Fractional Calculus

Xue Dingyü · Bai Lu

Fractional Calculus

High-Precision Algorithms and Numerical Implementations

Springer

Xue Dingyü
Northeastern University
Shenyang, Liaoning, China

Bai Lu
School of Information Engineering
Shenyang University
Shenyang, Liaoning, China

ISBN 978-981-99-2069-3 ISBN 978-981-99-2070-9 (eBook)
https://doi.org/10.1007/978-981-99-2070-9

Translation from the Chinese Simplified language edition: "分数阶微积分学——数值算法与实现" by Xue Dingyü and Bai Lu. © Tsinghua University Press, Beijing 2023. Published by Tsinghua University Press, Beijing. All Rights Reserved.

Part of the translation was done with the help of an artificial intelligence machine translation tool. A subsequent human revision was done primarily in terms of content.

This Springer imprint is published by the registered company Springer Nature Singapore Pte Ltd.
The registered company address is: 152 Beach Road, #21-01/04 Gateway East, Singapore 189721, Singapore

Paper in this product is recyclable.

Preface

Fractional calculus and applications are rapidly developing research directions in science and engineering at present. This book provides a systematic introduction to fractional calculus, focusing on the use of computer tools to directly solve problems in the field of fractional calculus and applications. The structure and general idea of this book are as follows: ▶ Chap. 1 gives a review of the related fields; ▶ Chap. 2 introduces some special functions commonly used in this field; ▶ Chaps. 3 and 4 introduce the evaluation of fractional-order derivatives and integrals of known functions or sampling points; ▶ Chap. 5 introduces the use of specially designed filters to find the fractional-order derivatives and integrals of signals when the signals are unknown in advance. Chapters 3 to 5 can also be understood as offline and online methods for evaluating fractional-order derivatives and integrals. ▶ Chapter 6 introduces the analytical and numerical solutions of linear fractional-order differential equations (FODEs), ▶ Chap. 7 introduces the "command-driven" solution of nonlinear FODEs, ▶ Chap. 8 introduces the block diagram-based solution of FODEs, and ▶ Chap. 9 introduces the solution of special FODEs (including implicit ones, delayed differential equations, differential equation boundary value problems, and partial differential equations) that were previously difficult or impossible to solve. Each part in the book is equipped with MATLAB general solution functions written by the author, whereas a new FOTF Toolbox is released with the book. The readers can directly use these reusable codes to reproduce the results in the book, and more importantly, use them to creatively solve practical problems and explore new knowledge.

In 2015, I was invited by Prof. Li Changpin of the Department of Mathematics, Shanghai University, an internationally renowned scholar in the field of fractional calculus, to write a related monograph for his series "Fractional Calculus in Applied Science and Engineering". In 2017, my monograph Fractional-Order Control Systems: Fundamentals and Numerical Implementations was fortunately published as the first volume of the series in de Gruyter Publishing. The following year, the corresponding monograph in Chinese, Fractional-Order Calculus and Fractional-Order Control, was officially published by Science Press.

This book was completed in collaboration with Dr. Bai Lu, School of Information Engineering of Shenyang University. It incorporates many of our new results in recent years. In this book, there are many original research results which are published for the first time, including analytical solutions in fractional calculus, high-precision algorithms for high-order fractional-order derivatives, simulation and stability analysis of irrational systems, new FOTF Toolbox and FOTF Blockset, unified framework for solving FODEs, solution methods for fractional-order delay differential equations, solution methods for FODEs in boundary value problems, and time-fractional partial differential equations, benchmark problems for more types of algorithms for solving various FODEs, and so on.

In around 2000, I was encouraged and even persuaded by a long-time collaborator, Prof. YangQuan Chen, now at the University of California, Merced, to start research in the field of fractional-order control. But it was not until 2003, when I started working with Prof. Chen on the first edition of MATLAB Solutions for

Advanced Applied Mathematics Problems, that I really took the time to study the literature in this area and began to study fractional calculus, I began my research on numerical implementations in fractional calculus. In that work, a lot of work on fractional calculus computation, filter approximation, closed-form solution algorithms for linear FODEs, and block diagram-based solution of nonlinear FODEs were systematically introduced, and many of these codes and models are still widely used by the researchers in the field of fractional calculus. Therefore, I must first thank Prof. YangQuan Chen here.

We would also like to thank a number of prominent scholars and active researchers in the field of fractional calculus, including Profs. Igor Podlubny and Ivo Petráš of Technical University of Košice, Slovakia, Li Changpin of Shanghai University, Li Yan of Shandong University, Wang Yong of University of Science and Technology China, Lu Junguo of Shanghai Jiaotong University, Li Donghai of Tsinghua University, Yu Yongguang of Beijing Jiaotong University, Sun Guanghui of Harbin Institute of Technology, Chen Wen and Sun Hongguang of Hohai University, Zeng Caibin of South China University of Technology, Wang Chunyang of Changchun University of Technology, Wei Yiheng of Southeast University, Liu Dayan and Driss Boutat of INSA Centre Val de Loire, France (in no particular order), and others. I am also grateful to my former co-authors, namely, Profs. Blas Vinagre, Concepción Monje, and Vicente Feliu, when I published my monograph at Springer Publishing House in 2010, and my discussions and idea exchanges with them have generated many new ideas and research results in this field, which have enriched the content of this book.

My in-depth discussions with my colleagues at Northeastern University, especially Profs. Pan Feng, Chen Dali, and Zhang Xuefeng, have also brought much meaningful content to this book. I also thank my former students for their results and contributions to this book and related research, specifically Dr. Zhao Chunna for her contributions to the numerical computation of fractional calculus and differential equations, Drs. Zhao Chunna and Meng Li for their contributions to filter design, Drs. Zhao Chunna, Meng Li, MSc student Wang Weinan, Dr. Liu Lu, and Dr. Li Tingxue for their contributions to controller design, and also in other related areas Dr. Yang Yang, Dr. Zhang Yanzhu, Dr. Liu Yanmei, Dr. Chen Zhen, Dr. Chen Lanfeng and Ph.D. students Cui Xinshu, Liu Yitong and Wang Zhe for their contributions.

We thank the National Natural Science Foundation of China for the Natural Science Foundation projects (project numbers: 61174145 and 61673094) for the research work on this book.

Last but not least, I would like to thank my wife, Yang Jun, and my daughter, Xue Yang, for their great help and encouragement in my life and career. Without their encouragement and continued support, this book and my other books would not have come out successfully, and I would like to dedicate this book to them.

Xue Dingyü
Shenyang, China
November 2022

Contents

Introduction to Fractional Calculus

Contents

D. Y. Xue and L. Bai, *Fractional Calculus*, https://doi.org/10.1007/978-981-99-2070-9_1

1.1 Historic Review of Fractional Calculus

At the beginning of the development of the theory of classical calculus (called integer-order calculus in this book), the British scientist Isaac Newton and the German mathematician Gottfried Wilhelm Leibniz used different symbols for different orders of derivatives. For example, Newton used the notation $\dot{y}(x)$, $\ddot{y}(x)$ and $\dddot{y}(x)$, while Leibniz used the notation $\mathrm{d}^n y(x)/\mathrm{d}x^n$, where n is a positive integer. A natural question is how to extend n into fractions or even complex numbers. In a letter written by the French mathematician Marquis de l'Hôpital to Leibniz in 1695, he asked question "what would be the meaning if $n = 1/2$ in the $\mathrm{d}^n y(x)/\mathrm{d}x^n$ notation". In a letter dated 30 September 1695, Leibniz replied, "Thus it follows that $\mathrm{d}^{1/2}x$ will be equal to $x\sqrt{\mathrm{d}x : x}$. This is an apparent paradox from which, one day, useful consequences will be drawn" [1]. The question and answer between these two mathematicians is widely considered to be the beginning of fractional calculus.

In 1819, the French mathematician Sylvestre François Lacroix used the Gamma function to study the power function

$$\frac{\mathrm{d}^n}{\mathrm{d}x^n}x^m = \frac{\Gamma(m+1)}{\Gamma(m-n+1)}x^{m-n+1}, \quad m \geqslant n, \tag{1.1}$$

and found that

$$\frac{\mathrm{d}^{1/2}}{\mathrm{d}x^{1/2}}x = \frac{\Gamma(2)}{\Gamma(3/2)}\sqrt{x} = 2\sqrt{\frac{x}{\pi}}. \tag{1.2}$$

This conclusion is in full agreement with the later results of Riemann–Liouville fractional-order derivatives.

Now it seems that the derivative notation invented by Newton is not suitable for extension to the field of fractional calculus, while the notation invented by Leibniz can be used directly in fractional calculus.

More than three centuries have passed and until a few decades ago research in the field of fractional calculus has focused on purely mathematical theoretical aspects of the work. Some of the better historical reviews in the field of fractional calculus can be found in References [1] and [2]. In Reference [1], Profs. Kenneth Miller and Bertram Ross gave a good historical review of fractional calculus from a purely mathematical point of view, and in Reference [2], Keith Oldham and Jerome Spanier quoted the chronicles of fractional calculus from its inception to the year 1975, summarized by Prof. Ross. An overview of historical figures and their contributions in the field of fractional calculus is presented in Reference [3].

Starting from 1960, the study of fractional calculus began to be extended to the field of science and engineering. To solve the nonzero initial value problems in fractional calculus, Prof. Michele Caputo, an Italian scholar, proposed a new definition of fractional calculus, which was later called Caputo's definition [4]. He presented a dissipation model based on fractional-order derivatives with Prof. Francesco Mainardi [5], which established the foundation for the engineering applications of fractional calculus. In Japan, Prof. Shunji Manabe extended the study of non-integer-order to the application of control systems and introduced the concept of non-integer-order control systems [6]. In Slovakia, Prof. Igor Podlubny proposed the fractional-order PID controller [7]. The research group led by Prof. Alain Oustaloup in France proposed the concept and technique of fractional-order robust control and successfully applied it to suspension control in the automotive industry [8, 9], which is regarded as a milestone for the applications of fractional calculus in the real world.

Since about 2000, a number of monographs devoted to fractional calculus and its applications have appeared in various fields of specialization, among which the more influential ones are Prof. Podlubny's 1999 book on fractional-order differential equations and their applications in the field of automatic control [10], Prof. Rudolf Hilfer's 2000 book in the field of physics [11], Prof. Richard Magin's 2006 book in the field of bioengineering [12], and so on.

Several works have also been published in recent years on the theory and numerical computation in fractional calculus, such as the works of Prof. Kai Diethelm's in 2010 [13], Prof. Das's in 2010 [14], Prof. Uchaikin's in 2013 [15], and Prof. Li Changpin and Dr. Zeng Fanhai's in 2015 [16]. Also, Prof. Xue Dingyü's monograph in 2017 on numerical implementation of fractional-order systems and control [17].

In 2017, de Gruyter published the series "Fractional Calculus in Applied Sciences and Engineering", edited by Prof. Li Changpin, and the "Handbook of Calculus in Applied Sciences and Engineering", edited by Profs. Anatoly Kochubei and Yuri Luchko.

In the field of automatic control, some monographs have been published in recent years, such as those by Profs. Caponetto, Dongola, Fortuna, and Petráš in 2010 [18]; Profs. Monje, Chen, Vinagre, Xue, and Feliu in 2010 [19]; Prof. Petráš in 2011 [20]; Dr. Luo Ying and Prof. Chen YangQuan in 2012 [21]; and Prof. Oustaloup in 2014 [22]. Prof. Uchaikin gave a good introduction to the applications of fractional calculus in various fields in 2013 [23].

Chinese scholars have also published textbooks and monographs on fractional calculus and its applications. There is a special chapter on fractional calculus and its computation in the book published by Profs. Xue Dingyü and Chen YangQuan in 2004 [24]. The following works are related to the research of this book, including the books by Profs. Chen Wen, Sun Hongguang, and Li Xicheng in 2010 [25]; Prof. Wang Jifeng in 2010 [26]; Profs. Zhao Chunna, Li Yingshun, and Lu Tao in 2011 [27]; Profs. Wang Chunyang, Li Mingqiu, and Jiang Shuhua in 2014 [28]; Profs. Li Wen and Zhao Huimin in 2014 [29]; Profs. Xiaozhong Liao and Zhe Gao in 2016 [30]; Profs. Qiang Wu and Jianhua Huang in 2016 [31]; Prof. Xue Dingyü in 2018 [32]; and so on.

Chinese scholars in the field of fractional calculus and applications established a special committee on fractional-order systems and control under the Chinese Association of Automation in July 2018 [33]. Special sessions or special issues in the field of fractional-order systems have also appeared one after another in many international conferences or international journals.

It is important to note that the term "fractional-order" is a misused one. The correct name should be "non-integer-order" or even "arbitrary order", because the order can be irrational or even complex. In addition to fractions (rational numbers), for example, $\mathrm{d}^{\sqrt{2}}y(t)/\mathrm{d}t^{\sqrt{2}}$ can be considered as the $\sqrt{2}$nd-order derivative of the signal $y(t)$. The complex order is beyond the scope of this book. However, the term "fractional-order" has been used by the huge majority of researchers in the vast reference text. So the term will be used in this book, but in essence it includes irrational orders and even irrational system structures.

Integer-order calculus has a concise and clear physical meaning. For example, displacement, velocity, and acceleration can be used to explain the relationship between a signal and its integer-order derivatives well. However, the fractional calculus does not have such a concise and understandable physical interpretation, although many scholars have tried to do so. One meaningful interpretation was given by Prof. Podlubny as "a shadow moving on a fence" [34], but it still seems to lack such a concise interpretation as the integer-order calculus.

Examples are given below to demonstrate fractional calculus operations for commonly used functions.

▶ Example 1.1

Consider the sinusoidal signal $\sin t$. It is known that the first-order derivative of this signal is $\cos t$. If we then find the high-order derivative of this signal, the result will be nothing more than $\pm\sin t$ and $\pm\cos t$. No other signal can be derived. What will happen if we introduce the concept of fractional calculus?

Solutions From the famous Cauchy integral formula

$$\frac{\mathrm{d}^n}{\mathrm{d}t^n}\sin t = \sin\left(t + n\frac{\pi}{2}\right).$$

In fact, the above formula holds when n is any non-integer, so the following MATLAB statements can be used to draw the surface of the fractional-order derivatives of the function, as shown in Fig. 1.1.

```
>> n0=0:0.1:1.5; t=0:0.2:2*pi; Z=[];
   for n=n0, Z=[Z; sin(t+n*pi/2)]; end, surf(t,n0,Z)
```

It can be seen that, in addition to the four known results $\pm\sin t$ and $\pm\cos t$, other information can be obtained and the results are asymptotic. Therefore, the fractional-order derivative of a function may provide richer information than the integer-order ones. In practical applications, if the world is viewed from the perspective of fractional calculus, it may reveal more things that were invisible from the perspective of integer-order calculus. ◀

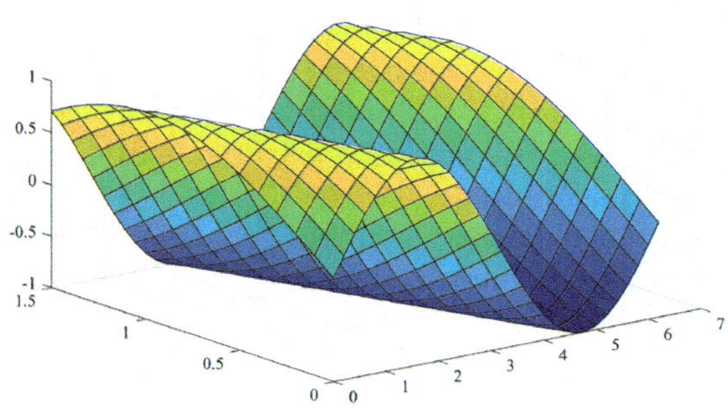

Fig. 1.1 Representation of surfaces with different order derivatives

1.2 Fractional Calculus Phenomena and Modeling Examples in Nature

There are many examples about the applications of fractional calculus in References [11, 12, 14]. Some relevant typical examples are listed here, which often cannot be described well in the framework of integer-order calculus. They must be described with the help of fractional calculus. Thus, fractional-order phenomena are actually ubiquitous.

▶ Example 1.2

In polymer materials and elastic material research, according to the suggestions in Reference [11], the rheological constitutive equation should be more precisely described as fractional-order differential equations (FODEs)

$$\sigma(t) + \tau^{\alpha-\beta}\frac{\mathrm{d}^{\alpha-\beta}\sigma(t)}{\mathrm{d}t^{\alpha-\beta}} = E\tau^{\alpha}\frac{\mathrm{d}^{\alpha}}{\mathrm{d}t^{\alpha}}\sigma(t),$$

where $0 < \alpha, \beta < 1$. ◀

▶ Example 1.3

Consider the driving-point impedance problem of a semi-infinite length lossy transmission line, whose standard voltage equation satisfies the integer-order partial differential equation. The boundary value conditions are known

$$\frac{\partial v(x,t)}{\partial t} = \alpha\frac{\partial^2 v(x,t)}{\partial x^2}, \quad v(0,t) = v_{\mathrm{I}}(t), \quad v(\infty,t) = 0.$$

After a series of direct mathematical formulations [14], the voltage–current equation for the impedance at the driving point can be derived as the following FODE with zero initial value:

$$i(t) = \frac{1}{R\sqrt{\alpha}}\frac{\mathrm{d}^{1/2}v(t)}{\mathrm{d}t^{1/2}} \quad \text{or} \quad v(t) = R\sqrt{\alpha}\frac{\mathrm{d}^{-1/2}i(t)}{\mathrm{d}t^{-1/2}}.$$

◀

▶ Example 1.4

The Bagley–Torvik equation [35, 36] describing the vibration of an oscillator in a viscoelastic medium is

$$A\frac{\mathrm{d}^2}{\mathrm{d}t^2}y(t) + B\frac{\mathrm{d}^{\alpha}}{\mathrm{d}t^{\alpha}}y(t) + Cy(t) = C(t+1). \tag{1.3}$$

The initial value conditions are $y(0) = y'(0) = 1$, with solution $y(t) = t + 1$. ◀

▶ Example 1.5

Ionic polymer metal composite (IPMC) is a new type of smart material that has a wide range of applications in areas such as robotics actuators and artificial muscles. In order to identify the model of IPMC, a set of frequency domain response data can be measured experimentally. The identification can be attempted by the well-established method of linear system model identification, but Reference [18] shows that no good identification model can be obtained in the framework of integer-order system. If the idea of fractional calculus is introduced, it may result in the following identification model:

$$G(s) = \frac{340}{s^{0.756}(s^2 + 3.85s + 5880)^{1.15}}.$$

It is clear that the identified model is a special form of fractional-order model. ◀

▶ Example 1.6

In a standard heat diffusion process, the temperature of a heat source rod at coordinate x can be directly described by the one-dimensional linear partial differential equation given below:

$$\frac{\partial c}{\partial t} = K\frac{\partial^2 c}{\partial x^2}.$$

If a constant temperature source C_0 is added at $x = 0$, the Laplace transform expression for the temperature under thermal diffusion can be deduced [37]

$$c(x, s) = \frac{C_0}{s}\mathrm{e}^{-x\sqrt{s/k}}.$$

◀

▶ Example 1.7

The memristor (resistor with memory) was pointed out as the fourth basic circuit element (the first three are the familiar and physically present resistor, capacitor, and inductor) by Prof. Chua in 1971 [38]. In 2008, researchers claimed to have found such a missing element [39]. Since the fractional calculus has the capabilities of describing memory, its resistance can be expressed in the fractional calculus sense as follows:

$$R_\mathrm{m} = \left[R_\mathrm{in}^{\alpha+1} \mp 2kR_\mathrm{d}\int_0^t \frac{v(\tau)}{(t-\tau)^{1-\alpha}}\mathrm{d}\tau\right]^{1/(\alpha+1)},$$

where the integral function is the basic expression of the fractional calculus [40]. ◀

In controller design, it is often necessary to deliberately introduce fractional-order actions in order to achieve a certain control performance index. An example of a controller is given below.

▶ Example 1.8

Consider the fractional-order quantitative feedback theory (QFT) given in the literature [19]. The controller model is as follows, which contains fractional-order operations.

$$G_c(s) = 1.8393\left(\frac{s+0.011}{s}\right)^{0.96}\left(\frac{8.8\times 10^{-5}s+1}{8.096\times 10^{-5}s+1}\right)^{1.76}\frac{1}{(1+s/0.29)^2}.$$

◂

In integer-order calculus, the signal in the system exhibits the form of an exponential function, while it can be observed in real life that certain phenomena exhibit results that may be power functions of time, often referred to as power-law phenomenon. This phenomenon is easier to understand under the fractional calculus framework. With the perspective of fractional calculus, one can better understand the complex world [41].

1.3 Historic Review of Fractional Calculus Computations

1.3.1 Numerical Computing in Fractional Calculus

With the maturity of computer technology, the theory of numerical computation in fractional calculus has developed rapidly. The first problem that was extensively studied was how to evaluate fractional-order derivatives and integrals numerically. Many classical algorithms emerged. A concise approach is to remove the limit sign from the definition of Grünwald–Letnikov fractional calculus, which leads to an approximate formula. References [10, 42] proved that this method has $O(h)$ accuracy. For most functions, the Grünwald–Letnikov definition is equivalent to the Riemann–Liouville definition, so this method can also be applied to evaluate the Riemann–Liouville fractional-order derivatives and integrals. Profs. Meerschaert and Tadjeran proposed the shifted approximation formula, which shifts the discrete points backward, so as to increase the convergence speed of the formula [43].

Prof. Lubich and co-workers, in their study of the Abel–Volterra integral equation, proposed the linear fractional-order multi-step method [44, 45], which introduced higher accuracy formulas for evaluating fractional-order derivatives and integrals. Reference [44] proved that the accuracy of this algorithm is not only related to the order of the generating function, but also affected by the initial value conditions of the original function. The linear fractional-order multi-step method is a classical algorithm that has had a great influence on the development of fractional-order numerical algorithms. Prof. Podlubny introduced the fast Fourier transform-based method to evaluate w_j on this basis. This algorithm has a faster computational speed [10], although it is possible to introduce greater errors as a result.

The finite-part integral (FPI) method proposed by Prof. Kai Diethelm and co-workers [13, 46–48] allows the evaluation of not only Riemann–Liouville fractional-order derivatives and integrals, but also Caputo fractional-order derivatives. This method converts the fractional-order derivative or integral into a Hadamard integral by dividing equidistant grid points on the integration interval, finding a simple function to replace the original function at the grid points, and then deriving the computational formula. Reference [46] proves that this method has $O(h^2)$ accuracy.

Prof. Igor Podlubny proposed the matrix method [49], which represents fractional-order derivatives and integrals in the form of matrices. Applying this method, it is also possible to convert FODEs into matrix equations, which are easier to find the numerical solutions.

1.3.2 Numerical Computing in Fractional-Order Ordinary Differential Equations

With the vigorous development of fractional calculus theory, more and more FODEs appear in engineering, and how to solve FODEs becomes a hot topic of research. From the perspective of the demand of applied science, the solution of FODEs under Caputo's definition is widely concerned. For some simple FODEs, the analytical solutions can be found directly. Commonly used methods include the integral transform method [50, 51], the Green's function method [10, 52, 53], and the Adomian decomposition method [54]. However, for most of the FODEs, finding analytical solutions is an impractical matter. Therefore, numerical algorithms for solving FODEs have become a hot research topic in this field.

The Adams–Bashforth–Moulton algorithm [55, 56] is an effective method for solving one-term FODEs. This algorithm first derives the integral form of the equation, then applies the first-order extrapolation to calculate the predictor solution, then substitutes the predictor solution into the original equation and applies the second-order extrapolation to calculate the corrector solution. The Adams–Bashforth–Moulton algorithm has $O(h^2)$ accuracy. Even if the FODE is nonlinear, it can also be solved by this algorithm.

It is also possible to introduce Prof. Igor Podlubny's matrix algorithm into the numerical solution process of linear and nonlinear FODEs. It should be noted in particular that the matrix algorithm can also solve certain fractional-order implicit differential equations.

The finite partial integration algorithm proposed by Prof. Kai Diethelm and his collaborators can be used to solve the one-term FODEs [46]. This algorithm writes the solution of the original equation in the form of Hadamard integral, applies the interpolation algorithm to find a simple function to replace the integrand, numerically computes the integral to obtain the predictor solution, and substitutes the predictor solution into the original equation to calculate the corrector solution. Reference [46] proves that the finite partial integration algorithm has $O(h^2)$ accuracy. Based on the finite partial integration algorithm, Diethelm proposed the predictor–corrector (PECE) algorithm [48, 57, 58]. This algorithm first applies extrapolation to solve the predictor solution of the equation and then substitutes the predictor solution into the original equation to find the corrector solution. In Reference [59], FODEs with nonzero initial value conditions were investigated. An auxiliary function was introduced to transform the original equation into an equation with zero initial value condition before solving it efficiently. The iterative algorithm was embedded into a corrector process in References [17, 32], which substantially improved the solution accuracy of the predictor–corrector algorithm.

Chinese scholars have made many contributions in the numerical algorithms about fractional calculus, including the closed-form solutions of linear FODEs proposed by Prof. Xue Dingyü and co-workers [24, 60]; high-precision algorithms in fractional calculus [61, 62]; high-precision numerical algorithms for Caputo ordinary differential equations [17, 32]; block diagram-based unified modeling and simulation framework for Caputo ordinary differential equations [63]; the high-precision numerical algorithms for fractional calculus and differential equations proposed by Prof. Li Changpin and co-workers [64, 65]; the block-by-block algorithm proposed by Profs. Wang Ziqiang and Cao Junying [66]; and so on. The algorithm in Reference [65] has a high-precision accuracy. With this algorithm it is possible to evaluate the Caputo derivative of orders in the (0, 1) interval and to find the numerical solution of the equation. Reference [65] proves that this method has an accuracy of $O(h^{r+1-\alpha})$. In the block-by-block algorithm, the FODE is transformed into the Volterra integral equation, and the integral interval is divided into equally spaced intervals. The quadratic Lagrange basis function is applied to replace the integrand in each interval to derive the formula for solving the numerical solution of the original equation.

To evaluate the algorithms for solving FODEs, the authors of this book proposed a series of benchmark problems for Caputo equations [67, 68]. These problems are highly general and can be used for a fair comparison of various FODE algorithms. Based on a general modeling and simulation approach based on block diagrams, this book further explores fractional-order state space equations with orders larger than 1, implicit FODEs, fractional-order delay differential equations (FODDEs), and FODEs with known boundary value conditions. Many of them are extremely rare or even completely non-existent in the literature, but the general modeling and simulation methods presented in this book make it easy to derive high-precision numerical solutions.

1.3.3 Numerical Computing in Fractional-Order Partial Differential Equations

Fractional-order partial differential equations (FOPDEs) can be broadly classified into time-fractional partial differential equations, space-fractional partial differential equations, time–space-fractional partial differential equations, and so on. There are some mature methods for solving FOPDEs, including finite difference algorithm, Adomian decomposition method, and variational iteration method.

Prof. Liu Fawang and co-workers [69] and Prof. Sun Zhizhong and co-workers [70] have discussed finite difference methods in detail in their respective monographs, given difference format algorithms for different kinds of FOPDEs. The accuracy, stability, solvability, and convergence of each difference format are discussed. More results and literature on finite difference methods are also listed and the development of finite difference methods is summarized. The authors of this book give algorithms for the solution of the unified two time-fractional difference formats [71].

The decomposition method proposed by Prof. Adomian is a method to apply the series to find the approximate analytical solution of equation [72]. Some scholars have applied the Adomian decomposition method to solve FOPDEs [73, 74]. The Adomian decomposition method avoids the discretization of FODEs and thus is with less computational load, but this method requires the calculation of complicated fractional-order integrals, so it increases the computational difficulty.

Many numerical methods for solving other forms of FOPDEs, such as variational iterative methods, finite element methods, spectral methods, and so on, have also appeared in the literature. A summary of numerical algorithms for fractional calculus and FOPDEs is given in the works of Prof. Liu Fawang and co-workers [69], Prof. Guo Boling and co-workers [75], and Prof. Li Changpin and Dr. Zeng Fanhai [16], and interested readers can find the related literature.

1.4 Tools in Fractional Calculus and Fractional-Order Control

There are several MATLAB toolboxes that are widely used in the field of fractional calculus and fractional-order control. A comparative review of the commonly used toolboxes is given in Reference [76]. In fact, many of the tools compared in Reference [76] are only single MATLAB functions, among which only four can be called toolboxes. These toolboxes happen to be in the field of fractional-order control. A brief comparison is presented here in the chronological order of their introduction.

(1) **CRONE Toolbox** [8] is the result of the CRONE research group led by the famous French scholar Prof. Oustaloup. The work started around 1990 as a practical tool for solving fractional-order system identification and robust controller design. The disadvantage is that it is distributed in MATLAB pseudo-code encrypted form, and there is no way for the user to modify or extend any of the features of the toolbox. CRONE is abbreviation of a French words *commande robuste d'ordre non entier*, meaning non-integer-order robust control.

(2) **Ninteger Toolbox** [77] was developed by a Portuguese scholar, Prof. Valério, in 2001. The earliest version mainly implemented the CRONE controller, and in the more mature version 2.3, the toolbox provided a set of functions, models, and interfaces for the design and analysis of fractional-order system and controller design based on MATLAB and Simulink. The toolbox has two core functions, one is the identification of fractional-order systems and the other is the approximation of fractional-order systems with integer-order models. No new versions of this toolbox have been released since 2009.

(3) **FOTF Toolbox** [78, 79] is a MATLAB toolbox for fractional calculus and fractional-order control system research written by Prof. Xue Dingyü in China. It was first made public under the name of FOTF in 2006, and since 2004, the author has successively released many MATLAB functions and Simulink models for fractional calculus and control. In 2017, all the programs and models were rewritten to support the analysis and design of fractional-order multivariable systems in conjunction with the publication of the monographs [17, 32]. In addition, the underlying fractional calculus calculations are replaced by the high-precision algorithms proposed by the authors, which are usually many orders of magnitude more accurate than the existing ones, making the toolbox itself more efficient and reliable. A brief introduction and demonstration of the main functions of the toolbox are given in Reference [80]. With the writing and publication of this book, the FOTF Toolbox has undergone another major revision, which is more powerful and more suitable for practical applications. This book will introduce the theoretical knowledge of the system and the details of the toolbox software development based on this toolbox in detail. The code is also all open source, which is also useful for readers to learn the numerical implementation in the field of fractional-order systems.

(4) **The FOMCON Toolbox** [81] was developed by the Estonian scholar Prof. Aleksie Tepljakov in his master's research. The toolbox was developed initially by copying

and integrating the facilities of the FOTF Toolbox and the Ninteger Toolbox. Later, some of the procedures were rewritten under the framework established by the original authors to form a set of procedures and models for solving the identification, analysis, and design of fractional-order systems. Since the toolbox was based on an earlier version of the FOTF class, the toolbox is limited to solving problems for univariate fractional-order systems, and the new version provides limited functionality for solving problems in multivariate systems [82].

In addition, readers are advised to download user contributed toolboxes and utility functions that may be available from the "File Exchange" section of the MathWorks' website, but because of the varying levels of programming on this site, some tools may be of poor quality and may sometimes lead to incorrect results. Users should be especially careful when choosing to download toolboxes and utility functions.

1.5 Structures in the Book

1.5.1 Main Contents

This book systematically introduces the fundamentals related to numerical computational problems in the field of fractional calculus, provides readers with directly usable computer tools that can enhance their understanding of the content and improve the manipuility of individual topics.

In ▸ Chap. 1, a brief review of the development of fractional calculus and its application areas is given. An explanation of why the fractional calculus perspective on the world is introduced through some real-world fractional-order phenomena, and a summary of several internationally available MATLAB toolboxes for fractional calculus and fractional-order control are presented. This book will make extensive use of the FOTF Toolbox developed by the authors.

▸ Chapter 2 focuses on the definition, properties, and computation of various special functions commonly used in the field of fractional calculus, and provides a foundation for introducing the definition and computation of fractional calculus.

▸ Chapter 3 introduces various common definitions of fractional calculus, such as the Cauchy integral formula, the Grünwald–Letnikov definition, the Riemann–Liouville definition, and the Caputo definition; summarizes the relationships between the definitions; and gives concise analytical and numerical operations. The chapter takes the fractional-order integral as an example, its geometric interpretation is given.

▸ Chapter 4 addresses the problem of numerical computation in fractional calculus by proposing and implementing a series of numerical algorithms with high accuracy, several orders of magnitude higher than conventional algorithms, which can be considered as the basis of numerical computation for the subsequent contents of the book. This chapter also explores a new path to efficiently derive solutions to fractional-order derivative problems of higher orders.

▸ Chapter 5 introduces filter approximation methods for fractional-order behaviors and discusses various continuous and discrete filter design methods, including filter implementation methods for fractional-order operators, fractional-order transfer functions (FOTFs), irrational transfer functions, and so on. An online implementation method for fractional-order derivative of unknown signals is given, as well as suggestions for the selection of filter design parameters.

▶ Chapter 6 introduces the format of linear FODEs, discusses the analytical and numerical algorithms, proposes high-precision numerical solutions of linear FODEs with zero and nonzero initial values, and gives the stability assessment methods of linear fractional-order systems and irrational systems. In particular, this chapter also systematically investigates the simulation method of irrational systems based on numerical Laplace transform and its inverse, which can be extended to the study of feedback control of irrational systems.

▶ Chapter 7 discusses the numerical methods for nonlinear fractional-order systems, focusing mainly on the command-driven solution algorithm and MATLAB implementation. Firstly, the numerical methods for explicit FODEs and fractional-order state space equations are introduced, and the computer code of the traditional numerical algorithm is given; in addition, a high-precision numerical method for nonlinear explicit FODEs is proposed and implemented, which significantly improves the computational accuracy and efficiency of the algorithm.

▶ Chapter 8 continues the discussion on the numerical methods for nonlinear FODEs, focusing mainly on block diagram-based differential equation solving methods, which significantly extend the solution capabilities and improve the efficiency. This chapter gives a brief introduction to the FOTF Toolbox, introduces the practical blocks that can be used for Simulink modeling and simulation, and presents a general method based on integrator chains for modeling and solving the Riemann–Liouville and the Caputo equations, and evaluates the accuracy and efficiency of the methods. Such methods can theoretically be used to deal with the solution of nonlinear FODEs with arbitrary complexity.

▶ Chapter 9 introduces algorithms for solving special FODEs, including implicit FODEs that are difficult or even impossible to solve in the traditional sense, fractional-order delay differential equations with nonzero history functions and boundary value problems of FODEs, and so on. It also introduces numerical methods for solving time-fractional partial differential equations. This chapter provides the ideas and implementation for a comprehensive study about the solution of FODEs.

In order to test and evaluate the accuracy and the efficiency of the algorithms for solving FODEs, Appendix A designs and proves some benchmark problems composed of various FODEs, which can be used to fairly compare the accuracy, efficiency, and other superiority of various numerical algorithms. Appendix B gives the common special functions and Laplace transforms related to fractional calculus. Appendix C lists the function tables of the FOTF Toolbox for the readers' reference and review, so that the readers can better use the FOTF Toolbox to complete the research in the field of fractional-order systems.

1.5.2 Reading Suggestions

An important feature of this book is that each topic is supported by MATLAB codes and models written by the authors, so that the reader can also better understand the relevant contents of this book from the supporting codes. More importantly, the reader can directly use these codes to solve relevant problems and even creatively solve some unknown problems that may be associated with them. The FOTF Toolbox provided in this book can solve basic problems in fractional calculus operations and can be used throughout the process of modeling, analysis, and design of fractional-order systems.

The FOTF Toolbox can be downloaded for free at the following URL.

▶ http://cn.mathworks.com/matlabcentral/fileexchange/60874-fotf-toolbox

This book can also be used as a textbook and reference book for general readers to learn fractional calculus and fractional-order control. In particular, the following study suggestions are given for reference to the following three groups of readers, i.e., control engineers who want to introduce fractional-order control methods into their own research fields, general researchers who want to study fractional calculus and fractional-order control systematically, and researchers in the field of mathematics who want to study fractional calculus.

(1) For the control engineers who only want to introduce the concept of fractional-order control in their practical applications, they can skip all the theoretical contents of this book and only need to understand the basic definition of fractional calculus, and then learn the direct calculation in fractional calculus—for known functions and data, calculating directly by the functions of the MATLAB toolbox; for unknown signals, reconstructing its fractional-order derivative and integral by the filters of the MATLAB toolbox. Such readers should also learn the methods for solving linear and nonlinear FODEs, especially those based on the Simulink environment. They should also learn how to use the two classes FOTF and FOSS written by the authors and will find the process of modeling and analyzing linear fractional-order systems as easy and convenient as integer-order systems. With these tools as a basis, users can introduce the concept of fractional calculus into their practical work as they wish and try out a class of controllers that may yield better results.

(2) For non-mathematical researchers, they can make full use of the special writing format of this book, use the provided MATLAB program as the main tool to systematically learn the necessary knowledge in the field of fractional calculus and fractional-order control, and reproduce the results in the book to better understand the relevant technical contents. Further, one can make full use of such tools to solve similar problems directly, to investigate more complicated and unknown problems creatively, and try to apply fractional-order systems theory to one's own research field to find more useful results. The proofs of the theorems can be skipped directly when reading this book.

(3) For the researchers with mathematical backgrounds, the proofs of the theorems covered in this book are not very comprehensive. It is recommended to find the omitted parts from other related works or to give the proofs by yourself. The programming details and techniques given in this book are worthy of reference and learning. It is recommended that such readers fully study the MATLAB programming techniques in this book to improve their programming ability and increase the efficiency of their code implementation. In addition, the authors sincerely hope that the computational examples which are selected carefully in this book can be used as the comparisons for such readers to test their algorithms. A set of benchmark problems for comparing the performance of numerical algorithms for solving FODEs are also constructed in Appendix A. These benchmark problems can be solved directly by the algorithms given in the book, and many of them have already yielded numerical solutions with

considerable accuracy and speed. In addition, the reusable codes and models in this book can also reproduce the numerical solutions of these problems, and readers can also write codes for their own or others' algorithms to find the results which are better than those in this book in terms of speed, accuracy, and adaptability. Readers can also develop their own benchmark problems to challenge the algorithms in this book. As the old Chinese saying goes, "Try it to show you are a horse not a mule", and only through such fair comparisons and solutions can meaningful progress be made in this area of research in the right direction.

All the results in this book can be reproduced by the corresponding MATLAB statements given in the book and the open-source FOTF Toolbox. It is believed that readers can use this book and the toolbox to gain more knowledge in the field of fractional calculus.

References

1. Miller KS, Ross B (1993) An introduction to the fractional calculus and fractional differen-tial equations[M]. Wiley, New York
2. Oldham KB, Spanier J (1974) The fractional calculus: theory and applications of differentiation and integration to arbitrary order[M]. Academic Press, San Diego
3. Valério D, Machado JT, Kiryakova V (2014) Some pioneers of the applications of fractional calculus[J]. Fract Calc Appl Anal 17(2):552–578
4. Caputo M (1967) Linear models of dissipation whose Q is almost frequency independent II[J]. Geophys J Int 13(5):529–539
5. Caputo M, Mainardi F (1971) A new dissipation model based on memory mechanism [J]. Pure Appl Geophys 91(8):134–147
6. Manabe S (1960) The non-integer integral and its application to control systems (in Japanese)[J]. Jpn J Inst Electr Eng 80:589–597
7. Podlubny I (1999) Fractional-order systems and $PI^{\lambda}D^{\mu}$-controllers[J]. IEEE Trans Autom Control 44(1):208–214
8. La Oustaloup A (1991) Commande CRONE[M]. Hermès, Paris
9. La Oustaloup A (1995) Dérivation non entière: théorie, synthèse et applications[M]. Hermès, Paris
10. Podlubny I (1999) Fractional differential equations[M]. Academic Press, San Diego
11. Hilfer R (2000) Applications of fractional calculus in physics[M]. World Scientific, Singapore
12. Magin RL (2006) Fractional calculus in bioengineering[M]. Begell House Publishers, Redding
13. Diethelm K (2010) The analysis of fractional differential equations: an application-oriented exposition using differential operators of Caputo type[M]. Springer, New York
14. Das S (2011) Functional fractional calculus[M]. Springer, Berlin
15. Uchaikin VV (2013) Fractional derivatives for physicists and engineers—Volume I: background and theory[M]. Higher Education Press, Beijing
16. Li CP, Zeng FH (2015) Numerical methods for fractional calculus. CRC Press, Boca Raton
17. Xue DY (2017) Fractional-order control systems—fundamentals and numerical implementations [M]. De Gruyter, Berlin
18. Caponetto R, Dongola G, Fortuna L et al (2010) Fractional order systems: modeling and control applications[M]. World Scientific, Singapore
19. Monje CA, Chen YQ, Vinagre BM et al (2010) Fractional-order systems and controls: fundamentals and applications[M]. Springer, London
20. Petráš I (2011) Fractional-order nonlinear systems: modelling, analysis and simulation[M]. Higher Education Press, Beijing
21. Luo Y, Chen YQ (2012) Fractional-order motion control[M]. John Wiley & Sons, London
22. Oustaloup A (2014) Diversity and non-integer differentiation for system dynamics[M]. Wliey, London

23. Uchaikin VV (2013) Fractional derivatives for physicists and engineers—Volume II: applications. Higher Education Press, Beijing
24. Xue DY, Chen YQ (2004) MATLAB solutions of higher advanced applicational mathematical problems (in Chinese)[M]. Tsinghua University Press, Beijing
25. Chen W, Sun HG, Li XC (2010) Fractional-order derivative-based modeling in mechanics and engineering problems (in Chinese)[M]. Science Press, Beijing
26. Wang JF (2010) Control performance analysis of fractional-order systems (in Chinese)[M]. Electronic Industry Press, Beijing
27. Zhao CN, Li YS, Lu T (2011) Fractional-order system analysis and design (in Chinese)[M]. National Defense Industry Press, Beijing
28. Wang CY, Li MQ, Jiang SH et al (2014) Fractional-order control system design (in Chinese)[M]. National Defense Industry Press, Beijing
29. Li W, Zhao HM (2014) Fractional-order controller design methods and vibration suppression performance analysis (in Chinese)[M]. Science Press, Beijing
30. Liao XZ, Gao Z (2016) Fractional-order system robustness analysis and robust control (in Chinese)[M]. Science Press, Beijing
31. Wu Q, Zhuang JH (2016) Fractional calculus (in Chinese)[M]. Tsinghua University Press, Beijing
32. Xue DY (2018) Fractional calculus and fractional-order control (in Chinese)[M]. Science Press, Beijing
33. Chinese Association of Automation. Special committee of "Fractional-order systems and controls"[OL]. ► http://www.caa.org.cn/article/205/1107.html
34. Podlubny I (2001) Geometric and physical interpretations of fractional integration and differentiation[J]. Fract Calc Appl Anal 5(4):230–237
35. Bagley RL, Torvik PJ (1983) Fractional calculus—a different approach to the analysis of viscoelastically damped structures[J]. AIAA J 21(5):741–748
36. Torvik PJ, Bagley RL (1984) On the appearance of the fractional derivative in the behavior of real materials[J]. Trans ASME 51(4):294–298
37. Jesus IS, Tenreiro Machado JA, Barbosa RS (2010) Fractional dynamics and control of distributed parameter systems[J]. Comput Math Appl 59(5):1687–1694
38. Chua LO (1971) Memristor–the missing circuit element[J]. IEEE Trans Circuit Theory 18(5):507–519
39. Strukov DB, Snider GS, Stewart DR et al (2008) The missing memristor found[J]. Nat 453:80–83
40. Fouda ME, Radwan AG (2013) On the fractional-order memristor model[J]. J Fract Calc Appl 4(1): 1–7
41. West BJ (2016) Fractional calculus view of complexity: tomorrow's science[M]. CRC Press, Boca Raton
42. Vinagre BM, Chen YQ (2003) Petŕàs I. Two direct Tustin discretization methods for fractional-order differentiator/integrator[J]. J Frankl Inst 340(5):349–362
43. Meerschaert MM, Tadjeran C (2004) Finite difference approximations for fractional advection-dispersion flow equations[J]. J Comput Appl Math 172(1):65–77
44. Lubich C (1986) Discretized fractional calculus[J]. SIAM J Math Anal 17(3):704–719
45. Hairer E, Lubich C, Schlichte M (1985) Fast numerical solution of nonlinear Volterra convolution equations[J]. SIAM J Sci & Stat Comput 6(3):532–541
46. Diethelm K (1997) Generalized compound quadrature formulae for finite-part integrals[J]. IMA J Numer Anal 17(3):479–493
47. Diethelm K, Ford NJ (2004) Multi-order fractional differential equations and their numerical solution[J]. Appl Math Comput 154(3):621–640
48. Diethelm K (2008) An investigation of some nonclassical methods for the numerical approximation of Caputo-type fractional derivatives[J]. Numer Algorithms 47(4):475–490
49. Podlubny I (2000) Matrix approach to discrete fractional calculus[J]. Fract Calc Appl Anal 3(4):359–386
50. Brzeziński DW, Ostalczyk P (2016) Numerical calculations accuracy comparison of the inverse Laplace transform algorithms for solutions of fractional order differential equations[J]. Nonlinear Dyn 84(1):65–77
51. Uddin M, Ahmad S (2017) On the numerical solution of Bagley-Torvik equation via the Laplace transform[J]. Tbil Math J 10(1):279–284
52. Kelly JF, Mcgough RJ (2016) Approximate analytical time-domain Green's functions for the Caputo fractional wave equation[J]. J Acoust Soc Am 140(2):1039–1047

53. Akram G, Rasheed R (2017) Existence and uniqueness of nonlinear multi-order fractional differential equations via Green function[J]. Int J Appl Comput Math 3:3831–3856
54. Adomian G (1996) Solution of coupled nonlinear partial differential equations by decomposition[J]. Comput Math Appl 31(6):117–120
55. Diethelm K, Ford NJ, Freed AD (2004) Detailed error analysis for a fractional Adams method[J]. Numer Algorithms 36(1):31–52
56. Baskonus HM, Bulut H (2015) On the numerical solutions of some fractional ordinary differential equations by fractional Adams-Bashforth-Moulton method[J]. Open Math 13(1):547–556
57. Diethelm K, Ford NJ, Freed AD (2002) A predictor-corrector approach for the numerical solution of fractional differential equations[J]. Nonlinear Dyn 29(1):3–22
58. Diethelm K (2011) An efficient parallel algorithm for the numerical solution of fractional differential equations[J]. Fract Calc Appl Anal 14(3):475–490
59. Diethelm K, Ford NJ (2002) Analysis of fractional differential equations[J]. J Math Anal Appl 265(2):229–248
60. Zhao CN, Xue DY (2008) Closed-form solutions to fractional-order linear differential equations[J]. Front Electr Electron Eng China 3(2):214–217
61. Bal L, Xue DY (2018) A high-precision recursive algorithm for fractional-order calculus (in Chinese)[J]. J Northeast Univ (Nat Sci Ed) 39(4):604–608
62. Xue DY, Bai L (2017) Numerical algorithms for Caputo fractional-order differential equations[J]. Int J Control 90(6):1201–1211
63. Bai L, Xue DY (2018) Universal block diagram based modeling and simulation schemes for fractional-order control systems. ISA Trans 82:153–162
64. Cao JX, Li CP, Chen YQ (2015) High-order approximation to Caputo derivatives and Caputo-type advection-diffusion equations (II)[J]. Fract Calc Appl Anal 18(3):735–761
65. Li HF, Cao JX, Li CP (2016) High-order approximation to Caputo derivatives and Caputo-type advection-diffusion equations (II)[J]. J Comput Appl Math 299:159–175
66. Wang ZQ, Cao JY (2015) Numerical algorithms and their error analysis in fractional-order integro-differential equations (in Chinese)[M]. Xi'an Jiaotong University Press, Xi'an
67. Xue DY, Bai L (2017) Benchmark problems for Caputo fractional-order ordinary differential equations[J]. Fract Calc Appl Anal 20(5):1305–1312
68. Bai L, Xue DY (2022) Block diagram modeling and simulation for delay fractional-order differential equations[C]. In: Proceedings of Chinese Control and Decision Conference. Hefei, China, pp 137–142
69. Liu FW, Zhuang PH, Liu QX (2015) Numerical algorithms and their applications in fractional-order partial differential equations (in Chinese)[M]. Science Press, Beijing
70. Sun ZZ, Gao GH (2015) Finite difference algorithms in fractional-order differential equations (in Chinese)[M]. Science Press, Beijing
71. Bai L, Xue DY, Meng L (2021) A numerical algorithm for time-fractional partial differential equations (in Chinese)[J]. J Math Pract Theory 51(12):245–251
72. Adomian G (1994) Solving frontier problems of physics: the decomposition method[M]. Kluwer Academic Publishers, Boston
73. Yu Q, Liu FW, Anh V et al (2008) Solving linear and non-linear space-time fractional reaction-diffusion equations by the Adomian decomposition method[J]. Int J Numer Methods Eng 74(1):138–158
74. Jafari H, Daftardar-Gejji V (2006) Solving linear and nonlinear fractional diffusion and wave equations by Adomian decomposition[J]. Appl Math Comput 180(2):488–497
75. Guo BL, Pu XK, Huang FH (2011) Fractional-order partial differential equations and their numerical solutions (in Chinese)[M]. Science Press, Beijing
76. Li Z, Liu L, Dehghan S et al (2017) A review and evaluation of numerical tools for fractional calculus and fractional order control[J]. Int J Control 90(6):1165–1181
77. Valério D (2006) Ninteger v. 2.3 fractional control toolbox for MATLAB[R]. Universudade Téchica de Lisboa
78. Xue DY (2022) Computer-aided design of control systems with MATLAB (in Chinese)[M], 4th edn. Tsinghua University Press, Beijing
79. Chen YQ, Petráš I, Xue DY (2009) Fractional order control—a tutorial[C]. In: Proceedings of the American Control Conference. St Louis, pp 1397–1411
80. Xue DY (2019) FOTF Toolbox for fractional-order control systems. In: Petráš I (ed) Handbook of fractional calculus with applications. Volume 6: applications in control[M]. De Gruyter, Berlin

81. Tepljakov A (2011) Fractional-order calculus based identification and control of linear dynamic systems[D]. Tallinn University of Technology, Tallinn
82. Tepljakov A, Petlenkov E, Belikov J (2019) FOMCON toolbox for modeling, design and implementation of fractional-order control systems. Petŕš. I. Handbook of fractional calculus with applications. Volume 6: applications in control[M]. De Gruyter, Berlin

Commonly Used Special Functions: Definitions and Computing

Contents

D. Y. Xue and L. Bai, ***Fractional Calculus***, https://doi.org/10.1007/978-981-99-2070-9_2

Special functions are dedicated mathematical functions invented by mathematicians. For instance, if the integrand is $f(x) = \mathrm{e}^{-x^2}$, the analytical solution to its indefinite integrals does not exist. Therefore, mathematicians invented a special function $\mathrm{erf}(x)$ to express the integral expression. This special function is regarded as the analytical solution of the integral problem. In different applications, many other such special functions are invented. For instance, Gamma function, Beta function, and so on. Many special functions are invented for indefinite integral and differential equation problems.

In addition, for linear differential equations with constant coefficients, the mathematical form of its analytical equation is a weighted sum of exponential functions. Since fractional calculus is directly extended from integer-order calculus, the low-level functions in the analytical solutions are the Mittag-Leffler functions, to be presented in detail in this chapter. In fact, the role of Mittag-Leffler functions in fractional calculus is similar to the exponential functions in integer-order calculus. Knowledge of these special functions will be helpful in learning the theory and applications of fractional calculus.

Some special functions closely related to fractional calculus are discussed in this chapter. Computer solutions to these functions are also presented. In ▶ Sect. 2.1, concentrations are made on error functions and complementary error functions. In ▶ Sects. 2.2 and 2.3, Gamma and Beta functions are introduced. In ▶ Sects. 2.4 and 2.5, Dawson and hypergeometric functions are introduced, and MATLAB-based solutions of these special functions are presented. In ▶ Sect. 2.6, various Mittag-Leffler functions are introduced, and evaluation of Mittag-Leffler functions and their integer-order derivatives are discussed.

2.1 Error and Complementary Error Functions

In classical calculus, the integrand e^{-t^2} is usually regarded non-integrable. The integral of this function is actually used in many fields. For instance, in probability theory and statistics, probability density function of normal distribution is such a function. Therefore, the integral of such a function is expected in solving probability distribution problems. Therefore, a special function is needed to denote such an integral. The mathematician then invented the error function.

Definition 2.1 Error function

The mathematical form of the error function, also known as erf function, is

$$\mathrm{erf}(z) = \frac{2}{\sqrt{\pi}} \int_0^z \mathrm{e}^{-t^2} \mathrm{d}t = \frac{1}{\sqrt{\pi}} \int_{-z}^z \mathrm{e}^{-t^2} \mathrm{d}t. \tag{2.1}$$

It is seen that error function satisfies the following formulas:

$$\mathrm{erf}(0) = \frac{2}{\sqrt{\pi}} \int_0^0 \mathrm{e}^{-t^2} \mathrm{d}t = 0 \tag{2.2}$$

$$\mathrm{erf}(\infty) = \frac{2}{\sqrt{\pi}} \int_0^\infty \mathrm{e}^{-t^2} \mathrm{d}t = 1, \quad \mathrm{erf}(-\infty) = -1. \tag{2.3}$$

Theorem 2.1 Infinite series representations
Error function can also be described by an infinite series

$$\mathrm{erf}(z) = \frac{2}{\sqrt{\pi}} \sum_{k=0}^{\infty} \frac{(-1)^k}{(2k+1)} \frac{z^{2k+1}}{k!}, \tag{2.4}$$

with convergent range $|z| < \infty$.

Proof Integrating each term in the Taylor series expansion of the exponential function, it is found that

$$\mathrm{erf}(z) = \frac{2}{\sqrt{\pi}} \int_0^z \sum_{k=0}^{\infty} \frac{(-t^2)^k}{k!} \mathrm{d}t = \frac{2}{\sqrt{\pi}} \sum_{k=0}^{\infty} \frac{(-1)^k}{(2k+1)} \frac{z^{2k+1}}{k!},$$

where the convergent field of Taylor series function is $|z| < \infty$. □

It is obvious that $\dfrac{\mathrm{d}}{\mathrm{d}z}\mathrm{erf}(z) = \dfrac{2}{\sqrt{\pi}}\mathrm{e}^{-z^2}$. The high-order derivative of the error function is

$$\frac{\mathrm{d}^n}{\mathrm{d}z^n}\mathrm{erf}(z) = \frac{\mathrm{d}^{n-1}}{\mathrm{d}z^{n-1}} \frac{2}{\sqrt{\pi}} \mathrm{e}^{-z^2} = \frac{2(-1)^n}{\sqrt{\pi}} \mathrm{H}_{n-1}(z) \mathrm{e}^{-z^2}, \tag{2.5}$$

where $\mathrm{H}_n(z)$ is the Hermite polynomial, satisfying the following recursive formula:

$$\mathrm{H}_0(z) = 1, \; \mathrm{H}_n(z) = 2z\mathrm{H}_{n-1}(z) - \mathrm{H}'_{n-1}(z), \; n = 1, 2, \cdots \tag{2.6}$$

Function `erf()` provided in MATLAB can be used to evaluate error functions, with the syntax `y=erf(z)`, where, if z is a numeric vector, vector y returns the error function of all the points in vector z. If z is a symbolic variable, the returned y is the symbolic expression of the error function.

Definition 2.2 Complementary errorfunction

The mathematical form of complimentary error function is

$$\mathrm{erfc}(z) = \frac{2}{\sqrt{\pi}} \int_z^\infty \mathrm{e}^{-t^2} \mathrm{d}t. \tag{2.7}$$

It can be seen that the relationship between the two functions is

$$\mathrm{erfc}(z) = 1 - \mathrm{erf}(z). \tag{2.8}$$

Function `erfc()` provided in MATLAB can be used to evaluate the complementary error function erfc(z), with the syntax `y=erfc(z)`. It should be noted that these two functions can only deal real z arguments. If z is a complex one, numerical integrals should be used, which will be demonstrated later through examples.

▶ Example 2.1

Draw the curves of error and complementary error functions.

Solutions The following commands can be used to draw error and complementary error functions, as shown in Fig. 2.1.

```
>> x=-5:0.1:5; y1=erf(x); y2=erfc(x); plot(x,y1,x,y2,'--')
```

If symbolic expressions are used, the following MATLAB can be issued, and the plots obtained are the same.

```
>> syms z, fplot([erf(z),erfc(z)],[-5,5])      % the same results found
```

◀

▶ Example 2.2

Compute complex error function erf$(1+\mathrm{j}2)$ with numerical method.

Solutions The `erf()` function in the current version of MATLAB does not support complex arguments. Numerical integrals should be employed to evaluate such error functions. Anonymous function can be used to describe the integrand. Note that, dot operations must be used. The numerical integral solution can be obtained with `integral()` function, which yields erf$(1+\mathrm{j}2) = -0.5366 - \mathrm{j}5.0491$.

```
>> f=@(t)2*exp(-t.^2)/sqrt(pi); I=integral(f,0,1+2i)
   I=vpa(erf(sym(1+2i)))      % direct computing insymbolic framework
```

Under symbolic framework, a much more accurate solution can be found, which is $I = -0.53664356577856503399179555931419 - 5.0491437034470346695430369586141\mathrm{j}$. It is seen that the numerical solution obtained above is also very accurate. It is usually adequate to use numerical method for this type of problems. ◀

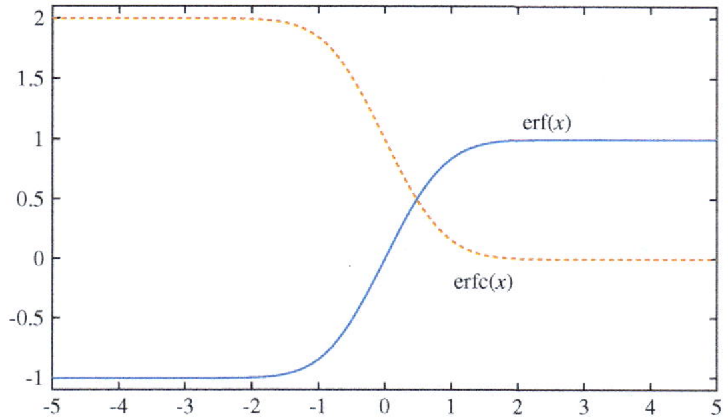

Fig. 2.1 Error and complementary error functions

2

2.2 Gamma Functions

Gamma function is the fundamental in fractional calculus, since many of the fractional calculus definitions involve Gamma functions. In this section, the definition of Gamma function is proposed first, followed by the properties and computations of Gamma functions.

2.2.1 Definition and Properties of Gamma Functions

The mathematical concepts of factorials were established in very old days. The notation $n!$ can be used to represent the factorial of n, if n is a nonnegative integer. What happens if $n = x$, while x is not a nonnegative integer? The concept of factorials should be extended. In the eighteenth century, Swiss mathematician Leonhard Euler found an integral expression.

$$\Gamma(x+1) = \int_0^1 (-\ln t)^x \mathrm{d}t = \int_0^\infty \mathrm{e}^{-t} t^x \mathrm{d}t, \tag{2.9}$$

which is a generalized function of factorials. Further generalization to the formula, for instance, let x be any complex variable z, the following formal definition of Gamma function is proposed.

Definition 2.3 Gamma function

Gamma function is defined by the following infinite integral:

$$\Gamma(z) = \int_0^\infty \mathrm{e}^{-t} t^{z-1} \mathrm{d}t. \tag{2.10}$$

Theorem 2.2 A important property of Gamma function

A very important property of Gamma function is

$$\Gamma(z+1) = z\Gamma(z). \tag{2.11}$$

Proof Consider the integration by parts method in calculus

$$\int u(t)\mathrm{d}v(t) = u(t)v(t) - \int v(t)\mathrm{d}u(t). \tag{2.12}$$

Assume that $u(t) = \mathrm{e}^{-t}$ and $v(t) = t^z/z$, it is found that $\mathrm{d}u(t) = \mathrm{e}^{-t}\mathrm{d}t$ and $\mathrm{d}v(t) = t^{z-1}\mathrm{d}t$. It follows that

$$\Gamma(z)=\int_0^\infty u(t)\mathrm{d}v(t)=\frac{1}{z}\,\mathrm{e}^{-t}t^z\Big|_0^\infty+\frac{1}{z}\int_0^\infty \mathrm{e}^{-t}t^z\mathrm{d}t=\frac{1}{z}\,\Gamma(z+1).$$

Therefore, the property in (2.11) can be proved. □

Corollary 2.1 *As a specific example, for a nonnegative integer z, (2.11) can be used directly to compute the factorial.*

$$\Gamma(z+1)=z\Gamma(z)=z(z-1)\Gamma(z-1)=\cdots=z! \tag{2.13}$$

Proof It is obvious that $\Gamma(1)=-\mathrm{e}^{-t}\Big|_0^\infty=1$. Therefore, considering Theorem 2.2, Corollary 2.1 can be proved immediately. □

Gamma functions can be regarded as the interpolations of factorials. On the other hand, Gamma function can also be considered as the extension of factorials on z in the complex domain. If z is a negative integer, $\Gamma(z+1)$ tends alternatively to $+\infty$ and $-\infty$.

In MATLAB, function `y=gamma(x)` can be used to compute directly Gamma functions. If $\boldsymbol{x}$ is a vector, the returned $\boldsymbol{y}$ is also a vector. In addition, $\boldsymbol{x}$ can be matrices or other data types. The structure of the returned $\boldsymbol{y}$ is the same as the $\boldsymbol{x}$ argument. Note that, the function `gamma()` can only be used to deal with real input arguments. The evaluation of complex Gamma functions will be discussed later.

▶ Example 2.3

Draw the Gamma function in the interval $x\in(-5,5)$.

Solutions The curve of Gamma function can be obtained with the following commands, as shown on Fig. 2.2. Since the values of Gamma function at $z=0,-1,-2,\cdots$ tend to infinity, function `ylim()` can be called to restrict the range of y-axis, such that the curves obtained are meaningful.

```
>> a=-5:0.002:5; plot(a,gamma(a)), ylim([-15,15])
   hold on, v=[1:4]; plot(v,gamma(v),'o'), hold off
```

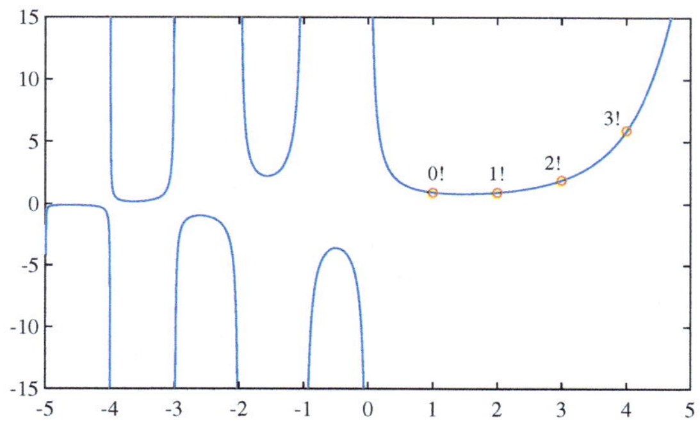

Fig. 2.2 Gamma function curves

The same curves can be found with symbolic expressions, and the asymptotes are also drawn.

```
>> syms x; fplot(gamma(x),[-5,5]), ylim([-15,15])
```

◀

For certain specific values of z, Gamma function values can be found.

$$\Gamma\left(\frac{1}{2}\right)=\sqrt{\pi},\ \Gamma\left(\frac{3}{2}\right)=\frac{\sqrt{\pi}}{2},\ \Gamma\left(\frac{5}{2}\right)=\frac{3\sqrt{\pi}}{4},\ \Gamma\left(\frac{7}{2}\right)=\frac{15\sqrt{\pi}}{8},\cdots \tag{2.14}$$

▶ Example 2.4

Prove (2.14) with MATLAB.

Solutions The Gamma function values in (2.14) can be obtained directly with

```
>> syms t z; Gam=int(exp(-t)*t^(z-1),t,0,inf);
   I1=subs(Gam,z,sym(1/2)), I2=subs(Gam,z,sym(3/2))
   I3=subs(Gam,z,sym(5/2)), I4=subs(Gam,z,sym(7/2))
```

Of course, `gamma()` function can also be used to prove (2.14).

```
>> I1=gamma(sym(1/2)), I2=gamma(sym(3/2))
   I3=gamma(sym(5/2))), I4=gamma(sym(7/2))
```

◀

Theorem 2.3 Gamma function formulas

More generally, if z is not an integer, Gamma function can be extended to

$$\Gamma\left(z+\frac{1}{m}\right)=(2\pi)^{(m-1)/2}m^{1/2-mz}\,\Gamma(mz). \tag{2.15}$$

Corollary 2.2 *It can be seen from Theorem 2.3 that, if $m=2$, then*

$$\Gamma\left(z+\frac{1}{2}\right)=\frac{\sqrt{\pi}\,\Gamma(2z+1)}{2^{2z}\,\Gamma(z+1)}. \tag{2.16}$$

Theorem 2.4 Gamma function and Legendre formulas

Gamma function satisfies the following Legendre formulas:

$$\Gamma(z)\,\Gamma\left(z+\frac{1}{2}\right)=2^{1-2z}\sqrt{\pi}\,\Gamma(2z). \tag{2.17}$$

Proof Through a series formulations with (2.16), the theorem can be proved.

$$\Gamma\left(z+\frac{1}{2}\right)=\frac{\sqrt{\pi}\,\Gamma(2z+1)}{2^{2z}\Gamma(z+1)}=\frac{\sqrt{\pi}\,2z\,\Gamma(2z)}{2^{2z}z\,\Gamma(z)}=2^{1-2z}\sqrt{\pi}\,\frac{\Gamma(2z)}{\Gamma(z)}.$$

□

In addition, Gamma function satisfies the following equation:

$$\Gamma(z)\Gamma(1-z)=\frac{\pi}{\sin\pi z}. \tag{2.18}$$

These equations can be derived manually [1] or with MATLAB symbolic expressions. It can be found that

$$\Gamma(z)\Gamma(-z)=\frac{-\pi}{z\sin\pi z},\quad \Gamma(1+z)\Gamma(1-z)=\frac{\pi z}{\sin\pi z}. \tag{2.19}$$

Theorem 2.5 Gamma function property
Gamma function satisfies the following equation:

$$\Gamma\left(\frac{1}{2}+z\right)\Gamma\left(\frac{1}{2}-z\right)=\frac{\pi}{\cos\pi z},\quad \lim_{z\to\infty}\frac{z^a\Gamma(z)}{\Gamma(z+a)}=1,\ \ \mathbb{R}(a)>0, \tag{2.20}$$

where $\mathbb{R}(a)$ means to extract the real part of a.

▶ Example 2.5

Shown (2.18)~(2.20) with symbolic computing in MATLAB.

Solutions The Gamma function can be defined by a symbolic expression. Then function `subs()` can be used in variable substitution, so as to show that the above formulas are satisfied. Note that, under new versions of MATLAB, I_{22} cannot be simplified.

```
>> syms t z; Gam=int(exp(-t)*t^(z-1),t,0,inf);
   I1=simplify(Gam*subs(Gam,z,1-z))
   I21=simplify(Gam*subs(Gam,z,-z))
   I22=simplify(subs(Gam,z,1+z)*subs(Gam,z,1-z))
   I3=simplify(subs(Gam,z,1/2+z)*subs(Gam,z,1/2-z))
   syms a positive; I4=limit(z^a*Gam/subs(Gam,z,a+z),z,inf)
```

If `gamma()` function is called, the same results can be found under symbolic framework.

```
>> syms z; I1=simplify(gamma(z)*gamma(1-z))
   I21=simplify(gamma(z)*gamma(-z))
   I3=simplify(gamma(1/2+z)*gamma(1/2-z))
   syms a positive; I4=limit(z^a*gamma(z)/gamma(a+z),z,inf)
```

◀

2

▶ Example 2.6

Some integral problems can be expressed by Gamma functions. Find the analytical solutions of the following definite integrals:

$$I_1 = \int_0^{\pi/2} \sin^{2m-1} t \cos^{2n-1} t \,\mathrm{d}t,\quad I_2 = \int_0^{\infty} t^{x-1} \cos t \,\mathrm{d}t,\ x > 0.$$

Solutions The following commands can be used to solve the two integral problems:

```
>> syms m n t z; syms x positive
   I1=int(sin(t)^(2*m-1)*cos(t)^(2*n-1),t,0,sym(pi)/2)
   I2=simplify(int(t^(x-1)*cos(t),t,0,inf))
```

It is found that

$$I_1 = \frac{\Gamma(n)\Gamma(m)}{2\Gamma(n+m)},\quad I_2 = \frac{2^{x-1}\sqrt{\pi}\,\Gamma(x/2)}{\Gamma((1-x)/2)}.$$

◀

2.2.2 Complex Gamma Functions

MATLAB function `gamma(z)` can only be used to evaluate Gamma functions with real arguments. In many applications, for instance, in fractional-order integral expressions, Gamma functions with complex arguments are involved. It is necessary to extend the `gamma()` functions for the Gamma functions with complex arguments. It is obvious that numerical integrals can be used with the following MATLAB function:

```
function y=gamma_c(z)
  if isreal(z), y=gamma(z);          %only real arguments involved
  else, f=@(t)exp(-t).*t.^(z-1); %otherwise use numerical integral
     y=integral(f,0,inf,'ArrayValued',true);
end, end
```

With the `gamma()` function under symbolic framework, the Gamma functions with complex arguments can be evaluated.

▶ Example 2.7

Compute Gamma function values $\Gamma(2+2\mathrm{j})$ and $\Gamma(2-2\mathrm{j})$.

Solutions With the new `gamma_c()` function, it is found directly that $\Gamma(2\pm 2\mathrm{j}) = 0.1123 \pm 0.3236\mathrm{j}$. It is seen that, Gamma functions with complex arguments can be evaluated directly. If symbolic framework is adopted, it is found that the Gamma function values are $0.11229424234632617354 34 \pm 0.32361288550192725686823\mathrm{j}$.

```
>> z=[2+2i, 2-2i]; I=gamma_c(z), I2=vpa(gamma(sym(z)))
```

◀

2.2.3 Other Forms of Gamma Functions

Apart from the integral definitions, Gamma functions can also be represented in other forms.

Theorem 2.6 Infinite product descriptions of Gamma functions
Gamma function can be expressed in infinite product form.

$$\Gamma(z) = \frac{1}{z} e^{-\gamma z} \prod_{n=1}^{\infty} \left(\frac{n}{n+z}\right) e^{z/n}, \tag{2.21}$$

where $\gamma \approx 0.57721566490153286$ is the Euler constant.

Theorem 2.7 Limit representation of Gamma functions
Gamma functions can also be represented in limit form.

$$\Gamma(z) = \lim_{n\to\infty} \frac{n^z n!}{z(z+1)(z+2)\cdots(z+n)}. \tag{2.22}$$

It is seen that the poles of Gamma functions are at $z = 0, -1, -2, \cdots$.

It is a pity that if infinite product is used to compute Gamma functions, the computational load is even heavier. Therefore, these theorems are not suitable in computing Gamma functions, and the methods are not adopted in this book.

2.2.4 Incomplete Gamma Functions

Definition 2.4 Incomplete Gamma function

Incomplete Gamma function is defined as

$$\Gamma(z, \alpha) = \frac{1}{\Gamma(\alpha)} \int_0^z e^{-t} t^{\alpha-1} dt, \quad z \geqslant 0. \tag{2.23}$$

Function `gammainc()` can be used to evaluate incomplete Gamma functions in MATLAB. The syntax of the function is y=`gammainc`(z, α).

▶ Example 2.8

Draw the incomplete Gamma functions for different values of α.

Solutions The following commands can be used to draw directly the incomplete Gamma function curve, as shown in Fig. 2.3.

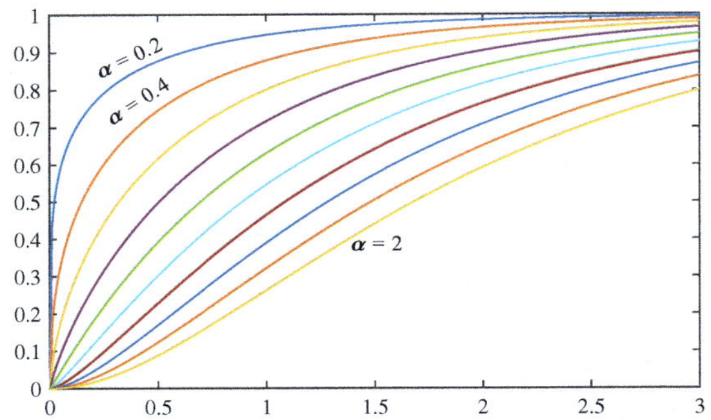

Fig. 2.3 The curve of the incomplete Gamma function

```
>> x=0:0.01:3; b=0.2:0.2:2;
   for a=b, plot(x,gammainc(x,a)); hold on; end
   hold off
```

◀

2.3 Beta Functions

The definitions and properties in Beta functions are presented in this section. Other forms of Beta functions are also presented.

2.3.1 Definition and Properties of Beta Functions

Beta function is a special function which is quite similar to Gamma function. It is also closely related to Gamma function.

Definition 2.5 Beta function definition

The definition of Beta function is

$$\mathrm{B}(z,m)=\int_0^1 t^{z-1}(1-t)^{m-1}\mathrm{d}t,\quad \mathbb{R}(m)>0,\ \mathbb{R}(z)>0. \tag{2.24}$$

Theorem 2.8 The relationship between Gamma and Beta functions

The relationship between Gamma and Beta functions is

$$\mathrm{B}(m,z)=\frac{\Gamma(m)\Gamma(z)}{\Gamma(m+z)}. \tag{2.25}$$

Proof It is known from Definition 2.5 that

$$\begin{aligned}\Gamma(m)\Gamma(z) &= \int_0^\infty \mathrm{e}^{-u}u^{m-1}\mathrm{d}u \int_0^\infty \mathrm{e}^{-v}v^{z-1}\mathrm{d}v \\ &= \int_0^\infty \int_0^\infty \mathrm{e}^{-(u+v)}u^{m-1}v^{z-1}\mathrm{d}u\mathrm{d}v.\end{aligned} \tag{2.26}$$

Define new variables $u = f(w, t) = wt$ and $v = g(w, t) = w(1 - t)$, where $w \in [0, \infty)$ and $t \in [0, 1]$. Substituting them into (2.26), it is found that

$$\Gamma(m)\Gamma(z) = \int_0^\infty \int_0^1 \mathrm{e}^{-w}(wt)^{m-1}\big[w(1-t)\big]^{z-1}\big|\boldsymbol{J}(w,t)\big|\mathrm{d}t\mathrm{d}w, \tag{2.27}$$

where $\big|\boldsymbol{J}(w,t)\big|$ is the absolute value of the Jacobian determinant of $u = f(w, t)$ and $v = g(w, t)$.

$$\boldsymbol{J}(w,t) = \begin{vmatrix} \partial u/\partial w & \partial u/\partial t \\ \partial v/\partial w & \partial v/\partial t \end{vmatrix} = \begin{vmatrix} t & w \\ 1-t & -w \end{vmatrix} = -w.$$

Substituting $\big|\boldsymbol{J}(w,t)\big| = w$ into (2.27), it is found

$$\Gamma(m)\Gamma(z) = \int_0^\infty \mathrm{e}^{-w}w^{z+m-1}\mathrm{d}w \int_0^1 t^{m-1}(1-t)^{z-1}\mathrm{d}t = \Gamma(m+z)\mathrm{B}(m,z).$$

Hence the theorem is proved. □

Corollary 2.3 *It is known from Theorem 2.8 that*

$$\mathrm{B}(z,m) = \mathrm{B}(m,z). \tag{2.28}$$

It can be seen from the above property that, the integral value I_1 in Example 2.6 can be further simplified to $\mathrm{B}(m,n)/2$. Function `y=beta(z,m)` provided in MATLAB can be used to evaluate Beta function, where z is the given vector. It is a pity that the function can only be used to evaluate nonnegative input arguments z and $\boldsymbol{m}$.

Even though negative parameters appear in Beta functions, Theorem 2.8 can still be used to find Beta function with `y=gamma(z).*gamma(m)./gamma(z+m)`.

2.3.2 Complex Beta Functions

Only real arguments m and z are supported in the Beta function here. If Beta functions with complex arguments are expected, the following MATLAB function can be written, such that the Beta function can be evaluated with numerical integrals.

```
function y=beta_c(z,m)
   if isreal(z), y=beta(z,m);
   else, f=@(t)t.^(z-1).*(1-t).^(m-1);
```

```
      y=integral(f,0,1,'ArrayValued',true);
end, end
```

2

▶ Example 2.9

Draw Beta functions under different values of m.

Solutions Selecting, respectively, $m = 1$ and $m = 3$, the extended Beta functions are drawn, as shown in Fig. 2.4.

```
>> x=-3:0.01:3; m=1; y1=gamma(x).*gamma(m)./gamma(x+m);
   m=3; y2=gamma(x).*gamma(m)./gamma(x+m);
   plot(x,y1,x,y2,'--'); ylim([-20 20])
```

◀

▶ Example 2.10

Draw the contour plots for the Beta functions.

Solutions A set of mesh grids can be generated, and the Beta function values at the mesh grids can be evaluated. With function `contour()`, the contour plots of Beta functions can be obtained, as shown in Fig. 2.5.

```
>> [x,y]=meshgrid(0:0.05:3); v=beta(x,y); contour(x,y,v,100)
```

◀

Direct extending Definition 2.5, the multivariate Beta function is introduced.

Definition 2.6 Multivariate Beta function

The mathematical form of multivariate Beta function is

$$B(\alpha_1, \alpha_2, \cdots, \alpha_n) = \frac{\Gamma(\alpha_1)\Gamma(\alpha_2)\cdots\Gamma(\alpha_n)}{\Gamma(\alpha_1 + \alpha_2 + \cdots + \alpha_n)}. \tag{2.29}$$

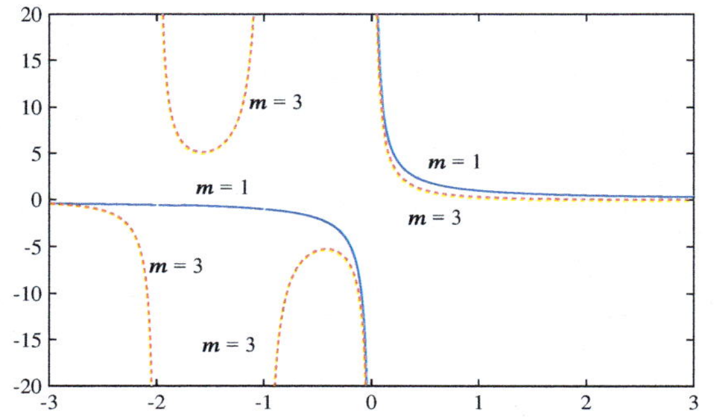

Fig. 2.4 The curves of extended Beta functions for $m = 1$ and $m = 3$

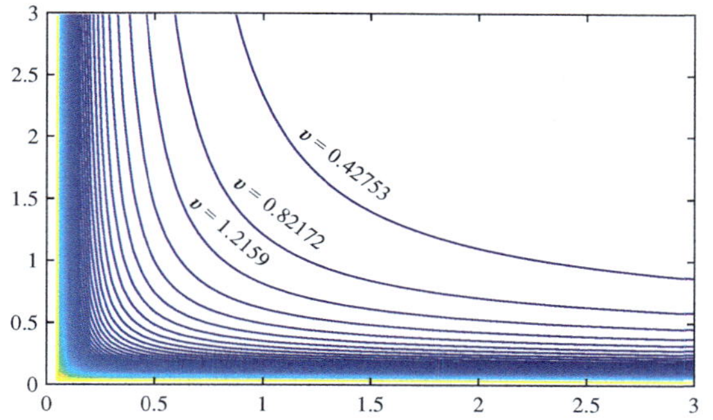

Fig. 2.5 Contour plots of Beta functions

2.3.3 Incomplete Beta Functions

Definition 2.7 Incomplete Beta function

Similar to the incomplete Gamma function, normalized incomplete Beta function is also defined.

$$B_x(z,m) = \frac{1}{B(z,m)} \int_0^x t^{z-1}(1-t)^{m-1}\mathrm{d}t,\ \mathbb{R}(m) > 0,\ \mathbb{R}(z) > 0, \tag{2.30}$$

where $0 \leqslant x \leqslant 1$.

Incomplete Beta functions can be evaluated directly with the MATLAB function `y=betainc(x,z,m)`.

2.4 Dawson Functions

Definition 2.8 Dawson function

Dawson function is defined as

$$\mathrm{daw}(z) = \mathrm{e}^{-z^2} \int_0^z \mathrm{e}^{\tau^2}\mathrm{d}\tau. \tag{2.31}$$

Dawson function is an odd function, i.e., $\mathrm{daw}(z) = -\mathrm{daw}(-z)$. Dawson function can also be written as

$$\mathrm{daw}(z) = -\frac{\mathrm{j}\sqrt{\pi}}{2}\mathrm{e}^{-z^2}\mathrm{erf}(\mathrm{j}z). \tag{2.32}$$

2

Theorem 2.9 Dawson function derivatives
The first-order derivative of Dawson function is

$$\frac{d}{dz}\mathrm{daw}(z) = 1 - 2z\,\mathrm{daw}(z). \tag{2.33}$$

Proof The theorem can be proved directly by the equation in (2.31).

$$\frac{d}{dz}\mathrm{daw}(z) = -2z\,e^{-z^2}\int_0^z e^{\tau^2}d\tau + e^{-z^2}e^{z^2} = 1 - 2z\,\mathrm{daw}(z).$$

□

Dawson function can also be directly evaluated through infinite series [2].

$$\mathrm{daw}(z) = z - \frac{2}{3}z^3 + \frac{4}{15}z^5 + \cdots = z\sum_{k=0}^{\infty}\frac{(-2z^2)^k}{(2k+1)!!}, \tag{2.34}$$

where $(2k+1)!! = (2k+1)(2k-1)(2k-3)\cdots 3\times 1$.

In the MuPAD engine under MATLAB symbolic framework, an interface to the low-level function `dawson()` can be called, with the syntax `y=feval(symengine,'dawson',x)`. The command can be used to compute Dawson functions. The returned argument y is a symbolic expression. If numerical solutions are expected, variable substitution should be made with `subs()` function. The function evaluation is demonstrated through examples.

▶ Example 2.11

Draw the Dawson functions $\mathrm{daw}(x)$ and $\mathrm{daw}(\sqrt{x})$.

Solutions The curves of the two Dawson functions can be obtained, as shown in Fig. 2.6.

```
>> syms x positive; y1=feval(symengine,'dawson',x);
   y2=feval(symengine,'dawson',sqrt(x)); fplot([y1,y2],[0,5])
```

◀

▶ Example 2.12

It is known from the first-order derivative of Dawson function $f(x)$ is $f'(x) = 1 - 2xf(x)$. Find high-order derivative formulas.

Solutions The symbolic function $f(x)$ is declared under MATLAB first. Then variable substitution method can be used to derive high-order derivatives.

```
>> syms x f(x); f1=1-2*x*f; f2=f1;
   for i=1:5
         f2=diff(f2,x); f2=expand(subs(f2,diff(f,x),f1))
   end
```

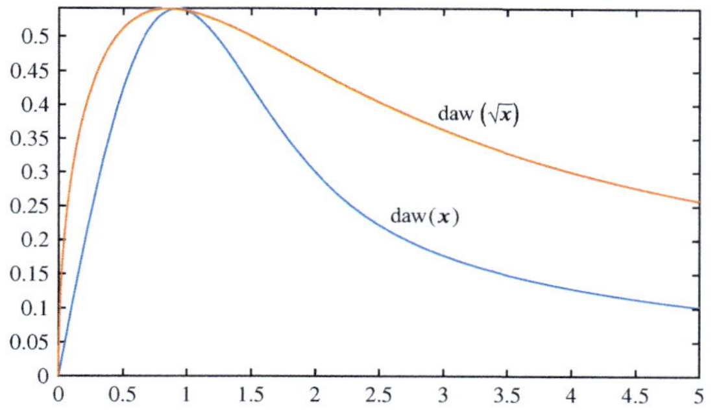

Fig. 2.6 Dawson function curves

A series of derivative functions for Dawson function are

$$\begin{aligned}
\mathrm{daw}''(x) &= -2x + (-2+4x^2)\,\mathrm{daw}(x)\\
\mathrm{daw}'''(x) &= -4+4x^2 + (12x-8x^3)\,\mathrm{daw}(x)\\
\mathrm{daw}^{(4)}(x) &= 20x - 8x^3 + (12-48x^2+16x^4)\,\mathrm{daw}(x)\\
\mathrm{daw}^{(5)}(x) &= 32-72x^2+16x^4 + (-120x+160x^3-32x^5)\,\mathrm{daw}(x)\\
\mathrm{daw}^{(6)}(x) &= -264x+224x^3-32x^5+(-120+720x^2-480x^4+64x^6)\,\mathrm{daw}(x).
\end{aligned}$$

◀

2.5 Hypergeometric Functions

Definition 2.9 Hypergeometric function

The general form of a hypergeometric function is [3]

$$\begin{aligned}
&{}_p\mathrm{F}_q(a_1, a_2, \cdots, a_p; b_1, b_2, \cdots, b_q; z)\\
&= \frac{\Gamma(b_1)\Gamma(b_2)\cdots\Gamma(b_q)}{\Gamma(a_1)\Gamma(a_2)\cdots\Gamma(a_p)} \sum_{k=0}^{\infty} \frac{\Gamma(a_1+k)\Gamma(a_2+k)\cdots\Gamma(a_p+k)}{\Gamma(b_1+k)\Gamma(b_2+k)\cdots\Gamma(b_q+k)} \frac{z^k}{k!},
\end{aligned} \tag{2.35}$$

where b_i cannot be non-positive integers. If $p \leqslant q$, the function converges for all the values of z. If $p = q+1$, the function converges when $|z| < 1$. If $p > q+1$, the function diverges for all the z.

Definition 2.10 Another form of hypergeometric functions

Hypergeometric functions are also defined as

$${}_p\mathrm{F}_q(a_1, a_2, \cdots, a_p; b_1, b_2, \cdots, b_q; z) = \sum_{k=0}^{\infty} \frac{(a_1)_k(a_2)_k\cdots(a_p)_k}{(b_1)_k(b_2)_k\cdots(b_q)_k} \frac{z^k}{k!}, \tag{2.36}$$

where $(\gamma)_k$ is the Pochhammer symbol.

2

Definition 2.11 Pochhammer symbol

Pochhammer symbol is defined as

$$(\gamma)_k = \gamma(\gamma+1)(\gamma+2)\cdots(\gamma+k-1) = \frac{\Gamma(k+\gamma)}{\Gamma(\gamma)}. \tag{2.37}$$

Pochhammer symbol is also known as the rising factorial. In the Pochhammer symbol, if $\gamma = 1$, then

$$(1)_k = \frac{\Gamma(k+1)}{\Gamma(1)} = k! \tag{2.38}$$

In many hypergeometric functions, two special forms are widely used in fractional calculus. These two functions are, respectively, Kummer hypergeometric function and Gauss hypergeometric function.

Definition 2.12 Kummer hypergeometric function

If $p = q = 1$, the hypergeometric function is Kummer hypergeometric function, also known as confluent hypergeometric function, whose definition is

$${}_1\mathrm{F}_1(a; b; z) = \frac{\Gamma(b)}{\Gamma(a)} \sum_{k=0}^{\infty} \frac{\Gamma(a+k)}{\Gamma(b+k)\,k!} z^k. \tag{2.39}$$

Kummer hypergeometric function is often denoted as $\mathrm{M}(a, b, z)$. If $b > a$, the function can be described as

$${}_1\mathrm{F}_1(a; b; z) = \frac{\Gamma(b)}{\Gamma(a)\Gamma(b-a)} \int_0^1 t^{a-1}(1-t)^{b-a-1}\mathrm{e}^{zt}\mathrm{d}t. \tag{2.40}$$

Theorem 2.10 The derivative of hypergeometric function

The first-order derivative of hypergeometric function ${}_1\mathrm{F}_1\,(a; b; z)$ satisfies

$$\frac{\mathrm{d}}{\mathrm{d}z}{}_1\mathrm{F}_1\,(a; b; z) = \frac{a}{c}\,{}_1\mathrm{F}_1(a+1; c+1; z). \tag{2.41}$$

Proof The following formulation leads directly to proof of the theorem, where variable substitution $k' = k - 1$ is made.

$$\begin{aligned}\frac{\mathrm{d}}{\mathrm{d}z}{}_1\mathrm{F}_1\,(a; b; z) &= \frac{\Gamma(c)}{\Gamma(a)} \sum_{k=1}^{\infty} \frac{\Gamma(a+k)}{\Gamma(c+k)\,k!} kz^{k-1} = \frac{\Gamma(c)}{\Gamma(a)} \sum_{k=1}^{\infty} \frac{\Gamma(a+k)}{\Gamma(c+k)\,(k-1)!} z^{k-1} \\ &= \frac{\Gamma(c)}{\Gamma(a)} \sum_{k'=0}^{\infty} \frac{\Gamma(a+k'+1)}{\Gamma(c+k'+1)} \frac{z^{k'}}{k'!} = \frac{a}{c} \frac{\Gamma(c+1)}{\Gamma(a+1)} \sum_{k=0}^{\infty} \frac{\Gamma(a+k+1)}{\Gamma(c+k+1)} \frac{z^k}{k!}.\end{aligned}$$

□

Definition 2.13 Gauss hypergeometric function

If $p = 2$ and $q = 1$, the function is referred to as Gauss hypergeometric function, whose definition is

$$_2\mathrm{F}_1(a, b; c; z) = \frac{\Gamma(c)}{\Gamma(a)\Gamma(b)} \sum_{k=0}^{\infty} \frac{\Gamma(a+k)\Gamma(b+k)}{\Gamma(c+k)\,k!} z^k. \tag{2.42}$$

If c is a nonnegative integer, the function converges for all the values of z.

Theorem 2.11 The relationship between two hypergeometric functions

The relationship between two hypergeometric functions is

$$_1\mathrm{F}_1(a; c; z) = \lim_{b\to\infty} {_2\mathrm{F}_1}\left(a, b; c; \frac{z}{b}\right). \tag{2.43}$$

Proof Consider the expression on the right-hand side of the equation

$$_2\mathrm{F}_1\left(a, b; c; \frac{z}{b}\right) = \frac{\Gamma(c)}{\Gamma(a)\,\Gamma(b)} \sum_{k=0}^{\infty} \frac{\Gamma(a+k)\,\Gamma(b+k)}{\Gamma(c+k)\,k!} \left(\frac{z}{b}\right)^k$$

$$= \frac{\Gamma(c)}{\Gamma(a)} \sum_{k=0}^{\infty} \frac{\Gamma(a+k)\,\overbrace{(b+k)(b+k-1)\cdots(b+1)}^{k\text{ term}}}{\Gamma(c+k)\,b^k\,k!} z^k.$$

Taking the limit for $b \to \infty$, the k terms in the numerator cancel the b^k term in the denominator, which complete the proof. □

Function `hypergeom()` is provided in MATLAB Symbolic Math Toolbox, which can be used in the evaluation of hypergeometric function ${_p\mathrm{F}_q}(a_1, a_2, \cdots, a_p; b_1, b_2, \cdots, b_q; z)$, with syntax `y=hypergeom([a1,a2,···,ap],[b1,b2,···,bq],z)`.

Hypergeometric function ${_1\mathrm{F}_1}(a, b; z)$ can be evaluated with `y=hypergeom(a,b,z)` function, and hypergeometric function ${_2\mathrm{F}_1}(a, b; c; z)$ can be evaluated with the syntax `y=hypergeom([a,b],c,z)`.

▶ Example 2.13

Draw the Gauss hypergeometric function ${_2\mathrm{F}_1}\,(1.5, -1.5; 1/2; (1 - \cos x)/2)$.

Solutions Hypergeometric function ${_2\mathrm{F}_1}\,(1.5, -1.5; 1/2; (1 - \cos x)/2)$ can be simplified to $\cos 1.5x$. If function `hypergeom()` is used, the hypergeometric function curve can be obtained, as shown in Fig. 2.7. It can be seen that the curve agrees well with $\cos 1.5x$.

```
>> syms x, f1=cos(1.5*x);
   f2=hypergeom([1.5,-1.5],0.5,0.5*(1-cos(x)));
   fplot([f1,f2],[-pi,pi])
```

◀

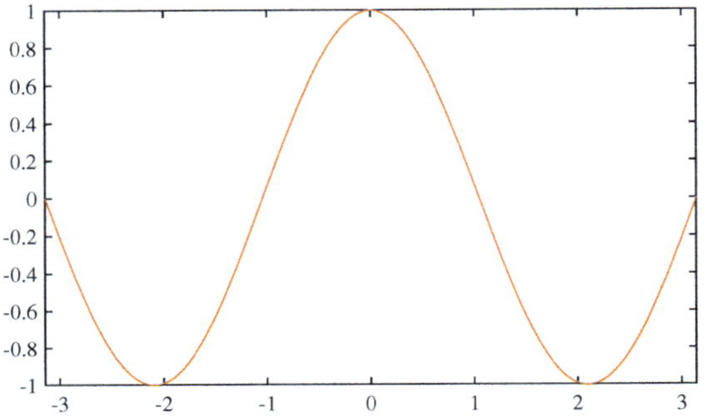

Fig. 2.7 The hypergeometric function in Example 2.13

Theorem 2.12 Complementary error function
Complementary error function can be converted into a hypergeometric function [3]

$$\mathrm{j}^n\mathrm{erfc}(z)=\mathrm{e}^{-z^2}\left[\frac{{}_1\mathrm{F}_1\left((n+1)/2;\,1/2;\,z^2\right)}{2^n\Gamma(1+n/2)}-\frac{z\,{}_1\mathrm{F}_1\left(n/2+1;\,3/2;\,z^2\right)}{2^{n-1}\Gamma((n+1)/2)}\right]. \tag{2.44}$$

Theorem 2.13 Incomplete Gamma function
Incomplete Gamma function can be converted into a hypergeometric function

$$\Gamma(x,a)=\frac{1}{a}x^a\mathrm{e}^{-x}\,{}_1\mathrm{F}_1(1;\,1+a;\,x)=\frac{1}{a}x^a\,{}_1\mathrm{F}_1(a;\,1+a;\,-x). \tag{2.45}$$

2.6 Mittag-Leffler Functions

Mittag-Leffler function is a direct extension of exponential functions. The simplest form of Mittag-Leffler function was proposed by Swedish mathematician Magnus Gustaf (Gösta) Mittag-Leffler (1846–1927) in 1903 [4]. The function is also known as one-parameter Mittag-Leffler function. Mittag-Leffler functions are later extended to the cases with two or more parameters. The role of Mittag-Leffler functions in fractional calculus is as important as exponential functions in integer-order systems. The definitions and computation methods of various Mittag-Leffler functions are presented in this section.

2.6.1 One-Parameter Mittag-Leffler Functions

Taylor series expansion of exponential function e^z is presented first.

$$e^z = \sum_{k=0}^{\infty} \frac{z^k}{k!} = \sum_{k=0}^{\infty} \frac{z^k}{\Gamma(k+1)}. \tag{2.46}$$

If the $k+1$ term in the Gamma function is changed to $\alpha k+1$, one-parameter Mittag-Leffler function can be introduced. It can be seen that Mittag-Leffler function is a direct extension of exponential functions.

Definition 2.14 One-parameter Mittag-Leffler function

The mathematical form of one-parameter Mittag-Leffler function is

$$E_\alpha(z) = \sum_{k=0}^{\infty} \frac{z^k}{\Gamma(\alpha k+1)}, \tag{2.47}$$

where $\alpha \in \mathbb{C}$, $\mathbb{C}$ is the complex set. The convergent condition of an infinite series is $\mathbb{R}(\alpha) > 0$.

In other words, exponential function e^z is a special case of one-parameter Mittag-Leffler function, $E_1(z) = e^z$. In addition, it is easily seen that

$$E_2(z) = \sum_{k=0}^{\infty} \frac{z^k}{\Gamma(2k+1)} = \sum_{k=0}^{\infty} \frac{(\sqrt{z})^{2k}}{(2k)!} = \cosh\sqrt{z} \tag{2.48}$$

$$E_{1/2}(z) = \sum_{k=0}^{\infty} \frac{z^k}{\Gamma(k/2+1)} = e^{z^2}\left[1+\operatorname{erf}(z)\right] = e^{z^2}\operatorname{erfc}(-z). \tag{2.49}$$

Function `symsum()` provided in the Symbolic Math Toolbox can be used to compute infinite series. Therefore, it can be seen from (2.47) that Mittag-Leffler function can be computed analytically with the following MATLAB function. The function here can also be used to compute multi-parameter Mittag-Leffler functions. One-parameter Mittag-Leffler function with parameter α can be evaluated with function f=`mittag_leffler(`α,z`)`.

```
function f=mittag_leffler(v,z)
   arguments, v(1,:), z(1,1); end
   v=sym([v,1,1,1]); a=v(1); b=v(2); c=v(3); q=v(4); syms k;
   f=symsum(pochhammer(c,k*q)*z^k/gamma(a*k+b)/gamma(k+1),k,0,inf);
end
```

2

► Example 2.14

Symbolic method can be used to derive the Mittag-Leffler function formulas.

Solutions The following commands can be used to derive the commonly used Mittag-Leffler functions. Note that, the earlier versions of MATLAB, for instance, R2008a, used Maple as its symbolic engine, so that correct results can be found. While in the new versions, some of the formulas cannot be found. The old versions are recommended here. If α is set to $\alpha = 1/3, 3, 4, 5, \cdots$, the following statements can be used to evaluate Mittag-Leffler functions.

```
>> syms z; Ia=mittag_leffler(1,z)
   Ib=mittag_leffler(2,z), Ic=mittag_leffler(1/2,z)
   I1=mittag_leffler(1/3,z), I2=mittag_leffler(3,z)
   I3=mittag_leffler(4,z), I4=mittag_leffler(5,z)
```

The following results can be found.

$$\mathrm{E}_{1/3}(z) = -\frac{\mathrm{e}^{z^3}}{2\pi\,\Gamma(2/3)}\left[-6\pi\,\Gamma\left(\frac{2}{3}\right)+\sqrt{3}\Gamma^2\left(\frac{2}{3}\right)\Gamma\left(\frac{1}{3},z^3\right)+2\pi\,\Gamma\left(\frac{2}{3},z^3\right)\right]$$

$$\mathrm{E}_3(z) = \frac{1}{3}\,\mathrm{e}^{\sqrt[3]{z}}+\frac{2}{3}\,\mathrm{e}^{-\sqrt[3]{z}/2}\cos\frac{\sqrt{3}}{2}\sqrt[3]{z}$$

$$\mathrm{E}_4(z) = \frac{1}{4}\,\mathrm{e}^{\sqrt[4]{z}}+\frac{1}{4}\,\mathrm{e}^{-\sqrt[4]{z}}+\frac{1}{2}\cos\sqrt[4]{z}$$

$$\mathrm{E}_5(z) = \frac{\mathrm{e}^{\sqrt[5]{z}}}{5}+\frac{2}{5}\mathrm{e}^{\sqrt[5]{z}\cos\pi/5}\cos\left(\sqrt[5]{z}\sin\frac{2\pi}{5}\right)+\frac{2}{5}\mathrm{e}^{-\sqrt[5]{z}\cos\pi/5}\cos\left(\sqrt[5]{z}\sin\frac{\pi}{5}\right).$$

If the new versions of MATLAB are used, I_1 cannot be evaluated. Some of the functions can be expressed as hypergeometric functions.

$$I_2={}_0\mathrm{F}_2\left(;\frac{1}{3},\frac{2}{3};\frac{z}{27}\right),\ I_3={}_0\mathrm{F}_3\left(;\frac{1}{4},\frac{1}{2},\frac{3}{4};\frac{z}{256}\right),\ I_4={}_0\mathrm{F}_4\left(;\frac{1}{5},\frac{2}{5},\frac{3}{5},\frac{4}{5};\frac{z}{3125}\right).$$

◄

It can be found that

$$\mathrm{E}_n(z) = {}_0\mathrm{F}_{n-1}\left(;\frac{1}{n},\frac{2}{n},\cdots,\frac{n-1}{n};\frac{z}{n^n}\right). \tag{2.50}$$

It can be seen from (2.47) that, accumulative methods can be used to find the numerical solutions of Mittag-Leffler functions. The following MATLAB function can be written, whose syntax is `y=ml_func(α,z)`. In fact, the function can also be used to evaluate multi-parameter Mittag-Leffler function, and its nth derivatives.

```
function f=ml_func(aa,z,n,eps0) % Mittag-Leffler function computing
   arguments
      aa(1,:) double, z double
      n(1,1){mustBeNonnegativeInteger}=0, eps0(1,1)=eps;
   end
   aa=[aa,1,1,1]; a=aa(1); b=aa(2); c=aa(3); q=aa(4);
   f=0; k=0; fa=1;
```

```
if n==0                               % Mittag-Leffler function
   while norm(fa,1)>=eps0             % if not convergent, continue
      fa=prod(c:(c+q*k-1))/gamma(k+1)/gamma(a*k+b)*z.^k;
      f=f+fa; k=k+1;
   end
   if ~isfinite(f)                    % if not convergent
     if c*q==1, N=round(-log10(eps0));
        f=mlf(a,b,z(:),N); f=reshape(f,size(z));
     else, error('Error: truncation method failed');
     end, end
else
   aa(2)=aa(2)+n*aa(1); aa(3)=aa(3)+aa(4)*n;     % 3/4 pars
   f=pochhammer(c,q*n)*ml_func(aa,z,0,eps0);     % derivatives
end, end
```

Remark Numerical solutions of Mittag-Leffler functions

(1) Support Mittag-Leffler functions with up to four parameters.

(2) If the accumulated value tends to infinity, the current algorithm fails to find the numerical solution. The solver `mlf()` [5] by Prof. Podlubny is adopted automatically. Compared with `ml_func()` function, the speed is much lower.

(3) The embedded `mlf()` function cannot evaluate Mittag-Leffler functions with more than two parameters.

(4) The function `mustBeNonnegativeInteger()`, see FOTF Toolbox. □

▶ Example 2.15

Draw the Mittag-Leffler functions $E_\alpha(z)$ for different values of α.

Solutions For different values of α, the following loop structure can be used to draw the Mittag-Leffler functions, as shown in Fig. 2.8. It can be seen that, the curve of $\alpha = 1$ is the same as the one in e^x.

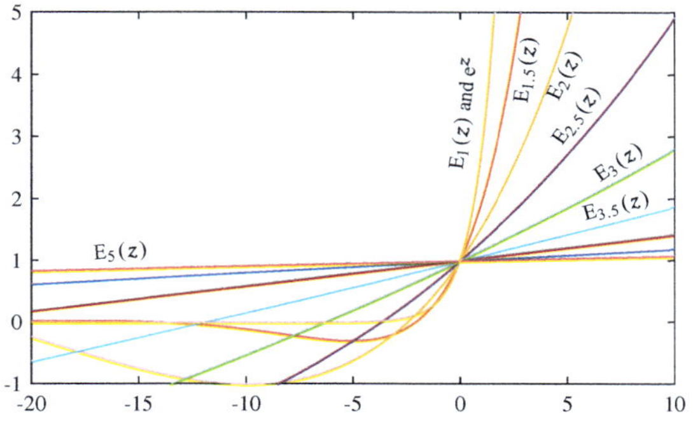

Fig. 2.8 The curve of one-parameter Mittag-Leffler function

```
>> a=[1:0.5:5]; t=[-20:0.1:10]; Y=[];
   for a=[1:0.5:5], Y=[Y; ml_func(a,t)]; end
   plot(t,Y), line(t,exp(t)); ylim([-1 5])
```

◄

2.6.2 Two-Parameter Mittag-Leffler Functions

Consider again the definition of one-parameter Mittag-Leffler function in (2.47). If 1 in the Gamma function is replaced by a variable β, two-parameter Mittag-Leffler function can be introduced.

Definition 2.15 Two-parameter Mittag-Leffler function

The definition of two-parameter Mittag-Leffler function is

$$\mathrm{E}_{\alpha,\beta}(z)=\sum_{k=0}^{\infty}\frac{z^k}{\Gamma(\alpha k+\beta)}, \tag{2.51}$$

where $\alpha,\beta\in\mathbb{C}$, and the infinite series is convergent for $z\in\mathbb{C}$ is $\mathbb{R}(\alpha)>0$ and $\mathbb{R}(\beta)>0$.

If $\beta=1$, the two-parameter Mittag-Leffler function can be reduced to the one-parameter Mittag-Leffler function, i.e.,

$$\mathrm{E}_{\alpha,1}(z)=\mathrm{E}_{\alpha}(z). \tag{2.52}$$

In other words, one-parameter Mittag-Leffler function is a special case of two-parameter Mittag-Leffler functions. Apart from the special case, it can be extended to the case for other Mittag-Leffler functions.

Some of the two-parameter Mittag-Leffler functions are

$$\mathrm{E}_{1,2}(z)=\sum_{k=0}^{\infty}\frac{z^k}{\Gamma(k+2)}=\frac{1}{z}\sum_{k=0}^{\infty}\frac{z^{k+1}}{(k+1)!}=\frac{\mathrm{e}^z-1}{z} \tag{2.53}$$

$$\mathrm{E}_{1,3}(z)=\sum_{k=0}^{\infty}\frac{z^k}{\Gamma(k+3)}=\sum_{k=0}^{\infty}\frac{z^k}{(k+2)!}=\frac{1}{z^2}\sum_{k=0}^{\infty}\frac{z^{k+2}}{(k+2)!}=\frac{\mathrm{e}^z-1-z}{z^2}. \tag{2.54}$$

More generally [1]

$$\mathrm{E}_{1,m}(z)=\sum_{k=0}^{\infty}\frac{z^k}{\Gamma(k+m)}=\frac{1}{z^{m-1}}\sum_{k=0}^{\infty}\frac{z^{k+m-1}}{(k+m-1)!}=\frac{1}{z^{m-1}}\left(\mathrm{e}^z-\sum_{k=0}^{m-2}\frac{z^k}{k!}\right). \tag{2.55}$$

In addition, it can be seen that

$$\mathrm{E}_{2,2}(z)=\sum_{k=0}^{\infty}\frac{z^k}{\Gamma(2k+2)}=\frac{1}{\sqrt{z}}\sum_{k=0}^{\infty}\frac{(\sqrt{z})^{2k+1}}{(2k+1)!}=\frac{\sinh\sqrt{z}}{\sqrt{z}} \tag{2.56}$$

$$\mathrm{E}_{2,1}(z^2)=\sum_{k=0}^{\infty}\frac{z^{2k}}{\Gamma(2k+1)}\sum_{k=0}^{\infty}\frac{z^{2k}}{(2k)!}=\cosh z \tag{2.57}$$

$$\mathrm{E}_{2,2}(z^2)=\sum_{k=0}^{\infty}\frac{z^{2k}}{\Gamma(2k+2)}=\frac{1}{z}\sum_{k=0}^{\infty}\frac{z^{2k+1}}{(2k+1)!}=\frac{\sinh z}{z}. \tag{2.58}$$

Since two-parameter Mittag-Leffler functions are also represented in infinite series form, the new function `mittag_leffler()` can also be used to find the expressions for parameters (α, β) with f=`mittag_leffler(`$[\alpha,\beta]$`,`z`)`.

▶ Example 2.16

Some one- or two-parameter Mittag-Leffler functions can be derived with MATLAB. Derive $\mathrm{E}_4(z)$, $\mathrm{E}_{4,5}(z)$, $\mathrm{E}_{5,6}(z)$, and $\mathrm{E}_{1/2,4}(z)$.

Solutions The following MATLAB commands can be used to derive directly some of the commonly used Mittag-Leffler functions. It can be seen that with symbolic computing, the formulas (2.53), (2.54), (2.56), and (2.58) can be evaluated.

```
>> syms z, I1=mittag_leffler([1,2],z), I2=mittag_leffler([2,2],z)
   I3=mittag_leffler([1,3],z), I4=mittag_leffler([4,1],z)
   I5=mittag_leffler([4,5],z), I6=mittag_leffler([5,6],z)
   I7=mittag_leffler([1/2,4],z)
   t=-3:0.1:2; t=t(t~=0); I71=subs(I7,z,t); y=ml_func([1/2,4],t);
   plot(t,y,t,I71,'--'), ylim([0,1.2])  % numerical and analytical solutions
   I8=mittag_leffler([2,1],z^2), I9=mittag_leffler([2,2],z^2)
```

and the results are

$$\mathrm{E}_4(z)=\frac{1}{4}\mathrm{e}^{\sqrt[4]{z}}+\frac{1}{4}\mathrm{e}^{-\sqrt[4]{z}}+\frac{1}{2}\cos\sqrt[4]{z}$$

$$\mathrm{E}_{4,5}(z)=-\frac{1}{4}+\frac{1}{4z}\left(\mathrm{e}^{\sqrt[4]{z}}+\mathrm{e}^{-\sqrt[4]{z}}+\mathrm{e}^{\mathrm{j}\sqrt[4]{z}}+\mathrm{e}^{-\mathrm{j}\sqrt[4]{z}}\right)$$

$$\mathrm{E}_{5,6}(z)=-\frac{1}{z}+\frac{\mathrm{e}^{\sqrt[5]{z}}}{5z}\left[1+\mathrm{e}^{(-1)^{2/5}}+\mathrm{e}^{(-1)^{4/5}}+\mathrm{e}^{-(-1)^{1/5}}+\mathrm{e}^{-(-1)^{3/5}}\right]$$

$$\mathrm{E}_{1/2,4}(z)=\frac{\mathrm{e}^{z^2}}{z^6}-\frac{1}{z^6}-\frac{1}{z^4}-\frac{1}{2z^2}+\frac{z\mathrm{e}^{z^2}\operatorname{erf}(z)}{z^7}-\frac{8}{15\sqrt{\pi z}}-\frac{4}{3\sqrt{\pi z^3}}-\frac{2}{\sqrt{\pi z^5}}.$$

◀

The numerical solutions of two-parameter Mittag-Leffler function can be evaluated with `ml_func()` function, with the syntax y=`ml_func(`$[\alpha,\beta]$`,`z`)`. Analytical and numerical solutions can be drawn for the $\mathrm{E}_{1/2,4}(z)$ function, as shown in Fig. 2.9. The two curves are exactly the same.

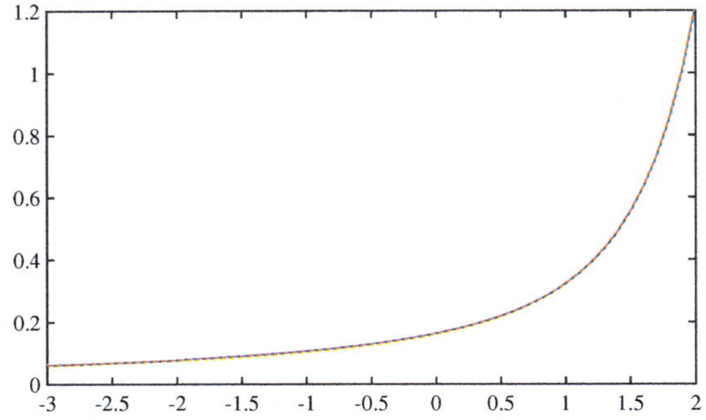

Fig. 2.9 Numerical and analytical solutions of $\mathrm{E}_{1/2,4}(z)$

Note that the expected analytical solutions cannot be found with new versions of MATLAB. For instance, I_4, I_5, and I_6 can only be expressed as hypergeometric functions with new versions. The curves can be drawn with `fplot()` function.

$$I_4 = {}_0\mathrm{F}_3\left(;\frac{1}{4},\frac{1}{2},\frac{3}{4};\frac{z}{256}\right),\ I_5=\frac{I_4}{z}-\frac{1}{z},\ I_6=\frac{1}{z}\,{}_0\mathrm{F}_4\left(;\frac{1}{5},\frac{2}{5},\frac{3}{5},\frac{4}{5};\frac{z}{3125}\right)-\frac{1}{z}.$$

It can be seen that

$$\mathrm{E}_{n,n+1}(z)=\frac{1}{z}\mathrm{E}_n(z)-\frac{1}{z}=\frac{1}{z}\,{}_0\mathrm{F}_{n-1}\left(;\frac{1}{n},\frac{2}{n},\cdots,\frac{n-1}{n};\frac{z}{n^n}\right)-\frac{1}{z}. \tag{2.59}$$

Corollary 2.4 *It is found from (2.59) that*

$$\mathrm{E}_{\alpha,\alpha+1}(z)=\frac{1}{z}\mathrm{E}_\alpha(z)-\frac{1}{z}. \tag{2.60}$$

Proof The corollary can be directly proved by the Mittag-Leffler function definition.

$$\begin{aligned}\mathrm{E}_{\alpha,\alpha+1}(z)&=\sum_{k=0}^{\infty}\frac{z^k}{\Gamma(\alpha k+\alpha+1)}=\frac{1}{z}\sum_{k=0}^{\infty}\frac{z^{k+1}}{\Gamma(\alpha(k+1)+1)}\\&=\frac{1}{z}\sum_{k=-1}^{\infty}\frac{z^{k+1}}{\Gamma(\alpha(k+1)+1)}-\frac{1}{z}\\&=\frac{1}{z}\sum_{k'=0}^{\infty}\frac{z^{k'}}{\Gamma(\alpha k'+1)}-\frac{1}{z}=\frac{1}{z}\mathrm{E}_\alpha(z)-\frac{1}{z}.\end{aligned} \tag{2.61}$$

□

► Example 2.17

Draw the curves of Mittag-Leffler functions $\mathrm{E}_{1,2}(x)$ and $\mathrm{E}_{3/2,1/2}(x)$.

Solutions The following commands can be used to draw the two curves, as shown in Fig. 2.10.

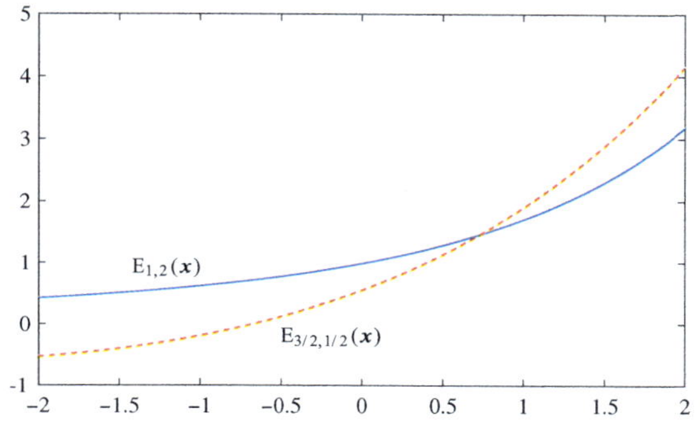

Fig. 2.10 The curves of two-parameter Mittag-Leffler functions

```
>> x=-2:0.1:2; y1=ml_func([1,2],x); y2=ml_func([3/2,1/2],x);
   plot(x,y1,x,y2,'--')
```

◀

Theorem 2.14 Mittag-Leffler function properties
Two properties of two-parameter Mittag-Leffler functions.

$$\begin{aligned} E_{\alpha,\beta}(z) + E_{\alpha,\beta}(-z) &= 2E_{2\alpha,\beta}(z^2) \\ E_{\alpha,\beta}(z) - E_{\alpha,\beta}(-z) &= 2zE_{2\alpha,\alpha+\beta}(z^2). \end{aligned} \tag{2.62}$$

Proof The two formulas can be directly proved.

$$\begin{aligned} E_{\alpha,\beta}(z) + E_{\alpha,\beta}(-z) &= \sum_{k=0}^{\infty} \frac{z^k + (-z)^k}{\Gamma(\alpha k + \beta)} = \sum_{k'=0}^{\infty} \frac{2z^{2k'}}{\Gamma(2\alpha k' + \beta)} \\ &= 2\sum_{k'=0}^{\infty} \frac{(z^2)^{k'}}{\Gamma(2\alpha k' + \beta)} = 2E_{2\alpha,\beta}(z^2) \end{aligned}$$

$$\begin{aligned} E_{\alpha,\beta}(z) - E_{\alpha,\beta}(-z) &= \sum_{k=0}^{\infty} \frac{z^k - (-z)^k}{\Gamma(\alpha k + \beta)} = \sum_{k'=0}^{\infty} \frac{2z^{2k'+1}}{\Gamma((2k'+1)\alpha + \beta)} \\ &= 2z\sum_{k'=0}^{\infty} \frac{(z^2)^{k'}}{\Gamma(2k'\alpha + \alpha + \beta)} = 2zE_{2\alpha,\alpha+\beta}(z^2). \end{aligned}$$

□

2

▶ Example 2.18

If z is complex, draw the surface of the real part of the function $E_{0.8,0.9}(z)$ [6].

Solutions For complex variable z, function $E_{0.8,0.9}(z)$ is a complex function. Mesh grids can be generated in the x-y-plane. Then complex function values can be evaluated. The real and imaginary parts can be extracted, such that the surface of the real part can be obtained, as shown in Fig. 2.11. The function values are restricted to the interval $[-3, +3]$ [6]. With such a method, the surface of the imaginary part can also be found.

```
>> [x y]=meshgrid(-6:0.2:6); z=x+sqrt(-1)*y;
   L=ml_func([0.8,0.9],z); L1=real(L); L2=imag(L);
   ii=find(L1>3); L1(ii)=3; ii=find(L1<-3); L1(ii)=-3;
   surf(x,y,L1), axis([-6 6 -6 6 -3 3])
```

◀

2.6.3 Multi-Parameter Mittag-Leffler Functions

Under certain applications, multi-parameter Mittag-Leffler function may be encountered [7].

Definition 2.16 Three- or four-parameter Mittag-Leffler functions

Three- or four-parameter Mittag-Leffler functions are [7]

$$E_{\alpha,\beta}^{\gamma}(z)=\sum_{k=0}^{\infty}\frac{(\gamma)_k}{\Gamma(\alpha k+\beta)}\frac{z^k}{k!},\quad E_{\alpha,\beta}^{\gamma,q}(z)=\sum_{k=0}^{\infty}\frac{(\gamma)_{kq}}{\Gamma(\alpha k+\beta)}\frac{z^k}{k!},\tag{2.63}$$

where $\alpha,\beta,\gamma\in\mathbb{C}$. For any $z\in\mathbb{C}$, the convergent condition for the infinite series is $\mathbb{R}(\alpha)>0$, $\mathbb{R}(\beta)>0$, $\mathbb{R}(\gamma)>0$ and $q\in\mathbb{N}$. $\mathbb{N}$ is the set of positive integers. $(\gamma)_k$ is the Pochhammer symbol, see Definition 2.11.

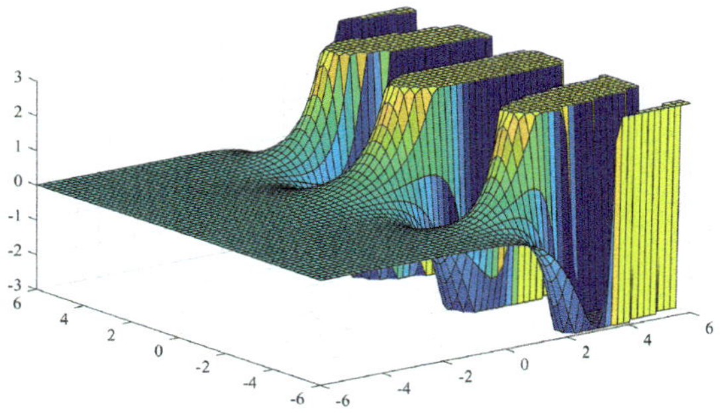

Fig. 2.11 Surface of the real part of Mittag-Leffler function $E_{0.8,0.9}(z)$

Definition 2.17 Multi-parameter Mittag-Leffler functions

The general form of multi-parameter Mittag-Leffler function is [1]

$$\mathrm{E}_{(\alpha_1,\cdots,\alpha_m),\beta}(z_1,\cdots,z_m)=\sum_{k=0}^{\infty}\sum_{\substack{l_1+\cdots+l_m=k\\ l_1>0,\cdots,l_m>0}}\frac{(k;l_1,\cdots,l_m)\prod_{i=1}^{m}z_i^{l_i}}{\Gamma\left(\beta+\sum_{i=1}^{m}\alpha_i l_i\right)},\tag{2.64}$$

where

$$(k;l_1,\cdots,l_m)=\frac{k!}{l_1!\cdots l_m!}.\tag{2.65}$$

Theorem 2.15 Special cases in Mittag-Leffler functions

Two-parameter Mittag-Leffler function is a special case of the one with three parameters, while the one with three parameters is a special case of the one with four parameters.

$$\mathrm{E}_{\alpha,\beta}(z)=\mathrm{E}_{\alpha,\beta}^{1}(z),\ \ \mathrm{E}_{\alpha,\beta}^{\gamma}(z)=\mathrm{E}_{\alpha,\beta}^{\gamma,1}(z).\tag{2.66}$$

Proof Substituting (2.38) into (2.63), it is found that

$$\mathrm{E}_{\alpha,\beta}^{1}(z)=\sum_{k=0}^{\infty}\frac{(1)_k}{\Gamma(\alpha k+\beta)}\frac{z^k}{k!}=\sum_{k=0}^{\infty}\frac{z^k}{\Gamma(\alpha k+\beta)}=\mathrm{E}_{\alpha,\beta}(z).$$

In addition, it can be seen from (2.63) that if $q=1$, then

$$\mathrm{E}_{\alpha,\beta}^{\gamma,1}(z)=\mathrm{E}_{\alpha,\beta}^{\gamma}(z).$$

□

2.6.4 The Relationship Between Mittag-Leffler and Hypergeometric Functions

The relationship between four-parameter Mittag-Leffler functions and hypergeometric functions is presented in Reference [8]. It reveals at least the relationship when α is an integer. It should be noted that the definition of four-parameter Mittag-Leffler functions in [8] is different from the one in this book, improvement is made to the theorem, where the relationship theorem of a three-parameter function is proposed.

Lemma 2.1 *A property of rising factorial is [8]*

$$(\beta)_{nk} = n^{nk}\prod_{r=1}^{n}\left(\frac{\beta+r-1}{n}\right)_k = (n^n)^k\left(\frac{\beta}{n}\right)_k\left(\frac{\beta+1}{n}\right)_k\cdots\left(\frac{\beta+n-1}{n}\right)_k. \quad (2.67)$$

Theorem 2.16 The relationship between Mittag-Leffler and hypergeometric functions

If $\alpha = n \in \mathbb{N}$, the following functions converge for all $z \in \mathbb{C}$.

$$\mathrm{E}_{n,\beta}^{\gamma}(z) = \frac{1}{\Gamma(\beta)}\,{}_1\mathrm{F}_n\left(\gamma;\ \frac{\beta}{n}, \frac{\beta+1}{n}, \cdots, \frac{\beta+n-1}{n};\ \frac{z}{n^n}\right). \quad (2.68)$$

Proof It is known from Definition 2.11 that $\Gamma(nk+\beta) = \Gamma(\beta)(\beta)_{kn}$. Therefore, it is found from the definition of three-parameter Mittag-Leffler function that

$$\mathrm{E}_{n,\beta}^{\gamma}(z) = \sum_{k=0}^{\infty}\frac{(\gamma)_k}{\Gamma(nk+\beta)}\frac{z^k}{k!} = \frac{1}{\Gamma(\beta)}\sum_{k=0}^{\infty}\frac{(\gamma)_k}{(\beta)_{kn}}\frac{z^k}{k!}.$$

With Lemma 2.1, the theorem can be proved. Since n is a positive integer, the function converges.

$$\mathrm{E}_{n,\beta}^{\gamma}(z) = \frac{1}{\Gamma(\beta)}\sum_{k=0}^{\infty}\frac{(\gamma)_k}{\left(\frac{\beta}{n}\right)_k\left(\frac{\beta+1}{n}\right)_k\cdots\left(\frac{\beta+n-1}{n}\right)_k}\frac{z^k}{(n^n)^k k!}.$$

□

If $\beta = \gamma = 1$, the one-parameter Mittag-Leffler function, $\mathrm{E}_n(z)$, satisfies (2.50).

2.6.5 Derivatives of Mittag-Leffler Functions

Integer-order derivatives of Mittag-Leffler functions are often encountered in fractional calculus. There are many properties in the derivatives. In this section, the integer-order derivatives of Mittag-Leffler functions are presented, and the derivative functions of a general four-parameter Mittag-Leffler function are discussed.

Theorem 2.17 Derivatives of Mittag-Leffler functions
The integer-order derivative of two-parameter Mittag-Leffler function, $E_{\alpha,\beta}(z)$, can be expressed as

$$\frac{d^n}{dz^n}E_{\alpha,\beta}(z) = \sum_{k=0}^{\infty}\frac{(k+n)!}{k!\,\Gamma(\alpha k+\alpha n+\beta)}z^k. \tag{2.69}$$

The first-order derivative of Mittag-Leffler function $E_{\alpha,\beta}(z)$ with respect to z is

$$\frac{d}{dz}E_{\alpha,\beta}(z) = \sum_{k=1}^{\infty}\frac{kz^{k-1}}{\Gamma(\alpha k+\beta)}, \tag{2.70}$$

where the term $k = 0$ has been eliminated because its value is zero. Denote $k' = k - 1$, the above equation can be rewritten as

$$\frac{d}{dz}E_{\alpha,\beta}(z) = \sum_{k'=0}^{\infty}\frac{(k'+1)z^{k'}}{\Gamma(\alpha k'+\alpha+\beta)} = \sum_{k=0}^{\infty}\frac{(k+1)z^k}{\Gamma(\alpha k+\alpha+\beta)}. \tag{2.71}$$

Taking the first-order derivative to the above formula again, we get

$$\frac{d^2}{dz^2}E_{\alpha,\beta}(z) = \sum_{k=0}^{\infty}\frac{(k+1)(k+2)z^k}{\Gamma(\alpha k+2\alpha+\beta)}. \tag{2.72}$$

The nth-order derivative of the function can be derived by taking the derivative one at a time

$$\begin{aligned}\frac{d^n}{dz^n}E_{\alpha,\beta}(z) &= \sum_{k=0}^{\infty}\frac{(k+n)(k+n-1)\cdots(k+1)}{\Gamma(\alpha k+\alpha n+\beta)}z^k\\ &= \sum_{k=0}^{\infty}\frac{(k+n)!}{k!\,\Gamma(\alpha k+\alpha n+\beta)}z^k.\end{aligned} \tag{2.73}$$

Proof The formal proof of the theorem is given below by mathematical induction. Equation (2.71) has been shown to hold for $n = 1$. Now assume that the theorem also holds for $n = m$, that is,

$$\frac{d^m}{dz^m}E_{\alpha,\beta}(z) = \sum_{k=0}^{\infty}\frac{(k+m)!}{k!\,\Gamma(\alpha k+\alpha m+\beta)}z^k.$$

The above equation is then evaluated for the first-order derivative of z. It can be seen that the term at $k = 0$ is eliminated, and

$$\frac{d^{m+1}}{dz^{m+1}}E_{\alpha,\beta}(z) = \sum_{k=1}^{\infty}\frac{(k+m)!\,k}{k!\,\Gamma(\alpha k+\alpha m+\beta)}z^{k-1}.$$

Denote $k = k' + 1$, the above equation can be rewritten as

$$\frac{\mathrm{d}^{m+1}}{\mathrm{d}z^{m+1}} \mathrm{E}_{\alpha,\beta}(z) = \sum_{k'=0}^{\infty} \frac{(k'+1+m)!(k'+1)}{(k'+1)!\,\Gamma(\alpha(k'+1)+\alpha m+\beta)} z^{k'}.$$

For eliminating the numerator and denominator of the $(k'+1)$ term, and note that $k'+1+m$ is $k'+(m+1)$, and $\alpha(k'+1)+\alpha m$ as $\alpha k'+\alpha(m+1)$, and then replacing k' with k, we find

$$\frac{\mathrm{d}^{m+1}}{\mathrm{d}z^{m+1}} \mathrm{E}_{\alpha,\beta}(z) = \sum_{k=0}^{\infty} \frac{(k+(m+1))!}{k!\,\Gamma(\alpha k+\alpha(m+1)+\beta)} z^{k}.$$

That is, the theorem holds when $n = m+1$. From this, the theorem can be proved by mathematical induction. □

Theorem 2.18 Properties of two-parameter Mittag-Leffler functions
The following condition holds for two-parameter Mittag-Leffler functions:

$$\mathrm{E}_{\alpha,\beta}(z) = \beta \mathrm{E}_{\alpha,\beta+1}(z) + \alpha z \frac{\mathrm{d}}{\mathrm{d}z} \mathrm{E}_{\alpha,\beta+1}(z). \tag{2.74}$$

Proof If the expressions on the right-hand side of the equal sign are rearranged, it can be seen that the theorem holds.

$$\begin{aligned} \text{right-hand side} &= \beta \sum_{k=0}^{\infty} \frac{z^k}{\Gamma(\alpha k+\beta+1)} + \alpha z \sum_{k=0}^{\infty} \frac{k z^{k-1}}{\Gamma(\alpha k+\beta+1)} \\ &= \sum_{k=0}^{\infty} \frac{(\beta+k\alpha)z^k}{\Gamma(\alpha k+\beta+1)} = \sum_{k=0}^{\infty} \frac{z^k}{\Gamma(\alpha k+\beta)} = \mathrm{E}_{\alpha,\beta}(z). \end{aligned}$$

□

Theorem 2.19 Derivative formula of Mittag-Leffler functions
The nth-order derivative of a four-parameter Mittag-Leffler function $\mathrm{E}_{\alpha,\beta}^{\gamma,q}(z)$ satisfies the following formula [7]:

$$\frac{\mathrm{d}^n}{\mathrm{d}z^n} \mathrm{E}_{\alpha,\beta}^{\gamma,q}(z) = (\gamma)_{qn} \mathrm{E}_{\alpha,\beta+n\alpha}^{\gamma+qn,q}(z). \tag{2.75}$$

Proof The first-order derivative of four-parameter Mittag-Leffler function $\mathrm{E}_{\alpha,\beta}^{\gamma,q}(z)$ satisfies the following equation:

$$\begin{aligned}\frac{\mathrm{d}}{\mathrm{d}z}\mathrm{E}_{\alpha,\beta}^{\gamma,q}(z)&=\sum_{k=1}^{\infty}\frac{(\gamma)_{kq}}{\Gamma(\alpha k+\beta)}\frac{kz^{k-1}}{k!}\\&=\sum_{k=1}^{\infty}\frac{(\gamma)_{kq}}{\Gamma(\alpha k+\beta)}\frac{z^{k-1}}{(k-1)!}=\sum_{k=0}^{\infty}\frac{(\gamma)_{(k+1)q}}{\Gamma(\alpha k+\alpha+\beta)}\frac{z^k}{k!}.\end{aligned}\tag{2.76}$$

It can be seen that for any integer q, we have

$$(\gamma)_{(k+1)q}=\frac{\Gamma(kq+q+\gamma)}{\Gamma(\gamma)}=\frac{\Gamma(kq+\gamma+q)}{\Gamma(\gamma+q)}\frac{\Gamma(\gamma+q)}{\Gamma(\gamma)}=(\gamma+q)_{kq}(\gamma)_q,$$

where $(\gamma)_q$ is uncorrelated with k. Substituting it into (2.76) shows that

$$\frac{\mathrm{d}}{\mathrm{d}z}\mathrm{E}_{\alpha,\beta}^{\gamma,q}(z)=(\gamma)_q\sum_{k=0}^{\infty}\frac{(\gamma+q)_{kq}}{\Gamma(\alpha k+\alpha+\beta)}\frac{z^k}{k!}=(\gamma)_q\mathrm{E}_{\alpha,\beta+\alpha}^{\gamma+q,q}(z).\tag{2.77}$$

It can be seen that the theorem holds for $n=1$. In fact, the above formulation can be avoided, since the theorem holds for $n=0$. Mathematical induction is still used to prove the theorem. Suppose that for $n=k$, the theorem holds, with

$$\frac{\mathrm{d}^k}{\mathrm{d}z^k}\mathrm{E}_{\alpha,\beta}^{\gamma,q}(z)=(\gamma)_{qk}\mathrm{E}_{\alpha,\beta+k\alpha}^{\gamma+kq,q}(z).$$

Taking the first-order derivative of the above equation and using the result in (2.77) show that

$$\frac{\mathrm{d}^{k+1}}{\mathrm{d}z^{k+1}}\mathrm{E}_{\alpha,\beta}^{\gamma,q}(z)=\frac{\mathrm{d}}{\mathrm{d}z}(\gamma)_{qk}\mathrm{E}_{\alpha,\beta+k\alpha}^{\gamma+kq,q}(z)=(\gamma)_{q(k+1)}\mathrm{E}_{\alpha,\beta+(k+1)\alpha}^{\gamma+(k+1)q,q}(z).$$

This implies that the theorem still holds for $n=k+1$, so it is proved by mathematical induction. □

Theorem 2.20 Derivatives of two-parameter Mittag-Leffler functions
The nth-order derivative of a two-parameter Mittag-Leffler function satisfies

$$\frac{\mathrm{d}^n}{\mathrm{d}z^n}\mathrm{E}_{\alpha,\beta}(z)=n!\,\mathrm{E}_{\alpha,\beta+n\alpha}^{n+1}(z).\tag{2.78}$$

Proof Consider Theorem 2.19, since $\mathrm{E}_{\alpha,\beta}(z)=\mathrm{E}_{\alpha,\beta}^{1,1}(z)$, where $\gamma=1$ and $q=1$, it can be proved immediately.

$$\frac{\mathrm{d}^n}{\mathrm{d}z^n}\mathrm{E}_{\alpha,\beta}(z) = (1)_n\mathrm{E}_{\alpha,\beta+n\alpha}^{n+1,1}(z) = n!\,\mathrm{E}_{\alpha,\beta+n\alpha}^{n+1}(z). \tag{2.79}$$

□

Theorem 2.21 Some properties of Mittag-Leffler functions
There are various properties in the derivatives of Mittag-Leffler functions. For instance, for any positive integer n, the following equation holds:

$$\mathrm{E}_{\alpha,\beta}^{\gamma,q}(z) - \mathrm{E}_{\alpha,\beta}^{\gamma-1,q}(z) = \frac{z}{q}\mathrm{E}_{\alpha,\alpha+\beta}^{\gamma,q+1}(z) + \frac{z^2}{q}\frac{\mathrm{d}}{\mathrm{d}z}\mathrm{E}_{\alpha,\alpha+\beta}^{\gamma,q+1}(z)$$

$$\mathrm{E}_{\alpha,\beta}^{\gamma,q}(z) - \mathrm{E}_{\alpha,\beta}^{\gamma,q-1}(z) = \frac{z}{1-q}\frac{\mathrm{d}}{\mathrm{d}z}\mathrm{E}_{\alpha,\beta}^{\gamma,q}(z) \tag{2.80}$$

$$\frac{\mathrm{d}^n}{\mathrm{d}z^n}\left[z^{\beta-1}\mathrm{E}_{\alpha,\beta}^{\gamma,q}(\omega z^\alpha)\right] = z^{\beta-n-1}\mathrm{E}_{\alpha,\beta-n}^{\gamma,q}(\omega z^\alpha),\ \ \mathbb{R}(\beta-n)>0 \tag{2.81}$$

$$\frac{\mathrm{d}^n}{\mathrm{d}z^n}\left[z^{q-1}\mathrm{E}_{\alpha,\beta}^{\gamma,q}(z)\right] = (\delta-n)_n z^{q-1-n}\mathrm{E}_{\alpha,\beta}^{\gamma,q-n}(z). \tag{2.82}$$

Proof Taking (2.81) as an example, the theorem is proved here. The theorem can be proved directly by the Mittag-Leffler function definition.

$$\begin{aligned}\frac{\mathrm{d}^n}{\mathrm{d}z^n}\left[z^{\beta-1}\mathrm{E}_{\alpha,\beta}^{\gamma,q}(\omega z^\alpha)\right] &= \sum_{k=0}^{\infty}\frac{(\gamma)_{kq}\omega^k}{\Gamma(\alpha k+\beta-n)}\frac{z^{\alpha k+\beta-n-1}}{k!}\\ &= z^{\beta-n-1}\sum_{k=0}^{\infty}\frac{(\gamma)_{kq}}{\Gamma(\alpha k+\beta-n)}\frac{\omega^k z^{\alpha k}}{k!}\\ &= z^{\beta-n-1}\mathrm{E}_{\alpha,\beta-n}^{\gamma,q}(\omega z^\alpha).\end{aligned}$$

The proofs of the other formulas can be found in References [7] and [8]. □

2.6.6 Numerical Evaluation of Mittag-Leffler Functions and Their Derivatives

In fact, the `ml_func()` function given earlier evaluates the function values of the one- to four-parameter Mittag-Leffler functions and directly finds the nth-order derivative of the Mittag-Leffler functions. The function is called in the form

```
f=ml_func(α,z,n,ϵ0)             %nth- order derivative of E_α(z)
f=ml_func([α,β],z,n,ϵ0)         %nth- order derivative of E_{α,β}(z)
f=ml_func([α,β,γ],z,n,ϵ0)       %nth- order derivative of E^γ_{α,β}(z)
f=ml_func([α,β,γ,q],z,n,ϵ0)     %nth- order derivative of E^{γ,q}_{α,β}(z)
```

where ϵ_0 is the error tolerance, when the absolute value of a term is smaller than the error tolerance, the subsequent terms are truncated and the summation ends. The returned vector $\boldsymbol{f}$ is the nth-order derivative of the corresponding Mittag-Leffler function. The last two arguments can be omitted and their default values are $n = 0$ and ϵ_0 =eps. If only the Mittag-Leffler function is needed, then set $n = 0$.

Remark Limitations of the solver

(1) The cumulative truncation algorithm used here has the advantage of being fast, but the method sometimes diverges, then the program automatically calls the embedded Prof. Podlubny's reliable `mlf()` function [5]. Sometimes it is slow, and due to its own limitations, the `mlf()` function can only solve one- or two-parameter Mittag-Leffler function and cannot compute their derivatives.

(2) Because the derivatives of some Mittag-Leffler functions cannot be solved by `ml_func()` function, it is recommended to use the method of numerical derivatives, and it is recommended to use the high-precision numerical derivative function `num_diff()`, which will be introduced in ▶ Chap. 4.

(3) Since the `gamma()` function called by the `ml_func()` function can only handle real variables, it should be replaced by `gamma_c()` when dealing with complex variables, but this case is not considered in this book. □

▶ Example 2.19

Plot the Mittag-Leffler functions $\mathrm{E}_{1,1}^{(5)}(x)$ and $\mathrm{E}_{\sqrt{2},1.3}^{(2)}(x)$.

Solutions It can be seen that the former has an analytical solution of e^x, but the latter has no analytical solutions. The curves of the two functions can be directly obtained with the following statements, as shown in ◘ Fig. 2.12, where the e^x curve is also superimposed, which can be seen to coincide exactly with the first Mittag-Leffler curve, with a maximum error of 3.5527×10^{-15}.

```
>> x=0:0.001:2;
   y1=ml_func([1,1],x,5); y2=ml_func([sqrt(2),1.3],x,2);
   plotyy(x,[y1; exp(x)],x,y2), max(abs(y1-exp(x)))
```

◀

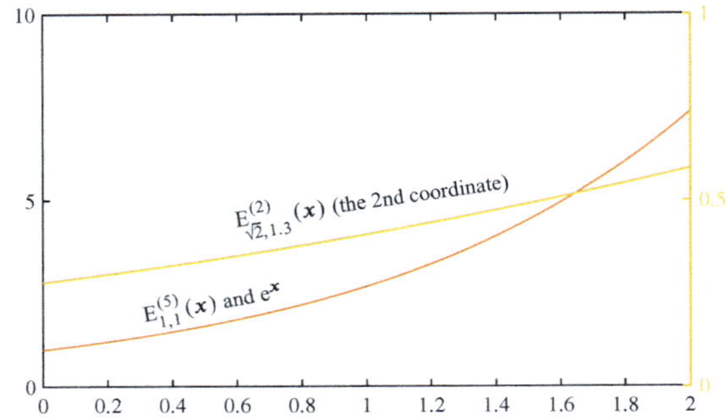

Fig. 2.12 Mittag-Leffler function curves

2.7 Exercises

(1) Prove $\Gamma(\alpha+1) = \alpha!$ using theoretical derivations and MATLAB symbolic computing, where α is a nonnegative integer.

(2) Using MATLAB's symbolic computing, prove that

$$\int_0^\infty e^{-t^2}dt = \frac{\sqrt{\pi}}{2}.$$

(3) Write a program that can calculate the Gamma functions based on the expression of the Gamma function given in Definition 2.3, and apply the written program to calculate $\Gamma(0.5)$, $\Gamma(1.5)$, and $\Gamma(2.5)$.

(4) Pochhammer symbol $(x)_n$ can be calculated by the symbolic function `pochhammer(x,n)`. If n is a given integer, is there more concise numerical method to directly calculate $(x)_n$?

(5) Apply the symbolic computing in MATLAB to prove the following equations:

$$\text{(a) } \Gamma(z)\Gamma(1-z) = \frac{\pi}{\sin \pi z}, \quad \text{(b) } \Gamma(z)\Gamma(-z) = \frac{-\pi}{z \sin \pi z},$$

$$\text{(c) } \Gamma(0.5+z)\Gamma(0.5-z) = \frac{\pi}{\cos \pi z}.$$

(6) Apply the symbolic computing in MATLAB to prove the following equations:

$$\int_0^{\pi/2} \sin^\alpha x \cos^\beta x dx = \frac{1}{2}\frac{\Gamma((\alpha+1)/2)\,\Gamma((\beta+1)/2)}{\Gamma((\alpha+\beta+2)/2)}, \quad (\alpha > -1, \beta > -1).$$

(7) The Euler constant γ is defined as

$$\gamma = \lim_{n\to\infty}\left[\left(1+\frac{1}{2}+\frac{1}{3}+\cdots+\frac{1}{n}\right) - \ln n\right]. \tag{2.83}$$

Write a piece of MATLAB code, try a larger value of n, and compute numerically the Euler constant γ. What is the minimum number of terms that should be added if you want to keep the first 10 significant digits correct?

(8) Write a piece of MATLAB code that can calculate the Gamma functions from the limit definition formula of the Gamma function given in Theorem 2.7. Apply the code to calculate $\Gamma(0.5)$, $\Gamma(1.5)$, and $\Gamma(2.5)$.

(9) Draw the surface of the Beta function $\mathrm{B}(x, y)$ in the interval $1 \leqslant x \leqslant 10$, $1 \leqslant y \leqslant 10$.

(10) Prove the following equation using theoretical formulations or the symbolic computing function of MATLAB.

$$\text{(a) } \mathrm{E}_{1,2}(z) = \frac{\mathrm{e}^z - 1}{z}, \quad \text{(b) } \mathrm{E}_{1,3}(z) = \frac{\mathrm{e}^z - 1 - z}{z^2},$$

$$\text{(c) } \mathrm{E}_{2,1}(z^2) = \cosh z, \quad \text{(d) } \mathrm{E}_{2,2}(z^2) = \frac{1}{z}\sinh z.$$

(11) Use the MATLAB functions provided in this chapter to plot the curve of the function in Exercise 2.7 on the interval [0, 10] and apply the curve to check the correctness of the above results.

(12) Write a MATLAB program to prove Theorem 2.9 of this chapter. Plot the curves of the expressions of Theorem 2.9 on the interval [−5, 5], and see whether the equal sign holds.

(13) Prove that the hypergeometric function satisfies

$$\text{(1) } {}_1\mathrm{F}_1\,(a; a; x) = \mathrm{e}^x, \quad \text{(2) } {}_2\mathrm{F}_1\,(a, 1; 1; z) = \frac{1}{(1-z)^a},$$

$$\text{(3) } {}_2\mathrm{F}_1\,(1, 1; 2; -z) = \frac{1}{z}\ln(z+1), \quad \text{(4) } {}_2\mathrm{F}_1\left(\frac{1}{2}, 1; \frac{3}{2}; z^2\right) = \frac{1}{2z}\ln\frac{z+1}{1-z},$$

$$\text{(5) } {}_2\mathrm{F}_1\left(\frac{1}{2}, \frac{1}{2}; \frac{3}{2}; z^2\right) = \frac{1}{z}\arcsin z.$$

(14) Use the `mittag_leffler()` function to find concise formulas of $\mathrm{E}_1(z^2)$, $\mathrm{E}_2(z+1)$, $\mathrm{E}_{1/2}(\sqrt{z})$, and $\mathrm{E}_{1/3}(z+5)$. If necessary, you can try MATLAB R2008a and earlier versions.

(15) Use `ml_func()` function to plot the functions $\mathrm{E}_{2/3,1/4}(-z^2)$, $\mathrm{E}_{3,\sqrt{2}}(-\sqrt{z})$, and $\mathrm{E}_{\pi,\mathrm{e}}(-z)$ on the interval [0, 10].

(16) Apply the derivative formula of the Mittag-Leffler function to calculate the first-order derivative of $\mathrm{E}_{1,1.5}(z)$ with respect to z. Then use `ml_func()` function and MATLAB's symbolic computing to plot the curve of this first-order derivative on the interval [0, 10], and apply the plotted curve to check the correctness of the derivative expression.

(17) Compute the following Mittag-Leffler functions:

$$\text{(a) } \mathrm{E}_{2,6}(z^8), \quad \text{(b) } \mathrm{E}_{2,6}(-z^8), \quad \text{(c) } \mathrm{E}_{1,6}(z^8),$$

$$\text{(d) } \mathrm{E}_{1,6}(-z^8), \quad \text{(e) } \mathrm{E}_{1,9}(z^8) + \mathrm{E}_{1,9}(-z^8).$$

(18) Use MATLAB commands to plot the curves of the expression on each side of the following equation on the interval [0, 10], and apply the plotted curves to check whether the equation holds.

$$\frac{\mathrm{d}}{\mathrm{d}t}\left(t^{1.5}\mathrm{E}_{1,2.5}(t)\right) = t^{0.5}\mathrm{E}_{1,1.5}(t).$$

(19) The following equation is cited from Reference [1]. Use MATLAB commands to plot the expressions on both sides of the equal sign on the interval [−1, 1]. Apply the plotted curve to check whether the equal sign holds.

$$\mathrm{E}_{1/2,1}(z) = \mathrm{e}^{z^2}\mathrm{erfc}(-z).$$

(20) Let $\alpha = 2$ and $\beta = 2.5$. Use the Mittag-Leffler function code given in this chapter, verify the following equations numerically:

$$\text{(a)}\,\mathrm{E}_{\alpha,\beta}(x)+\mathrm{E}_{\alpha,\beta}(-x)=2\mathrm{E}_{2\alpha,\beta}\left(x^2\right),\quad \text{(b)}\,\mathrm{E}_{\alpha,\beta}(x)-\mathrm{E}_{\alpha,\beta}(-x)=2x\mathrm{E}_{2\alpha,\alpha+\beta}\left(x^2\right),$$

$$\text{(c)}\,\mathrm{E}_{\alpha,\beta}(x) = \frac{1}{\beta}+x\mathrm{E}_{\alpha,\alpha+\beta}(x),\quad \text{(d)}\,\mathrm{E}_{\alpha,\beta}(x) = \beta\mathrm{E}_{\alpha,\beta+1}(x)+\alpha x\frac{\mathrm{d}}{\mathrm{d}x}\mathrm{E}_{\alpha,\beta+1}(x).$$

(21) Two functions `ml_func()` and `mlf()` are provided in FOTF Toolbox. Try some α and β combinations in the Mittag-Leffler functions, and compare the execution efficiencies of these two functions.

(22) Prove (2.50) by Theorem 2.16.

(23) Prove the other three formulas in Theorem 2.21.

References

1. Podlubny I (1999) Fractional differential equations[M]. Academic Press, San Diego
2. Magin RL (2006) Fractional calculus in bioengineering[M]. Begell House Publishers, Redding
3. Abramowita M, Stegun IA (1970) Handbook of mathematical functions with formulas, graphs and mathematical tables[M], 9th edn. United States Department of Commerce, National Bureau of Standards, Washengton D C
4. Mittag-Leffler GM (1903) Sur la vonvelle fonction $\mathrm{E}_\alpha(x)$[J]. Comptes Rendus de l'Académie des Sciences Paris 137:554–558
5. Podlubny I Mittag-Leffler function [OL]. ► http://ww2.mathworks.cn/matlabcentral/fileexchange/8738-mittag-leffler-function
6. Seybold HJ, Hilfer R (2005) Numerical results for the generalized Mittag-Leffler function[J]. Fract Calc Appl Anal 8(2):127–139
7. Shukla AK, Prajapati JC (2007) On a generalization of Mittag-Leffler function and its properties[J]. J Math Anal Appl 336(2):797–811
8. Salim TO (2009) Some properties relating to the generalized Mittag-Leffler function[J]. Adv Appl Math Anal 4(1):21–30

Definitions and Numerical Evaluations of Fractional Calculus

Contents

D. Y. Xue and L. Bai, *Fractional Calculus*, https://doi.org/10.1007/978-981-99-2070-9_3

As previously noted, fractional calculus dates back to the days when Newton and Leibniz first invented traditional calculus. In the absence of a unified and widely accepted definitions, fractional calculus did not progress well in its early development. It was not until the middle of the nineteenth century that some famous scholars, including the French mathematician Liouville (1809–1882) in 1834, the German mathematician Riemann (1826–1866) in 1847, the Czech mathematician Grünwald (1838–1920) in 1867, and the Russian mathematician Letnikov (1837–1888) in 1868 proposed meaningful definitions of fractional calculus.

Later, the definitions of Liouville and Riemann were unified into the Riemann–Liouville definition, while the definitions of Grünwald and Letnikov were integrated into the Grünwald–Letnikov definition. These two definitions have been widely used in the field of fractional calculus and were later shown to be equivalent in engineering applications. Both definitions are suitable for describing fractional calculus problems with zero initial values.

For the functions with nonzero initial values, the definition proposed by the Italian mathematician Caputo in 1967 is very useful, especially in the study of fractional-order differential equations (FODEs), where the Riemann–Liouville and Grünwald–Letnikov definitions require that the initial values of the fractional-order derivatives of the signal be known, whereas the Caputo definition requires that the value of the signal and its integer-order derivatives are known at the initial moment, which is closer to practical applications.

In addition to these widely used definitions, there are other definitions of fractional-order derivatives in the literatures, such as the Erdélyi–Kober definition, Hadamard definition, Marchaud definition, Riesz definition, Riesz–Miller definition, Miller–Ross definition, Weyl definition, and so on. These definitions are not involved in this book, but it is possible to extend the numerical algorithms presented in this book to other definitions.

The main problem studied in this chapter is how to evaluate the fractional-order derivatives or integrals of the original function $f(t)$, if the function or its sample points are known. Later chapters will also describe how to construct fractional-order derivative and integral signals of a signal if the original function is unknown.

A unified operator ${}_{t_0}\mathscr{D}_t^{\alpha}$ for fractional-order derivatives and integrals will be introduced here and widely used in this book, where α is restricted to real numbers, t is the independent variable, and t_0 is the lower bound of that variable.

Definition 3.1 Unified fractional-order operator

The unified fractional-order operator ${}_{t_0}\mathscr{D}_t^{\alpha}$ is defined as

$$
{}_{t_0}\mathscr{D}_t^{\alpha} f(t) = \begin{cases} \dfrac{\mathrm{d}^{\alpha}}{\mathrm{d}t^{\alpha}} f(t), & \alpha > 0 \\ f(t), & \alpha = 0 \\ \displaystyle\int_{t_0}^{t} f(\tau)\mathrm{d}\tau^{-\alpha}, & \alpha < 0. \end{cases} \tag{3.1}
$$

Remark Unified fractional-order operator

(1) If $\alpha \geqslant 0$ and $t_0 = 0$, the notation t_0 can be omitted. If the independent variable is t and there are no other variables, t can also be omitted.

(2) If $\alpha > 0$, ${}_{t_0}\mathscr{D}_t^{\alpha}$ operator is the αth-order derivative with respect to independent variable t; if $\alpha = 0$, the original signal is returned; if $\alpha < 0$, the $-\alpha$th-order integral is returned.

(3) If α is a complex number, its real part determines whether derivative or integral is defined, which is beyond the scope of this book. □

▶ Section 3.1 of this chapter begins with an extension of the integer-order Cauchy integral formula to give the fractional-order Cauchy integral formula, but the definition is not considered in subsequent chapters. In ▶ Sects. 3.2 and ▶ 3.3, respectively, the traditional calculus definition is directly extended by introducing the Grünwald–Letnikov and Riemann–Liouville definitions in fractional calculus. Numerical algorithms for these definitions are given, with general solution functions implemented in MATLAB. In ▶ Sect. 3.4, the concept of fractional calculus is extended to functions with nonzero initial values. The definition of Caputo fractional-order derivatives is then proposed. In ▶ Sect. 3.5, the relationship between the various fractional calculus definitions is explained. The numerical algorithms of fractional-order operators based on the Caputo definition are introduced. In ▶ Sect. 3.6, the properties of fractional calculus are presented, and attempts to illustrate the physical interpretation of fractional-order integrals are made.

It is important to note that

(1) The methods for evaluating fractional-order derivatives given in this chapter are only conventional algorithms for the approximate evaluation in fractional calculus, which are generally of low accuracy. ▶ Chapter 4 will comprehensively introduce the high-precision numerical algorithms and their MATLAB implementations for fractional-order derivatives of functions under various definitions.

(2) The numerical solution algorithm presented here is proposed when the original function or its sample points are known. If the original function is unknown, the construction of its fractional-order derivative signals can be found in ▶ Chap. 5.

3.1 Fractional-Order Integral Formula

In the field of complex analysis, there is a very important Cauchy integral formula. It converts problems with high-order derivatives of known complex functions $f(z)$ into integral problems with closed curves. This section will explore the extension of the Cauchy integral formula to the fractional-order Cauchy integral formula for fractional-order derivatives, and present the properties of the corresponding fractional-order operators.

3.1.1 Cauchy Integral Formula

First, let us observe the integer-order derivative representation of the Cauchy integral formula.

$$\frac{\mathrm{d}^n}{\mathrm{d}t^n} f(t) = \frac{n!}{2\pi \mathrm{j}} \oint_{\mathrm{C}} \frac{f(\tau)}{(\tau - t)^{n+1}} \mathrm{d}\tau, \tag{3.2}$$

where C is a smooth closed curve, and function $f(t)$ is a single-valued analytic one in the region.

If n is replaced by a non-integer γ, there will be an isolated singularity at the point $t = \tau$. This singularity should be removed when calculating the closed-path integral, and the factorial should be replaced by the Gamma function. Then, (3.2) can be directly extended to deal with fractional-order derivatives.

Definition 3.2 Cauchy integral definition of fractional-order derivatives

The Cauchy integral definition of fractional-order derivatives of $f(t)$ is

$$\mathscr{D}_t^{\gamma} f(t) = \frac{\Gamma(\gamma + 1)}{2\pi \mathrm{j}} \oint_{\mathrm{C}} \frac{f(\tau)}{(\tau - t)^{\gamma+1}} \mathrm{d}\tau, \tag{3.3}$$

where the order γ is an arbitrary real number.

3.1.2 Derivative and Integral Formulas for Commonly Used Functions

In traditional calculus, it is well known that the nth-order derivative of trigonometric functions can be calculated as follows:

$$\frac{\mathrm{d}^n}{\mathrm{d}t^n} \sin at = a^n \sin\left(at + \frac{n\pi}{2}\right) \tag{3.4}$$

$$\frac{\mathrm{d}^n}{\mathrm{d}t^n} \cos at = a^n \cos\left(at + \frac{n\pi}{2}\right). \tag{3.5}$$

If integer n is replaced by any real number α, the above definition can be directly extended to yield the fractional-order derivatives of trigonometric functions.

$$\mathscr{D}_t^{\alpha} \sin at = a^{\alpha} \sin\left(at + \frac{\alpha\pi}{2}\right) \tag{3.6}$$

$$\mathscr{D}_t^{\alpha} \cos at = a^{\alpha} \cos\left(at + \frac{\alpha\pi}{2}\right). \tag{3.7}$$

In addition, the fractional-order derivative of an exponential function can be found from the following equation:

$$\mathscr{D}_t^{\alpha} \mathrm{e}^{\lambda t} = \lambda^{-\alpha} \mathrm{E}_{1,1-\alpha}(\lambda t). \tag{3.8}$$

The nth-order derivative of a power function t^m can be expressed as

$$\frac{\mathrm{d}^n}{\mathrm{d}t^n} t^m = m(m-1)\cdots(m-n+1)\, t^{m-n} = \frac{m!}{(m-n)!} t^{m-n}. \tag{3.9}$$

By replacing the factorials with the Gamma functions, the equation can be directly extended to fractional-order derivatives as follows:

$$\mathscr{D}_t^{\alpha} t^m = \frac{\Gamma(m+1)}{\Gamma(m-\alpha+1)} t^{m-\alpha}, \quad m > -1. \tag{3.10}$$

Here, m is no longer restricted to integers, it can also be non-integers.

Reference [1] also presents formulas for the fractional-order derivatives of more commonly used functions. The interested reader is referred to that work. However, the definition itself is also rather demanding in that it requires $f(t)$ to be a single-valued analytic function in a closed region. The definition of fractional calculus based on the Cauchy integral formula will not be explored further with this book.

3.2 Definition and Numerical Evaluation of Grünwald–Letnikov Integrals and Derivatives

The Grünwald–Letnikov definition of fractional calculus is one of the most widely used definitions in fractional calculus. A general representation of the integer-order derivative is given first in this section, and then it is extended to give the Grünwald–Letnikov definition of the fractional-order derivatives. This section also gives some numerical algorithms based on this definition. The advantages and disadvantages of various algorithms are compared with examples.

3.2.1 Formulations in High-Order Integer-Order Derivatives

Before presenting the definition of Grünwald–Letnikov fractional-order derivatives, let us observe the integer-order derivative formulas. Consider first the formula for the first-order derivative of the function $f(t)$.

$$\frac{\mathrm{d}^1}{\mathrm{d}t^1} f(t) = \lim_{h\to 0} \frac{1}{h} \big[f(t) - f(t-h)\big]. \tag{3.11}$$

The second-order derivative formula can be easily derived such that

$$\frac{\mathrm{d}^2}{\mathrm{d}t^2} f(t) = \lim_{h\to 0} \frac{1}{h^2} \big[f(t) - 2f(t-h) + f(t-2h)\big]. \tag{3.12}$$

Use the above method repeatedly, the nth-order derivative of the function can be derived as

$$\frac{\mathrm{d}^n}{\mathrm{d}t^n} f(t) = \lim_{h\to 0} \frac{1}{h^n} \sum_{j=1}^{n} (-1)^j \binom{n}{j} f(t-jh), \tag{3.13}$$

where the binomial expression can be written as

$$(1-z)^n = \sum_{j=0}^{n} (-1)^j \binom{n}{j} z^j = \sum_{j=0}^{n} \frac{(-1)^j n!}{j!\,(n-j)!} z^j = \sum_{j=0}^{n} w_j z^j, \tag{3.14}$$

and the binomial coefficients can be computed directly from the following formula:

$$w_j = (-1)^j \binom{n}{j} = (-1)^j \frac{n!}{j!\,(n-j)!},\ j = 0, 1, 2, \ldots, n. \tag{3.15}$$

3.2.2 Definition of Grünwald–Letnikov Fractional-Order Derivatives

The above formula for the nth-order derivative can be easily extended directly to the case of non-integer $n = \alpha$. Unlike the integer-order ones, the binomial expression is no longer a sum of finite terms, but takes the form of an infinite series, that is,

$$(1-z)^\alpha = \sum_{j=0}^{\infty} (-1)^j \binom{\alpha}{j} z^j = \sum_{j=0}^{\infty} w_j z^j. \tag{3.16}$$

Therefore, the expanded binomial coefficients become

$$w_j = (-1)^j \binom{\alpha}{j} = \frac{(-1)^j \Gamma(\alpha+1)}{\Gamma(j+1)\Gamma(\alpha-j+1)},\ j = 0, 1, 2, \ldots \tag{3.17}$$

Comparing the binomial coefficients in (3.15) with those in (3.17), it can be seen that the latter simply replaces the factorials of the former with more general Gamma functions, thus expanding its scope of applications.

Suppose that when $t \leqslant t_0$, the value of the function $f(t)$ is zero, then the sum of infinite terms can be converted into the sum of finite terms, so that the Grünwald–Letnikov fractional-order derivative formula can be introduced.

Definition 3.3 Grünwald–Letnikov definition

The αth-order Grünwald–Letnikov derivative of function $f(t)$ is

$$ {}^{\mathrm{GL}}_{t_0}\mathscr{D}^{\alpha}_{t} f(t) = \lim_{h \to 0} \frac{1}{h^\alpha} \sum_{j=0}^{[(t-t_0)/h]} (-1)^j \binom{\alpha}{j} f(t-jh), \tag{3.18}$$

where $[\,\cdot\,]$ takes the nearest integer.

Remark Grünwald–Letnikov definition

(1) The GL notation in the upper left corner of the operator indicates the Grünwald–Letnikov definition, which can be omitted when there is no conflict.

(2) It can be seen that integer-order derivatives only use the current and the few recent finite steps of the function values, while fractional-order derivatives involve all the function values since t_0. Therefore, the fractional-order derivative can be considered to have memory.

(3) If α is an integer, (3.18) is exactly the same as (3.13).

(4) The same definition applies to derivatives for $\alpha > 0$ and integrals for $\alpha < 0$. Also, if $\alpha = 0$, it follows from the definition that ${}_{t_0}^{\rm GL}\mathscr{D}_t^0 f(t) = f(t)$.

(5) The Grünwald–Letnikov definition satisfies Definition 3.1. □

Later, various methods for evaluating binomial coefficients will be presented, and the numerical evaluation of the Grünwald–Letnikov fractional-order derivatives will be introduced.

3.2.3 Numerical Evaluation of Grünwald–Letnikov Fractional-Order Derivatives and Integrals

The limit operation in (3.18) can be ignored, if the chosen step size h is small enough, such that the fractional-order derivatives and integrals under the Grünwald–Letnikov definition can be evaluated directly from the following equation:

$$ {}_{t_0}^{\rm GL}\mathscr{D}_t^\alpha f(t) \approx \frac{1}{h^\alpha} \sum_{j=0}^{[(t-t_0)/h]} w_j f(t-jh), \tag{3.19}$$

where the binomial coefficients can be evaluated directly from (3.17). Thus, it seems natural to implement the following algorithm and write the MATLAB solution function.

Algorithm 3.1 Direct evaluation of fractional-order derivatives
(1) Compute the samples from the function $f(t)$ and compose the vector $\boldsymbol{f}$.
(2) Compute the binomial coefficients from (3.17).
(3) Compute directly the fractional-order derivatives from (3.18).

Based on Algorithm 3.1, the following MATLAB function can be written:

```
function df=glfdiff0(f,t,gam)
   arguments, f(:,1), t(:,1) double, gam(1,1) double, end
   [f,h,n]=fdiffcom(f,t); J=0:(n-1); a0=f(1);
   if a0~=0 && gam>0, dy(1)=sign(a0)*Inf; end
   w=gamma(gam+1)*(-1).^J./gamma(J+1)./gamma(gam-J+1);
   for i=2:n, dy(i,1)=w(1:i)*f(i:-1:1)/h^gam; end
end
```

The syntax of the function is $f_1 =$ `glfdiff0`(f,t,α), where $\boldsymbol{t}$ is an equally spaced time column vector and α is the order, where f can be either a column vector of function samples of function $f(t)$ at each point of $\boldsymbol{t}$ or a function handle of $f(t)$. In other words, the original function can be represented by an anonymous function at this point. The numerical solution of the fractional-order derivative is returned in the column vector of variables $\boldsymbol{f}_1$. If the $\boldsymbol{t}$-vector is not a column vector, the function will automatically convert it into a column vector.

The next-level utility function `fdiffcom()` is used in this function, which reads

```
function [y,h,n]=fdiffcom(f,t)
   y=f; if strcmp(class(f),'function_handle'), y=f(t); end
   h=t(2)-t(1); n=length(t);
end
```

Unfortunately, such an algorithm has obvious defects: it directly involves the value of the Gamma function of large numbers, and $\Gamma(172)$ and the subsequent terms have become infinite terms `Inf` under the MATLAB double-precision data structure. Therefore, the influence of the latter terms is completely ignored, resulting in large errors in the final evaluation results. Thus, the evaluation of fractional-order derivatives requires a more reliable method.

The biggest problem with the above algorithm is that it uses the Gamma function when evaluating the binomial coefficients, thus leading to unavoidable errors. If the Gamma function is avoided by introducing a more efficient method for the binomial coefficients, more reliable results can be obtained. The recursive algorithm for the binomial coefficients is given below.

Theorem 3.1 Recursive formula for binomial coefficients
A recursive formula for computing binomial coefficients w_j is as follows:

$$w_0 = 1, \quad w_j = \left(1 - \frac{\alpha+1}{j}\right) w_{j-1}, \quad j = 1, 2, \ldots \tag{3.20}$$

Proof It is seen from (3.17) that, $w_0 = 1$. Taking the ratio of the current term w_j to the previous one w_{j-1} in the binomial coefficient formula, and employing the property $\Gamma(x+1) = x\Gamma(x)$, the following formula can be derived:

$$\begin{aligned}\frac{w_j}{w_{j-1}} &= \frac{(-1)^j\Gamma(\alpha+1)}{\Gamma(j+1)\Gamma(\alpha-j+1)} \bigg/ \frac{(-1)^{j-1}\Gamma(\alpha+1)}{\Gamma(j)\Gamma(\alpha-j+2)} \\ &= -\frac{\Gamma(j)\Gamma(\alpha-j+2)}{\Gamma(j+1)\Gamma(\alpha-j+1)} = -\frac{\cancel{\Gamma(j)}(\alpha-j+1)\cancel{\Gamma(\alpha-j+1)}}{j\cancel{\Gamma(j)}\cancel{\Gamma(\alpha-j+1)}} \\ &= -\frac{\alpha-j+1}{j} = 1 - \frac{\alpha+1}{j}.\end{aligned}$$

The recursive formula in (3.20) can be proved by multiplying both sides of the equation by w_{j-1} at the same time. □

In this way, the binomial coefficients are recursively evaluated using (3.20), and the fractional-order derivative of a given function can be evaluated directly from (3.19). Since the recursive algorithm successfully avoids the direct computation of the Gamma function, it solves the problem in Algorithm 3.1. It can be shown that the accuracy of the following algorithm reaches $O(h)$ [2].

Algorithm 3.2 The Grünwald–Letnikov derivatives and integrals

(1) Compute the samples of the given signal and construct the vector $\boldsymbol{f}$.

(2) Compute the binomial coefficients w_j recursively from (3.20).

(3) Evaluate the fractional-order derivatives or integrals from (3.19).

Based on the above algorithm, the functions for the numerical evaluation of Grünwald–Letnikov derivatives can be written in MATLAB language as follows:

```
function dy=glfdiff(y,t,gam)
   arguments, y(:,1), t(:,1) double, gam(1,1) double, end
   [y,h,n]=fdiffcom(y,t); w=[1,zeros(1,n-1)];
   a0=y(1); dy(1)=0;
   if a0~=0 && gam>0, dy(1)=sign(a0)*Inf; end
   for j=2:n, w(j)=w(j-1)*(1-(gam+1)/(j-1)); end
   for i=1:length(t), dy(i,1)=w(1:i)*y(i:-1:1)/h^gam; end
end
```

The syntax of the function is y_1 = `glfdiff(`y`,`t`,`γ`)`, and the arguments are exactly the same as the ones in function `glfdiff0()`. Compared with the `glfdiff0()` function, only the $\boldsymbol{w}$-vector evaluation statement is modified here, nothing else is changed. If γ is negative, then this function can directly evaluate the $-\gamma$th-order integral of the original function.

▶ Example 3.1

Find the 0.75th-order Cauchy derivative and the Grünwald–Letnikov derivative of a constant and, in addition, the 0.75th-order integral of the constant under these two definitions.

Solutions In the integer-order calculus, the derivative of each order of the constant is zero and its first-order integral is a straight line. It is natural for people to wonder what the fractional-order derivatives and integrals of the constant really is.

In fact, the constant 1 can be expressed as a power function $f(t) = 1 = t^0$, so that the fractional-order derivative and the integral are satisfied by (3.10), respectively.

$$\mathscr{D}_t^{0.75} y(t) = \frac{\Gamma(1)\, t^{-0.75}}{\Gamma(1-0.75)} = \frac{t^{-0.75}}{\Gamma(0.25)}, \quad \mathscr{D}_t^{-0.75} y(t) = \frac{\Gamma(1)\, t^{0.75}}{\Gamma(1+0.75)} = \frac{t^{0.75}}{\Gamma(1.75)}.$$

The Cauchy definition of the constant and the 0.75th-order derivative of the Grünwald–Letnikov definition and the integral can be derived from the following statement, as shown in ◘ Fig. 3.1. It can be seen that the resulting two sets of curves are in good agreement.

```
>> t=0:0.001:1; y=ones(size(t));
   y1=glfdiff(y,t,0.75); y1a=t.^(-0.75)/gamma(0.25);
   y2=glfdiff(y,t,-0.75); y2a=t.^(0.75)/gamma(1.75);
   y2a=gamma(1)*t.^(-0.75)/gamma(1-0.75);
   plot(t,y1,t,y1a,'--',t,y2,t,y2a,'--'), ylim([0 5])
```

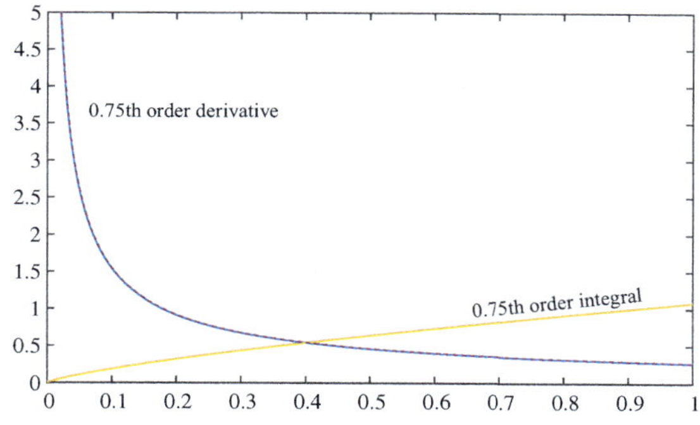

Fig. 3.1 0.75th-order derivative and integral of a constant

Choosing different orders $\gamma = 0$, 1/2, 3/4, 1 and 5/4, the derivatives of each order of the function are shown in Fig. 3.2. It can be seen that the results are identical to the original function and its first-order derivatives when γ is equal to the integers 0 and 1.

```
>> t=0:0.001:0.5; y=ones(size(t)); gam=0:0.25:1.25;
   for g=gam, y1=glfdiff(y,t,g); plot(t,y1), hold on; end
   ylim([-3 5]), hold off
```

The function `glfdiff()` can also be used to evaluate fractional-order integrals directly if the orders of integral are set to 0, 1/4, 1/2, 3/4, 1, and 5/4, i.e., if the order vector is set to $\boldsymbol{\gamma} = [0, -1/4, -1/2, -3/4, -1, -5/4]$, then to plot the fractional-order integral by the following statements, as shown in Fig. 3.3. It can be seen that the results obtained when $\gamma = 0$ and $\gamma = -1$ are exactly the same as the original function and the first-order integral.

```
>> t=0:0.01:1.5; y=ones(size(t)); gam=-1.25:0.25:0;
   for g=gam, y1=glfdiff(y,t,g); plot(t,y1), hold on; end
```

◀

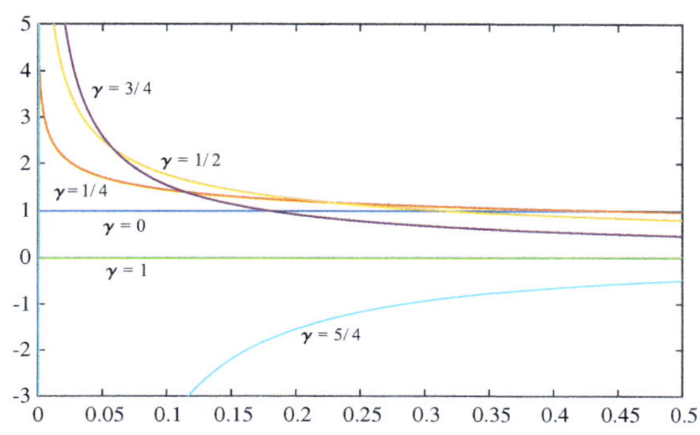

Fig. 3.2 Fractional-order derivatives of the constant

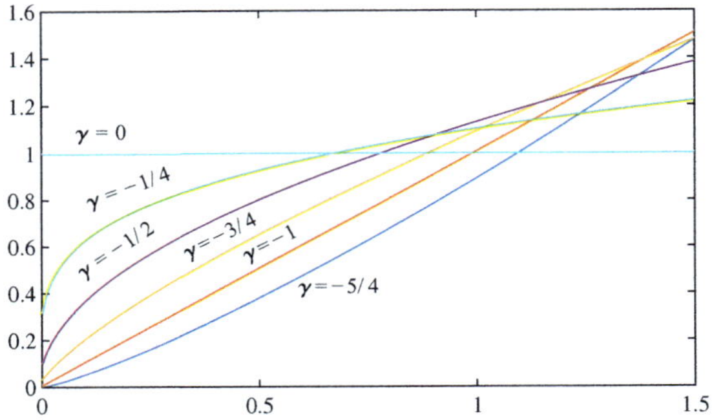

Fig. 3.3 Fractional-order integral curves of different orders

▶ Example 3.2

Select different step sizes to find the 0.5th-order derivative of the step function and evaluate the accuracy of the solution.

Solutions By choosing the step sizes of $h = 0.0001$, 0.001, 0.01 and 0.1, the following statements can be used to find the 0.5th-order derivatives of the original function. Compare them with the theoretical value to find their errors, the results are listed in Table 3.1. It can be seen that the theoretical accuracy of $O(h)$ is correct.

```
>> t0=0.2:0.2:1; y0=gamma(1)*t0(:).^(-0.5)/gamma(1-0.5);
   t=0:0.0001:1; y=ones(size(t)); y1=glfdiff(y,t,0.5);
   t=0:0.001:1;  y=ones(size(t)); y2=glfdiff(y,t,0.5);
   t=0:0.01:1;   y=ones(size(t)); y3=glfdiff(y,t,0.5);
   t=0:0.1:1;    y=ones(size(t)); y4=glfdiff(y,t,0.5);
   y1=y1(2001:2000:10001); y2=y2(201:200:1001);
   y3=y3(21:20:101); y4=y4(3:2:11);
   T=[[t0, t0]; [y1 y2 y3 y4]', [y0-y1 y0-y2 y0-y3 y0-y4]']
```

◀

In the previous example, since the analytical expression of the fractional-order derivative is known, the computational error can be evaluated compared to the exact

Table 3.1 Errors of fractional-order derivatives at different step sizes

Step sizes	The derivatives at t_k				The errors at t_k			
h	0.2	0.4	0.6	0.8	0.2	0.4	0.6	0.8
y_0	1.2616	0.8921	0.7284	0.6308				
0.0001	1.2615	0.8920	0.7284	0.6308	7.9×10^{-5}	2.8×10^{-5}	1.5×10^{-5}	7.1×10^{-6}
0.001	1.2608	0.8918	0.7282	0.6307	0.000786	0.000279	0.000152	7.1×10^{-5}
0.01	1.2537	0.8893	0.7269	0.6298	0.0079	0.00278	0.00152	0.00070
0.1	1.1859	0.8647	0.7134	0.6210	0.07571	0.02738	0.015	0.00701

value. In general applications where the exact value is unknown, the result can be obtained by choosing a certain step size and then recalculating it by choosing a smaller step size to see whether the two agree. If they agree, the results are reliable; if not, the step size should be reduced again and recalculated until a consistent result is obtained.

► Example 3.3

Now consider the given function $f(t) = \mathrm{e}^{-t}\sin(3t+1)$, $t \in (0, \pi)$, where the initial value of the function is not zero. Evaluate the fractional-order derivative of this signal.

Solutions By choosing the step sizes of $h = 0.01$, $h = 0.001$, and $h = 0.0005$, respectively, the 0.5th-order Grünwald–Letnikov derivative of the known signal can be evaluated from the following statements, as shown in Fig. 3.4. It can be seen that several curves match well, except for a very small region near the moment $t = 0$, indicating that the result is less accurate at $h = 0.01$. The results for $h = 0.001$ and $h = 0.0005$ coincide almost exactly.

```
>> f=@(t)exp(-t).*sin(3*t+1); % use anonymous function for the given one
   t=0:0.001:pi; dy=glfdiff(f,t,0.5); % taking derivatives
   t1=0:0.01:pi; dy1=glfdiff(f,t1,0.5);
   t2=0:0.0005:pi; dy2=glfdiff(f,t2,0.5);
   plot(t,dy,t1,dy1,'--',t2,dy2,':'), axis([0,pi,-1.5 6])
```

By choosing different derivative orders $\gamma = 0, 1/4, 1/2, 3/4, 1, 5/4$, the fractional-order derivatives of the function can be evaluated by the following statements, as shown in Fig. 3.5.

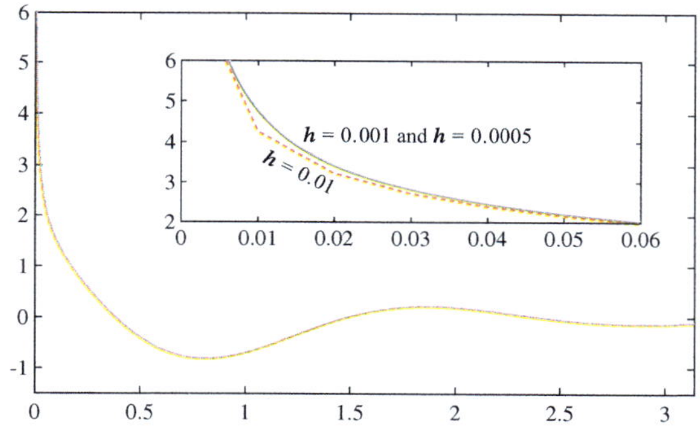

Fig. 3.4 Comparison of results with different step sizes

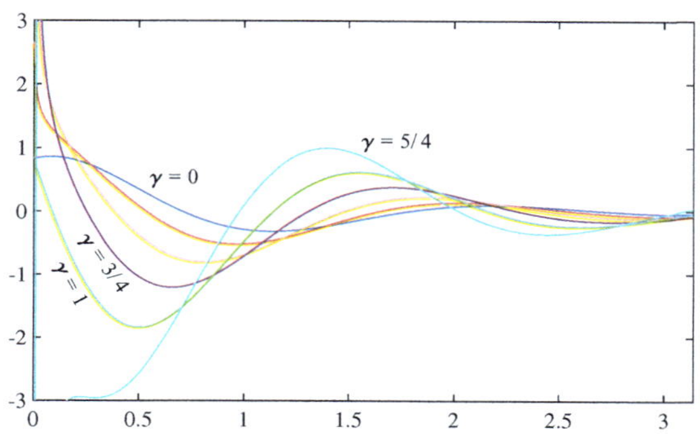

Fig. 3.5 Fractional-order derivative curves of different orders

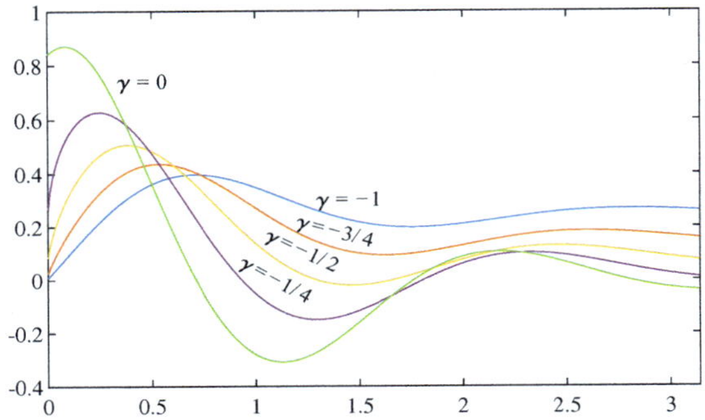

Fig. 3.6 Fractional-order integral curves of different orders

```
>> t=0:0.01:pi; y=exp(-t).*sin(3*t+1); gam=0:0.25:1.25;
   for g=gam, y1=glfdiff(y,t,g); plot(t,y1); hold on; end
   axis([0 pi, -3 3]), hold off
```

The following statements also directly yield the fractional-order integrals at different orders, as shown in Fig. 3.6. It can be seen that the fractional-order integrals of this function contain richer information than the integer-order ones.

```
>> for gam=-1:0.25:0
     y1=glfdiff(y,t,gam); plot(t,y1); hold on
   end
   xlim([0 pi]), hold off
```

◀

► Example 3.4

Find the 0.75th-order derivative of $f(t) = \sin(3t + 1)$ using the Cauchy definition and the Grünwald–Letnikov definition, respectively, and compare their results.

Solutions From the Cauchy integral formula given in (3.6), the 0.75th-order derivative of the function is ${}_0\mathscr{D}_t^{0.75}f(t) = 3^{0.75}\sin(3t+1+0.75\pi/2)$, while the numerical solution of the Grünwald–Letnikov definition can be found by the `glfdiff()` function. The two curves are shown in Fig. 3.7.

```
>> t=0:0.01:pi;
   y=sin(3*t+1); y1=3^0.75*sin(3*t+1+0.75*pi/2);
   y2=glfdiff(y,t,0.75); plot(t,y1,t,y2,'--'), axis([0 pi -4 8])
```

Comparing these two derivative curves, it can be seen that under the Cauchy definition there is no sudden jump at $t = 0$. This is because the function $y(t)$ at $t \leqslant 0$ is also considered to satisfy $f(t) = \sin(3t + 1)$, while in the Grünwald–Letnikov definition, the initial value at $t < 0$ is assumed to be zero. ◀

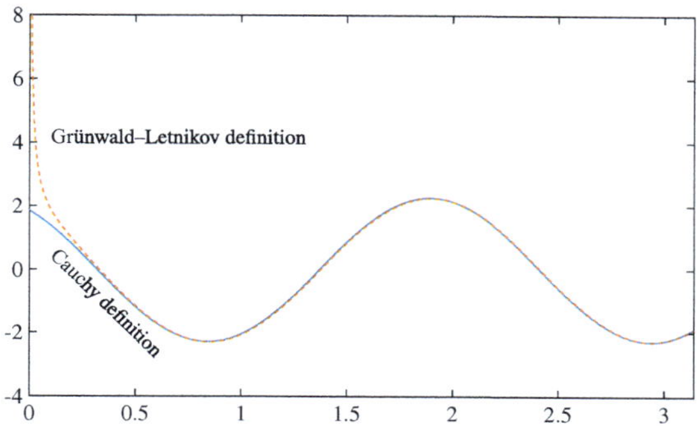

Fig. 3.7 Fractional-order derivative curves under different definitions

3.2.4 Podlubny's Matrix Algorithm

The matrix algorithm proposed by Prof. Igor Podlubny can also be used to evaluate the Grünwald–Letnikov fractional-order derivatives for a given function [3] with the following algorithm.

Algorithm 3.3 Matrix algorithm for Grünwald–Letnikov derivatives
(1) Construct column vector $\boldsymbol{f}$ from the samples of the given function.
(2) Compute the binomial coefficients w_j recursively from (3.20).
(3) Matrix form from (3.19) can be applied to evaluate the fractional-order derivatives.

$$\mathscr{D}^{\alpha} f = \frac{1}{h^{\alpha}} \begin{bmatrix} w_0 & & & & & \\ w_1 & w_0 & & & & \\ w_2 & w_1 & w_0 & & & \\ \ddots & \ddots & \ddots & \ddots & & \\ w_{N-1} & \ddots & w_2 & w_1 & w_0 & \\ w_N & w_{N-1} & \ddots & w_2 & w_1 & w_0 \end{bmatrix} \begin{bmatrix} f_0 \\ f_1 \\ f_2 \\ \vdots \\ f_{N-1} \\ f_N \end{bmatrix}. \tag{3.21}$$

Note that the matrix in (3.21) is actually a rotated Hankel matrix. The matrix algorithm can be implemented in MATLAB to find the fractional-order derivatives of a given function.

```
function dy=glfdiff_mat(y,t,gam)
   arguments, y(:,1), t(:,1) double, gam(1,1) double, end
   [y,h,n]=fdiffcom(y,t); w=[1,zeros(1,n-1)];
   for j=2:n, w(j)=w(j-1)*(1-(gam+1)/(j-1)); end
   dy=rot90(hankel(w(end:-1:1)))*y/h^gam;
end
```

It can be seen that, the method given here is mathematically simpler and more suitable for sequential derivatives of signals, i.e., the action $\mathscr{D}^{\alpha}\mathscr{D}^{\beta}\mathscr{D}^{\gamma}y(t)$ can be converted into a matrix product for computation. The limitation of this algorithm is that sometimes the total number of computation points N is too large for MATLAB to carry out such large-scale matrix operations.

3.2.5 Exploring Short-Time Memory Effects

In general, if the step size h is chosen too small or the value of $[(t-t_0)/h]$ is too large, the number of points involved in the summation of (3.19) will be particularly large, which may lead to a significant increase in the computational load and eventually produce no computational results. Therefore, a reasonable reduction in the number of points should be considered. In practical applications, it is not necessary to use all the points in the previous interval $[(t-t_0)/h]$, but the information in the nearest time interval $[t-L, t]$ reduces the number of computations [2].

$$ {}_{t_0}\mathscr{D}_t^{\alpha} f(t) \approx {}_{(t-L)}\mathscr{D}_t^{\alpha} f(t). \tag{3.22}$$

This method is also known as the short-time memory effect [2, 4]. Using this method, the Grünwald–Letnikov fractional-order derivative can be approximated as

$$ y(t) \approx \frac{1}{h^{\alpha}} \sum_{j=0}^{N(t)} w_j f(t-jh), \tag{3.23}$$

where

$$ N(t) = \min\left\{ \left[\frac{t-t_0}{h}\right], \frac{L}{h} \right\}, \tag{3.24}$$

where L is referred to as the memory duration. In this way, how to choose this memory duration L will become the most critical issue. Suppose the value of the function $f(t)$ in the time interval of interest (t_0, T) satisfies $|f(t)| \leqslant M$, then the visible approximation error is

$$ \begin{aligned} \Delta(t) &= \left| {}_{t_0}\mathscr{D}_t^{\alpha} f(t) - {}_{(t-L)}\mathscr{D}_t^{\alpha} f(t) \right| \\ &= \frac{1}{\Gamma(1-\alpha)} \int_{t_0}^{L} \frac{f'(\tau)}{(t-\tau)^{\alpha}} \mathrm{d}\tau \leqslant \frac{ML^{-\alpha}}{|\Gamma(1-\alpha)|}. \end{aligned} \tag{3.25}$$

If the expected error is less than a pre-specified positive number ε, i.e., $\Delta(t) < \varepsilon$, then the memory duration should be chosen as

$$ L \geqslant \left(\frac{M}{\varepsilon|\Gamma(1-\alpha)|} \right)^{1/\alpha}. \tag{3.26}$$

However, the formula may be too harsh in choosing the memory duration. For example, if the error tolerance $\varepsilon = 10^{-6}$ is chosen, the memory duration will be a very large value unless the error requirement is reduced. If a relatively large error tolerance $\varepsilon = 10^{-3}$ is chosen, the order $\alpha = 0.5$, and the upper bound of the function is $M = 1$, the required memory duration is $L = 318309$. If $\varepsilon = 10^{-4}$ is chosen, the memory duration $L = 31830988$. If the order is smaller, such as $\alpha = 0.1$, the memory duration will be astronomical, so when accuracy is required, (3.26) may be too conservative, and sometimes the short-time memory effect may not have practical significance.

Based on the idea of short-time memory, the `glfdiff()` function is rewritten to add the short-time memory function and write a new `glfdiff_mem()` function, whose syntax is y_1 = `glfdiff_mem(`y,t,γ,L_0`)`, where y, t, γ, and y_1 are exactly the same as in `glfdiff()`. The additional L_0 is the number of computed points, i.e., $L_0 = L/h$. If L_0 is not given, the result is the Grünwald–Letnikov fractional-order derivative.

```
function dy=glfdiff_mem(y,t,gam,L0)
   arguments, y(:,1), t(:,1), gam(1,1), L0=length(t); end
   [y,h,n]=fdiffcom(y,t); w=[1,zeros(1,n-1)]; dy(1)=0;
   for j=2:L0+1, w(j)=w(j-1)*(1-(gam+1)/(j-1)); end
   for i=1:n
      L=min([i,L0]); dy(i,1)=w(1:L)*y(i:-1:i-L+1)/h^gam;
end, end
```

It is also possible to consider setting the value following the $\boldsymbol{w}$ vector in matrix Algorithm 3.3 to zero, but this does not help to improve the speed of matrix operations. Examples will be given later to explore the practical application and the effect of the short-time memory effect.

▶ Example 3.5

In the formula for fractional-order derivatives, the binomial coefficient w_j shows the dependence of the derivative value on the previous function values. Draw the w_j curves for $\alpha = 0.1$ and $\alpha = 0.8$.

Solutions Using the recursive formula in (3.20), the w_j coefficients can be evaluated directly for different values of α to plot the curves, as shown in Fig. 3.8. It can be seen that the fractional-order derivatives are highly dependent on a few recent points and somewhat less dependent on the earlier points. For example, when $\alpha = 0.1$, $w_{100} = -5.9076 \times 10^{-4}$. However, the information from the previous points cannot be completely ignored.

```
>> t=0:0.1:3; a1=0.1; a2=0.8;
   w1=1; for j=2:length(t), w1(j)=w1(j-1)*(1-(a1+1)/(j-1)); end
   w2=1; for j=2:length(t), w2(j)=w2(j-1)*(1-(a2+1)/(j-1)); end
   plot(t,w1,t,w2,'--')
```

◀

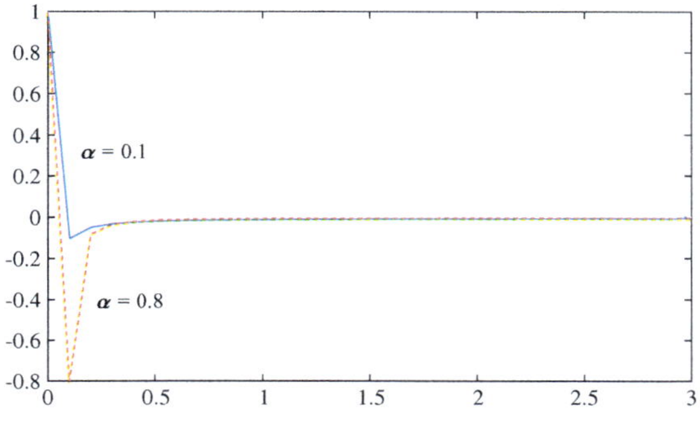

Fig. 3.8 Curves of binomial coefficients w_j at different orders of α

3

► Example 3.6

Consider again the function $f(t) = e^{-t}\sin(3t + 1)$ from Example 3.3. Use the short-time memory effect to find the 0.5th-order derivative of the function and explore the effect of the short-time memory effect approximation.

Solutions Choose a step size $h = 0.001$ and a tolerance error of $\varepsilon = 10^{-3}$. Since the maximum value of the function is $M = 1$, the minimum memory duration calculated from (3.26) is $L = \left(1/(10^{-3}\Gamma(1-0.5))\right)^{1/0.5} = 3.2 \times 10^5$. Of course, such a requirement is somewhat too demanding.

A shorter memory duration L_0 was chosen to explore the approximation and the comparison results are given in Table 3.2. It can be seen that if the memory duration is chosen to be $L_0 = 1200$, a satisfactory approximation can basically be obtained. Because the total full duration is 3142, the memory duration of $L_0 = 1200$ reduces 2/3 of the computational load. The curves obtained are shown in Fig. 3.9.

Table 3.2 Relationship between memory duration and error when the step size is $h = 0.001$

Memory L_0	Error norm	Memory L_0	Error norm	Memory L_0	Error norm
100	0.71736	1100	0.063559	2100	0.028305
200	0.38972	1200	0.057211	2200	0.026644
300	0.26417	1300	0.051879	2300	0.025141
400	0.19737	1400	0.047346	2400	0.023777
500	0.15597	1500	0.043449	2500	0.022532
600	0.12789	1600	0.040070	2600	0.021099
700	0.10767	1700	0.037115	2700	0.018628
800	0.09246	1800	0.034513	2800	0.015194
900	0.08065	1900	0.032206	2900	0.011026
1000	0.07123	2000	0.030149	3000	0.006443

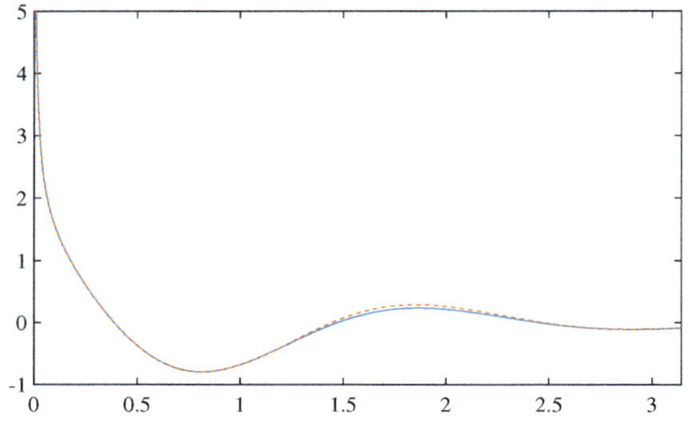

Fig. 3.9 Short-time memory effect at $L_0 = 1200$

```
>> t=0:0.001:pi; y=exp(-t).*sin(3*t+1);
   y0=glfdiff(y,t,0.5); L0=[100:100:3000]; T=[];
   for L=L0
     dy=glfdiff_mem(y,t,0.5,L); T=[T; L, norm(y0-dy,inf)];
   end
   dy=glfdiff_mem(y,t,0.5,1200);
   plot(t,y0,t,dy,'--'); axis([0 pi -1 5])
```

In the same example, if the termination time becomes 50 s, it takes 2.74 s to derive the fractional-order derivative using the ordinary Grünwald–Letnikov algorithm, while it takes only 0.21 s using a short-time memory algorithm with a memory duration of $L_0 = 1200$. In practice, the use of the short-time memory algorithm can be considered for particularly time-consuming applications, but the results obtained need to be tested.

```
>> t=0:0.001:50; y=exp(-t).*sin(3*t+1);
   tic, y0=glfdiff(y,t,0.5); toc
   tic, dy=glfdiff_mem(y,t,0.5,1200); toc
```

◀

A counterexample to the short-time memory effect is given below.

▶ Example 3.7

Find the fractional-order integral of a step function using a short-time memory method.

Solutions The 0.75th-order integral of the step signal can be derived directly from the Cauchy integral formula $t^{0.75}/\Gamma(1+0.75)$. Choosing a step size of $h = 0.001$, a total of 10 s is required to generate 10000 data points. Now try a larger memory duration, such as $L_0 = 4000$, the following statements can be used to evaluate the fractional-order integral directly, as shown in Fig. 3.10. It can be seen that even with such a large memory duration, a large computational error is still produced, indicating that the short-time memory algorithm is not suitable for use in this example and that the full information must be used.

```
>> t=0:0.001:10; y=ones(size(t));
   y0=gamma(1)*t.^(0.75)/gamma(1+0.75);  %Cauchy integral formula
```

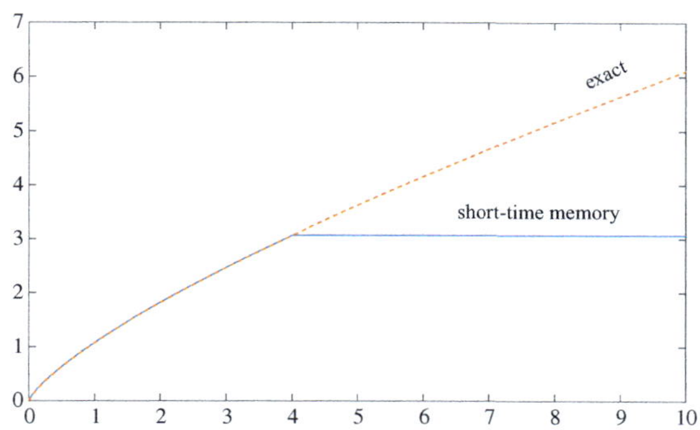

Fig. 3.10 Failed short-time memory effect

```
dy=glfdiff_mem(y,t,-0.75,4000);          % short-time memory effect
dy1=glfdiff(y,t,-0.75); plot(t,dy,t,dy1,'--');
```

◀

3

As can be seen from Example 3.7, although many people believe that the short-time memory effect is an efficient approximation, in fact it is not. This is because there are still risks in using the method, and sometimes even wrong or misleading results can be obtained. Since (3.26) is too demanding to ensure less expected accuracy requirements, it is also often the case that the calculated L is much larger than the actual number of points involved. Therefore, unless it is absolutely necessary, do not use this algorithm, but consider a better evaluation method.

3.3 Definition and Evaluation of Riemann–Liouville Derivatives and Integrals

The Grünwald–Letnikov definition given earlier, although favorable for numerical evaluations, is not easy to handle in terms of theoretical derivation or proof of formulas. Other mathematical definitions should also be considered. For example, the Riemann–Liouville definition, which is extended from integer-order integrals and so on, will be introduced in this section.

3.3.1 High-Order Integer-Order Integral Formulas

Consider first the integer-order integral. It is clear that the first-order integral of a given function $f(t)$ can be expressed as

$$\frac{\mathrm{d}^{-1}}{\mathrm{d}t^{-1}}f(t) = \int_0^t f(\tau)\mathrm{d}\tau. \tag{3.27}$$

Taking integral once again, the second-order integral derived for the original function is as follows:

$$\frac{\mathrm{d}^{-2}}{\mathrm{d}t^{-2}}f(t) = \int_0^t \int_0^\tau f(\tau)\mathrm{d}\tau\mathrm{d}t = \int_0^t f(\tau)(t-\tau)\mathrm{d}\tau. \tag{3.28}$$

Similarly, the nth-order integral can be formulated as

$$\begin{aligned}\frac{\mathrm{d}^{-n}}{\mathrm{d}t^{-n}}f(t) &= \underbrace{\int_0^t \int_0^\tau \cdots \int_0^\tau}_{n\text{ fold}} f(\tau)\,\underbrace{\mathrm{d}\tau_1\mathrm{d}\tau_2\cdots\mathrm{d}\tau}_{n\text{ fold}} \\ &= \frac{1}{(n-1)!}\int_0^t f(\tau)(t-\tau)^{n-1}\mathrm{d}\tau = \frac{1}{(n-1)!}\int_0^t \frac{f(\tau)}{(t-\tau)^{1-n}}\mathrm{d}\tau.\end{aligned} \tag{3.29}$$

3.3.2 Definitions of Riemann–Liouville Fractional-Order Derivatives and Integrals

If the integer order n is replaced by the real number α, then the Riemann–Liouville fractional-order integral formula is given directly as follows.

Definition 3.4 Riemann–Liouville integral definition

The αth-order Riemann–Liouville integral of function $f(t)$ is defined as

$$ {}_{t_0}^{\mathrm{RL}}\mathscr{D}_t^{-\alpha} f(t) = \frac{1}{\Gamma(\alpha)} \int_{t_0}^{t} \frac{f(\tau)}{(t-\tau)^{1-\alpha}} \mathrm{d}\tau, \tag{3.30} $$

where RL represents Riemann–Liouville definition, and it can be omitted if there is no conflict.

The Riemann–Liouville definition is one of the more widely used definitions in the field of fractional calculus, where the subscripts at both sides of the operator $\mathscr{D}$ are the lower and upper bounds of the integral [5].

Fractional-order derivative cannot be defined directly by replacing α in the integral with $-\alpha$, but should be given based on the definition of the integral. Now consider βth-order fractional-order derivative. If $n-1 < \beta \leqslant n$, denote $n = \lceil \beta \rceil$, then

$$ {}_{t_0}^{\mathrm{RL}}\mathscr{D}_t^{\beta} f(t) = \frac{\mathrm{d}^n}{\mathrm{d}t^n} \Big[{}_{t_0}^{\mathrm{RL}}\mathscr{D}_t^{-(n-\beta)} f(t) \Big]. \tag{3.31} $$

Based on the definition of the integral, a formal definition of the Riemann–Liouville fractional-order derivative can be given.

Definition 3.5 Riemann–Liouville derivative formula

The Riemann–Liouville derivative of function $f(t)$ is defined as

$$ {}_{t_0}^{\mathrm{RL}}\mathscr{D}_t^{\beta} f(t) = \frac{1}{\Gamma(n-\beta)} \frac{\mathrm{d}^n}{\mathrm{d}t^n} \int_{t_0}^{t} \frac{f(\tau)}{(t-\tau)^{1+\beta-n}} \mathrm{d}\tau, \tag{3.32} $$

where $n-1 < \beta \leqslant n$ and $n = \lceil \beta \rceil$.

3.3.3 Riemann–Liouville Derivative and Integral Formulas for Commonly Used Functions

Consider first the Riemann–Liouville fractional-order derivatives of the power and exponential functions at the lower bound $t_0 = 0$.

Theorem 3.2 Fractional-order derivative of power function
For power function $f(t)=t^{\mu}$, with $\mu>-1$, ${}_{0}^{\mathrm{RL}}\mathscr{D}_t^{\alpha}f(t)$ satisfies [2]

$$ {}_{0}^{\mathrm{RL}}\mathscr{D}_t^{\alpha}t^{\mu}=\frac{\Gamma(\mu+1)}{\Gamma(\mu+1-\alpha)}t^{\mu-\alpha}. \tag{3.33}$$

Theorem 3.3 Fractional-order derivative of exponential function
The Riemann–Liouville derivative of exponential function of $f(t)=\mathrm{e}^{\lambda t}$ is

$$ {}_{0}^{\mathrm{RL}}\mathscr{D}_t^{\alpha}\mathrm{e}^{\lambda t}=t^{-\alpha}\mathrm{E}_{1,1-\alpha}(\lambda t). \tag{3.34}$$

Proof The Taylor expansion of the exponential function $\mathrm{e}^{\lambda t}$ is

$$\mathrm{e}^{\lambda t}=\sum_{k=0}^{\infty}\frac{(\lambda t)^k}{k!}=\sum_{k=0}^{\infty}\frac{\lambda^k t^k}{\Gamma(k+1)}. \tag{3.35}$$

Using the properties of (3.33), the Riemann–Liouville fractional-order derivative can be proved by computing term by term the fractional-order derivatives on the right-hand side of (3.35).

$$\begin{aligned} {}_{0}^{\mathrm{RL}}\mathscr{D}_t^{\alpha}\mathrm{e}^{\lambda t}&=\sum_{k=0}^{\infty}{}_{0}^{\mathrm{RL}}\mathscr{D}_t^{\alpha}\frac{\lambda^k t^k}{\Gamma(k+1)}=\sum_{k=0}^{\infty}\frac{\lambda^k\Gamma(k+1)t^{k-\alpha}}{\Gamma(k+1)\Gamma(k+1-\alpha)}\\ &=t^{-\alpha}\sum_{k=0}^{\infty}\frac{(\lambda t)^k}{\Gamma(k+1-\alpha)}=t^{-\alpha}\mathrm{E}_{1,1-\alpha}(\lambda t). \end{aligned} \tag{3.36}$$

□

Theorem 3.4 Riemann–Liouville derivatives of commonly used functions
Riemann–Liouville derivatives of commonly used functions are [2]

$$ {}^{\mathrm{RL}}\mathscr{D}_t^{\alpha}H(t-a)=\frac{(t-a)^{-\alpha}}{\Gamma(1-\alpha)}H(t-a) \tag{3.37}$$

$$ {}_{0}^{\mathrm{RL}}\mathscr{D}_t^{\alpha}H(t-a)f(t)=H(t-a)\,{}_{a}^{\mathrm{RL}}\mathscr{D}_t^{\alpha}f(t) \tag{3.38}$$

$$ {}_{0}^{\mathrm{RL}}\mathscr{D}_t^{\alpha}\delta(t)=\frac{t^{-\alpha-1}}{\Gamma(-\alpha)} \tag{3.39}$$

$$ {}_{0}^{\mathrm{RL}}\mathscr{D}_t^{\alpha}\delta^{(n)}(t)=\frac{t^{-\alpha-n-1}}{\Gamma(-\alpha-n)} \tag{3.40}$$

$$ {}_{0}^{\rm RL}\mathscr{D}_t^{\alpha} \cosh\sqrt{\lambda t} = t^{-\alpha} \mathrm{E}_{2,1-\alpha}\left(\lambda t^2\right) \tag{3.41} $$

$$ {}_{0}^{\rm RL}\mathscr{D}_t^{\alpha} \frac{\sinh\sqrt{\lambda t}}{\sqrt{\lambda t}} = t^{1-\alpha} \mathrm{E}_{2,2-\alpha}\left(\lambda t^2\right) \tag{3.42} $$

$$ {}_{-\infty}^{\rm RL}\mathscr{D}_t^{\alpha} \mathrm{e}^{\lambda t} = \lambda^{\alpha} \mathrm{e}^{\lambda t} \tag{3.43} $$

$$ {}_{-\infty}^{\rm RL}\mathscr{D}_t^{\alpha} \mathrm{e}^{\lambda t+\mu} = \lambda^{\alpha} \mathrm{e}^{\lambda t+\mu},\ \lambda > 0 \tag{3.44} $$

$$ {}_{-\infty}^{\rm RL}\mathscr{D}_t^{\alpha} \sin\lambda t = \lambda^{\alpha} \sin\left(t + \alpha\pi/2\right) \tag{3.45} $$

$$ {}_{-\infty}^{\rm RL}\mathscr{D}_t^{\alpha} \cos\lambda t = \lambda^{\alpha} \cos\left(t + \alpha\pi/2\right). \tag{3.46} $$

In the above theorem, $H(t-a)$ is the Heaviside function, i.e., the step function, defined as

$$ H(t-a) = \begin{cases} 1, & t \geqslant a \\ 0, & \text{otherwise.} \end{cases} \tag{3.47} $$

Function $\delta(t)$ is an impulsive function, also known as Dirac function, defined as

$$ \text{when } t \neq 0,\ \delta(t) = 0, \text{ and } \int_{-\infty}^{\infty} \delta(t)\mathrm{d}t = 1. \tag{3.48} $$

The Dirac function is the first-order derivative of the Heaviside function.

In addition, introducing the intermediate variables $r = \sqrt{\lambda^2 + \mu^2}$ and $\tan\varphi = \mu/\lambda$, it also follows that

$$ {}_{-\infty}^{\rm RL}\mathscr{D}_t^{\alpha} \mathrm{e}^{\lambda t} \sin\mu t = r^{\alpha} \mathrm{e}^{\lambda t} \sin\left(\mu t + \alpha\varphi\right) \tag{3.49} $$

$$ {}_{-\infty}^{\rm RL}\mathscr{D}_t^{\alpha} \mathrm{e}^{\lambda t} \cos\mu t = r^{\alpha} \mathrm{e}^{\lambda t} \cos\left(\mu t + \alpha\varphi\right). \tag{3.50} $$

It is important to note that, only few analytical solutions of Riemann–Liouville fractional-order derivatives exist for a few classes of functions.

From the Riemann–Liouville definition, the analytical solution of the fractional-order derivative of the general function $f(t)$ is difficult to find because the variable t appears simultaneously in the integral boundary and the integrand. Only numerical methods can be used to evaluate the Riemann–Liouville fractional-order derivative. Of course, the numerical solution is also not effective.

3.3.4 Initial Time Translation Properties

In the previous descriptions of fractional-order derivatives, some formulas involve an initial moment $t_0 = 0$. If one wants to generalize the initial moment of all these formulas to t_0, one should make the necessary transformations to the interval of interest. In this section, a theorem will be introduced that allows the reader to make the corresponding transformations.

Recall the definition of the Riemann–Liouville fractional-order derivative:

$$ {}^{\mathrm{RL}}_{t_0}\mathscr{D}^{\alpha}_{t} f(t) = \frac{1}{\Gamma(n-\alpha)} \frac{\mathrm{d}^n}{\mathrm{d}t^n} \int_{t_0}^{t} \frac{f(\tau)}{(t-\tau)^{1+\alpha-n}} \mathrm{d}\tau. \tag{3.51} $$

Theorem 3.5 Translation formula
If ${}^{\mathrm{RL}}_{0}\mathscr{D}^{\alpha}_{t} f(t)$ is known, the Riemann–Liouville derivative is written as

$$ {}^{\mathrm{RL}}_{t_0}\mathscr{D}^{\alpha}_{t} f(t) = {}^{\mathrm{RL}}_{0}\mathscr{D}^{\alpha}_{s} f(s+t_0), \tag{3.52} $$

where $s = t - t_0$.

Proof Denote $\tau = \theta + t_0$, then $\mathrm{d}\tau = \mathrm{d}\theta$ and $\theta \in [0, t - t_0]$. Substituting them into the right-hand side of (3.51), then

$$ \frac{1}{\Gamma(n-\alpha)} \frac{\mathrm{d}^n}{\mathrm{d}t^n} \int_{t_0}^{t} \frac{f(\tau)}{(t-\tau)^{1+\alpha-n}} \mathrm{d}\tau = \frac{1}{\Gamma(n-\alpha)} \frac{\mathrm{d}^n}{\mathrm{d}t^n} \int_{0}^{t-t_0} \frac{f(\theta+t_0)}{(t-t_0-\theta)^{1+\alpha-n}} \mathrm{d}\theta. \tag{3.53} $$

Let $s = \tau + t_0$, then $\mathrm{d}t = \mathrm{d}s$, so $t - t_0 = s$. Substituting them into the right-hand side of (3.53), then

$$ \frac{1}{\Gamma(n-\alpha)} \frac{\mathrm{d}^n}{\mathrm{d}t^n} \int_{0}^{t-t_0} \frac{f(\theta+t_0)}{(t-t_0-\theta)^{1+\alpha-n}} \mathrm{d}\theta = \frac{1}{\Gamma(n-\alpha)} \frac{\mathrm{d}^n}{\mathrm{d}s^n} \int_{0}^{s} \frac{f(\theta+t_0)}{(s-\theta)^{1+\alpha-n}} \mathrm{d}\theta. \tag{3.54} $$

Let $F(\theta) = f(\theta + t_0)$, and the transformation of (3.54) can prove the theorem directly.

$$ \frac{1}{\Gamma(n-\alpha)} \frac{\mathrm{d}^n}{\mathrm{d}s^n} \int_{0}^{s} \frac{F(\theta)}{(s-\theta)^{1+\alpha-n}} \mathrm{d}\theta = {}^{\mathrm{RL}}_{0}\mathscr{D}^{\alpha}_{s} F(s) = {}^{\mathrm{RL}}_{0}\mathscr{D}^{\alpha}_{s} f(s+t_0). \tag{3.55} $$

□

► Example 3.8

The fractional-order derivative of the exponential function is known to be ${}^{\mathrm{RL}}_{0}\mathscr{D}^{\alpha}_{t} \mathrm{e}^{\lambda t} = t^{-\alpha}\mathrm{E}_{1,1-\alpha}(\lambda t)$. Derive the Riemann–Liouville derivative at lower bound t_0.

Solutions According to Theorem 3.5, let $s = t - t_0$, and after variable substitution we can derive

$$ \begin{aligned} {}^{\mathrm{RL}}_{t_0}\mathscr{D}^{\alpha}_{t} \mathrm{e}^{\lambda t} &= {}^{\mathrm{RL}}_{0}\mathscr{D}^{\alpha}_{s} \mathrm{e}^{\lambda(s+t_0)} = \mathrm{e}^{\lambda t_0}\, {}^{\mathrm{RL}}_{0}\mathscr{D}^{\alpha}_{s} \mathrm{e}^{\lambda s} = \mathrm{e}^{\lambda t_0} s^{-\alpha}\mathrm{E}_{1,1-\alpha}(\lambda s) \\ &= \mathrm{e}^{\lambda t_0}(t-t_0)^{-\alpha}\mathrm{E}_{1,1-\alpha}\big(\lambda(t-t_0)\big). \end{aligned} \tag{3.56} $$

Assuming $\lambda = -1$, $t_0 = 0.5$ and $\alpha = 0.7$, the Grünwald–Letnikov derivative can be evaluated directly from the following statements and the exact solution can be calculated using (3.56), as shown in ◘ Fig. 3.11. It can be seen that, the two curves in this way overlap

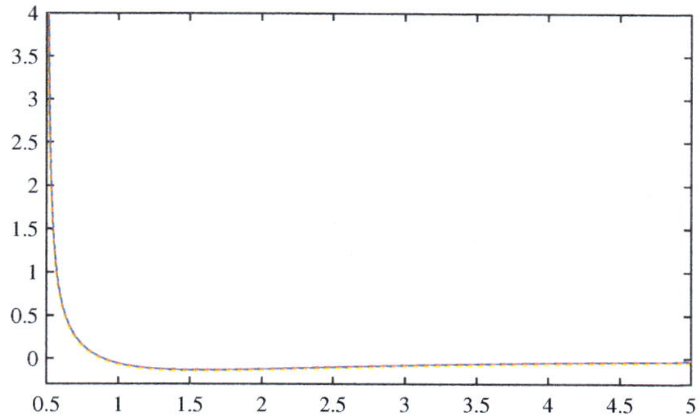

Fig. 3.11 Comparison of the Grünwald–Letnikov derivative with the translation formula

exactly, indicating that the analytical equation derived from the translation formula is correct.

```
>> lam=-1; t0=0.5; alfa=0.7;
   t=t0:0.01:5; u=exp(lam*t); y1=glfdiff(u,t,alfa);
   y2=exp(lam*t0)*(t-t0).^(-alfa).*ml_func([1,1-alfa],lam*(t-t0));
   plot(t,y1,t,y2,'--')
```

◀

3.3.5 Numerical Evaluation of Riemann–Liouville Derivatives and Integrals

A numerical algorithm for evaluating Riemann–Liouville derivatives is presented in this section. If the step size is h, the following procedures can be used to evaluate the αth-order derivatives.

Algorithm 3.4 Numerical algorithm for Riemann–Liouville derivatives
(1) Construct the vector $\boldsymbol{f} = \left[f(0), f(h), f(2h), \cdots\right]^{\mathrm{T}}$.
(2) Let $n = \lceil\alpha\rceil$, compute the binomial coefficients $w_i = ht_i{}^{n-\alpha}/\Gamma(n-\alpha+1)$.
(3) Compute the following vector $\boldsymbol{s}$:

$$s_j = \sum_{i=1}^{j} w_{j-i} f_i - \frac{1}{2} w_1 f_j - \frac{1}{2} w_j f_1. \tag{3.57}$$

(4) Under step size h, compute $(n+1)$th-order difference to the vector $\boldsymbol{s}$.

The MATLAB implementation of the above algorithm is given below. In fact, the authors have added other elements to the implementation of the algorithm so that fractional- and integer-order integrals can also be computed directly.

```
function [dy,t]=rlfdiff(y,t,r)
   arguments, y(:,1), t(:,1) double, r(1,1) double, end
   [y,h,n]=fdiffcom(y,t); dy=zeros(n,1); dy1=dy;
   if r>-1, m=ceil(r)+1; p=m-r; y3=t.^(p-1);
   elseif r==-1, yy=0.5*(y(1:n-1)+y(2:n)).*diff(t);
      for i=2:n, dy(i)=dy(i-1)+yy(i-1); end, return
   else, m=-r; y3=t.^(m-1); end
   for i=1:n, dy1(i)=y(i:-1:1).'*(y3(1:i)); end
   if r>-1, dy=diff(dy1,m)/(h^(m-1))/gamma(p); t=t(1:n-m);
   else, dy=dy1*h/gamma(m); end
end
```


The syntax of the function is $[\boldsymbol{y}_1, \boldsymbol{t}] = \texttt{rlfdiff}(\boldsymbol{y}, \boldsymbol{t}, \gamma)$, which is similar to the `glfdiff()` function, except that it returns two vectors $\boldsymbol{y}_1$ and $\boldsymbol{t}$. Because the function uses the MATLAB `diff()` function internally, the actual length of the vector $\boldsymbol{y}_1$ may be smaller than the length of the vector $\boldsymbol{y}$.

▶ Example 3.9

The analytical solution of the 0.75th-order derivative of the step function is known to be $\mathscr{D}_t^{0.75} y(t) = t^{-0.75}/\Gamma(0.25)$. Compare the accuracy of numerical solutions of Grünwald–Letnikov and Riemann–Liouville fractional-order derivatives.

Solutions Choosing the step size $h = 0.01$, the following statements evaluate the exact value of the 0.75th-order derivative of the step function and the numerical solution of the Grünwald–Letnikov and Riemann–Liouville fractional-order derivatives, as shown in Fig. 3.12. It can be seen that the directly evaluated Riemann–Liouville fractional-order derivatives have large errors. It is not recommended to use this function to evaluate the Riemann–Liouville fractional-order derivatives in practical applications.

```
>> t0=0:0.01:3; y0=t0.^-0.75/gamma(0.25);
   h=0.01; t1=0:h:3; y=ones(size(t1));
   y1=glfdiff(y,t1,0.75); [y2,t2]=rlfdiff(y,t1,0.75);
   plot(t0,y0,t1,y1,'--',t2,y2,':'), ylim([0 6])
```

◀

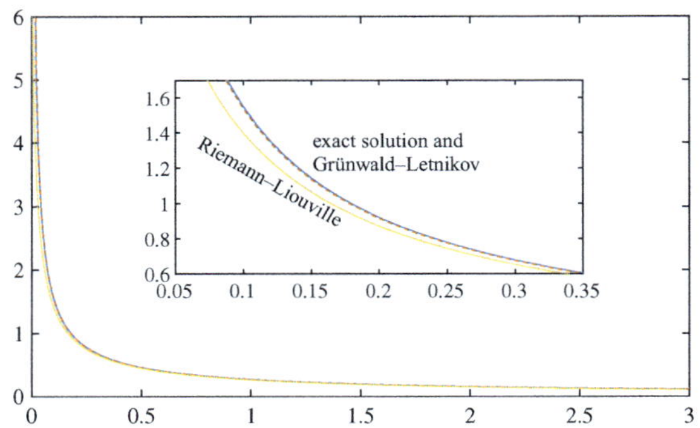

Fig. 3.12 0.75th-order Riemann–Liouville derivatives of the step function

3.3.6 Symbolic Computing in Riemann–Liouville Derivatives

With MATLAB symbolic arithmetic functions, it is possible to directly implement Definitions 3.4 and 3.5 on Riemann–Liouville fractional-order integral and derivative. Analytical computational functions can be written as follows:

```
function dy=riemannsym(f,alpha,t0,t)
   arguments
      f(1,1), alpha(1,1), t0(1,1)=0, t(1,1)=symvar(f)
   end
   syms tau; alpha=sym(alpha);
   switch class(f)
      case 'sym', F=subs(f,t,tau);
      case 'symfun', F=f(tau);
      otherwise, error('f must be a symbolic expression')
   end
   if alpha<0, alpha=-alpha;
      dy=1/gamma(alpha)*int(F/(t-tau)^(1-alpha),tau,t0,t);
   elseif alpha==0, dy=f;
   else, n=ceil(alpha);
      f1=int(F/(t-tau)^(1+alpha-n),tau,t0,t);
      dy=diff(f1,t,n)/gamma(n-alpha);
   end, dy=simplify(dy);
end
```

where f is a symbolic expression or a symbolic function, α is the order of derivative. If its value is negative, Riemann–Liouville integral is computed. Argument t_0 is the lower bound, while t is the symbolic variable. From this, the analytical solution of the Riemann–Liouville derivatives and integrals can be calculated for certain functions.

3.4 Caputo Fractional Calculus Definition

In the previously introduced Grünwald–Letnikov and Riemann–Liouville definitions, if the initial conditions are nonzero, sometimes the description of the differential equation may be difficult. In the study and simulation of systems with nonzero initial conditions, it is necessary to introduce another definition, the Caputo definition [6].

3.4.1 Definition of Caputo Derivatives and Integrals

Definition 3.6 Caputo fractional-order derivative

The Caputo fractional-order derivative is defined as

$$ {}_{t_0}^{\mathrm{C}}\mathscr{D}_t^{\alpha} y(t) = \frac{1}{\Gamma(m-\alpha)} \int_{t_0}^{t} \frac{y^{(m)}(\tau)}{(t-\tau)^{1+\alpha-m}} \mathrm{d}\tau, \tag{3.58} $$

where $m = \lceil \alpha \rceil$ is an integer. The function $y(t)$ and its ith-order derivative are continuous, $i = 1, 2, \ldots, m$ [6].

Definition 3.7 Caputo fractional-order integral

Caputo fractional-order integral is defined as

$${}_{t_0}^{\mathrm{C}}\mathscr{D}_t^{-\gamma} y(t) = \frac{1}{\Gamma(\gamma)} \int_{t_0}^{t} \frac{y(\tau)}{(t-\tau)^{1-\gamma}} \mathrm{d}\tau. \tag{3.59}$$

It can be seen that the definition of the Caputo fractional-order integral is in full agreement with the Riemann–Liouville definition.

3.4.2 Commonly Used Caputo Derivative Formulas

As an example, consider first the problem of the Caputo fractional-order derivative of the exponential function $f(t) = \mathrm{e}^{\lambda t}$.

Theorem 3.6 Caputo derivative of exponential function
Denote $m = \lceil \alpha \rceil$, $\gamma = m - \alpha$, the Caputo fractional-order derivative of exponential function $f(t) = \mathrm{e}^{\lambda t}$ is

$${}_{0}^{\mathrm{C}}\mathscr{D}_t^{\alpha} \mathrm{e}^{\lambda t} = \lambda^m t^{\gamma} \mathrm{E}_{1,\gamma+1}(\lambda t). \tag{3.60}$$

Proof Since $\alpha > 0$, $m = \lceil \alpha \rceil$, $\gamma = m - \alpha$, and the Riemann–Liouville fractional-order integral of the exponential function can be found from Theorem 3.3, which is rewritten as

$${}_{0}^{\mathrm{RL}}\mathscr{D}_t^{-\gamma} \mathrm{e}^{\lambda t} = t^{\gamma} \mathrm{E}_{1,1+\gamma}(\lambda t). \tag{3.61}$$

Since the definitions of Caputo and Riemann–Liouville fractional-order integrals are identical, after the following derivation, it can be found that the Caputo fractional-order derivative of the exponential function satisfies the conclusion of the theorem.

$${}_{0}^{\mathrm{C}}\mathscr{D}_t^{\alpha} \mathrm{e}^{\lambda t} = {}_{0}^{\mathrm{RL}}\mathscr{D}_t^{-\gamma} \left(\frac{\mathrm{d}^m}{\mathrm{d}t^m} \mathrm{e}^{\lambda t} \right) = \lambda^m \, {}_{0}^{\mathrm{RL}}\mathscr{D}_t^{-\gamma} \mathrm{e}^{\lambda t} = \lambda^m t^{\gamma} \mathrm{E}_{1,1+\gamma}(\lambda t).$$

□

It is important to note that, the property used in the first step above is Theorem 3.18, which will be introduced later.

Theorem 3.7 Caputo derivatives of commonly used functions
Caputo fractional-order derivatives of commonly used functions are [2]

$$
{}_0^{\mathrm{C}}\mathscr{D}_t^{\alpha} t^{\lambda} = \begin{cases} 0, & \lambda \text{ is a natural number, and } \lceil\alpha\rceil > \lambda \\ \dfrac{\Gamma(\lambda+1)}{\Gamma(\lambda+1-\alpha)} t^{\lambda-\alpha}, & \text{otherwise} \end{cases} \tag{3.62}
$$

$$
{}_0^{\mathrm{C}}\mathscr{D}_t^{\alpha} (t+c)^{\lambda} = \frac{\Gamma(\lambda+1)}{\Gamma(\lambda+1-m)} \frac{c^{\lambda-m-1} t^{\gamma}}{\Gamma(\gamma+1)} \mathrm{E}_{1,\gamma+1}^{m-\lambda}(-t/c), \; c > 0 \tag{3.63}
$$

$$
{}_0^{\mathrm{C}}\mathscr{D}_t^{\alpha} \sin\lambda t = \begin{cases} \dfrac{\lambda^m \mathrm{j}(-1)^{m/2} t^{\gamma}}{2\Gamma(\gamma+1)} \left[-F(\mathrm{j}\lambda t) + F(-\mathrm{j}\lambda t)\right], & m \text{ is even} \\ \dfrac{\lambda^m (-1)^{(m-1)/2} t^{\gamma}}{2\Gamma(\gamma+1)} \left[F(\mathrm{j}\lambda t) + F(-\mathrm{j}\lambda t)\right], & m \text{ is odd} \end{cases} \tag{3.64}
$$

$$
{}_0^{\mathrm{C}}\mathscr{D}_t^{\alpha} \cos\lambda t = \begin{cases} \dfrac{\lambda^m (-1)^{m/2} t^{\gamma}}{2\Gamma(\gamma+1)} \left[F(\mathrm{j}\lambda t) + F(-\mathrm{j}\lambda t)\right], & m \text{ is even} \\ \dfrac{\lambda^m \mathrm{j}(-1)^{(m-1)/2} t^{\gamma}}{2\Gamma(\gamma+1)} \left[F(\mathrm{j}\lambda t) - F(-\mathrm{j}\lambda t)\right], & m \text{ is odd,} \end{cases} \tag{3.65}
$$

where, λ is real, $F(x) = {}_1\mathrm{F}_1(1; \gamma+1; x)$, ${}_1\mathrm{F}_1(\cdot)$ is a hypergeometric function defined in ▶ Sect.2.5.

Corollary 3.1 *The formulas for the fractional-order derivatives of trigonometric functions can also be written as*

$$
{}_0^{\mathrm{C}}\mathscr{D}_t^{\alpha} \sin\lambda t = \frac{(\mathrm{j}\lambda)^m}{2\mathrm{j}} t^{\gamma} \left[\mathrm{E}_{1,\gamma+1}(\mathrm{j}\lambda t) - (-1)^m \mathrm{E}_{1,\gamma+1}(-\mathrm{j}\lambda t)\right] \tag{3.66}
$$

$$
{}_0^{\mathrm{C}}\mathscr{D}_t^{\alpha} \cos\lambda t = \frac{(\mathrm{j}\lambda)^m}{2} t^{\gamma} \left[\mathrm{E}_{1,\gamma+1}(\mathrm{j}\lambda t) + (-1)^m \mathrm{E}_{1,\gamma+1}(-\mathrm{j}\lambda t)\right]. \tag{3.67}
$$

Proof The corollary can be proved directly by the famous Euler formula.

$$
\sin\lambda t = \frac{1}{2\mathrm{j}}\left(\mathrm{e}^{\mathrm{j}\lambda t} - \mathrm{e}^{-\mathrm{j}\lambda t}\right), \quad \cos\lambda t = \frac{1}{2}\left(\mathrm{e}^{\mathrm{j}\lambda t} + \mathrm{e}^{-\mathrm{j}\lambda t}\right). \tag{3.68}
$$

□

Let $m = 0$ and $\gamma = -\alpha$, the Caputo fractional-order integral of commonly used functions can be evaluated directly using the above formula. The numerical evaluation of the Caputo fractional-order derivatives and integrals will be described later.

3.4.3 Symbolic Computing in Caputo Calculus

In fact, the Caputo derivatives of some functions can be computed analytically by computer. From Definitions 3.6, the following MATLAB functions can be written:

```
function dy=caputosym(f,alpha,t0,t)
   arguments
      f(1,1), alpha(1,1), t0(1,1)=0, t(1,1)=symvar(f)
   end
   if alpha<=0, dy=riemannsym(f,alpha,t0,t); return; end
   syms tau; alpha=sym(alpha); m=ceil(alpha);
   switch class(f)
      case 'sym', F=diff(subs(f,t,tau),m);
      case 'symfun', F=diff(f(tau),m);
      otherwise, error('f must be a symbolic expression')
   end
   dy=1/gamma(m-alpha)*int(F/(t-tau)^(1+alpha-m),tau,t0,t);
   dy=simplify(dy);
end
```


The function internally identifies whether the given function is a symbolic variable or a symbolic function, and handles it separately depending on the situations.

It is important to note that, since low-level functions in MuPAD are used in the function, the analytical solutions of many functions cannot be derived. In addition, since the MuPAD kernel does not have special functions such as the Mittag-Leffler function, the results sometimes do not exactly agree with those in this book. In such cases, the correctness of the results can be verified by comparing the curves.

▶ Example 3.10

Use symbolic facilities to compute ${}_0^{\rm C}\mathscr{D}_t^{0.5}(2t+1)$ and ${}_0^{\rm C}\mathscr{D}_t^{0.4}(2t+1)$.

Solutions The current version of the caputosym() function can only handle genuine polynomials, not pseudo-polynomials.

```
>> syms t positive, f(t)=2*t+1; %or with f=2*t+1 command
   R1=caputosym(f,sym(0.5),0,t), R2=caputosym(f,sym(0.4),0,t)
```

The results obtained are $R_1 = 4\sqrt{t/\pi}$ and $R_2 = 10t^{0.6}/(3\Gamma(0.6))$. ◀

▶ Example 3.11

For the exponential function $f(t) = {\rm e}^{-3t}$, compute ${}_0^{\rm C}\mathscr{D}_t^{4.3}f(t)$.

Solutions For the problem given here, it follows from Theorem 3.6 that $m=5$, $\gamma=0.7$, and $\lambda=-3$, so the solution of ${}_0^{\rm C}\mathscr{D}_t^{4.3}f(t)$ is $-243t^{-0.7}{\rm E}_{1,1.7}(-3t)$. Now try to find its Caputo derivative using the caputosym() function.

```
>> syms t positive, f=exp(-3*t); F(t)=caputosym(f,sym(4.3),0,t)
```

The results obtained are extremely tedious and can be collated by hand to obtain

$$F(t) = \frac{81\sqrt[10]{27}}{4\Gamma(0.7)}{\rm e}^{-3t}\left(\Gamma(0.7)-\Gamma(0.7,-3t)\right)\left(\sqrt{10-2\sqrt{5}}+\left(1+\sqrt{5}\right){\rm i}\right).$$

As can be seen, the result here is much more complicated than the analytical solution introduced earlier. However, by drawing the curves for the interval $t \in (0,5)$, it can be seen that the curves of the two coincide exactly, thus indicating that the results obtained by the symbolic computing are also correct.

```
>> t0=[0:0.01:5]'; y0=F(t0); y0=double(y0);
   y1=-243*t0.^(0.7).*ml_func([1,1.7],-3*t0);
   plot(t0,real(y0),t0,y1,'--')       %omit the tiny imaginary elements
```

◄

3.5 The Relationship Among Different Fractional Calculus Definitions

Several different definitions in fractional calculus were presented earlier, and they are closely related to each other, except for the Cauchy integral formula. Some definitions are equivalent to each other under certain conditions. Some definitions are interconvertible under certain conditions. This section describes the interrelationship between the different definitions. Numerical evaluation of Caputo derivatives is discussed.

3.5.1 The Relationship Between Grünwald–Letnikov and Riemann–Liouville Definitions

It is stated in Ref. [2] that, the Grünwald–Letnikov definition is equivalent to the Riemann–Liouville definition for a large number of functions. The relationship between the two is given here specifically.

Theorem 3.8 Grünwald–Letnikov integrals
If function $f(t)$ is continuous in the interval (t_0, T), and is $(n-1)$th-order differentiable, while $f^{(n)}(t)$ is integrable, then if $0 \leqslant m-1 < \alpha < m \leqslant n$, Grünwald–Letnikov fractional-order derivative can be written in the following integral form [2]:

$$ {}_{t_0}^{\mathrm{GL}}\mathscr{D}_t^{\alpha} f(t) = \sum_{j=0}^{m-1} \frac{f^{(j)}(t_0)(t-t_0)^{j-\alpha}}{\Gamma(1+j-\alpha)} + \frac{1}{\Gamma(m-\alpha)} \int_{t_0}^{t} \frac{f^{(m)}(\tau)\mathrm{d}\tau}{(t-\tau)^{1+\alpha-m}}. \tag{3.69} $$

Theorem 3.9 Equivalent definitions
If function $f(t)$ is continuous in the interval (t_0, T), and is $(n-1)$th-order differentiable, while $f^{(n)}(t)$ is integrable, then if $0 \leqslant m-1 < \alpha < m \leqslant n$, then Grünwald–Letnikov definition is equivalent to Riemann–Liouville definition.

Proof The proof of the theorem can be found in Ref. [2]. Taking the mth-order derivative of the right-hand side of (3.69) yields

$$\frac{d^m}{dt^m}\left[{}_{t_0}^{\mathrm{GL}}\mathscr{D}_t^{\alpha}f(t)\right]=\sum_{j=0}^{m-1}\frac{f^{(j)}(t_0)(t-t_0)^{m+j-\alpha}}{\Gamma(1+m+j-\alpha)}+\frac{1}{\Gamma(2m-\alpha)}\int_{t_0}^{t}\frac{f^{(m)}(\tau)d\tau}{(t-\tau)^{\alpha-2m+1}}. \tag{3.70}$$

If $m \geqslant 1$, then the highest order in the first term of the above equation is t^{m-1}, so the term will vanish completely when finding the mth-order derivative. Using the integration by parts method to find the mth-order integral, then

$$ {}_{t_0}^{\mathrm{GL}}\mathscr{D}_t^{\alpha}f(t)=\frac{1}{\Gamma(m-\alpha)}\int_{t_0}^{t}\frac{f^{(m)}(\tau)d\tau}{(t-t_0)^{1+\alpha-m}}={}_{t_0}^{\mathrm{RL}}\mathscr{D}_t^{\alpha}f(t). \tag{3.71}$$

It can be seen that the result is in full agreement with the Riemann–Liouville definition. □

3.5.2 The Relationship Between Caputo and Riemann–Liouville Definitions

If $0 < \alpha < 1$, it is known that $m = 1$. Therefore, (3.58) can be rewritten as

$$ {}_{t_0}^{\mathrm{C}}\mathscr{D}_t^{\alpha}y(t)=\frac{1}{\Gamma(1-\alpha)}\int_{t_0}^{t}\frac{y'(\tau)}{(t-\tau)^{\alpha}}d\tau. \tag{3.72}$$

If the initial value of $y(t)$ is nonzero, it is found by comparing the Caputo and Riemann–Liouville definitions that

$$ {}_{t_0}^{\mathrm{C}}\mathscr{D}_t^{\alpha}y(t)={}_{t_0}^{\mathrm{RL}}\mathscr{D}_t^{\alpha}\big(y(t)-y(t_0)\big), \tag{3.73}$$

where the fractional-order derivative of constant $y(t_0)$ can be written as

$$ {}_{t_0}^{\mathrm{RL}}\mathscr{D}_t^{\alpha}y(t_0)=\frac{y(t_0)}{\Gamma(1-\alpha)}(t-t_0)^{-\alpha}. \tag{3.74}$$

It can be seen that, the relationship between Caputo and Riemann–Liouville fractional-order derivatives is

$$ {}_{t_0}^{\mathrm{C}}\mathscr{D}_t^{\alpha}y(t)={}_{t_0}^{\mathrm{RL}}\mathscr{D}_t^{\alpha}y(t)-\frac{y(t_0)}{\Gamma(1-\alpha)}(t-t_0)^{-\alpha}. \tag{3.75}$$

More generally, if the order α satisfies $\alpha > 0$, the relationship between Caputo and the Riemann–Liouville definitions can be described by the following theorem.

Theorem 3.10 Relation between Caputo and Riemann–Liouville definitions
Denote $m = \lceil\alpha\rceil$, then the relationship between Caputo and Riemann–Liouville fractional-order derivative definitions is

$$ {}_{t_0}^{\mathrm{C}}\mathscr{D}_t^{\alpha} y(t) = {}_{t_0}^{\mathrm{RL}}\mathscr{D}_t^{\alpha} y(t) - \sum_{k=0}^{m-1} \frac{y^{(k)}(t_0)}{\Gamma(k-\alpha+1)}(t-t_0)^{k-\alpha}. \tag{3.76}$$

Remark Caputo fractional-order derivative formula

(1) It can be seen that $0 \leqslant \alpha \leqslant 1$ is just a special case of the above relation.

(2) As introduced earlier, if $\alpha < 0$, the Riemann–Liouville fractional-order integral is identical to the Caputo fractional-order integral. The two definitions are not distinguished in practical applications.

3.5.3 Numerical Evaluations of Caputo Derivatives and Integrals

From the description in ▶ Sect. 3.5.1, it is clear that the difference between Caputo and the Riemann–Liouville or Grünwald–Letnikov definitions lies in the treatment of initial values. The difference between the two can be evaluated directly from (3.76). In this way, the code for the evaluation of the Caputo fractional-order derivative ${}_{t_0}^{\mathrm{C}}\mathscr{D}_t^{\alpha} y(t)$ can be written on the basis of the function `glfdiff()`.

```
function dy=caputo(y,t,gam,y0,L)
   arguments, y(:,1), t(:,1), gam(1,1), y0, L(1,1)=10, end
   dy=glfdiff(y,t,gam);
   if gam>0, q=ceil(gam); if gam<=1, y0=y(1); end
      for k=0:q-1
         dy=dy-y0(k+1)*t.^(k-gam)./gamma(k+1-gam);
      end
      yy1=interp1(t(L+1:end),dy(L+1:end),t(1:L),'spline');
      dy(1:L)=yy1;
end, end
```

The syntax of the function is y_1 = `caputo(`$\boldsymbol{y}$,$\boldsymbol{t}$,α,$\boldsymbol{y}_0$,L`)`. If $\alpha \leqslant 0$, the Grünwald–Letnikov fractional-order integral is return; if $\alpha < 1$, $y(t_0)$ is extracted from $\boldsymbol{y}$ automatically, such that the Caputo derivative is returned; if $\alpha > 1$, $\boldsymbol{y}_0$ vector should be provided such that $\boldsymbol{y}_0 = \left[y(t_0), y'(t_0), \ldots, y^{(q-1)}(t_0)\right]$, $q = \lceil\alpha\rceil$.

Remark Caputo fractional-order derivatives and integrals

(1) In the actual calculation, the first few initial points of the returned Caputo derivative will have large errors, so that the value of the first L points is recalculated using the method of spline interpolation, the default value of L is 10. If the order is increased, the value of L should also be increased.

(2) Due to the use of spline interpolation in the first few points, sometimes the results have very large errors. This function is not very practical, and not recommended for real use. A better method and its implementation will be given later.

► Example 3.12

Consider the fractional-order derivative problem for the sinusoidal function in Example 3.4, with $f(t) = \sin(3t+1)$. Find its 0.3th-, 1.3th-, and 2.3th-order Caputo fractional-order derivatives.

Solutions It can be seen that the initial value of the function at $t = 0$ is $\sin 1$, such that the difference between the two definitions is $d(t) = t^{-0.3}\sin 1/\Gamma(0.7)$. The curve of the Grünwald–Letnikov and Caputo definitions can be drawn with the following statements, and difference function $d(t)$ is also plotted, as shown in Fig. 3.13.

```
>> t=0:0.01:pi; y=sin(3*t+1); d=t.^(-0.3)*sin(1)/gamma(0.7);
   y1=glfdiff(y,t,0.3); y2=caputo(y,t,0.3,sin(1));
   plot(t,y1,t,y2,'--',t,d,':'), axis([0 pi -3 3])
```

To compute ${}_0^{\rm C}\mathscr{D}_t^{2.3}y(t)$, the initial values of $y'(0)$ and $y''(0)$ are also needed, which can be obtained from symbolic computing. The result can be transformed back to a vector of double precision. The 1.3th- and 2.3th-order derivative curves of the function can be derived. To obtain a better approximation, the values of the first few points of the fractional-order derivatives are reconstructed by means of spline interpolation. The resulting curves are shown in Fig. 3.14.

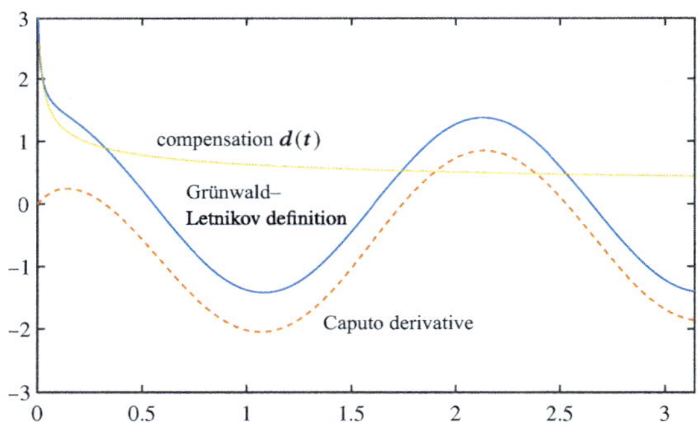

Fig. 3.13 0.3th-order derivative

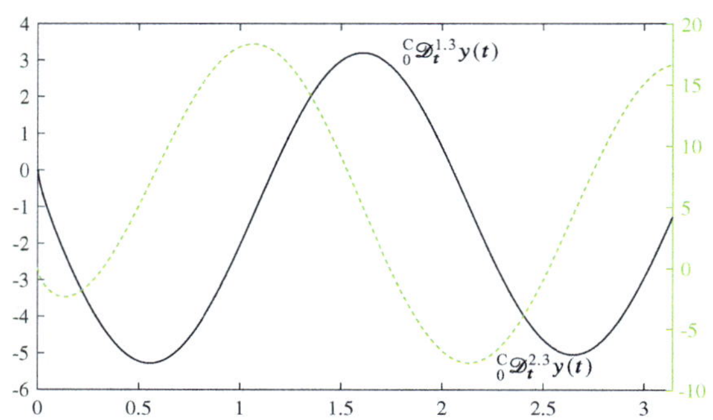

Fig. 3.14 Double vertical axes curves of 1.3th- and 2.3th-order derivatives

```
>> syms t; y(t)=sin(3*t+1); y00=sin(1);
   y1=diff(y); y10=double(y1(0)); y2=diff(y1); y20=double(y2(0));
   t=0:0.01:pi; y=@(t)sin(3*t+1); y1=caputo(y,t,1.3,[y00 y10],10);
   y2=caputo(y,t,2.3,[y00,y10,y20],30);
   yyaxis left, plot(t,y1), yyaxis right, plot(t,y2,'--')
```

◀

In fact, the solution function given here has two pitfalls.

(1) The interpolation results used here introduce other errors.

(2) In practical applications, only the samples of $y(t)$ can be provided, there is usually no information about the initial values of the derivatives. Therefore, the function given here is not suitable for computing Caputo derivatives, and is not recommended for actual use. A numerical function with high-precision algorithm will be introduced in ► Chap. 4.

3.6 Properties and Geometrical Interpretations of Fractional Calculus

Similar to the integer-order calculus, the fractional calculus has many properties of its own. Some common properties in fractional calculus are introduced first. A physical and geometric interpretation of a simple fractional-order integral is illustrated. However, there is no widely accepted physical and geometric interpretation in fractional calculus so far.

3.6.1 Properties of Fractional Calculus

Theorem 3.11 Linear property—superposition principle
Fractional-order operator is linear, i.e., for any constants a and b, there exists

$$\mathscr{D}_t^{\alpha}\big[af(t)+bg(t)\big]=a\,\mathscr{D}_t^{\alpha}f(t)+b\,\mathscr{D}_t^{\alpha}g(t). \tag{3.77}$$

Proof Consider here first the proof that the Grünwald–Letnikov fractional-order operator is linear, noting that $\tau=t-jh$, $\Delta={}_{t_0}^{\mathrm{GL}}\mathscr{D}_t^{\alpha}\big[af(t)+bg(t)\big]$, the linear property can be proved with the following formulas:

$$\begin{aligned}\Delta &= \lim_{h\to 0}\frac{1}{h^\alpha}\sum_{j=0}^{[(t-t_0)/h]}(-1)^j\binom{\alpha}{j}\big[af(\tau)+bg(\tau)\big]\\ &= a\lim_{h\to 0}\frac{1}{h^\alpha}\sum_{j=0}^{[(t-t_0)/h]}(-1)^j\binom{\alpha}{j}f(\tau)+b\lim_{h\to 0}\frac{1}{h^\alpha}\sum_{j=0}^{[(t-t_0)/h]}(-1)^j\binom{\alpha}{j}g(\tau)\\ &= a\,{}_{t_0}^{\mathrm{GL}}\mathscr{D}_t^\alpha f(t)+b\,{}_{t_0}^{\mathrm{GL}}\mathscr{D}_t^\alpha g(t).\end{aligned}$$

3

For Riemann–Liouville definition, denote $\Delta = {}_{t_0}^{\mathrm{RL}}\mathscr{D}_t^\alpha\big[af(t)+bg(t)\big]$, the following formulation also shows that it is linear.

$$\begin{aligned}\Delta &= \frac{1}{\Gamma(k-\alpha)}\frac{\mathrm{d}^k}{\mathrm{d}t^k}\int_{t_0}^t\frac{af(\tau)+bg(\tau)}{(t-\tau)^{1+\alpha-k}}\mathrm{d}\tau\\ &= \frac{a}{\Gamma(k-\alpha)}\frac{\mathrm{d}^k}{\mathrm{d}t^k}\int_{t_0}^t\frac{f(\tau)}{(t-\tau)^{1+\alpha-k}}\mathrm{d}\tau+\frac{b}{\Gamma(k-\alpha)}\frac{\mathrm{d}^k}{\mathrm{d}t^k}\int_{t_0}^t\frac{g(\tau)}{(t-\tau)^{1+\alpha-k}}\mathrm{d}\tau\\ &= a\,{}_{t_0}^{\mathrm{RL}}\mathscr{D}_t^\alpha f(t)+b\,{}_{t_0}^{\mathrm{RL}}\mathscr{D}_t^\alpha g(t).\end{aligned}$$

It is also possible to prove that the Caputo operator is linear. □

In addition to the linear property, there are other properties in fractional calculus. Some commonly used properties will be listed here without proof. The proofs of these properties can be found in other references.

Theorem 3.12 Unified fractional- and integer-order operators
If $\alpha = n$ is an integer, the fractional-order operator and integer-order one are identical, and ${}_0\mathscr{D}_t^0 f(t) = f(t)$.

Theorem 3.13 Leibniz property
In integer-order calculus, the two functions $f(t)$ and $g(t)$ are nth-order differentiable, the product is also nth-order differentiable and satisfies the following properties:

$$\frac{\mathrm{d}^n}{\mathrm{d}t^n}\big[f(t)g(t)\big] = \sum_{k=0}^n\binom{n}{k}f^{(k)}(t)g^{(n-k)}(t). \tag{3.78}$$

If n is not an integer, the above equation also applies. Proof see Reference [2].

Theorem 3.14 Analytic property
If $f(t)$ is analytic, then ${}_{t_0}\mathscr{D}_t^\alpha f(t)$ is also analytic for t and α.

Theorem 3.15 Commutative law for sequential integrals
Sequential fractional-order integrals satisfy commutative law, and

$$ {}_{t_0}^{\mathrm{RL}}\mathscr{D}_t^{-\alpha}\left[{}_{t_0}^{\mathrm{RL}}\mathscr{D}_t^{-\beta}f(t)\right] = {}_{t_0}^{\mathrm{RL}}\mathscr{D}_t^{-\beta}\left[{}_{t_0}^{\mathrm{RL}}\mathscr{D}_t^{-\alpha}f(t)\right] = {}_{t_0}^{\mathrm{RL}}\mathscr{D}_t^{-(\alpha+\beta)}f(t). \tag{3.79}$$

Theorem 3.16 Commutative law not satisfied for sequential derivatives
If $\alpha, \beta > 0$, and $f(t)$ has rth-order continuous derivatives, where $r = \max(\lceil\alpha\rceil, \lceil\beta\rceil)$, sequential fractional-order derivatives can be computed by [2]

$$ {}_{t_0}^{\mathrm{RL}}\mathscr{D}_t^{\alpha}\left[{}_{t_0}^{\mathrm{RL}}\mathscr{D}_t^{\beta}f(t)\right] = {}_{t_0}^{\mathrm{RL}}\mathscr{D}_t^{\alpha+\beta}f(t) - \sum_{k=0}^{n-1}\frac{(t-t_0)^{k-\alpha-n}}{\Gamma(1+k-n-\alpha)}\,{}_{t_0}^{\mathrm{RL}}\mathscr{D}^{k+\beta-n}f(t)\Big|_{t_0}, \tag{3.80}$$

where $n = \lceil\beta\rceil$, and sequential fractional-order derivatives do not satisfy commutative law.

Corollary 3.2 (Sequential derivatives) *If n is an integer, the following sequential fractional-order derivatives satisfy:*

$$ \frac{\mathrm{d}^n}{\mathrm{d}t^n}\left[{}_{t_0}^{\mathrm{RL}}\mathscr{D}_t^{\alpha}f(t)\right] = {}_{t_0}^{\mathrm{RL}}\mathscr{D}_t^{n+\alpha}f(t) \tag{3.81}$$

$$ {}_{t_0}^{\mathrm{RL}}\mathscr{D}_t^{\alpha}\left[\frac{\mathrm{d}^n}{\mathrm{d}t^n}f(t)\right] = {}_{t_0}^{\mathrm{RL}}\mathscr{D}_t^{n+\alpha}f(t) - \sum_{k=0}^{n-1}\frac{(t-t_0)^{k-\alpha-n}}{\Gamma(1+k-n-\alpha)}f^{(k)}(t_0). \tag{3.82}$$

Theorem 3.17 Sequential derivatives with zero initial conditions
If function $f(t)$ is at rest at $t = t_0$, i.e., the initial values of the function and all its fractional-order derivatives are zero, the sequential fractional-order derivatives satisfy the commutative law, and

$$ {}_{t_0}^{\mathrm{RL}}\mathscr{D}_t^{\alpha}\left[{}_{t_0}^{\mathrm{RL}}\mathscr{D}_t^{\beta}f(t)\right] = {}_{t_0}^{\mathrm{RL}}\mathscr{D}_t^{\beta}\left[{}_{t_0}^{\mathrm{RL}}\mathscr{D}_t^{\alpha}f(t)\right] = {}_{t_0}^{\mathrm{RL}}\mathscr{D}_t^{\alpha+\beta}f(t). \tag{3.83}$$

Theorem 3.18 Modeling theorem I
If $n = \lceil\gamma\rceil$ and $y(t)$ is nth-order differentiable, then

$$ {}_{t_0}^{\mathrm{C}}\mathscr{D}_t^{\gamma}y(t) = {}_{t_0}^{\mathrm{RL}}\mathscr{D}_t^{\gamma-n}\left[y^{(n)}(t)\right]. \tag{3.84}$$

Proof The theorem is proved directly from Theorem 3.10 and Corollary 3.2. □

This is a very important theorem and will be one of the theoretical foundations of the method for block diagram-based framework for solving the Caputo equations, to be presented later.

3.6.2 Geometrical Interpretations of Fractional Integrals

Integer-order calculus has a good physical and geometric interpretation. If a function represents the displacement of a mass, its first-order derivative represents the instantaneous velocity of that mass. If the function represents the spatial position in a curve, its first-order derivative represents the tangent direction.

In contrast, although fractional calculus is also a discipline with a long history of more than 300 years, there is no widely accepted geometrical and physical interpretation in the definitions of fractional calculus to date. Prof. Igor Podlubny has given a reasonable geometrical and physical interpretation of the Riemann–Liouville fractional-order integral [7, 8], and that interpretation will be presented here.

First, recall the Riemann–Liouville fractional-order integral definition

$$ {}_0^{\mathrm{RL}}\mathscr{D}_t^{-\gamma} f(t) = \frac{1}{\Gamma(\gamma)} \int_0^t \frac{f(\tau)}{(t-\tau)^{1-\gamma}} \mathrm{d}\tau. \tag{3.85} $$

An auxiliary function can be introduced

$$ g(\tau) = \frac{1}{\Gamma(\gamma+1)} \left[t^{\gamma} - (t-\tau)^{\gamma} \right]. \tag{3.86} $$

In this way, the fractional-order integral can be written directly as a Stieltjes integral in the form

$$ {}_0^{\mathrm{RL}}\mathscr{D}_t^{-\gamma} f(t) = \int_0^t f(\tau) \mathrm{d}g(\tau). \tag{3.87} $$

Now choose three axes τ, $g_t(\tau)$, and $f(\tau)$ to establish a three-dimensional coordinate system. Function $f(t)$ can be represented under the τ-$f(\tau)$-coordinate system, the $g(\tau)$-$f(\tau)$ coordinate system, and the whole 3D coordinate system, respectively. In this way, the area enclosed under the $g(\tau)$-$f(\tau)$ coordinate system can be regarded as the fractional-order integral of the $f(t)$ function, while the area under the τ-$f(\tau)$-coordinate system is the first-order integral of the function. The authors have written a MATLAB function `fence_shadow()` to draw the graphs in the 3D coordinate system as follows:

```
function fence_shadow(t,f,gam,key)
   arguments, t(1,:)
      f, gam(1,1), key(1,1){mustBeMember(key,[1,2,3])};
   end
   x=t; z=f(x); tn=x(end);
   switch key
      case 1, y=(tn^gam-(tn-x).^gam)/gamma(1+gam);
```

```
            axis([minmax(x),minmax(y),minmax(z)]), hold on
            for i=1:length(x)-1
               ii=[i,i,i+1,i+1]; x0=x(ii); z0=[0 z(i:i+1) 0];
               y0=zeros(1,4); patch(x0,y0,z0,'c'),
               y0=y(ii); patch(x0,y0,z0,'g')
               x0=zeros(1,4); patch(x0,y0,z0,'r')
            end, view(-37.5000,30)
         case {2,3}, axis([minmax(x),minmax(z)]), hold on
            for i=1:length(x)-1
               x1=0:0.01:x(i+1); tn=x1(end);
               y=(tn^gam-(tn-x1).^gam)/gamma(1+gam);
               if key==2, plot(x1,y);
               else, z=f(x1);
               plot(y,z,[y(end),y(end)],[z(end),0]);
      end, end, end, hold off
   end
```

The syntax of the function is `fence_shadow(t, f, γ, key)`, where $\boldsymbol{t}$ is the sample vector of time instances, f is the function handle, and γ is the order of the integral. The argument key is an identifier, corresponding to the three graphs in Ref. [7].

Here the function $g(\tau)$ can be regarded as the observer's own time coordinate system, which is the time coordinate system for the physical time t distortion. If the velocity of the object's motion is observed to be $f(t)$ under the observer's own time coordinate system $g(t)$, then the displacement of the object observed by the observer is the αth-order integral of the $f(t)$ function.

▶ Example 3.13

Given a function $f(t) = t + 0.5 \sin t$ and $t \in (0, 10)$. Show the geometric interpretation of its 0.75th-order Riemann–Liouville integral.

Solutions According to the above description, the vector $\boldsymbol{t}$ and the original function $f(t)$ can be generated so that the geometric interpretation of the Riemann–Liouville fractional-order integral can be obtained, as shown in Fig. 3.15. It can be seen that the graph is in fact constructed from filled color blocks.

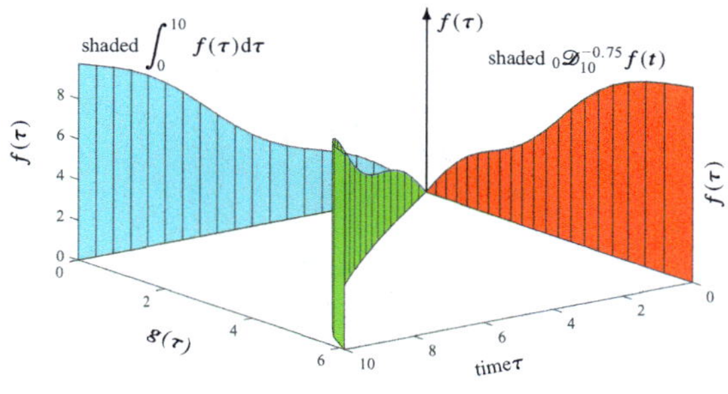

Fig. 3.15 Geometric interpretation of fractional-order integral

```
>> t=0:0.5:10; f=@(t)t+sin(t)/2;
   clf, fence_shadow(t,f,0.75,1), grid
   set(gca,'xdir','reverse'), set(gca,'ydir','reverse')
```

It can be seen that the shaded area on the left-hand side wall is the first-order integral of the function $f(t)$, while the shaded area on the right-hand side wall is the desired fractional-order integral. If `key` is chosen to be 2 and 3, other geometrical interpretations of the diagram can be derived as shown in Figs. 3.16 and 3.17, respectively.

```
>> clf, fence_shadow(t,f,0.75,2)
   set(gca,'xdir','reverse','ydir','reverse'), ylim([0 2*pi])
   figure, fence_shadow(t,f,0.75,3), xlim([0 2*pi])
```

If the geometric interpretation in Fig. 3.15 graphically describes the fractional-order integral as a shadow on a fence, the length of the fence shadow changes as τ increases. Figure 3.16 depicts the process of changing the length of the fence shadow, i.e., the projection on the bottom surface τ-$g_t(\tau)$, while Fig. 3.17 gives the process of changing the projection on the wall surface $g_t(\tau)$-$f(\tau)$. ◄

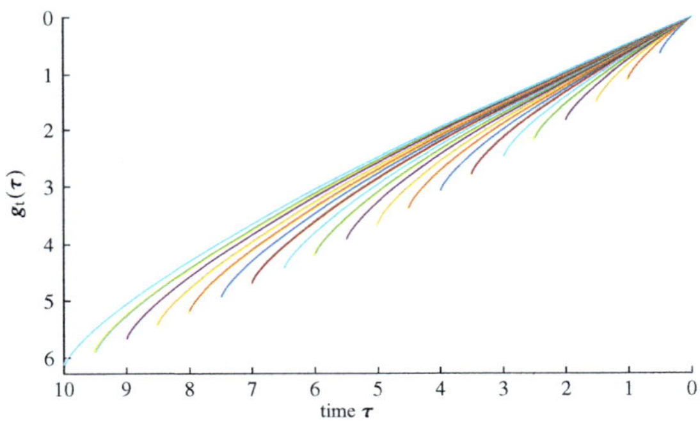

Fig. 3.16 Variation of the shadow length on the fence

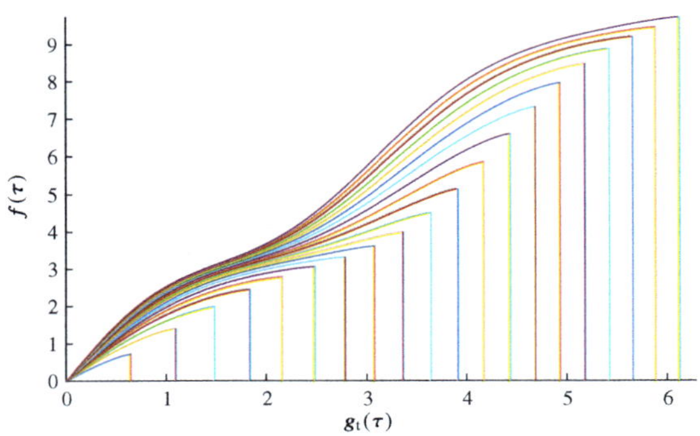

Fig. 3.17 Variation of the shadow of the fence on the wall

Although Prof. Igor Podlubny's geometric interpretation describes well the fractional-order integral, more effort is needed to arrive at a more understandable geometric and physical interpretation in fractional calculus under various definitions.

3.7 Exercises

(1) Let the function $u(t) = C$, C is an arbitrary constant. Compute the expressions for ${}_0^{\rm RL}\mathscr{D}_t^{0.5}u(t)$ and ${}_0^{\rm RL}\mathscr{D}_t^{-0.5}u(t)$ according to Definitions 3.4 and 3.5, with the symbolic facilities of MATLAB.

(2) Use numerical method to verify the formula

$$ {}_a^{\rm GL}\mathscr{D}_t^{\alpha}(t-a)^{\nu} = \frac{\Gamma(\nu+1)}{\Gamma(\nu+1-\alpha)}(t-a)^{\nu-\alpha}, $$

where a is an arbitrary constant. If $\alpha < 0$, $\nu > -1$, while if $\alpha > 0$, $\nu > \lfloor\alpha\rfloor$.

(3) Find an example of (3.28) and use symbolic facilities to prove the example.

$$ \int_0^t\int_0^{\tau}\sin\tau{\rm d}\tau{\rm d}t = \int_0^t (t-\tau)\sin\tau{\rm d}\tau. $$

(4) Generate randomly a set of values of λ and α, and verify the following equation numerically:

$$ {}_0^{\rm RL}\mathscr{D}_t^{\alpha}{\rm e}^{\lambda t} = t^{-\alpha}{\rm E}_{1,1-\alpha}(\lambda t). $$

(5) Let function $u(t)$ continuous in its domain of definition, then its Riemann–Liouville fractional-order integral satisfies the following equation. Choose a function $u(t)$ to satisfy the conditions, using numerical methods to verify the following equation:

$$ {}_0^{\rm RL}\mathscr{D}_t^{-\alpha}\left[{}_0^{\rm RL}\mathscr{D}_t^{-\beta}u(t)\right] = {}_0^{\rm RL}\mathscr{D}_t^{-\beta}\left[{}_0^{\rm RL}\mathscr{D}_t^{-\alpha}u(t)\right] = {}_0^{\rm RL}\mathscr{D}_t^{-\alpha-\beta}u(t), $$

where $\alpha>0$ and $\beta>0$.

(6) According to its definition, the Riemann–Liouville fractional-order derivative of the function $u(t)$ can be written in the following form:

$$ {}_0^{\rm RL}\mathscr{D}_t^{\alpha}u(t) = \frac{{\rm d}^n}{{\rm d}t^n}\left[{}_0^{\rm RL}\mathscr{D}_t^{\alpha-n}u(t)\right], $$

where $\alpha > 0$ and $n = \lceil\alpha\rceil$. Choose a $u(t)$, use numerical method to verify the above equation.

(7) If function $f(t)$ has Riemann–Liouville fractional-order derivatives and integrals of any order, then the following equation is satisfied [9]:

$$ {}_0^{\rm RL}\mathscr{D}_t^{\alpha}[tf(t)] = t\,{}_0^{\rm RL}\mathscr{D}_t^{\alpha}f(t) + \alpha\,{}_0^{\rm RL}\mathscr{D}_t^{\alpha-1}f(t), $$

where $\alpha > 0$. Choose a function $f(t)$ satisfying the above conditions, and verify it numerically.

(8) Prove the following equations using the symbolic facilities in MATLAB [9]:

(a) $\mathscr{L}\left[{}_0^{\mathrm{RL}}\mathscr{D}_t^{-\alpha}t^{\nu}\right] = \dfrac{\Gamma(\nu+1)}{s^{\alpha+\nu+1}}$, $(\alpha > 0, \nu > -1)$,

(b) $\mathscr{L}\left[{}_0^{\mathrm{RL}}\mathscr{D}_t^{-\alpha}\mathrm{e}^{at}\right] = \dfrac{1}{s^{\alpha}(s-a)}$, $(\alpha > 0)$,

(c) $\mathscr{L}\left[{}_0^{\mathrm{RL}}\mathscr{D}_t^{-\alpha}\cos at\right] = \dfrac{1}{s^{\alpha-1}(s^2+a^2)}$, $(\alpha > 0)$.

(9) According to the Caputo fractional-order derivative definition, the Caputo fractional-order derivative of the function $u(t)$ can be written in the following form:

$${}_0^{\mathrm{C}}\mathscr{D}_t^{\alpha}u(t) = {}_0^{\mathrm{RL}}\mathscr{D}_t^{\alpha-n}\left[u^{(n)}(t)\right],$$

where $\alpha > 0$ and $n = \lceil\alpha\rceil$. Choose a function $u(t)$. Verify numerically the above equation.

(10) According to Theorem 3.6, the following equation can be derived:

$${}_0^{\mathrm{C}}\mathscr{D}_t^{1.3}\mathrm{e}^{-0.5t} = (-0.5)^2\, t^{0.7}\, \mathrm{E}_{1,1.7}(-0.5t)\,.$$

Use `caputo()` function and function `ml_func()` provided in ▶ Chap. 2 to verify the above equation.

(11) It is known from Example 3.1 that the 0.5th-order Grünwald–Letnikov derivative of a constant is not zero. What is the 0.5th-order Caputo derivative of the constant?

(12) It is known from Theorem 3.9 that, if the function $f(t)$ satisfies the conditions of the theorem, then the Grünwald–Letnikov fractional-order integral and derivative of the function $f(t)$ are equivalent to the Riemann–Liouville ones. Choose some functions that satisfy the conditions and apply the functions `glfdiff()` and `rlfdiff()` to verify this conclusion.

(13) According to Theorem 3.10, the following equation can be obtained:

$${}_0^{\mathrm{C}}\mathscr{D}_t^{1.5}\sin t = {}_0^{\mathrm{RL}}\mathscr{D}_t^{1.5}\sin t + \frac{1}{\Gamma(0.5)\, t^{0.5}}.$$

Verify the above equation with numerical methods.

(14) Apply the functions `rlfdiff()` and `caputo()` to evaluate ${}_0^{\mathrm{RL}}\mathscr{D}_t^{\alpha}t^2$ and ${}_0^{\mathrm{C}}\mathscr{D}_t^{\alpha}t^2$, taking $\alpha = 0.5, 1, 1.5, 2, 2.5$, respectively. Compare the Riemann–Liouville fractional-order derivatives and Caputo ones of the same order.

(15) The following equation can be obtained from Theorem 3.2:

$${}_0^{\mathrm{RL}}\mathscr{D}_t^{-0.5}t^{0.5} = \frac{\sqrt{\pi}}{2}t.$$

Verify the above equation with `rlfdiff()` function. Select different step sizes to validate the above equation, and find the relationship between the step sizes and computation errors.

(16) Show the following equation, and validate them numerically with functions `caputo()` and `rlfdiff()`.

$$ {}_{0}^{\mathrm{C}}\mathscr{D}_{t}^{1.5}t^{2}\mathrm{e}^{-t}={}_{0}^{\mathrm{RL}}\mathscr{D}_{t}^{-0.5}\left[\sum_{k=0}^{2}\frac{2!}{k!(2-k)!}\left(\frac{\mathrm{d}^{k}}{\mathrm{d}t^{k}}t^{2}\right)\left(\frac{\mathrm{d}^{2-k}}{\mathrm{d}t^{2-k}}\mathrm{e}^{-t}\right)\right]. $$

(17) Prove the following equation, and validate it numerically with functions `caputo()` and `rlfdiff()`.

$$ {}_{0}^{\mathrm{C}}\mathscr{D}_{t}^{\alpha}\sin t={}_{0}^{\mathrm{RL}}\mathscr{D}_{t}^{\alpha-\lceil\alpha\rceil}\sin\left(t+\frac{\pi}{2}\lceil\alpha\rceil\right), $$

where $\alpha>0$.

References

1. Oldham KB, Spanier J (1974) The fractional calculus: theory and applications of differentiation and integration to arbitary order [M]. Academic, San Diego
2. Podlubny I (1999) Fractional differential equations [M]. Academic, San Diego
3. Podlubny I (2000) Matrix approach to discrete fractional calculus[J]. Fract Calc Appl Anal 3(4):359–386
4. Dorčák L (1994) Numerical models for simulation the fractional-order control systems[R]. UEF-04-94. The Academy of Sciences Institute of Experimental Physics, Kosice, Slovak Republi
5. Hilfer R (2000) Applications of fractional calculus in physics[M]. World Scientific, Singapore
6. Caputo M (1967) Linear models of dissipation whose Q is almost frequency independent II[J]. Geophys J Int 13(5):529–539
7. Podlubny I (2001) Geometric and physical interpretations of fractional integration and differentiation[J]. Fract Calc Appl Anal 5(4):230–237
8. Bai L, Xue D Y, Meng L (2020) Geometric interpretation for Riemann-Liouville fractional-order integral[C]. In: Proceedings of the Chinese control and decision conference. Hefei, China, pp 3225–3230
9. Guo BL, Pu XK, Huang FH (2011) Fractional-order partial differential equations and their numerical solutions (in Chinese)[M]. Science Press, Beijing

High-Precision Numerical Algorithms and Implementation in Fractional Calculus

Contents

D. Y. Xue and L. Bai, *Fractional Calculus*, https://doi.org/10.1007/978-981-99-2070-9_4

The accuracy of the algorithm described earlier is at the $O(h)$ level, also known as the first-order algorithm. The computational error is closely related to the step size h. If h is large, the computational error is also large. For example, imprecisely, if $h = 0.01$, the computational error is almost 0.01. If there is an algorithm with $O(h^2)$, called a second-order algorithm, it is possible to obtain a computational error of 0.0001, while a fourth-order algorithm $O(h^4)$ may bring the error down to $0.01^4 = 10^{-8}$. It follows that if we want to obtain a numerical solution with high accuracy, we need to increase the order of the algorithm.

To achieve better computational accuracy, polynomial expansions can be considered instead of binomial expansions in the recursive formula of the Grünwald–Letnikov derivatives. Reference [1] gives the generating function for the computational accuracy $O(h^p)$, for an order $p \leq 6$. High-order generating functions exist. This chapter discusses the construction of the generating functions and presents a method for the high-precision numerical evaluation of the Grünwald–Letnikov and Caputo fractional-order derivatives, with $O(h^p)$ accuracy.

A method for constructing generating functions for any integer p is presented first in ► Sect. 4.1. Theoretically, generating functions can be constructed with arbitrary accuracy. A recursive algorithm is proposed in ► Sect. 4.2, based on the generating functions for functions with zero initial values only. The algorithm can be used to find the Grünwald–Letnikov derivatives and integrals of known functions. The numerical algorithm based on fast Fourier transform is commented. In fact, the method will lead to huge errors for nonzero initial value problems. A signal decomposition method is explored in ► Sect. 4.3 to obtain high-precision Grünwald–Letnikov derivatives and integrals for signals with nonzero initial values. The accuracies of the algorithm are demonstrated by examples. As will be shown in the examples, such an algorithm may be more than a dozen orders of magnitude more accurate than the existing $O(h)$ algorithms. In ► Sect. 4.4, a high-precision algorithm is presented for Caputo derivatives and demonstrates its advantages through dedicated benchmark problems. In ► Sect. 4.5, the problem is also explored for evaluating numerically derivatives of even higher fractional orders, in conjunction with a high-precision integer-order derivative algorithm.

4.1 Generating Function Construction for Arbitrary Integer Orders

The previously described computational function `glfdiff()` based on the Grünwald–Letnikov definition appears to be a faithful implementation of the Grünwald–Letnikov definition. But in fact it is not. The largest source of error in this implementation is the elimination of the limit operator. The direct approximation of the summation of the function by binomials, which leads to an error of $O(h)$ in this algorithm. To better approximate the Grünwald–Letnikov definition, it is important to keep some more terms other than the binomial term information, to compensate for the effect of artificial cancellation of the limit operations.

Let us first look at how to find the first-order derivative using the difference algorithms. Obviously, by directly removing the limit operation from the definition of derivative, the approximation of the first-order derivative can be found directly by the backward-difference method as follows:

$$y'_k = \lim_{h\to 0} \frac{y(t) - y(t-h)}{h} \approx \frac{y_k - y_{k-1}}{h}. \tag{4.1}$$

In addition to the backward-difference method, forward- and central-difference methods are often found in the literatures. Since the latter two require the information of y_{k+i} (the value of the y function after the current point) in calculating y'_k, these two difference methods are not considered in this book. The error of the backward-difference method can be derived by combining the Taylor series expressions:

4

$$\begin{aligned}\frac{y(t) - y(t-h)}{h} &= \frac{f(t) - f(t) + hf'(t) - h^2 f''(t)/2 + \cdots}{h}\\ &= f'(t) - \frac{h}{2} f''(t) + \cdots\end{aligned} \tag{4.2}$$

It can be seen that, omitting the limit operation, the error of the algorithm is $-hf''(t)/2+\cdots$, which is mathematically known as $O(h)$ order algorithm. If one wants to obtain $O(h^2)$ order accuracy, then the following backward-difference formula can be introduced:

$$f'_k \approx \frac{1}{h}\left(\frac{3}{2}y_k - 2y_{k-1} + \frac{1}{2}y_{k-2}\right). \tag{4.3}$$

It can be shown that the error is now $-h^2 f'''(t)/3+\cdots$, hence the $O(h^2)$ algorithm.

▶ Example 4.1

Show that the backward-difference algorithm of (4.3) is an $O(h^2)$ algorithm. Prove that the following equation is an $O(h^3)$ one:

$$f'_k \approx \frac{1}{h}\left(\frac{11}{6}y_k - 3y_{k-1} + \frac{3}{2}y_{k-2} - \frac{1}{3}y_{k-3}\right).$$

Solutions Imitating the method of (4.2), these expressions can be proved, but here the algorithms are proved directly using MATLAB's symbolic facilities. Symbolic function $f(x)$ can be defined first, function `taylor()` can then be called to expand it. For the two formulas here, it is sufficient to expand to five terms and then substitute them into the two backward-difference formulas, and subtract $f'(x)$ from the result before simplifying it.

```
>> syms h x f(x)
   F1=taylor(f(x-h),h,'Order',5); F2=taylor(f(x-2*h),h,'Order',5);
   F3=taylor(f(x-3*h),h,'Order',5);
   H1=simplify((3*f/2-2*F1+F2/2)/h-diff(f))
   H2=simplify((11*f/6-3*F1+3*F2/2-F3/3)/h-diff(f))
```

The following errors are obtained. Obviously, both the errors satisfy the accuracy requirement.

$$H_1 = \frac{h^2}{12}\left(-4f'''(x) + 3hf^{(4)}(x)\right),\ H_2 = -h^3 f^{(4)}(x).$$

◀

How can we obtain the above combination of coefficients? Reference [2] gives a table of coefficients of polynomials (Table 4.1). The formula for generating the

Table 4.1 Low-order generating functions

Order p	Generating functions $g_p(z)$	Order p	Generating functions $g_p(z)$
1	$1-z$	4	$\frac{25}{12}-4z+3z^2-\frac{4}{3}z^3+\frac{1}{4}z^4$
2	$\frac{3}{2}-2z+\frac{1}{2}z^2$	5	$\frac{137}{60}-5z+5z^2-\frac{10}{3}z^3+\frac{5}{4}z^4-\frac{1}{5}z^5$
3	$\frac{11}{6}-3z+\frac{3}{2}z^2-\frac{1}{3}z^3$	6	$\frac{147}{60}-6z+\frac{15}{2}z^2-\frac{20}{3}z^3+\frac{15}{4}z^4-\frac{6}{5}z^5+\frac{1}{6}z^6$

coefficients of z-polynomials can be summarized from the table. The following theorem is given here without proof.

Theorem 4.1 pth-order generating functions
To pursuit the high-precision behavior expected in Ref. [1], the generating function is defined as a pth-order polynomial.

$$g_p(z) = \sum_{k=1}^{p} \frac{1}{k}(1-z)^k. \tag{4.4}$$

Therefore, to achieve $O(h^p)$ precision behavior for $p = 1, 2, \cdots, 6$, the generating function table (Table 4.1) is proposed by Lubich [3].

With the generating functions given in Table 4.1, it is straightforward to write out the integer-order numerical derivative formula. For example, if $p = 4$, then the expression is written as

$$y_k' = \frac{1}{h}\left(\frac{25}{12}y_k - 4y_{k-1} + 3y_{k-2} - \frac{4}{3}y_{k-3} + \frac{1}{4}y_{k-4}\right). \tag{4.5}$$

Unfortunately, the table only lists a number of mathematical expressions for generating functions, and does not give rules and methods for the construction of general pth-order generating functions. If one expects to obtain generating functions for polynomials of arbitrary order, the following theorem should be introduced.

Theorem 4.2 pth-order generating functions

The pth-order generating function $g_p(z)$ can be expressed in the polynomial

$$g_p(z) = \sum_{k=0}^{p} g_k z^k, \tag{4.6}$$

where coefficients g_k can be evaluated directly from

$$\begin{bmatrix} 1 & 1 & 1 & \cdots & 1 \\ 1 & 2 & 3 & \cdots & p+1 \\ 1 & 2^2 & 3^2 & \cdots & (p+1)^2 \\ \vdots & \vdots & \vdots & \ddots & \vdots \\ 1 & 2^p & 3^p & \cdots & (p+1)^p \end{bmatrix} \begin{bmatrix} g_0 \\ g_1 \\ g_2 \\ \vdots \\ g_p \end{bmatrix} = - \begin{bmatrix} 0 \\ 1 \\ 2 \\ \vdots \\ p \end{bmatrix}. \tag{4.7}$$

Proof It is seen from (4.4) and (4.6) that

$$\sum_{k=0}^{p} g_k z^k = \sum_{k=1}^{p} \frac{1}{k}(1-z)^k. \tag{4.8}$$

Substituting $z = 1$ into (4.8), we have

$$\sum_{k=0}^{p} g_k = 0. \tag{4.9}$$

Multiplying both sides of (4.8) by z and taking the first-order derivative with respect to z, then

$$\sum_{k=0}^{p} (k+1) g_k z^k = \sum_{k=1}^{p} \frac{1}{k}(1-z)^k - z\sum_{k=1}^{p}(1-z)^{k-1}. \tag{4.10}$$

Substituting $z = 1$ again into (4.10), we have

$$\sum_{k=0}^{p} (k+1) g_k = -1.$$

Multiplying both sides of (4.10) by z and taking the first-order derivative with respect to z, then

$$\sum_{k=0}^{p} (k+1)^2 g_k z^k = \sum_{k=1}^{p} \frac{1}{k}(1-z)^k - 3z\sum_{k=1}^{p}(1-z)^{k-1} + z^2 \sum_{k=2}^{p} \frac{1}{k-1}(1-z)^{k-2}. \tag{4.11}$$

Substituting $z = 1$ into (4.11), we have

$$\sum_{k=0}^{p} (k+1)^2 g_k = -2.$$

Repeating the above process, the following linear equation can be established:

$$\begin{cases} g_0+g_1+g_2+\cdots+g_p=0 \\ g_0+2g_1+3g_2+\cdots+(p+1)g_p=-1 \\ g_0+2^2g_1+3^2g_2+\cdots+(p+1)^2g_p=-2 \\ \vdots \\ g_0+2^pg_1+3^pg_2+\cdots+(p+1)^pg_p=-p. \end{cases}$$

The theorem is proved from the matrix form of (4.7). □

Based on this theorem, a table of generating functions for higher orders of p can be derived, as shown in Table 4.2, where, again, the result for $p=6$ is given because its first term 147/60 has been correctly simplified to 49/20.

The MATLAB implementation of the above algorithm is

```
function g=genfunc(p)
   arguments, p(1,1) {mustBePositiveInteger}, end
   a=1:p+1; A=rot90(vander(a)); g=(1-a)/sym(A');
end
```

To ensure the accuracies of the generating function coefficients, symbolic facilities are used. The function syntax is g=genfunc(p), where p is the order, while the returned symbolic vector $\boldsymbol{g}$ contains the coefficients of generating functions.

The above generating function corresponds to an integer-order generating function. The actual generating function used for the αth-order derivative is $g_p^\alpha(z)$. Unlike the integer-order calculus, such a generating function is an infinite series and requires the calculation of many values of w_i, as in the case of the binomial coefficients introduced earlier.

Table 4.2 Higher order generating functions

p	**Generating functions** $g_p(z)$
6	$\frac{49}{20}-6z+\frac{15}{2}z^2-\frac{20}{3}z^3+\frac{15}{4}z^4-\frac{6}{5}z^5+\frac{1}{6}z^6$
7	$\frac{363}{140}-7z+\frac{21}{2}z^2-\frac{35}{3}z^3+\frac{35}{4}z^4-\frac{21}{5}z^5+\frac{7}{6}z^6-\frac{1}{7}z^7$
8	$\frac{761}{280}-8z+14z^2-\frac{56}{3}z^3+\frac{35}{2}z^4-\frac{56}{5}z^5+\frac{14}{3}z^6-\frac{8}{7}z^7+\frac{1}{8}z^8$
9	$\frac{7129}{2520}-9z+18z^2-28z^3+\frac{63}{2}z^4-\frac{126}{5}z^5+14z^6-\frac{36}{7}z^7+\frac{9}{8}z^8-\frac{1}{9}z^9$
10	$\frac{7381}{2520}-10z+\frac{45}{2}z^2-40z^3+\frac{105}{2}z^4-\frac{252}{5}z^5+35z^6-\frac{120}{7}z^7+\frac{45}{8}z^8-\frac{10}{9}z^9+\frac{1}{10}z^{10}$
11	$\frac{83711}{27720}-11z+\frac{55}{2}z^2-55z^3+\frac{165}{2}z^4-\frac{462}{5}z^5+77z^6-\frac{330}{7}z^7+\frac{165}{8}z^8-\frac{55}{9}z^9+\frac{11}{10}z^{10}-\frac{1}{11}z^{11}$

4.2 Trials on High-Precision Algorithms for Grünwald–Letnikov Derivatives

This section gives methods for evaluating fractional-order derivatives based on generating functions and introduces a method for evaluating them based on Fourier transforms. However, it must be pointed out that these direct implementations have significant drawbacks and cannot be practically applied. Compensations are required for the algorithm to become truly practical.

Recall the definition of Grünwald–Letnikov fractional-order derivatives

$$ {}_{t_0}^{\mathrm{GL}}\mathscr{D}_t^{\alpha} f(t) = \lim_{h\to 0} \frac{1}{h^{\alpha}} \sum_{j=0}^{[(t-t_0)/h]} (-1)^{\alpha} \binom{\alpha}{j} f(t-jh). \tag{4.12} $$

If the binomial coefficients are replaced by the polynomial coefficients with $O(h^p)$ ones, high-precision algorithms can be introduced.

Theorem 4.3 pth-order approximation methods
High-precision Grünwald–Letnikov derivatives can be approximated with

$$ {}_{t_0}^{\widetilde{\mathrm{GL}}}\mathscr{D}_t^{\alpha} f(t) = \frac{1}{h^{\alpha}} \sum_{k=0}^{[(t-t_0)/h]} w_k^{(\alpha,p)} f(t-kh), \tag{4.13} $$

where the coefficient vector $\boldsymbol{w}^{(\alpha,p)} = \left[w_0^{(\alpha,p)}, w_1^{(\alpha,p)}, w_2^{(\alpha,p)}, \cdots\right]$ can be applied in Lubich's linear multi-step methods. If there are no conflicts, the (α, p) superscripts can be omitted. The accuracy of the algorithm is $O(h^p)$.

Here, the vector $\boldsymbol{w}$ contains the coefficients of the Taylor series expansion of the generating function. If a first-order generating function is chosen, the coefficient vector $\boldsymbol{w}$ can be calculated recursively as follows:

$$ w_0 = 1, \quad w_k = \left(1 - \frac{\alpha+1}{k}\right) w_{k-1}. \tag{4.14} $$

This equation is actually the recursive formula for the binomial coefficients introduced earlier.

Theorem 4.4 Error tolerances
If $f(t) = t^{\nu-1}$, $(\nu > 0)$, the coefficient of pth-order Taylor series expansion is $\boldsymbol{w}$[3], then

$$ {}_{t_0}^{\mathrm{RL}}\mathscr{D}_t^{\alpha} f(t) = \frac{1}{h^\alpha} \sum_{k=0}^{[(t-t_0)/h]} w_k f(t-kh) + O(h^\nu) + O(h^p). \tag{4.15} $$

4.2.1 An FFT-Based Algorithm

From the previous analysis, it is clear that the Taylor series expansion algorithm is not suitable for computing vector $\boldsymbol{w}$ in practical applications. Prof. Podlubny introduced a computation method based on fast Fourier transform (FFT) [1]. The algorithm will be presented here, and the error of the algorithm will be demonstrated by a computational example. The causes of the error will be analyzed.

Suppose the coefficients of the Taylor series expansion of the generating function $g_p^\alpha(z)$ are $\boldsymbol{w} = [w_0, w_1, \cdots]$, then Taylor series expansion can be written as

$$ g_p^\alpha(z) = \sum_{k=0}^{\infty} w_k z^k. \tag{4.16} $$

By replacing the variable z in (4.16) by $e^{-j\varphi}$, then

$$ g_p^\alpha(e^{-j\varphi}) = \sum_{k=0}^{\infty} w_k e^{-jk\varphi}. \tag{4.17} $$

Definition 4.1 Inner product formula

If the inner product of functions $e^{jn\varphi}$ and $e^{-jm\varphi}$ is

$$ \langle e^{jn\varphi}, e^{-jm\varphi} \rangle = \frac{1}{2\pi} \int_0^{2\pi} e^{jn\varphi} e^{-jm\varphi} d\varphi = \begin{cases} 0, & m \neq n \\ 1, & m = n, \end{cases} \tag{4.18} $$

the two functions are orthogonal ones.

If we calculate the inner product of the two functions as $\langle g_p^\alpha(e^{-j\varphi}), e^{jk\varphi} \rangle$, then it is found that

$$ w_k = \frac{1}{2\pi} \int_0^{2\pi} g_p^\alpha(e^{-j\varphi}) e^{jk\varphi} d\varphi. \tag{4.19} $$

Equation (4.19) shows that, the vector $\boldsymbol{w}$ is the Fourier coefficients of the function $g_p^\alpha(e^{-j\varphi})$, such that the coefficient vector $\boldsymbol{w}$ can be efficiently computed by the FFT algorithm. Based on such an idea, the algorithm for computing the vector $\boldsymbol{w}$ can be given as follows.

Algorithm 4.1 FFT-based method
(1) Compute sample vector $\boldsymbol{f}$ in step size h from the original function $y = f(t)$.
(2) Choose p, compute fractional-order generating function $g_p^\alpha(z)$.
(3) Evaluate coefficients w_k through FFT in (4.19).
(4) Evaluate fractional-order derivatives or integrals from (4.13).

Based on the above algorithm, the following MATLAB functions can be written:

```
function dy=glfdiff_fft(y,t,gam,p)
   arguments
      y(:,1), t(:,1) double, gam(1,1) double
      p(1,1){mustBePositiveInteger}=5
   end
   [y,h,n]=fdiffcom(y,t); dy=zeros(n,1);
   g=double(genfunc(p)); T=2*pi/(n-1);
   if y(1)~=0 && gam>0, dy(1)=sign(y(1))*Inf; end
   tt=[0:T:2*pi]; F=g(1); f1=exp(1i*tt); f0=f1;
   for i=2:p+1, F=F+g(i)*f1; f1=f1.*f0; end
   w=real(fft(F.^gam))*T/2/pi;
   for k=2:n, dy(k)=w(1:k)*y(k:-1:1)/h^gam; end
end
```

Since the FFT calculation is used in Algorithm 4.1, the resulting Fourier coefficients are not very accurate, which may lead to large computational errors, as demonstrated here in Example 4.2.

▶ Example 4.2

Compute $y_1(t) = {}_0^{\rm RL}\mathscr{D}_t^{0.6}{\rm e}^{-t}$ numerically within [0, 5].

Solutions The analytical solution of fractional-order derivative is $y_1(t) = t^{-0.6}{\rm E}_{1,0.4}(-t)$, where ${\rm E}(\cdot)$ is the two-parameter Mittag-Leffler function. From the following statements, the analytical and numerical solutions of the fractional-order derivative problem can be completed and drawn as shown in Fig. 4.1. It can be seen that the difference between the analytical and numerical solutions is too large, and the resulting fractional-order derivatives appear as erroneous results.

```
>> t=0:0.001:5; y0=t.^-0.6.*ml_func([1,0.4],-t,0,eps);
   h=0.01; t1=0:h:5; y1=exp(-t1); dy1=glfdiff_fft(y1,t1,0.6,6);
   h=0.001; t2=0:h:5; y2=exp(-t2); dy2=glfdiff_fft(y2,t2,0.6,6);
   plot(t,y0,t1,dy1,'--',t2,dy2,':'), ylim([-1 4])
```

◀

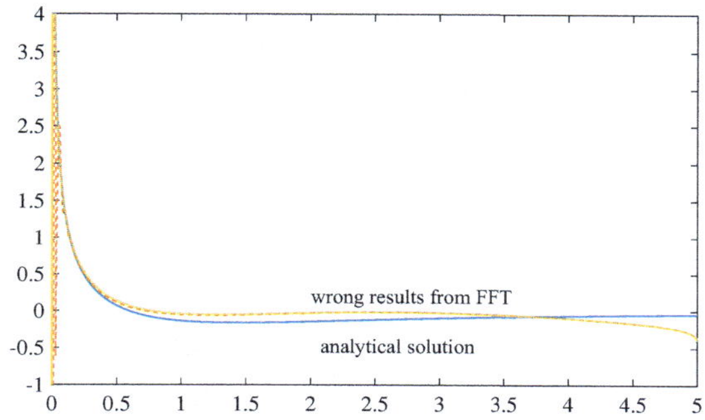

Fig. 4.1 Erroneous results from the FFT-based algorithm

It can be seen that Algorithm 4.1 is fast enough; however, if different step sizes $h = 0.01$ and $h = 0.001$ are chosen, and $p = 4$, there is always a huge error from the initial moment. The reason for this is that Algorithm 4.1 does not evaluate the vector $\boldsymbol{w}$ accurately enough, which leads to a large error in the evaluation of the fractional-order derivative in (4.13). In a word, the FFT-based algorithm is not suitable for high-precision evaluation of fractional-order derivatives.

4.2.2 A Recursive Formula for Generating Function Coefficients

The previous analysis shows that the computational error of Algorithm 4.1 is very large, so an efficient numerical algorithm that can really achieve high-precision fractional-order derivatives is needed. The important thing in the algorithm is how to find the coefficient vector $\boldsymbol{w}$.

Consider the right-hand side of (4.16) in the infinite series with respect to the variable z. Since the generating function $g_p^\alpha(z)$ is a differentiable function, the first-order derivative of z at both sides of (4.16) shows that

$$\frac{\mathrm{d}}{\mathrm{d}z}g_p^\alpha(z) = \frac{\mathrm{d}}{\mathrm{d}z}\sum_{k=0}^{\infty} w_k z^k = \sum_{k=0}^{\infty} w_k \frac{\mathrm{d}}{\mathrm{d}z} z^k. \tag{4.20}$$

Processing both sides of the equal sign yields

$$\alpha g_p^{\alpha-1}(z)\frac{\mathrm{d}g_p(z)}{\mathrm{d}z} = \sum_{k=1}^{\infty} k w_k z^{k-1}. \tag{4.21}$$

Multiplying both sides of (4.21) by $g_p(z)$, we have

$$\alpha g_p^\alpha(z)\frac{\mathrm{d}g_p(z)}{\mathrm{d}z} = g_p(z)\sum_{k=1}^{\infty} k w_k z^{k-1}. \tag{4.22}$$

Substituting $g_p^\alpha(z) = \sum_{k=0}^{\infty} w_k z^k$ into the left-hand side of (4.22), it is found

$$\alpha \frac{\mathrm{d}g_p(z)}{\mathrm{d}z} \sum_{k=0}^{\infty} w_k z^k = g_p(z) \sum_{k=1}^{\infty} k w_k z^{k-1}. \tag{4.23}$$

Theorem 4.5 Fractional-order generating function
If the fractional-order generating function $g_p^\alpha(z)$ is expressed as

$$g_p^\alpha(z) = \left(g_0 + g_1 z + \cdots + g_p z^p\right)^\alpha, \tag{4.24}$$

its Taylor series can be written as

$$g_p^\alpha(z) = \sum_{k=0}^{\infty} w_k z^k. \tag{4.25}$$

Therefore, when $k = 0$, $w_0 = g_0^\alpha$, while if $k > 0$, the subsequent coefficients w_k can be evaluated recursively from

$$w_k = -\frac{1}{g_0} \sum_{i=1}^{p} g_i \left(1 - i\frac{1+\alpha}{k}\right) w_{k-i}, \tag{4.26}$$

where if $k < 0$, then $w_k = 0$.

Proof Let $z = 0$ in (4.24), it can be proved that $w_0 = g_0^\alpha$.
Rewriting (4.24) and extending the sum to the $(-\infty, \infty)$ range, then

$$\left(g_0 + g_1 z + \cdots + g_p z^p\right)^\alpha = \sum_{k=-\infty}^{\infty} w_k z^k. \tag{4.27}$$

Clearly, if $k < 0$, then $w_k = 0$.
Taking the first-order derivative with respect to z on both sides of (4.27), we get

$$\alpha\left(g_1 + 2g_2 z + \cdots + p g_p z^{p-1}\right)\left(g_0 + g_1 z + \cdots + g_p z^p\right)^{\alpha-1} = \sum_{k=-\infty}^{\infty} k w_k z^{k-1}. \tag{4.28}$$

Both sides of (4.28) are simultaneously multiplied by $\left(g_0 + g_1 z + \cdots + g_p z^p\right)$, which shows that

$$\alpha\left(g_1 + \cdots + p g_p z^{p-1}\right)\left(g_0 + \cdots + g_p z^p\right)^\alpha = \left(g_0 + \cdots + g_p z^p\right) \sum_{k=-\infty}^{\infty} k w_k z^{k-1}. \tag{4.29}$$

Substituting (4.27) into (4.29), we can get

$$\alpha\left(g_1+\cdots+pg_pz^{p-1}\right)\sum_{k=-\infty}^{\infty} w_kz^k = \left(g_0+\cdots+g_pz^p\right)\sum_{k=-\infty}^{\infty} kw_kz^{k-1}. \tag{4.30}$$

With the translation law $z^d w_k = w_{k-d}$, the left-hand side of (4.30) is written as

$$\sum_{k=-\infty}^{\infty} \alpha\left(g_1w_k+2g_2w_{k-1}+\cdots+pg_pw_{k-p+1}\right)z^k. \tag{4.31}$$

and its right-hand side is written as

$$\sum_{k=-\infty}^{\infty} \left[(k+1)g_0w_{k+1}+kg_1w_k+\cdots+(k-p+1)g_pw_{k-p+1}\right]z^k. \tag{4.32}$$

Comparing (4.31) and (4.32) the coefficients with the same power of z, we can conclude that

$$\alpha\left(g_1w_k+\cdots+pg_pw_{k-p+1}\right) = (k+1)g_0w_{k+1}+\cdots+(k-p+1)g_pw_{k-p+1}. \tag{4.33}$$

Shifting (4.33) back one step, i.e., by letting $k=k-1$, the equation becomes

$$g_0kw_k+g_1(k-1-\alpha)w_{k-1}+\cdots+g_p(k-p-p\alpha)w_{k-p}=0.$$

If $k\neq 0$, it follows that

$$w_k = -\frac{1}{g_0}\left[g_1\left(1-\frac{1+\alpha}{k}\right)w_{k-1}+g_2\left(1-2\frac{1+\alpha}{k}\right)w_{k-2}+\cdots+ g_p\left(1-p\frac{1+\alpha}{k}\right)w_{k-p}\right],$$

for $w_k=0$ for $k<0$. The above is a recursive formula, so the theorem is proved.

□

In the actual implementation, conditions such as "$w_k=0$ for $k<0$" may lead to negative subscripts, so from a numerical implementation point of view, the programming should be based on the following corollary.

Corollary 4.1 *If the generating function and its Taylor series expansion are still represented by (4.24) and (4.25), then the coefficient w_k of the Taylor series can be derived recursively from the following equation:*

$$w_k = -\frac{1}{g_0}\sum_{i=1}^{p} g_i\left(1-i\frac{1+\alpha}{k}\right)w_{k-i},\ \ \text{if}\ k=p,p+1,p+2,\cdots \tag{4.34}$$

4

For $m = 1, 2, \cdots, p-1$, the coefficients of the first few terms are

$$w_m = -\frac{1}{g_0}\sum_{i=1}^{m-1} g_i\left(1 - i\frac{1+\alpha}{m}\right) w_{m-i}, \tag{4.35}$$

and $w_0 = g_0^\alpha$.

The recursive formulas for $p = 1, 2, \cdots, 6$ are given in Table 4.3, and $k < p$ initial terms are shown in Table 4.4. Of course, this can be achieved directly by implementing the equations in Corollary 4.1.

Algorithm 4.2 Direct recursive algorithm

(1) Choose a step size h, compute the samples $\boldsymbol{f}$ from function $y = f(t)$.

(2) Choose order p to construct generating function $g_p^\alpha(z)$, create vector $\boldsymbol{w}$ from Corollary 4.1.

(3) Evaluate fractional-order derivatives directly from (4.13).

Table 4.3 Recursive formulas of different orders

Order p	Recursive formulas for w_k, where $k = p, p+1, \cdots$
1	$w_k = \left(1 - \frac{1+\alpha}{k}\right) w_{k-1}$
2	$w_k = \frac{1}{3}\left[4\left(1 - \frac{1+\alpha}{k}\right) w_{k-1} - \left(1 - 2\frac{1+\alpha}{k}\right) w_{k-2}\right]$
3	$w_k = \frac{1}{11}\left[18\left(1 - \frac{1+\alpha}{k}\right) w_{k-1} - 9\left(1 - 2\frac{1+\alpha}{k}\right) w_{k-2} + 2\left(1 - 3\frac{1+\alpha}{k}\right) w_{k-3}\right]$
4	$w_k = \frac{1}{25}\left[48\left(1 - \frac{1+\alpha}{k}\right) w_{k-1} - 36\left(1 - 2\frac{1+\alpha}{k}\right) w_{k-2} + 16\left(1 - 3\frac{1+\alpha}{k}\right) w_{k-3} - \left(1 - 4\frac{1+\alpha}{k}\right) w_{k-4}\right]$
5	$w_k = \frac{1}{137}\left[300\left(1 - \frac{1+\alpha}{k}\right) w_{k-1} - 300\left(1 - 2\frac{1+\alpha}{k}\right) w_{k-2} + 200\left(1 - 3\frac{1+\alpha}{k}\right) w_{k-3} - 75\left(1 - 4\frac{1+\alpha}{k}\right) w_{k-4} + 12\left(1 - 5\frac{1+\alpha}{k}\right) w_{k-5}\right]$
6	$w_k = \frac{1}{147}\left[360\left(1 - \frac{1+\alpha}{k}\right) w_{k-1} - 450\left(1 - 2\frac{1+\alpha}{k}\right) w_{k-2} + 400\left(1 - 3\frac{1+\alpha}{k}\right) w_{k-3} - 225\left(1 - 4\frac{1+\alpha}{k}\right) w_{k-4} + 72\left(1 - 5\frac{1+\alpha}{k}\right) w_{k-5} - 10\left(1 - 6\frac{1+\alpha}{k}\right) w_{k-6}\right]$

Table 4.4 Initial terms when $k < p$

p	**Initial terms** w_k
1	$w_0 = 1$
2	$w_0 = (3/2)^\alpha$
	$w_1 = -4\alpha w_0/3$
3	$w_0 = (11/6)^\alpha$
	$w_1 = -18\alpha w_0/11$
	$w_2 = 9\alpha w_0/11 + 9(1-\alpha)w_1/11$
4	$w_0 = (25/12)^\alpha$
	$w_1 = -48\alpha w_0/25$
	$w_2 = (36\alpha w_0 + 24(1-\alpha)w_1)/25$
	$w_3 = (-16\alpha w_0 - 12(1-2\alpha)w_1 + 16(2-\alpha)w_2)/25$
5	$w_0 = (137/60)^\alpha$
	$w_1 = -300\alpha w_0/137$
	$w_2 = (300\alpha w_0 + 150(1-\alpha)w_1)/137$
	$w_3 = (-200\alpha w_0 - 100(1-2\alpha)w_1 + 100(2-\alpha)w_2)/137$
	$w_4 = (75\alpha w_0 + 50(1-3\alpha)w_1 - 75(2-2\alpha)w_2 + 75(3-\alpha)w_3)/137$
6	$w_0 = (49/20)^\alpha$
	$w_1 = -120\alpha w_0/49$
	$w_2 = (150\alpha w_0 + 60(1-\alpha)w_1)/49$
	$w_3 = (-400\alpha w_0 - 150(1-2\alpha)w_1 + 120(2-\alpha)w_2)/147$
	$w_4 = (225\alpha w_0 + 100(1-3\alpha)w_1 + 112.5(2-2\alpha)w_2 + 90(3-\alpha)w_3)/147$
	$w_5 = (-72\alpha w_0 - 45(1-4\alpha)w_1 + 80(2-3\alpha)w_2 - 90(3-2\alpha)w_3 + 72(4-\alpha)w_4)/147$

Based on the above algorithm, the following MATLAB function can be written:

```
function dy=glfdiff2(y,t,gam,p)
   arguments
      y(:,1), t(:,1) double, gam(1,1) double
      p(1,1){mustBePositiveInteger}=5
   end
   g=double(genfunc(p)); [y,h,n]=fdiffcom(y,t);
   w=get_vecw(gam,n,g); dy=zeros(n,1);
   for i=1:n, dy(i)=w(1:i)*y(i:-1:1)/h^gam; end
end
```

The authors wrote the supporting function get_vecw() to retrieve the vector $\boldsymbol{w}$, which requires the user to provide the generator function $\boldsymbol{g}$ vector. Many more functions will call this supporting function later.

```
function w=get_vecw(gam,n,g)
   p=length(g)-1; b=1+gam; g0=g(1);
   w=zeros(1,n); w(1)=g(1)^gam;
   for m=2:p, M=m-1; A=b/M;
      w(m)=-[g(2:m).*((1-A):-A:(1-b))]*w(M:-1:1).'/g0;
   end
   for k=p+1:n, M=k-1; A=b/M;
      w(k)=-[g(2:(p+1)).*...
            ((1-A):-A:(1-p*A))]*w(M:-1:(k-p)).'/g0;
end, end
```

▶ Example 4.3

Solve again the problem of Example 4.2, which is to find the 0.6th-order derivative of the exponential function e^{-t}.

Solutions The solution can be found by trying to find the 0.6th-order derivatives of the original function at $p = 1$, 5, and 6, as shown in ■ Fig. 4.2. It can be seen that when p is large, there is a strong oscillation at the initial moment, although the error is smaller when t is larger. Obviously, the calculation results obtained here are wrong. The reason for this is the forced assumption in Corollary 4.1 that the value of w_k is zero for $k < 0$, which leads to the overall effect of nonzero initial values.

```
>> t=[0:0.01:0.2]'; y=@(t)exp(-t);                %column vectors
   y0=t.^-0.6.*ml_func([1,0.4],-t,0,eps); %exact solutions
   y1=glfdiff2(y,t,0.6,1); y5=glfdiff2(y,t,0.6,5);
   y6=glfdiff2(y,t,0.6,6);
   plot(t,y0,t,y1,'--',t,y5,':',t,y6,'-')
```

◀

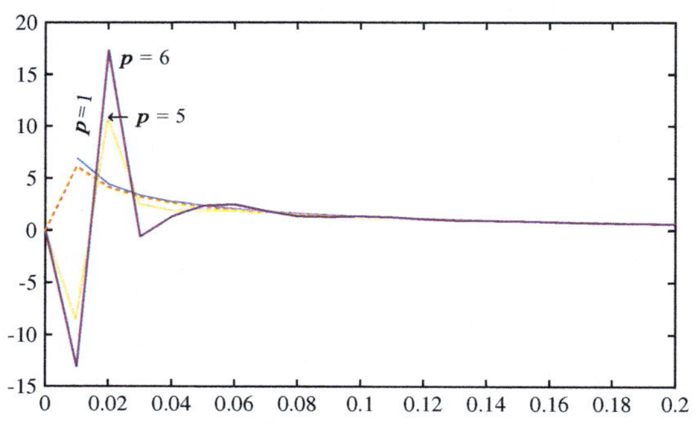

■ **Fig. 4.2** 0.6th-order derivatives of the function at different orders p

In fact, the algorithm given here is only suitable for functions with zero initial values. It cannot be used for functions with nonzero initial conditions, which require further exploration of specialized methods for solving the problems. In addition, the `glfdiff2()` function given here does not solve fractional-order integral problems. Therefore, it is not recommended to use `glfdiff2()` function directly in practical applications.

4.3 High-Precision Algorithm and Implementation for Grünwald–Letnikov Definition

The so-called high-precision algorithm attempted in the previous section has a fatal weakness, namely, if the function to be studied contains nonzero initial values in the function or its integer-order derivatives, it may lead to great errors, making the results far from the theoretical values. In this section, we will explore the compensation methods for nonzero initial value functions and construct practical high-precision algorithms.

4.3.1 Decomposition and Compensation for Nonzero Initial Value Functions

As can be seen by the previous examples, using the existing Algorithm 4.2 may lead to large errors in the initial moments. When $k < p$, there are no other initial values to use from if a high-precision algorithm is used. This is the direct cause of the large error. To solve such a problem, the authors define the following auxiliary function and propose an improved high-precision algorithm.

Theorem 4.6 Auxiliary function

In the numerical evaluation process for ${}_{t_0}^{\mathrm{RL}}\mathscr{D}_t^{\alpha} y(t)$, an auxiliary function can be introduced.

$$u(t) = \sum_{k=0}^{p} c_k (t-t_0)^k, \tag{4.36}$$

to ensure the first $p+1$ points in the signals $u(t)$ and $y(t)$ identical, the coefficients c_k in (4.36) should satisfy the following equation:

$$\begin{bmatrix} 1 & 0 & 0 & \cdots & 0 \\ 1 & h & h^2 & \cdots & h^p \\ 1 & 2h & (2h)^2 & \cdots & (2h)^p \\ \vdots & \vdots & \vdots & \ddots & \vdots \\ 1 & ph & (ph)^2 & \cdots & (ph)^p \end{bmatrix} \begin{bmatrix} c_0 \\ c_1 \\ c_2 \\ \vdots \\ c_p \end{bmatrix} = \begin{bmatrix} y(t_0) \\ y(t_0+h) \\ y(t_0+2h) \\ \vdots \\ y(t_0+ph) \end{bmatrix}, \tag{4.37}$$

where h is the chosen step size.

In this way, the original signal $y(t)$ can be decomposed into

$$y(t) = u(t) + v(t), \tag{4.38}$$

and

$${}^{\mathrm{RL}}_{t_0}\mathscr{D}_t^{\alpha} y(t) = {}^{\mathrm{RL}}_{t_0}\mathscr{D}_t^{\alpha} u(t) + {}^{\mathrm{RL}}_{t_0}\mathscr{D}_t^{\alpha} v(t), \tag{4.39}$$

where the signal $u(t)$ is the sum of the Heaviside and power functions, and the unified analytical solutions of the fractional-order derivatives of these input components are

$${}^{\mathrm{RL}}_{t_0}\mathscr{D}_t^{\alpha} (t-t_0)^k = \frac{\Gamma(k+1)}{\Gamma(k+1-\alpha)}(t-t_0)^{k-\alpha}, \quad k=0,1,2,\cdots \tag{4.40}$$

4.3.2 High-Precision Algorithm and Its Implementation

Since the first $p+1$ initial values of the two signals $u(t)$ and $y(t)$ are the same, the first $p+1$ initial values of $v(t)$ are zero. Thus, although the first $p+1$ points of Table 4.4 are missing in the evaluation of ${}^{\mathrm{RL}}_{t_0}\mathscr{D}_t^{\alpha} v(t)$, when $k < p$, the condition ${}^{\mathrm{RL}}_{t_0}\mathscr{D}_t^{\alpha} v(t_0+kh) = 0$ still holds, so that the impact of the first few missing terms in Table 4.4 is eliminated. In this way, the ${}^{\mathrm{RL}}_{t_0}\mathscr{D}_t^{\alpha} v(t)$ can be directly obtained using (4.40). In summary, a modification of Algorithm 4.2 is proposed [4].

Algorithm 4.3 Improved high-precision recursive algorithm

(1) Create an auxiliary function $u(t)$ from (4.36). Compute $v(t)=y(t)-u(t)$.
(2) Find the accurate solution of ${}^{\mathrm{RL}}_{t_0}\mathscr{D}_t^{\alpha} u(t)$ from (4.40).
(3) Use Algorithm 4.2 to compute the high-precision solution of ${}^{\mathrm{RL}}_{t_0}\mathscr{D}_t^{\alpha} v(t)$.
(4) Find the fractional-order derivative or integral from (4.39).

The following MATLAB function is written according to the above algorithm, where the supporting function `get_vecw()` introduced earlier is used in the calculation of the $\boldsymbol{w}$ vector.

```
function dy=glfdiff9(y,t,gam,p)
   arguments
      y(:,1), t(:,1) double, gam(1,1) double
      p(1,1){mustBePositiveInteger}=5
   end
   [y,h,n]=fdiffcom(y,t); u=0; du=0; r=(0:p)*h;
   R=sym(fliplr(vander(r))); c=double(R\y(1:p+1));
   for i=1:p+1, u=u+c(i)*t.^(i-1);
      du=du+c(i)*t.^(i-1-gam)*gamma(i)/gamma(i-gam);
   end
   v=y-u; g=double(genfunc(p)); w=get_vecw(gam,n,g);
   for i=1:n, dv(i,1)=w(1:i)*v(i:-1:1)/h^gam; end
```

```
    dy=dv+du; if abs(y(1))<1e-10, dy(1)=0; end
end
```

Remarks High-precision numerical calculus algorithm

(1) In actual high-precision operations, because of the limitation of the double-precision data structure used, the accuracy of $O(h^{p+1})$ is not guaranteed to be really higher than $O(h^p)$ for very large p-values. A verification method should be used to select the appropriate p.

(2) The functions with nonzero initial values must be decomposed into two signals, and the fractional-order derivatives are evaluated individually, with different methods. □

4.3.3 Testing and Assessment of the Algorithms

In this section, a test function with known analytical solutions of derivatives is given, different step sizes and p-values are chosen to find fractional-order derivatives. Computational accuracy and other specifications are examined.

► **Example 4.4**

Solve the problem in Example 4.2 again and assess the accuracy of the algorithm.

Solutions From the previous example, it can be seen that the FFT-based algorithm fails if the computation step $h = 0.01$ is chosen. The algorithm of Algorithm 4.2 has strong oscillations at the initial moment, yielding wrong results. Here the same step size is used, different p values are tried and the results of the new algorithm are compared with the analytical solution $y = t^{-0.6}\mathrm{E}_{1,0.4}(-t)$, the results of which are given in Table 4.5 (the resulting matrix $\boldsymbol{T}$ returns the full data of the table).

```
>> t10=[0.5:0.5:5]'; t=0:0.01:5;
   y10=t10.^-0.6.*ml_func([1,0.4],-t10,0,eps);
   f=@(t)exp(-t); ii=[51:50:length(t)]; T=t10;
   for p=1:6, y1=glfdiff9(f,t,0.6,p); T=[T, [y1(ii)-y10]]; end
```

It can be seen that the error at $p = 1$ is consistent with that of `glfdiff()`. The computational error is significantly reduced when a larger desired order p is chosen. Moreover, the algorithm here is particularly efficient, with a computational speed similar to that of `glfdiff()`. Although the step size $h = 0.01$ chosen here is already large from a conventional point of view, and the maximum accuracy given by this algorithm at $p = 6$ can still be as low as 10^{-13}, which is more than ten orders of magnitude higher than that of other existing algorithms.

Now try a larger step size of $h = 0.1$, and still measure the calculation errors of the high-precision algorithm for different values of p, as shown in Table 4.6.

```
>> t10=[0.5:0.5:5]';
   y10=t10.^-0.6.*ml_func([1,0.4],-t10,0,eps);
   t=0:0.1:5; y=exp(-t); ii=[6:5:51];
   for p=1:6, y1=glfdiff9(y,t,0.6,p); T=[T [y1(ii)-y10]]; end
```

4

Table 4.5 Computational errors for different orders p

t	$p=1$	$p=2$	$p=3$	$p=4$	$p=5$	$p=6$
0.5	−0.0018	1.19×10^{-5}	-8.89×10^{-8}	7.06×10^{-10}	-5.85×10^{-12}	5.30×10^{-14}
1	−0.0017	1.14×10^{-5}	-8.58×10^{-8}	6.84×10^{-10}	-5.69×10^{-12}	4.98×10^{-14}
1.5	−0.0011	1.00×10^{-5}	-7.52×10^{-8}	6.00×10^{-10}	-4.99×10^{-12}	4.74×10^{-14}
2	−0.0012	8.61×10^{-6}	-6.44×10^{-8}	5.14×10^{-10}	-4.28×10^{-12}	5.52×10^{-14}
2.5	−0.0011	7.38×10^{-6}	-5.52×10^{-8}	4.41×10^{-10}	-3.70×10^{-12}	-6.74×10^{-15}
3	−0.0009	6.39×10^{-6}	-4.78×10^{-8}	3.82×10^{-10}	-3.23×10^{-12}	1.11×10^{-14}
3.5	−0.0008	5.61×10^{-6}	-4.19×10^{-8}	3.35×10^{-10}	-2.93×10^{-12}	1.18×10^{-14}
4	−0.0007	4.99×10^{-6}	-3.73×10^{-8}	2.98×10^{-10}	-2.83×10^{-12}	-5.71×10^{-14}
4.5	−0.0006	4.50×10^{-6}	-3.36×10^{-8}	2.68×10^{-10}	-3.03×10^{-12}	8.69×10^{-14}
5	−0.0006	4.11×10^{-6}	-3.07×10^{-8}	2.45×10^{-10}	-3.46×10^{-12}	-8.91×10^{-13}

Table 4.6 Comparison of errors at large step size $h=0.1$

t	$p=1$	$p=2$	$p=3$	$p=4$	$p=5$	$p=6$
0.5	−0.01707	0.001033	-6.946×10^{-5}	4.526×10^{-6}	-1.979×10^{-7}	-3.068×10^{-9}
1	−0.01676	0.001080	-7.822×10^{-5}	5.981×10^{-6}	-4.731×10^{-7}	3.739×10^{-8}
1.5	−0.01474	0.000962	-7.058×10^{-5}	5.507×10^{-6}	-4.474×10^{-7}	3.726×10^{-8}
2	−0.01264	0.000827	-6.086×10^{-5}	4.771×10^{-6}	-3.900×10^{-7}	3.279×10^{-8}
2.5	−0.01083	0.000708	-5.208×10^{-5}	4.083×10^{-6}	-3.339×10^{-7}	2.809×10^{-8}
3	−0.00936	0.000611	-4.484×10^{-5}	3.510×10^{-6}	-2.865×10^{-7}	2.407×10^{-8}
3.5	−0.00820	0.000534	-3.907×10^{-5}	3.050×10^{-6}	-2.484×10^{-7}	2.081×10^{-8}
4	−0.00729	0.000473	-3.452×10^{-5}	2.688×10^{-6}	-2.183×10^{-7}	1.824×10^{-8}
4.5	−0.00656	0.000424	-3.092×10^{-5}	2.401×10^{-6}	-1.945×10^{-7}	1.621×10^{-8}
5	−0.00598	0.000386	-2.806×10^{-5}	2.174×10^{-6}	-1.756×10^{-7}	1.460×10^{-8}

Table 4.7 Comparison of errors at higher orders of p

t	$p=7$	$p=8$	$p=9$	$p=10$	$p=11$
0.5	-8.165×10^{-11}	-2.969×10^{-12}	-1.359×10^{-13}	-1.357×10^{-14}	9.714×10^{-17}
1	-2.914×10^{-9}	2.483×10^{-10}	-2.028×10^{-11}	7.797×10^{-13}	4.857×10^{-15}
1.5	-3.120×10^{-9}	2.479×10^{-10}	-2.098×10^{-11}	2.435×10^{-12}	-2.035×10^{-13}
2	-2.847×10^{-9}	2.530×10^{-10}	-1.629×10^{-11}	-4.066×10^{-13}	-1.541×10^{-13}
2.5	-2.445×10^{-9}	2.501×10^{-10}	-4.511×10^{-11}	6.793×10^{-12}	5.390×10^{-13}
3	-2.058×10^{-9}	1.519×10^{-10}	-6.358×10^{-12}	3.181×10^{-11}	-1.003×10^{-11}
3.5	-1.756×10^{-9}	4.491×10^{-11}	2.351×10^{-10}	-3.369×10^{-10}	8.919×10^{-11}
4	-1.547×10^{-9}	2.271×10^{-10}	-6.042×10^{-10}	9.424×10^{-10}	-3.980×10^{-10}
4.5	-1.394×10^{-9}	4.941×10^{-10}	-1.905×10^{-9}	7.291×10^{-9}	-3.496×10^{-9}
5	-1.258×10^{-9}	-2.166×10^{-10}	1.141×10^{-8}	-8.334×10^{-8}	9.951×10^{-8}

It can be seen that even with such a large step sizes, it is still possible to obtain a fairly high computational accuracy, which is simply not possible with other existing algorithms. Further increasing the precision p yields the error comparison shown in Table 4.7. Due to the limitations of the double-precision scheme, it is not always better to have a larger p. For this example, $p = 8$ is the best choice. If the choice of p is increased further, the final error is likely to increase. Therefore, we should consider increasing the value of p appropriately. For the case of unknown analytical solution, the user can consider comparing the error between two adjacent p-values, and if the error corresponding to two adjacent p-values at a certain p is the smallest, such a result can be accepted.

```
>> t10=[0.5:0.5:5]';
   y10=t10.^-0.6.*ml_func([1,0.4],-t10,0,eps);
   t=0:0.1:5; y=exp(-t); ii=[6:5:51]; T=[];
   for p=7:11, y1=glfdiff9(y,t,0.6,p); T=[T [y1(ii)-y10]]; end
```

◀

▶ Example 4.5

Find the 0.6th-order integral of the exponential function e^{-t} in the interval [0, 5].

Solutions The exact solution of 0.6th-order integral of the exponential function is known to be $t^{0.6}E_{1,1.6}(-t)$. The error between the numerical integral and the exact solution for different p is obtained by choosing the step size $h = 0.01$, which is shown in Table 4.8. It can be seen that the error of the numerical integral is similar to that of the numerical derivatives in Example 4.4.

Table 4.8 Comparison of the error of numerical integral at different p

t	$p=1$	$p=2$	$p=3$	$p=4$	$p=5$	$p=6$
0.5	0.00057	-3.61×10^{-6}	2.62×10^{-8}	-2.02×10^{-10}	1.62×10^{-12}	-1.32×10^{-14}
1	0.00146	-9.51×10^{-6}	7.01×10^{-8}	-5.51×10^{-10}	4.50×10^{-12}	-3.76×10^{-14}
1.5	0.00240	-1.57×10^{-5}	1.16×10^{-7}	-9.20×10^{-10}	7.57×10^{-12}	-6.43×10^{-14}
2	0.00330	-2.17×10^{-5}	1.61×10^{-7}	-1.27×10^{-9}	1.05×10^{-11}	-8.97×10^{-14}
2.5	0.00415	-2.73×10^{-5}	2.03×10^{-7}	-1.61×10^{-9}	1.33×10^{-11}	-1.13×10^{-13}
3	0.00494	-3.26×10^{-5}	2.42×10^{-7}	-1.92×10^{-9}	1.59×10^{-11}	-1.37×10^{-13}
3.5	0.00567	-3.74×10^{-5}	2.79×10^{-7}	-2.21×10^{-9}	1.83×10^{-11}	-1.58×10^{-13}
4	0.00635	-4.20×10^{-5}	3.12×10^{-7}	-2.48×10^{-9}	2.05×10^{-11}	-1.80×10^{-13}
4.5	0.00699	-4.62×10^{-5}	3.44×10^{-7}	-2.73×10^{-9}	2.27×10^{-11}	-2.03×10^{-13}
5	0.00759	-5.02×10^{-5}	3.74×10^{-7}	-2.97×10^{-9}	2.47×10^{-11}	-2.19×10^{-13}

```
>> t10=[0.5:0.5:5]';
   y10=t10.^0.6.*ml_func([1,1.6],-t10,0,eps);
   t=0:0.01:5; y=exp(-t); ii=[51:50:501]; T=t10;
   for p=1:6
       y1=glfdiff9(y,t,-0.6,p); T=[T [y1(ii)-y10]];
   end
```

◀

▶ Example 4.6

Try the new algorithm `glfdiff9()` given here to find the second-order derivative of the exponential function e^{-t}.

Solutions It is theoretically known that the second-order derivative of the exponential function e^{-t} is itself. Now choose different values of p for the numerical derivatives and obtain the results of the error at the step size $h = 0.01$, as shown in Table 4.9. It can be seen that the algorithm can also be used for high-precision integer-order numerical derivatives and still yield very high approximate accuracy. In addition, the length of the derivative vector is the same as the original function vector.

```
>> t10=[0.5:0.5:5]'; y10=exp(-t10); T=[];
   t=0:0.01:5; y=exp(-t); ii=[51:50:501];
   for p=1:6, y1=glfdiff9(y,t,2,p); T=[T [y1(ii)-y10]]; end
```

◀

Table 4.9 Computational errors of second-order derivatives of e^{-t} with different p

t	$p=1$	$p=2$	$p=3$	$p=4$	$p=5$	$p=6$
0.5	0.00610	-4.07×10^{-5}	3.07×10^{-7}	-2.43×10^{-9}	1.71×10^{-10}	6.60×10^{-11}
1	0.00370	-2.47×10^{-5}	1.86×10^{-7}	-1.48×10^{-9}	1.23×10^{-11}	-1.18×10^{-10}
1.5	0.00224	-1.50×10^{-5}	1.13×10^{-7}	-9.19×10^{-10}	3.42×10^{-10}	-7.17×10^{-11}
2	0.00136	-9.09×10^{-6}	6.85×10^{-8}	-5.91×10^{-10}	2.96×10^{-10}	7.21×10^{-11}
2.5	0.00082	-5.51×10^{-6}	4.15×10^{-8}	-5.08×10^{-10}	-7.35×10^{-11}	-2.28×10^{-9}
3	0.00050	-3.34×10^{-6}	2.54×10^{-8}	-1.43×10^{-9}	1.37×10^{-10}	-3.10×10^{-9}
3.5	0.00030	-2.03×10^{-6}	1.51×10^{-8}	-1.13×10^{-9}	2.86×10^{-10}	-8.86×10^{-9}
4	0.00018	-1.23×10^{-6}	9.45×10^{-9}	-2.51×10^{-9}	6.92×10^{-10}	-2.25×10^{-8}
4.5	0.00011	-7.46×10^{-7}	5.97×10^{-9}	-4.27×10^{-9}	5.35×10^{-9}	-3.77×10^{-8}
5	6.78×10^{-5}	-4.52×10^{-7}	4.84×10^{-9}	-7.35×10^{-9}	1.40×10^{-8}	-1.02×10^{-7}

4.3.4 Revisit to the Matrix Algorithm

Since there is a compensation term at the initial moment, (3.21) cannot be used directly to implement the high-precision algorithm. It is also necessary to introduce the auxiliary signal $u(t)$ shown in (4.36), whose coefficients can still be derived from (4.37). At this point, the Riemann–Liouville fractional-order derivative of the signal $f(t)$ can be written as

$$ {}_{t_0}^{\mathrm{RL}}\mathscr{D}^{\alpha} f(t) = {}_{t_0}^{\mathrm{RL}}\mathscr{D}^{\alpha} u(t) + {}_{t_0}^{\mathrm{RL}}\mathscr{D}^{\alpha} v(t). \tag{4.41} $$

The direct formula can be written in a matrix form for the high-precision Riemann–Liouville fractional-order derivative algorithm.

$$ {}_{t_0}^{\mathrm{RL}}\mathscr{D}^{\alpha} f(t) = \frac{1}{h^{\alpha}} \boldsymbol{W}\boldsymbol{v} + \sum_{k=0}^{p} \frac{\Gamma(k+1)c_k}{\Gamma(k+1-\alpha)} (t-t_0)^{k-\alpha}, \tag{4.42} $$

where $\boldsymbol{W}$ is a matrix composed of high-precision coefficients w_k, still in the form of (4.26). Vector $\boldsymbol{v}$ is vector evaluated from the signal $v(t) = y(t) - u(t)$. It can be seen that if the matrix algorithm is implemented, the resulting result will be exactly the same as Algorithm 4.3.

Substituting the command `for i=1:n, dv(i,1)=w(1:i)*v(i:-1:1)/h^gam; end` in `glfdiff9()` function into `dv=rot90(hankel(w(end:-1:1)))*v/h^gam`, the high-precision matrix function `glfdiff9_mat()` can be written. The syntax and results are identical to `glfdiff9()` function. The function reads as follows:

```
function dy=glfdiff9_mat(y,t,gam,p)
   arguments
      y(:,1), t(:,1) double, gam(1,1) double
      p(1,1){mustBePositiveInteger}=5
   end
   [y,h,n]=fdiffcom(y,t); u=0; du=0; r=(0:p)*h;
   R=sym(fliplr(vander(r))); c=double(R\y(1:p+1));
   for i=1:p+1, u=u+c(i)*t.^(i-1);
      du=du+c(i)*t.^(i-1-gam)*gamma(i)/gamma(i-gam);
   end
   v=y-u; g=double(genfunc(p)); w=get_vecw(gam,n,g);
   dv=rot90(hankel(w(end:-1:1)))*v/h^gam;
   dy=dv+du; if abs(y(1))<1e-10, dy(1)=0; end
end
```


4.4 High-Precision Algorithm for Caputo Derivatives

The major difference between the Grünwald–Letnikov and Caputo definitions is the treatment of initial values. This section analyzes the cause of this difference and gives a signal decomposition method.

4.4.1 The Algorithm and Its Implementation

As explained before, the Caputo fractional-order integral is fully equivalent to the Riemann–Liouville integral, so the Caputo fractional-order integral can be computed directly with the high-precision solver `glfdiff9()`. For the Caputo fractional-order derivatives, the following algorithm should be adopted to evaluate the Caputo fractional-order derivatives.

Algorithm 4.4 High-precision Caputo derivative algorithm
(1) Compute high-precision Grünwald–Letnikov derivative with `glfdiff9()`.
(2) Compute the compensating function from (3.76).
(3) Combine Grünwald–Letnikov derivative and compensating function to yield high-precision Caputo derivative.

Based on the above idea, the MATLAB function `caputo9()` can be written to compute the numerical solution of the high-precision Caputo fractional-order derivatives.

```
function dy=caputo9(y,t,gam,p)
   arguments, y(:,1), t(:,1) double, gam(1,1) double
      p(1,1) {mustBePositiveInteger}=5
   end
   if gam<0, dy=glfdiff9(y,t,gam,p); return; end
```

```
    h=t(2)-t(1); q=ceil(gam);
    r=max(p,q); R=sym(fliplr(vander((0:(r-1))'*h)));
    c=double(R\y(1:r)); u=0; du=0;
    for i=1:r, u=u+c(i)*t.^(i-1); end
    if q<r
       for i=(q+1):p
          du=du+c(i)*t.^(i-1-gam)*gamma(i)/gamma(i-gam);
    end, end
    v=y-u; dv=glfdiff9(v,t,gam,p); dy=dv+du;
end
```

4.4.2 Testing and Assessment of the Algorithm

As in the previous test of the Grünwald–Letnikov high-precision algorithm, the Caputo derivative is still computed here for functions with known analytical solutions. The actual accuracy of the solution algorithm is evaluated.

▶ Example 4.7

Evaluate the 0.6th-order Caputo derivative of the exponential function e^{-t}. Compare the accuracy under the step sizes $h = 0.01$ and $h = 0.1$.

Solutions From (3.60), since $\alpha = 0.6$, $\lambda = -1$, we have $q = \lceil \alpha \rceil = 1$, and $\gamma = q - \alpha = 0.4$, so that the analytical solution of the 0.6th-order Caputo fractional-order derivative of the exponential function is

$$y_0(t) = {}_0^{\mathrm{C}}\mathscr{D}_t^{\alpha} e^{-t} = \lambda^q t^{\gamma} \mathrm{E}_{1,\gamma+1}(\lambda t) = -t^{0.4}\mathrm{E}_{1,1.4}(-t).$$

Choosing a step size $h = 0.01$, the Caputo fractional-order derivatives for different values of p can be found by the following statement. Comparisons are made on the errors of the numerical solution in Table 4.10. It can be seen that the maximum error may be reduced as small as 10^{-13}, which is many orders of magnitude higher than the existing algorithms.

```
>> t0=[0.5:0.5:5]'; q=1; gam=q-0.6;
   t=0:0.01:5; y=exp(-t); T=t0;
   y0=-t0.^0.4.*ml_func([1,1.4],-t0,0,eps); ii=[51:50:501];
   for p=1:6
       y1=caputo9(y,t,0.6,p); T=[T [y1(ii)-y0]];
   end
```

If a larger step size $h = 0.1$ is used, the calculation error for each p choice can be measured from the following statements, as shown in Table 4.11.

```
>> t=[0:0.1:5]'; y=exp(-t); ii=[6:5:51]; T=t0;
   for p=4:9, y1=caputo9(y,t,0.6,p); T=[T [y1(ii)-y0]]; end
```

It can be seen that despite the choice of a large step size $h = 0.1$, the computational accuracy is still high, so the algorithm and implementation given here are efficient and of high precision. ◀

Table 4.10 Calculation errors for step size of $h = 0.01$

t	$p=1$	$p=2$	$p=3$	$p=4$	$p=5$	$p=6$
0.5	−0.0018	1.19×10^{-5}	-8.89×10^{-8}	7.07×10^{-10}	-5.85×10^{-12}	4.80×10^{-14}
1	−0.0017	1.15×10^{-5}	-8.59×10^{-8}	6.85×10^{-10}	-5.69×10^{-12}	4.85×10^{-14}
1.5	−0.0015	1.01×10^{-5}	-7.52×10^{-8}	6.00×10^{-10}	-4.99×10^{-12}	3.97×10^{-14}
2	−0.0012	8.61×10^{-6}	-6.45×10^{-8}	5.15×10^{-10}	-4.28×10^{-12}	4.26×10^{-14}
2.5	−0.0011	7.39×10^{-6}	-5.53×10^{-8}	4.41×10^{-10}	-3.70×10^{-12}	4.70×10^{-14}
3	−0.0009	6.40×10^{-6}	-4.78×10^{-8}	3.82×10^{-10}	-3.25×10^{-12}	1.18×10^{-14}
3.5	−0.0008	5.61×10^{-6}	-4.20×10^{-8}	3.35×10^{-10}	-2.96×10^{-12}	4.46×10^{-14}
4	−0.0007	4.99×10^{-6}	-3.73×10^{-8}	2.98×10^{-10}	-2.92×10^{-12}	3.52×10^{-14}
4.5	−0.0006	4.50×10^{-6}	-3.37×10^{-8}	2.68×10^{-10}	-2.72×10^{-12}	7.61×10^{-13}
5	−0.0006	4.11×10^{-6}	-3.07×10^{-8}	2.45×10^{-10}	-3.41×10^{-12}	-2.75×10^{-13}

Table 4.11 Calculation errors for step size of $h = 0.1$

t	$p=4$	$p=5$	$p=6$	$p=7$	$p=8$	$p=9$
0.5	4.53×10^{-6}	-1.98×10^{-7}	-3.07×10^{-9}	-8.17×10^{-11}	-2.97×10^{-12}	-1.30×10^{-13}
1	5.98×10^{-6}	-4.73×10^{-7}	3.74×10^{-8}	-2.91×10^{-9}	2.48×10^{-10}	-2.03×10^{-11}
1.5	5.51×10^{-6}	-4.47×10^{-7}	3.73×10^{-8}	-3.12×10^{-9}	2.48×10^{-10}	-2.10×10^{-11}
2	4.77×10^{-6}	-3.90×10^{-7}	3.28×10^{-8}	-2.85×10^{-9}	2.53×10^{-10}	-1.63×10^{-11}
2.5	4.08×10^{-6}	-3.34×10^{-7}	2.81×10^{-8}	-2.45×10^{-9}	2.50×10^{-10}	-4.51×10^{-11}
3	3.51×10^{-6}	-2.87×10^{-7}	2.41×10^{-8}	-2.06×10^{-9}	1.52×10^{-10}	-6.34×10^{-12}
3.5	3.05×10^{-6}	-2.48×10^{-7}	2.08×10^{-8}	-1.76×10^{-9}	4.49×10^{-11}	2.35×10^{-10}
4	2.69×10^{-6}	-2.18×10^{-7}	1.82×10^{-8}	-1.55×10^{-9}	2.27×10^{-10}	-6.05×10^{-10}
4.5	2.40×10^{-6}	-1.95×10^{-7}	1.62×10^{-8}	-1.39×10^{-9}	4.94×10^{-10}	-1.91×10^{-9}
5	2.17×10^{-6}	-1.76×10^{-7}	1.46×10^{-8}	-1.26×10^{-9}	-2.17×10^{-10}	1.14×10^{-8}

▶ Example 4.8

Solve again the problem in Example 3.12.

Solutions As can be seen from Example 3.12, since the first few calculated points show strong oscillations, it is possible to recalculate these points using spline interpolation in the

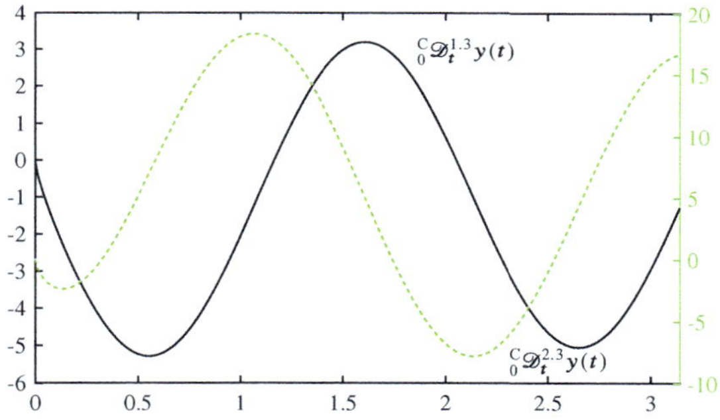

Fig. 4.3 Newly evaluated Caputo fractional-order derivatives

caputo() function, which inevitably introduces new or even unknown errors. A reliable numerical solution can be derived with the high-precision solver, as shown in Fig. 4.3.

```
>> t=0:0.01:pi; y=sin(3*t+1);
   y1=caputo9(y,t,1.3,5); y2=caputo9(y,t,2.3,5);
   yyaxis left, plot(t,y1), yyaxis right, plot(t,y2,'--')
```

It can be seen that with the new function, the information such as $y(0)$, $y'(0)$ and $y''(0)$ is no longer required. The method given here is more reliable and convenient. ◄

4.4.3 Solutions of a Benchmark Problem

In order to evaluate the accuracy and speed of numerical algorithms, the authors have designed a series of more challenging benchmark problems for testing numerical algorithms [5]. These benchmark problems are listed in Appendix A. The first of these problems is used to test the solution algorithm for the numerical Caputo integral problem. The benchmark problem is taken as an example in this section. The actual computational results based on the solution method are measured and assessed.

► Example 4.9

Evaluate the numerical Caputo integral $y(t) = {}_0^{\rm C}\mathscr{D}_t^{-0.7} f(t) + y(0)$ of the following piecewise function, where

$$f(t) = \begin{cases} \dfrac{1}{\Gamma(1.3)} t^{0.3}, & 0 \leqslant t \leqslant 1 \\ \dfrac{1}{\Gamma(1.3)} t^{0.3} - \dfrac{2}{\Gamma(2.3)} (t-1)^{1.3}, & t > 1, \end{cases} \tag{4.43}$$

where $0 \leqslant t \leqslant 2$. The initial value is $y(0) = 1$, and the exact solution of the problem is also a piecewise function

$$y(t) = \begin{cases} t+1, & 0 \leqslant t \leqslant 1 \\ t+1-(t-1)^2, & t > 1. \end{cases} \tag{4.44}$$

Solutions The following statements can be used to describe the original function and its solution function by anonymous functions. The function curves are shown in Fig. 4.4.

```
>> f=@(t)1/gamma(1.3)*t.^0.3-2/gamma(2.3)*(t-1).^1.3.*(t>1);
   y=@(t)t+1-(t-1).^2.*(t>1); fplot({f,y},[0,2])
```

Taking a step size $h = 0.01$, the exact values of the fractional-order integral can be evaluated from the following statements. The numerical integral results for $p = 1, 5, 6$, and 7 can be obtained as shown in Fig. 4.5, where the solid line is the exact value and the dashed lines are the numerical solutions for different values of p. For this specific example, the numerical integral at $p = 1$ is far from the exact value. As p increases, the numerical integral gradually converges to the exact value, with $p = 7$ being the closest to the exact one. If the value of p continues to increase, numerical instability may occur. It can be seen that for this benchmark problem, the traditional numerical integral algorithm may lead to great computational errors, while the high-precision ones yield more accurate results.

```
>> h=0.01; t=[0:h:2]'; p0=[1,5,6,7]; y0=f(t); fplot(y,[0,2])
   for p=p0, y1=caputo9(y0,t,-0.7,p)+1; line(t,y1); end
```

To further test the effect of the computational parameters on the accuracy, different step sizes h and orders p are tried. The resulting error paradigms are given in Table 4.12. Note that for the sake of fairness, the parametric index defined here is the norm of errors divided by the length of the vector. If $h = 0.001$, then p is too large and may cause the algorithm to fail.

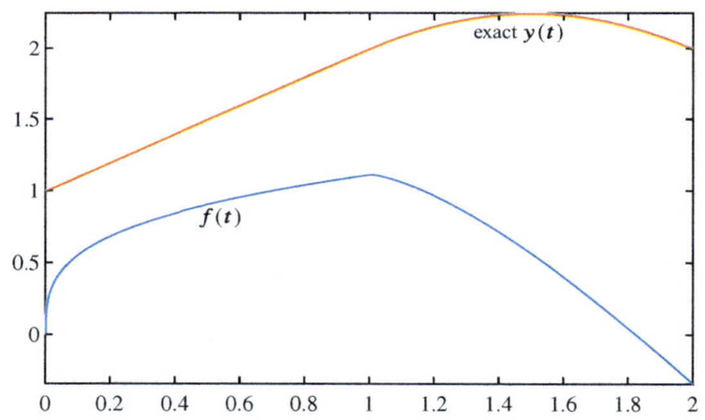

Fig. 4.4 Test function and solution function curves

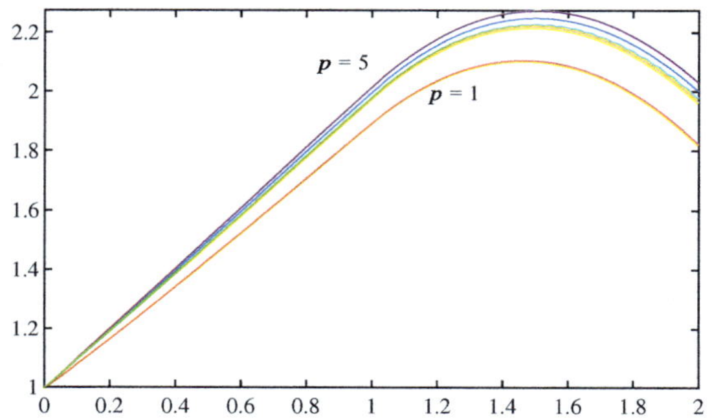

Fig. 4.5 Numerical integral results

Table 4.12 Error norms at different step sizes and orders p

Step sizes	$p=1$	$p=2$	$p=3$	$p=4$	$p=5$	$p=6$	$p=7$
0.001	0.0013	0.00065	0.00044	0.00032	0.00416	0.07026	4.7095×10^{10}
0.005	0.0046	0.00233	0.00159	0.00119	0.00096	0.00079	0.00371
0.01	0.0080	0.00400	0.00275	0.00203	0.00165	0.00134	0.00115
0.05	0.0288	0.01310	0.00939	0.00632	0.00545	0.00389	0.00370
0.1	0.0504	0.01981	0.01515	0.00901	0.00833	0.00503	0.00527

```
>> h=0.01; t=[0:h:2]'; T=h; y0=f(t); % the value 0.01 can be changed
   for p=1:7
      y1=caputo9(y0,t,-0.7,p)+1; T=[T,norm(y1-y(t))/length(y1)];
   end
```

The error norm is sometimes affected by the vector length and does not necessarily portray the actual error situation better. For example, a vector length of 201 at $h=0.01$ often has a larger norm than the calculated result for $h=0.1$ (vector length of 21). Other important measures, such as the maximum error, also need to be tested. If the above error calculation statement `norm(`y_1`-`$y(t)$`)/length(`t`)` is replaced by `max(abs(`y_1`-`$y(t)$`))`, the maximum error at each step size can be recorded, listed in Table 4.13. Considering these two error measures together, it can be found that the combination of parameters in Fig. 4.5 ($h=0.01, p=7$) yields the best results for this benchmark problem.◀

4.5 Computing of Higher Fractional-Order Derivatives

The algorithms presented earlier generally do not support higher orders of derivatives, such as the 3.6th-order derivative of a known function. Direct application of the previously presented methods may lead to erroneous results. In this section, the high-precision solver for the high-order integer-order derivative is studied first. Then the higher fractional-order derivatives of the result can be obtained.

Table 4.13 Maximum errors at different step sizes h and orders p

Step sizes	$p=1$	$p=2$	$p=3$	$p=4$	$p=5$	$p=6$	$p=7$
0.001	0.08815	0.04494	0.030568	0.021384	0.687540	14.009	2.6581×10^{13}
0.005	0.14418	0.07245	0.049399	0.037114	0.029971	0.025073	0.415880
0.01	0.17907	0.08861	0.060603	0.045222	0.036611	0.029983	0.026924
0.05	0.30488	0.13531	0.095335	0.067152	0.056391	0.042727	0.039269
0.1	0.39326	0.15229	0.112640	0.073125	0.064551	0.043361	0.041784

4.5.1 High-Precision Algorithms for Higher Integer-Order Derivatives

If one is looking for integer-order numerical derivative results with high-precision $O(h^p)$, one can also use the `num_diff()` function given in Ref. [6]:

```
[z,t]=num_diff(y,h,n,p)
```

where the nth-order numerical derivative of $\boldsymbol{y}$ is returned and $\boldsymbol{t}$ is the relative time vector. The actual time vector should be $\boldsymbol{t} = t_0+\boldsymbol{t}$, t_0 being the initial moment. Limited to the length of this book, the function's listing is not provided. Interested readers should refer to Ref. [6] for the theoretical description and code implementation, or refer to the source code of the function in the FOTF Toolbox of this book.

4

► Example 4.10

Solve again the second-order derivative problem in Example 4.6, using the integer-order forward-difference algorithm.

Solutions If the second-order derivative problem of $y(t) = \mathrm{e}^{-t}$ is solved again, using an integer-order numerical derivative algorithm, the absolute values of the resulting errors are shown in Table 4.14. It can be seen that the accuracy of such a numerical solution is much higher than the accuracy obtained by Example 4.6 (compare Table 4.9).

```
>> t10=[0.5:0.5:4.5]'; y10=exp(-t10);
   t=0:0.01:5; y=exp(-t); ii=[51:50:500]; T=[];
   for p=1:6, y1=num_diff(y,0.01,2,p); T=[T abs(y1(ii)-y10)]; end
```

Table 4.14 Errors of second-order derivatives with numerical derivative solver

t	$p=1$	$p=2$	$p=3$	$p=4$	$p=5$	$p=6$
0.5	0.0060	5.499×10^{-5}	4.974×10^{-7}	-4.545×10^{-9}	1.123×10^{-11}	2.353×10^{-12}
1	0.0036	3.335×10^{-5}	3.017×10^{-7}	2.750×10^{-9}	2.819×10^{-11}	3.769×10^{-12}
1.5	0.0022	2.023×10^{-5}	1.830×10^{-7}	1.661×10^{-9}	2.128×10^{-12}	1.119×10^{-11}
2	0.0013	1.227×10^{-5}	1.110×10^{-7}	1.004×10^{-9}	7.884×10^{-12}	2.342×10^{-11}
2.5	0.0008	7.442×10^{-6}	6.733×10^{-8}	6.162×10^{-10}	1.283×10^{-11}	2.005×10^{-11}
3	0.0005	4.514×10^{-6}	4.083×10^{-8}	3.734×10^{-10}	3.193×10^{-12}	4.177×10^{-13}
3.5	0.0003	2.738×10^{-6}	2.476×10^{-8}	2.266×10^{-10}	1.569×10^{-12}	1.206×10^{-12}
4	0.0001	1.660×10^{-6}	1.502×10^{-8}	1.350×10^{-10}	2.641×10^{-12}	5.971×10^{-12}
4.5	0.0001	1.007×10^{-6}	9.111×10^{-9}	8.224×10^{-11}	1.229×10^{-12}	3.519×10^{-12}

◄

▶ **Example 4.11**

If $x \in (1.5, 3.5)$, and the step size is $h = 0.02$, generate samples from

$$f(x) = \frac{1}{2}\ln(x+1) - \frac{1}{4}\ln(x^2 - x + 1) + \frac{1}{\sqrt{3}}\arctan\frac{2x-1}{\sqrt{3}},$$

and then evaluate the first- through the seventh-order numerical derivatives. Compare the results with exact ones, find the maximum errors with the solver.

Solutions If the accuracy p is chosen, the sample points can be generated directly from the following statements, and then the solutions of the numerical derivatives of the function at each order can be found from the sample points. Since the mathematical expression of the function is known, it is also possible to find the exact values at these sample points. In this way, the errors for each order of numerical derivatives are listed in Table 4.15. It can be seen that the results of the seventh-order numerical derivative are far from the exact values and cannot be used. Smaller values of h can be tried. The sixth-order derivatives are shown in Fig. 4.6. Only a very slight difference can be witnessed in the resulting curves. If one wants to obtain usable seventh-order numerical derivative results, consider reducing the step size.

```
>> syms x; h=0.02; x0=1.5:h:3.5;
   f(x)=log(1+x)/2-log(x^2-x+1)/4+atan((2*x-1)/sqrt(3))/sqrt(3);
   y0=double(f(x0));       % generate samples
   for n=1:7               % find derivative with numerical method
      [z,t]=num_diff(y0,h,n,6); t=1.5+t;     % derivative
      f1=diff(f,n); y1=double(f1(t));        % exact values
      norm(z-y1), max(abs(z-y1))             % errors
   end
   [z,t]=num_diff(y0,h,6,6); t=1.5+t;        % compute 6th order derivative
   f1=diff(f,6); y1=double(f1(t)); plot(t,y1,t,z,'--')
```

◀

4.5.2 Computing of Higher Fractional-Order Derivatives

Before introducing this section, an example for finding a higher fractional-order derivative is shown to demonstrate the failure process of the derivative algorithm introduced earlier. The reason is analyzed, and feasible alternative methods are proposed and demonstrated.

Table 4.15 Errors of derivatives of different orders

Derivative order	1	2	3	4	5	6	7
Error norm	9.47×10^{-10}	2.94×10^{-8}	4.45×10^{-7}	6.58×10^{-6}	0.0011	0.216	29.4
Maximum error	4.69×10^{-10}	1.28×10^{-8}	1.62×10^{-7}	1.93×10^{-6}	3.27×10^{-4}	0.049	7.32

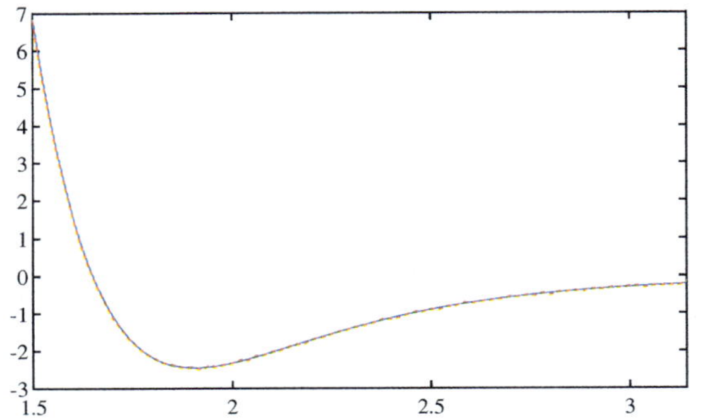

Fig. 4.6 Sixth-order derivative (solid line for exact, dashed line is for numerical)

▶ Example 4.12

For function $f(t) = \mathrm{e}^{-t}$, find its 3.6th-order derivative, ${}_0^{\mathrm{C}}\mathscr{D}_t^{3.6}f(t)$.

Solutions With Theorem 3.6, denote that $m = \lceil \alpha \rceil = 4$ and $\gamma = m - \alpha = 0.4$, the exact Caputo fractional-order derivative is given by

$$ {}_0^{\mathrm{C}}\mathscr{D}_t^{3.6}\mathrm{e}^{-t} = t^{0.4}\mathrm{E}_{1,1.4}(-t). \tag{4.45}$$

Consider using the `caputo9()` function to directly find the 3.6th-order derivative of the function. On comparing it with the exact solutions, the results are obtained, as shown in Fig. 4.7. Obviously, although a high-precision fractional-order derivative algorithm is used, the numerical derivatives thus derived deviate significantly from the exact values.

```
>> h=0.01; t=[0:h:1]'; f=@(t)exp(-t);
   y1=caputo9(f,t,3.6,4); y0=t.^0.4.*ml_func([1,1.4],-t);
   plot(t,y1,t,y0)
```

◀

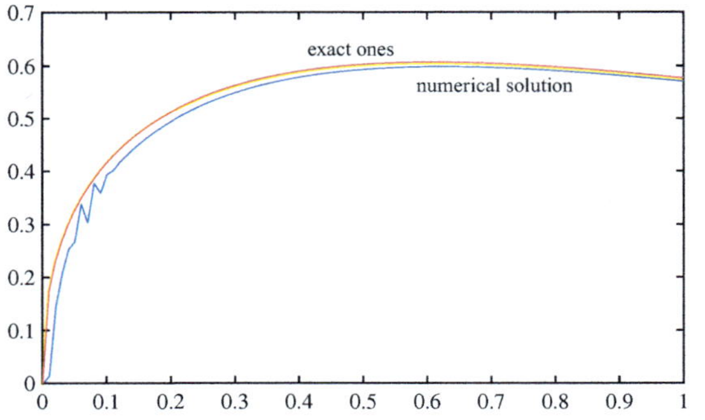

Fig. 4.7 Calculation results of using `caputo9()` function directly

Why is there such a large deviation in the calculation of higher fractional-order derivatives? The high-precision algorithm introduced in the previous sections is based on the generating functions given in Tables 4.1 and 4.2, which are derived by the first-order derivative formulas. In fact, in addition to the first-order derivatives, there are generating functions for high-order derivatives, see Table 4.16 [6].

Table 4.16 High-precision backward-difference algorithm coefficients

n	p	y_k	y_{k-1}	y_{k-2}	y_{k-3}	y_{k-4}	y_{k-5}	y_{k-6}	y_{k-7}	y_{k-8}
	$O(h)$	1	−1							
	$O(h^2)$	3/2	−2	1/2						
	$O(h^3)$	11/6	−3	3/2	−1/3					
1	$O(h^4)$	25/12	−4	3	−4/3	1/4				
	$O(h^5)$	137/60	−5	5	−10/5	5/4	−1/5			
	$O(h^6)$	49/20	−6	15/2	−20/3	15/4	−6/5	1/6		
	$O(h^7)$	363/140	−7	21/2	−35/3	35/4	−21/5	7/6	−1/7	
	$O(h^8)$	761/280	−8	14	−56/3	35/2	−56/5	14/3	−8/7	1/8
	$O(h)$	1	−2	1						
	$O(h^2)$	2	−5	4	−1					
	$O(h^3)$	35/12	−26/3	19/2	−14/3	11/12				
2	$O(h^4)$	15/4	−77/6	107/6	−13	61/12	−5/6			
	$O(h^5)$	203/45	−87/5	117/4	−254/9	33/2	−27/5	137/180		
	$O(h^6)$	469/90	−223/10	879/20	−949/18	41	−201/10	1019/180	−7/10	
	$O(h)$	1	−3	3	−1					
	$O(h^2)$	5/2	−9	12	−7	3/2				
3	$O(h^3)$	17/4	−71/4	59/2	−49/2	41/4	−7/4			
	$O(h^4)$	49/8	−29	461/8	−62	307/8	−13	15/8		
	$O(h^5)$	967/120	−638/15	3929/40	−389/3	2545/24	−268/5	1849/120	−29/15	
	$O(h)$	1	−4	6	−4	1				
	$O(h^2)$	3	−14	26	−24	11	−2			
4	$O(h^3)$	35/6	−31	137/2	−242/3	107/2	−19	17/6		
	$O(h^4)$	28/3	−111/2	142	−1219/6	176	−185/2	82/3	−7/2	
5	$O(h^4)$	27/2	−575/6	895/3	−1065/2	1790/3	−2581/6	195	−305/6	35/6
6	$O(h^3)$	39/4	−73	239	−447	1045/2	−391	183	−49	23/4

Obviously, the generating function given earlier is only a special case of Table 4.16 (the first part when $n = 1$). The generating functions corresponding to the tables of different order derivatives are only suitable for the evaluation of that order derivative, but not for the calculation of other order derivatives. The phenomenon of Example 4.10 is for the same reason, because Example 4.10 is calculated directly from the second-order numerical derivative formula, while Example 4.6 is calculated from the formula obtained by extending the first-order derivative, so the former is much more accurate.

Reference [6] provides the function `fdcoef()` for finding the coefficients of the difference algorithm, which is called in the form `c=fdcoef(n,n+p,α)`, where the integer n is the order of the derivative, p is the precision order of the algorithm, and $\boldsymbol{\alpha}$ is the subscript vector of y in Table 4.16. If you want to use the backward-difference algorithm, take `α=-(0:n+p)`, denoting $y_k, y_{k-1}, \cdots, y_{k-n-p}$. For example, the coefficients in the table of derivative order $n = 3$ and the algorithm $O(h^4)$ can be found directly by the following statements:

```
>> n=3; p=4; c=fdcoef(n,n+p,-(0:n+p))
```

The result obtained is $\boldsymbol{c} = [49/8, -29, 461/8, -62, 307/8, -13, 15/8]$, which is exactly the same as the coefficients given in the table. The corresponding formula for this coefficient vector is

$$y_k''' \approx \frac{1}{h^3}\left(\frac{49}{8}y_k - 29y_{k-1} + \frac{461}{8}y_{k-2} - 62y_{k-3} + \frac{307}{8}y_{k-4} - 13y_{k-5} + \frac{15}{8}y_{k-6}\right).$$

The accuracy of the algorithm is $O(h^4)$.

It can be seen that the function can be used to derive the coefficients for the arbitrary combinations of n and p.

If one wants to find a higher fractional-order derivative, for instance, ${}_0^{\mathrm{C}}\mathscr{D}_t^{3.6}y(t)$, two possible methods can be used.

(1) Based on the generating function of the high-order derivative, reconstruct the corresponding high-order derivative solving algorithm. For example, if we want to find the 3.6th-order derivative of a signal, we need to reconstruct the solution algorithm based on the generating function of orders 3 or 4 in Table 4.16. However, this means a large amount of work needed.

(2) First find out the corresponding integer-order numerical derivative. For example, first find out the fourth-order numerical derivative, and then call the function `glfdiff9()` for 0.4th-order integral (see Theorem 3.18), resulting in the required fractional-order derivative. Alternatively, first find out the third-order numerical derivative, and then call the function `caputo9()` for 0.6th-order derivative of the result, resulting in the required 3.6th-order Caputo derivative.

The effect of solving numerical higher fractional-order derivatives is practically tested by the following example.

▶ Example 4.13

Consider the function $f(t) = e^{-t}$ from Example 4.12. Compute again ${}_{0}^{\mathrm{C}}\mathscr{D}_t^{3.6} f(t)$.

Solutions The direct call of caputo9() function failed to solve the problem. Two ways are tried here to compute the 3.6th-order Caputo derivative again: The first method is to call the num_diff() function to find the fourth-order derivative of the original function, and then use Theorem 3.18 to call the glfdiff9() function to find the 0.4th-order Riemann–Liouville integral. The second method is to first find the third-order derivative and then find the 0.6th-order Caputo derivative of the result. The two results obtained are in perfect agreement with the exact curve (omitted) and no difference can be seen on the graph.

```
>> h=0.01; t=[0:h:1]'; f=@(t)exp(-t);
   y0=t.^0.4.*ml_func([1,1.4],-t); p=6;
   [z1,t1]=num_diff(f(t),h,4,p); y1=glfdiff9(z1,t1,-0.4,p);
   [z2,t2]=num_diff(f(t),h,3,p); y2=caputo9(z2,t2,0.6,p);
   plot(t,y0,t1,y1,t2,y2)
```

Due to the use of the num_diff() function, the resulting numerical vector of integer-order derivatives is shorter than the original function vector. Comparing the results with the theoretical values, it can be seen that the two methods yield maximum error values of 4.0337×10^{-5} and 3.4585×10^{-6}, respectively. The latter method yields a comparatively smaller error due to the slightly lower order in the integer-order numerical derivative.

```
>> max(abs(y0(1:length(y1))-y1))   % first 4th order then 0.4th order integral
   max(abs(y0(1:length(y2))-y2))   % first 3rd order, then 0.6th order derivative
```

◀

4.6 Exercises

(1) Write out the backward-difference formula corresponding to the last term in Table 4.16. Use the symbolic computing function provided by MATLAB to prove that the error of the formula for calculating the sixth-order derivative does satisfy $O(h^3)$.

(2) Use the symbolic computing function of MATLAB to calculate the accuracy of the following difference algorithm:

$$y_k' \approx \frac{1}{h}\left(y_k - y_{k-1}\right)$$

$$y_k' \approx \frac{1}{h}\left(\frac{3}{2}y_k - 2y_{k-1} + \frac{1}{2}y_{k-2}\right)$$

$$y_k' \approx \frac{1}{h}\left(\frac{11}{6}y_k - 3y_{k-1} + \frac{3}{2}y_{k-2} - \frac{1}{3}y_{k-3}\right).$$

(3) Use the function genfunc() to design a function to calculate the first-order derivative, and give examples of evaluations to test the effect of the calculation.

(4) Use MATLAB's symbolic computing to prove the following conclusion, where n and m are nonnegative integers.

$$\frac{1}{2\pi}\int_0^{2\pi} \mathrm{e}^{\mathrm{j}n\varphi}\mathrm{e}^{-\mathrm{j}m\varphi}\mathrm{d}\varphi = \begin{cases} 0, & m \neq n \\ 1, & m = n. \end{cases}$$

(5) Let the functions $y_1(t) = \mathrm{e}^{-t}$, $y_2(t) = \sin t$, $y_3(t) = \cos t$, with step size $h = 0.01$. Construct the $O(h^p)$ auxiliary functions of these functions according to Theorems 4.6, where p is taken as 1, 2, 3, 4, 5. Write MATLAB programs to implement the above functions.

(6) Verify the following equations using the functions `glfdiff9()` and `ml_func()`. Choose different step sizes and orders, and observe the impacts of step size and order on the calculation error.

$${}_0^{\mathrm{RL}}\mathscr{D}_t^{-0.5}\mathrm{e}^{-t} = t^{0.5}\mathrm{E}_{1,1.5}(-t)$$

$${}_0^{\mathrm{RL}}\mathscr{D}_t^{0.5}\mathrm{e}^{-t} = t^{-0.5}\mathrm{E}_{1,0.5}(-t).$$

(7) Use functions `glfdiff9()` and `ml_func()` to verify the following equation:

$${}_0^{\mathrm{RL}}\mathscr{D}_t^{-0.7}\cosh\sqrt{2}t = t^{0.7}\mathrm{E}_{2,1.7}(2t^2)$$

$${}_0^{\mathrm{RL}}\mathscr{D}_t^{0.7}\cosh\sqrt{2}t = t^{-0.7}\mathrm{E}_{2,0.3}(2t^2).$$

Choose different step sizes and orders and observe the impacts of step size and order on the calculation error.

(8) Consider the function $f(t) = \sqrt{t\sin t\sqrt{1-\mathrm{e}^{-t}}}$, defined in the domain of $t \in [0, \pi]$. Use the function `caputo9()` to calculate ${}_0^{\mathrm{C}}\mathscr{D}_t^{0.7}f(t)$, and use the function `glfdiff9()` to calculate ${}_0^{\mathrm{RL}}\mathscr{D}_t^{0.7}f(t)$. Compare the results of the two calculations to see if they are similar, and explain why.

(9) Apply the functions `glfdiff9()` and `glfdiff9_mat()` to calculate

$${}_0^{\mathrm{RL}}\mathscr{D}_t^{1.2}\cos\left(t^2 + \pi/3\right), \quad {}_0^{\mathrm{RL}}\mathscr{D}_t^{0.3}\sin\left(\sqrt{t} + \pi/4\right).$$

Compare whether the calculation results of the two algorithms are consistent.

(10) Use functions `glfdiff9()` and `caputo9()` to design two methods to compute ${}_0^{\mathrm{C}}\mathscr{D}_t^{1.5}\Gamma\left(\sin t + 2\right)$, and see whether the results are consistent.

(11) Use the function `glfdiff9()` to verify the following equation:

$${}_0^{\mathrm{RL}}\mathscr{D}_t^{-1.2}\mathrm{e}^{-t} = {}_0^{\mathrm{RL}}\mathscr{D}_t^{-0.5}\left({}_0^{\mathrm{RL}}\mathscr{D}_t^{-0.7}\mathrm{e}^{-t}\right).$$

(12) For the function $f(t) = \sin t/(t^2 + 4t + 3)$, evaluate ${}_0^{\mathrm{C}}\mathscr{D}_t^{1.7}f(t)$. Function `caputo9()` can be used in evaluation directly, or the expression for $f''(t)$ can be calculated by MATLAB's symbolic computing. Then applying the function `glfdiff9()` computes ${}_0^{\mathrm{RL}}\mathscr{D}_t^{-0.3}f''(t)$. Apply these two methods to calculate this expression and check the results of the calculation consistent or not.

(13) The original problem given in Appendix A is ${}_{0}^{\mathrm{C}}\mathscr{D}_{t}^{0.7}y(t)=f(t)$, while Example 4.9 converts it into a fractional-order integral problem: $y(t) = {}_{0}^{\mathrm{C}}\mathscr{D}_{t}^{-0.7}f(t) + y(0)$. Prove that such a conversion is correct.

(14) Use the functions `caputo9()` and `ml_func()` to verify the following equation, where $\alpha>0$, $m=\lceil\alpha\rceil$.

$$ {}_{0}^{\mathrm{C}}\mathscr{D}_{t}^{\alpha}\sin\lambda t=\frac{t^{m-\alpha}}{2\mathrm{j}}\left[(\mathrm{j}\lambda)^{m}\mathrm{E}_{1,1+m-\alpha}(\mathrm{j}\lambda t)-(-\mathrm{j}\lambda)^{m}\mathrm{E}_{1,1+m-\alpha}(-\mathrm{j}\lambda t)\right] $$

$$ {}_{0}^{\mathrm{C}}\mathscr{D}_{t}^{\alpha}\cos\lambda t=\frac{t^{m-\alpha}}{2}\left[(\mathrm{j}\lambda)^{m}\mathrm{E}_{1,1+m-\alpha}(\mathrm{j}\lambda t)+(-\mathrm{j}\lambda)^{m}\mathrm{E}_{1,1+m-\alpha}(-\mathrm{j}\lambda t)\right]. $$

(15) For the function $f(t)=\sqrt{2}\sin(t+\pi/4)$, prove that

$$ {}_{0}^{\mathrm{C}}\mathscr{D}_{t}^{4-\sqrt{\pi}}f(t)+{}_{0}^{\mathrm{C}}\mathscr{D}_{t}^{2-\sqrt{\pi}}f(t)=0. $$

(16) A function $y=f(t)$ at the sampling point takes the values shown in Table 4.17. Use the MATLAB functions provided in this chapter to calculate the values of ${}_{0}^{\mathrm{RL}}\mathscr{D}_{t}^{-2.3}f(t)$, ${}_{0}^{\mathrm{RL}}\mathscr{D}_{t}^{0.5}f(t)$, and ${}_{0}^{\mathrm{C}}\mathscr{D}_{t}^{1.3}f(t)$.

(17) Use the function `caputo9()` to calculate the following expressions:

$$ {}_{0}^{\mathrm{C}}\mathscr{D}_{t}^{\sqrt{2}}\mathrm{e}^{-0.1t}\sin(2t+\pi/3). $$

Choose different step sizes and orders and observe whether the calculation results are consistent.

(18) According to the Riemann–Liouville fractional-order integral definition, we can get

$$ {}_{0}^{\mathrm{RL}}\mathscr{D}_{t}^{-\pi}\mathrm{e}^{-t^2/2}=\frac{1}{\Gamma(\pi)}\int_0^t(t-\tau)^{\pi-1}\mathrm{e}^{-\tau^2/2}\mathrm{d}\tau. $$

The function `glfdiff9()` calculates the equation on the left, and then applies the numerical integral algorithm to calculate the equation on the right, comparing whether they are the same.

Table 4.17 Data for Exercise 4.6

t_i	0	0.1	0.2	0.3	0.4	0.5	0.6	0.7	0.8
y_i	0	2.2077	3.2058	3.4435	3.241	2.8164	2.3110	1.8101	1.3602
t_i	0.9	1.0	1.1	1.2					
y_i	0.9817	0.6791	0.4473	0.2768					

References

1. Podlubny I (1999) Fractional differential equations[M]. Academic, San Diego
2. Fornberg B (1988) Generation of finite difference formulas on arbitrarily spaced grids[J]. Math Comput 51(184):699–706
3. Lubich C (1986) Discretized fractional calculus[J]. SIAM J Math Anal 17(3):704–719
4. Bal L, Xue DY (2018) A high-precision recursive algorithm for fractional-order calculus (in Chinese)[J]. J Northeastern Univ (Natural Science Edition), 39(4):604–608
5. Xue DY, Bai L (2017) Benchmark problems for Caputo fractional-order ordinary differential equations[J]. Fract Calc Appl Anal 20(5):1305–1312
6. Xue DY (2021) Calculus problem solutions with MATLAB[M]. De Gruyter Press, Berlin

Approximations of Fractional-Order Operators and Systems

Contents

D. Y. Xue and L. Bai, *Fractional Calculus*, https://doi.org/10.1007/978-981-99-2070-9_5

Dynamical systems are the basis of mathematical models for describing many physical phenomena. From the point of view of system analysis and description, systems can usually be classified into linear and nonlinear systems. Starting from this chapter, the concept of systems will be introduced. The MATLAB Control System Toolbox based tools are introduced first in ► Sect. 5.1 for modeling and analyzing integer-order linear systems. The foundation for the comparison of system behaviors is then established and demonstrated in the section.

In the previous sections describing the numerical computation of the fractional-order derivative $\mathscr{D}^{\alpha} y(t)$, it has been assumed that the function $y(t)$ or its sampled values are known, so that the vector of sampled values of $\boldsymbol{y}$ is first constructed, before its fractional-order derivatives and integrals can be computed. If the fractional-order operator $\mathscr{D}^{\alpha}$ is driven by a previously unknown signal, as in a control system where the fractional-order plant is driven by the controller, then the previously discussed method cannot compute such a fractional-order action. In this case, it is usually necessary to build a device that can simulate the fractional-order behavior to accomplish such a task. With such a device, a signal can be fed into the device, and the output of the device is the desired αth order derivative signal. Of course, the same idea applies to the acquisition of fractional-order integral signals.

If one wants to design such a device, a more efficient and commonly used idea is to design a linear continuous integer-order filter. For example, a linear continuous transfer function model $G(s)$, such that its frequency response approximates the original fractional-order operator model as good as possible.

From the theoretical analysis, the Bode magnitude–frequency characteristics of the fractional-order operator s^{α} has a slope of 20α dB/dec, and the phase–frequency characteristic is a horizontal line with value $\alpha\pi/2$. The asymptote of the magnitude–frequency characteristic of the integer-order continuous transfer function model has a slope of $20k$ dB/dec, and k can only be an integer. Therefore, it is not possible to have an integer-order transfer function model that provides a consistent approximation to the fractional-order behavior over the entire frequency range. So, we have to settle for a filter that approximates the fractional-order behavior in a certain frequency band.

In the early studies of integer-order transfer function approximation, the main methods used were the continued fraction type of approximations, including the continued fraction approximation for the low- and high-frequency parts, the Carlson approximation and the Matsuda–Fujii approximation, and so on. A general introduction to these approximations and filter design is discussed in ► Sect. 5.2. From the present point of view, the continued fraction type of approximation is very ineffective and not very suitable for practical use, because there is no way to specify the frequency band of interest, which is crucial in the numerical simulation of fractional-order systems.

The filter approximation of fractional-order operators proposed by the French scholar Prof. Oustaloup and his colleagues opened a new era in the simulation of complicated fractional-order systems [1]. These filters allow the user to select actively the frequency band and order of interest, so as to approximate the fractional-order operator using an integer-order transfer function model. The design of such filters will be presented in detail in ► Sect. 5.3, and the improved forms of such filters are also presented. If each fractional-order operator in a fractional-order system or transfer function is replaced by this type of filter, the entire fractional-order system can be approximated by a high integer-order transfer function, or it can be approximated by a lower order optimal model reduction technique. A concrete implementation of these approximations is given in ▶ Sect.5.4.

For irrational systems that cannot be described by the standard form of the fractional-order transfer functions (FOTFs), for example, if the system contains a $p^{\gamma}(s)$ term, where $p(s)$ is an integer-order transfer function model, three types of approximation methods are presented in ▶ Sect. 5.5: the first is the time domain approximation of the $1/(\tau s+1)^{\nu}$ implicit model; the second is the design of filters by frequency response fitting; and the third is the Charef filter design techniques. The optimal Charef filter design algorithm and its implementation are also presented in ▶ Sect. 5.5.

In addition to the continuous filter approximation introduced earlier, The approximation of fractional-order operators with discrete filters or discretized filters [2, 3] are also studied in ▶ Sect. 5.6. However, from the perspective of frequency response fitting, the fitted model is often accompanied by extremely high gain at high frequencies, thus making such filters severely effective in amplifying noisy signals and not necessarily suitable for practical applications.

This book collectively refers to various high integer-order transfer functions derived from frequency response fitting as filters, since their role is to allow a function to yield the desired fractional-order operations after fitting the model.

5.1 Representation and Analysis of Linear Integer-Order Models

MATLAB's Control System Toolbox provides powerful support for model input and analysis of linear time-invariant (LTI) integer-order systems. As the underlying support for this chapter, the tools for modeling and analyzing linear integer-order systems in the MATLAB workspace are briefly described here.

5.1.1 Mathematical Model Input and Manipulations

LTI continuous integer-order systems are generally represented by transfer functions and state space equations. This section first introduces the mathematical forms of these two models, and then describes the methods for inputting these models using MATLAB.

Definition 5.1 Integer-order transfer function

The general form of integer-order LTI transfer function is

$$G(s)=\frac{b_1s^m+b_2s^{m-1}+\cdots+b_ms+b_{m+1}}{a_1s^n+a_2s^{n-1}+\cdots+a_ns+a_{n+1}}, \tag{5.1}$$

where $m\leq n$, also referred to as a physically realizable system. The case $m>n$ is not considered in this book.

Definition 5.2 Integer-order state space model

The general form of the integer-order LTI state space model is

$$\begin{cases} \boldsymbol{x}'(t) = \boldsymbol{A}\boldsymbol{x}(t) + \boldsymbol{B}\boldsymbol{u}(t) \\ \boldsymbol{y}(t) = \boldsymbol{C}\boldsymbol{x}(t) + \boldsymbol{D}\boldsymbol{u}(t), \end{cases} \tag{5.2}$$

where $\boldsymbol{x}(t)$, $\boldsymbol{u}(t)$, and $\boldsymbol{y}(t)$ are referred, respectively, as state variable, input, and output vectors. Matrix $\boldsymbol{A}$ is an $n \times n$ matrix, while $\boldsymbol{B}$, $\boldsymbol{C}$, and $\boldsymbol{D}$ are, respectively, $n \times p$, $q \times n$, and $q \times p$ matrices. Integer n is the number of state variables, also known as the order of the system; Integers p and q are, respectively, the numbers of input and output signals in the system.

The transfer function model of the system has two input methods.

(1) First, use `num=`$[b_1, b_2, \ldots, b_{m+1}]$, `den=`$[a_1, a_2, \ldots, a_{n+1}]$ commands to input the numerator and denominator polynomials of the transfer function, and then use `G=tf(num, den)` commands to construct the transfer function object.

(2) The Laplace operator s is defined by the command `s=tf('s')`, and then the transfer function model is input with the expression under MATLAB.

If the state space model is known, i.e., the $\boldsymbol{A}$, $\boldsymbol{B}$, $\boldsymbol{C}$, and $\boldsymbol{D}$ matrices are given, the state space model can be input by the command `G=ss(A, B, C, D)`.

If two models $G_1(s)$ and $G_2(s)$ are known to be connected in series, and these two models are represented by two variables G_1 and G_2 respectively, the overall model of the series system can be found directly by $G{=}G_2 * G_1$. If these two models are connected in parallel, the overall model can be found directly by $G{=}G_1 + G_2$. If these two models are connected with negative feedback structure, where the forward path is G_1 and the feedback path is G_2, the overall system can be found directly by the command `G = feedback(G1, G2)`. The overall model of the positive feedback system should be found with the command `G = feedback(G1, G2, 1)`.

5.1.2 Time and Frequency Domain Responses

If the model variable G of the system is known, the system can be analyzed regardless of the transfer function or state space model. For example, the `bode(G)`, `nyquist(G)`, and `nichols(G)` functions can draw the Bode, Nyquist, and Nichols plots of the system, respectively. The step or impulse response of the system can be plotted by the `step(G)` and `impulse(G)` functions. If the time vector $\boldsymbol{t}$ and the sampling point vector $\boldsymbol{u}$ of the excitation signal are known, the time response curve can be directly plotted by calling `lsim(G, u, t)` to directly plot the time domain response curve, or `y=lsim(G, u, t)` to obtain the time response data vector $\boldsymbol{y}$.

5.1.3 Modeling and Analysis of Linear Fractional-Order Systems

The author's FOTF Toolbox is developed on the basis of the modeling and analysis statement syntaxes discussed above. The system analysis functions and syntaxes are consistent with the functions introduced above as close as possible and can be called directly. The use of the FOTF Toolbox will be introduced later.

5.2 Some Approximation Methods with Continued Fractions

In early research, filters were often designed in the form of continued fraction expansions, and the continued fraction technique was often considered an effective tool for approximating certain nonlinear functions.

5

Definition 5.3 Continued fractions

The general form of continued fraction expansion for function $f(x)$ is

$$f(x)=b_1+\cfrac{(x-a)^{c_1}}{b_2+\cfrac{(x-a)^{c_2}}{b_3+\cfrac{(x-a)^{c_3}}{b_4+\cfrac{(x-a)^{c_4}}{b_5+\cfrac{(x-a)^{c_5}}{\cdots}}}}}, \tag{5.3}$$

where b_i are constants, c_i are rational numbers, and a is considered the reference point of expansion. For integer-order approximations, normally let $c_i = 1$.

5.2.1 Continued Fraction Approximation

It is theoretically impossible to make a direct continued fraction approximation to the fractional-order operator $s^{-\alpha}$. Therefore, the researchers make a continued fraction approximation to $G_1(s) = 1/(1+Ts)^\alpha$ and $G_2(s) = (1+1/s)^\alpha$ in the high- and low-frequency bands, respectively. When $\omega T \gg 1$ is approximated by $G_1(s)$ to $1/s^\alpha$; and $\omega \ll 1$, the integral operator is approximated by $G_2(s)$.

MATLAB's symbolic engine, MuPAD, provides the low-level function `contfrac()`, with which the authors have written an interface function that performs a continued fraction approximation to the given function.

```
function [F,r]=contfrac(f,s,n,a)
   arguments, f(1,1), s(1,1), n(1,1), a(1,1), end
   F=feval(symengine,'contfrac',...
           f,[inputname(2) '=' num2str(a)],n);
   if nargout==2
      r=feval(symengine,'contfrac::rational',F);
end, end
```

The syntax of the function is $[f_1, r]$=`contfrac`(f, s, n, a), where f is the symbolic expression to the given function, s is the independent variable, n is the stage number of the continued fractions, a is the reference point. The returned argument f_1 is the symbolic expression of the continued fraction, and r is the equivalent rational expression.

Unfortunately, the interface functions given here do not work properly because the low-level `contfrac()` function in recent versions of MuPAD no longer seems to support deriving continued fraction expressions for functions. If you want to derive continued fraction expressions, you can try earlier versions such as MATLAB 2018b, or use the Padé approximation technique instead.

▶ Example 5.1

Approximate the fractional-order integral operator $1/\sqrt{s}$ in the high- and low-frequency bands by continued fractions, respectively, and evaluate their fitting quality.

Solutions Select reference point $a = 2$, and number of stage $n = 9$, the fourth-order approximation model can be found.

```
>> syms s; T1=1/(1+s)^0.5; [c1,G2]=contfrac(T1,s,9,2)
```

The resulting continued fraction model is

$$c_1(s) \approx \frac{\sqrt{3}}{3} + \cfrac{s-2}{-6\sqrt{3} + \cfrac{s-2}{-\frac{2\sqrt{3}}{9} + \cfrac{s-2}{-54\sqrt{3} + \cfrac{s-2}{-\frac{2\sqrt{3}}{45} + \cfrac{s-2}{-150\sqrt{3} + \cfrac{s-2}{-\frac{2\sqrt{3}}{105} + \cdots}}}}}}.$$

The corresponding rational function approximation model is

$$G_2(s) = \frac{s^4 + 112s^3 + 1464s^2 + 4864s + 4240}{9\sqrt{3}s^4 + 288\sqrt{3}s^3 + 1944\sqrt{3}s^2 + 4032\sqrt{3}s + 2448\sqrt{3}}.$$

It can be seen that the approximate model derived here is not consistent with the model in Reference [4] shown below.

$$G_h(s) = \frac{0.3513s^4 + 1.405s^3 + 0.8433s^2 + 0.1574s + 0.008995}{s^4 + 1.333s^3 + 0.478s^2 + 0.064s + 0.002844}$$

$$G_l(s) = \frac{s^4 + 4s^3 + 2.4s^2 + 0.448s + 0.0256}{9s^4 + 12s^3 + 4.32s^2 + 0.576s + 0.0256}.$$

One possible reason is the different choice of reference points. Comparison of the Bode diagrams of the approximate model and the original model $1/\sqrt{s}$ can be derived from the following statements, as shown in Fig. 5.1.

```
>> w=logspace(-3,3); s=fotf('s'); G0=s^-0.5; s=tf('s');
   G2=(s^4+112*s^3+1464*s^2+4864*s+4240)/(9*3^(1/2)*s^4+...
      288*3^(1/2)*s^3+1944*3^(1/2)*s^2+4032*3^(1/2)*s+2448*3^(1/2));
   n=[0.3513 1.405 0.8433 0.1574 0.008995];
   d=[1 1.333 0.478 0.064 0.002844]; Gh=tf(n,d);
   n=[1 4 2.4 0.448 0.0256]; d=[9 12 4.32 0.576 0.0256];
   Gl=tf(n,d); H=bode(G0,w); bode(H,G2,Gh,Gl)
   bode(H,'-',G2,'--',Gh,':',Gl,'-.')
```

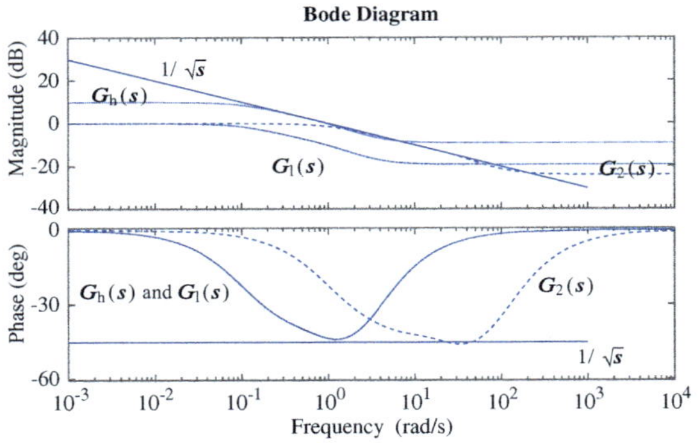

Fig. 5.1 Comparison of Bode plots for continued fraction fitting

It can be seen that the fit of model $G_2(s)$ is slightly better than that of $G_h(s)$, but it is worse in the low-frequency band and the fit is not very good from the whole frequency band. ◀

▶ Example 5.2

Compare the approximate effect of several filters on the 0.5th-order integral of the $f(t) = e^{-t}\sin(3t+1)$ signal with the calculated results of the Grünwald–Letnikov fractional-order integral.

Solutions The following statements can be used to derive the filtering results of these filters, which are evaluated by the `lsim()` function in MATLAB Control System Toolbox. The curves of the filtering results and the theoretical values are shown in Fig. 5.2. It can be seen that the difference between such filter results and the theoretical values is too large, so this filter cannot really be used for the simulation of fractional-order systems.

```
>> t=0:0.01:10; y=exp(-t).*sin(3*t+1);
   y0=glfdiff9(y,t,-0.5,5); y1=lsim(G2,y,t);
   y2=lsim(Gh,y,t); plot(t,y0,'-',t,y1,'--',t,y2,':')
```

◀

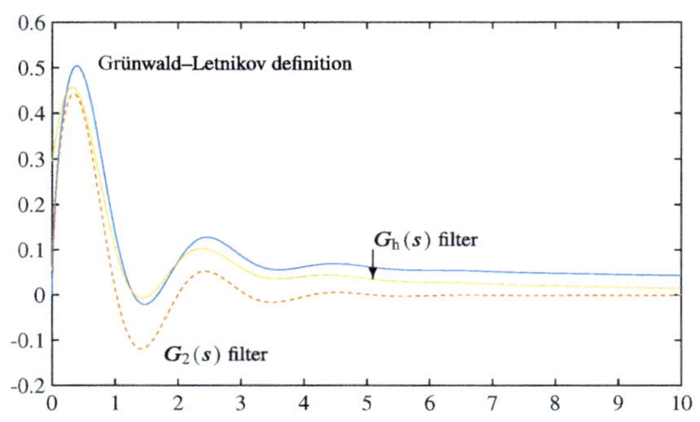

Fig. 5.2 Comparisons of 0.5th-order integral

5.2.2 Carlson Approximation

Another approximation method based on continued fractions is the Carlson method [4]. The target of the Carlson method is to find an integer-order approximation model $H(s)$ for a given integer-order base model $G(s)$ with a fractional power, i.e.

$$H(s) \approx G^{\alpha}(s). \tag{5.4}$$

Algorithm 5.1 Carlson filter design

(1) Input integer-order base model $G(s)$ and fractional power α.
(2) Let $q = \alpha$ and $m = q/2$, and initial fitting model $H_0(s) = 1$.
(3) Choose the number of iterations n.
(4) In each iteration, find the next model $H_i(s)$ from $H_{i-1}(s)$.

$$H_i(s) = H_{i-1}(s)\frac{(q-m)H_{i-1}^2(s) + (q+m)G(s)}{(q+m)H_{i-1}^2(s) + (q-m)G(s)}. \tag{5.5}$$

(5) The fitting model obtained directly is usually of extremely high order, with cancellable pole–zero pairs. Minimum realization can be made to the obtained model.

Based on the above algorithm, the following MATLAB function can be written to design the Carlson filter:

```
function H=carlson_fod(alpha,G,iter,epsx)
   arguments
      alpha(1,1) double, G(1,1)
      iter(1,1) {mustBePositiveInteger}=2, epsx(1,1)=1e-4;
   end
   q=alpha; m=q/2; H=1;
   for i=1:iter
      H=minreal(H*((q-m)*H^2+(q+m)*G)/((q+m)*H^2+(q-m)*G),epsx);
end, end
```

The syntax of the function is H=carlson_fod(α, G, n, ϵ), where G is the integer-order base model, α is the fractional power, n is the number of iterations, and H is the designed Carlson filter. Note that the actual order of the filter increases rapidly with the number of iterations, and in general $n = 2$ can be chosen. The user can also choose a larger error tolerance ϵ, for instance, taking a value of 10^{-4}, to do a minimum realization of the resulting approximate model.

▶ Example 5.3

Design a filter for $1/\sqrt{s}$ using Carlson's algorithm. Observe the frequency response fitting quality. Also, find the 0.5th-order integral of a given function $f(t) = \mathrm{e}^{-t}\sin(3t+1)$ and evaluate its accuracy.

Solutions For this specific problem, the base model is $G(s) = 1/s$, the power is $\alpha = 0.5$. The number of iterations is chosen to be 2, so that the Carlson filter can be designed.

```
>> s=tf('s'); G=1/s; alpha=0.5;
   H2=carlson_fod(alpha,G,2,1e-4)    % select a large error tolerance
```

The model derived here is equivalent to the one given in Reference [4] as follows.

$$H_2(s) = \frac{0.1111s^4 + 4s^3 + 14s^2 + 9.333s + 1}{s^4 + 9.333s^3 + 14s^2 + 4s + 0.1111}.$$

The frequency response fitting curve can be obtained by the following statements, as shown in Fig. 5.3. It can be seen that if the frequency interval of interest $(10^{-1}, 10^1)$ rad/s is chosen, the approximation of the filter is satisfactory.

```
>> G3=tf([1 36 126 84 9],[9 84 126 36 1]); w=logspace(-3,3);
   s=fotf('s'); H=bode(s^-0.5,w); bode(H,'-',H2,'--')
```

If these filters are used to evaluate the 0.5th-order integral of a given signal, the resulting results are shown in Fig. 5.4. As can be seen, the resulting filtering effect is almost identical to the theoretical value derived from the Grünwald–Letnikov definition.

```
>> t=0:0.01:10;
   y=exp(-t).*sin(3*t+1); y0=glfdiff9(y,t,-0.5,5);
   y1=lsim(H2,y,t); plot(t,y0,'-',t,y1,'--')
```

If the time interval is increased to $t \in (0, 100)$, a comparison of the time domain response fit of the filter with the theoretical value can be derived, as shown in Fig. 5.5.

```
>> t=0:0.01:100;
   y=exp(-t).*sin(3*t+1); y0=glfdiff9(y,t,-0.5,3);
   y1=lsim(H2,y,t); plot(t,y0,'-',t,y1,'--')
```

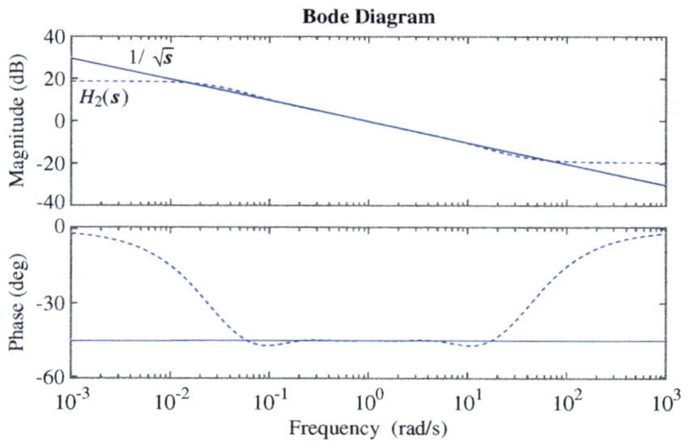

Fig. 5.3 Comparison of Bode diagrams of two filters

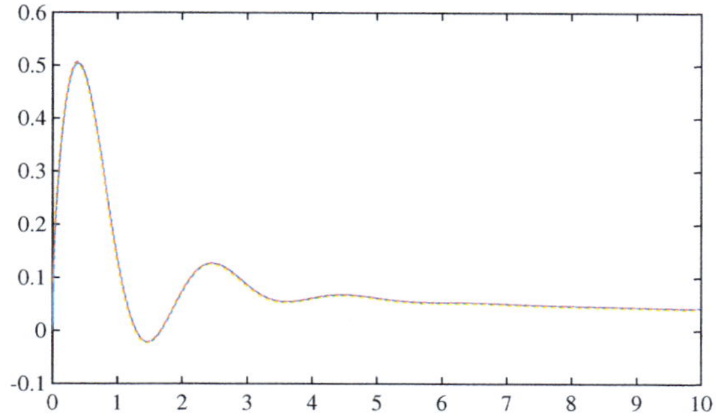

Fig. 5.4 Comparison of the calculation effect of 0.5th-order integral

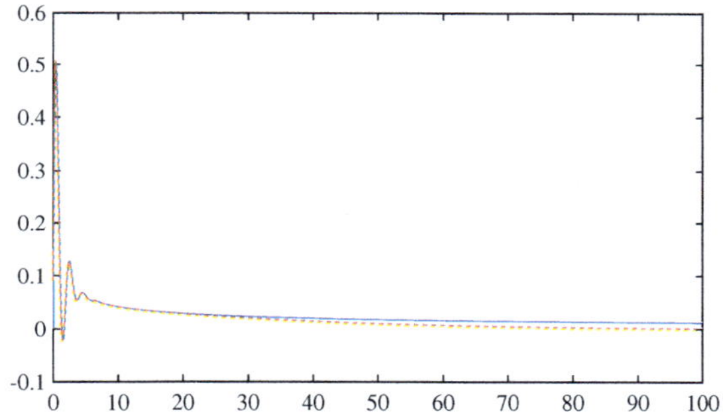

Fig. 5.5 The effect of 0.5th-order integral fitting for a larger time interval

However, the fitting effect then produces a large error at large t, which corresponds to the low-frequency band, and it can be seen that the error arises because of the poor fitting at the low-frequency band. Combined with the frequency response fitting in Fig. 5.3, it can be seen that the lower bound of the fitted region of 10^{-1} rad/s is not sufficient and should be extended to $\omega < 10^{-2}$ rad/s or even lower. Such a requirement is not possible in the Carlson filter design. A better algorithm that allows the user to choose the fitting interval is needed. ◄

5.2.3 Matsuda–Fujii Approximation

Matsuda and Fujii proposed an approximation method for approximating the irrational transfer function $G(s)$ with an integer-order transfer function by continued fraction technique [4, 5]. Unfortunately, the mathematical description of their original algorithm is wrong, and only the corrected algorithm is given here.

Algorithm 5.2 Corrected Matsuda–Fujii filter design method

(1) Choose a frequency vector ω_i, $i = 1, 2, \ldots, n$, and compute frequency response data $G_0(\mathrm{j}\omega_i)$.

(2) Choose an initial sequence $v_1(\omega_k) = |G_0(\mathrm{j}\omega_k)|$, $k = 1, 2, \ldots, n$.

(3) Make a loop for k: extract $a_k = v_k(\omega_k)$, and update $v_{k+1}(\omega)$ with

$$v_{k+1}(\omega_j) = \frac{\omega_j - \omega_k}{v_k(\omega_j) - v_k(\omega_k)}, \quad j = 1, 2, \ldots, n. \tag{5.6}$$

(4) Compute the coefficients a_k, $k = 1, 2, \ldots, n$, and construct Matsuda–Fujii filter $H(s)$.

$$H(s) = a_1 + \cfrac{s - \omega_1}{a_2 + \cfrac{s - \omega_2}{a_3 + \cfrac{s - \omega_3}{a_4 + \cfrac{s - \omega_4}{a_5 + \cdots}}}}. \tag{5.7}$$

5

MATLAB implementation of the Matsuda–Fujii filter algorithm is given below.

```
function G=matsuda_fod(G0,n,wb,wh)
   arguments
      G0,n,wb(1,1),wh(1,1){mustBeGreaterThan(wh,wb)}
   end
   if nargin==2, f=G0; w=n; n=length(w);
   else
      if isa(G0,'double'), s=fotf('s'); G0=s^G0; end
      w=logspace(log10(wb),log10(wh),n); f=mfrd(G0,w);
   end
   v=abs(f(:).'); s=tf('s'); n=length(w);
   for k=1:n, a(k)=v(k); v=(w-w(k))./(v-v(k)); end
   G=a(n); for k=n-1:-1:1, G=a(k)+(s-w(k))/G; end
end
```

Remarks Matsuda–Fujii filter design function

(1) The function has the following three syntaxes.

G=matsuda_fod(f,ω), % where frequency response data is needed

G=matsuda_fod(γ,n,ω_b,ω_h), % fit operator s^γ in the band $(\omega_\mathrm{b}, \omega_\mathrm{h})$

G=matsuda_fod(G_0,n,ω_b,ω_h), % n-point fitting to G_0 within $(\omega_\mathrm{b}, \omega_\mathrm{h})$

(2) Since ω_b and ω_h are the inner points within the frequency band of interest, the actual frequency band fitted is slightly larger than the specified $(\omega_\mathrm{b},\omega_\mathrm{h})$.

(3) To get a better fit, choose an odd number n, so that the actual filter order is $(n-1)/2$.

(4) Although the target of the fitting is the magnitude fit of the frequency response, the phase fit is also made automatically. □

▶ Example 5.4

Design a Matsuda–Fujii filter for $1/\sqrt{s}$ and assess the fitting quality.

Solutions If the frequency bands $(10^{-1}, 10^{1})$ rad/s and $(10^{-2}, 10^{2})$ rad/s are chosen, respectively, the Matsuda–Fujii filters can be designed directly, and the resulting Bode diagrams are compared as shown in Fig. 5.6.

```
>> s=fotf('s'); w=logspace(-3,3); G1=matsuda_fod(-0.5,9,1e-1,1e1)
   G2=matsuda_fod(-0.5,9,1e-2,1e2)  %design in large frequency band
   H=bode(s^-0.5); bode(H,G1,'--',G2,':',{1e-3,1e3})
```

The filters designed for the two frequency bands are identical to the ones given in Reference [4].

$$G_1(s) = \frac{0.0855s^4 + 4.876s^3 + 20.84s^2 + 13s + 1}{s^4 + 13s^3 + 20.84s^2 + 4.876s + 0.0855}$$

$$G_2(s) = \frac{0.04401s^4 + 8.142s^3 + 58.85s^2 + 30.76s + 1}{s^4 + 30.76s^3 + 58.85s^2 + 8.142s + 0.04401}.$$

The comparison of the frequency responses shows that the low-frequency fitting is close to that of the Carlson filter, and it can be said that the problem of low-frequency fitting has not been solved. The good thing is that the frequency band and order of Matsuda–Fujii filter can be artificially selected, so there is a lot of room for improvement in the actual frequency response fitting. For example, if the order of the filter can be increased and the fitted frequency band can be increased, the following statement directly designs an eighth-order filter (taking $n = 17$), which results in the Bode diagram shown in Fig. 5.7. It can be seen that the filter designed in this way has a significantly better quality, with a wider frequency band and a better fit of the Bode diagram.

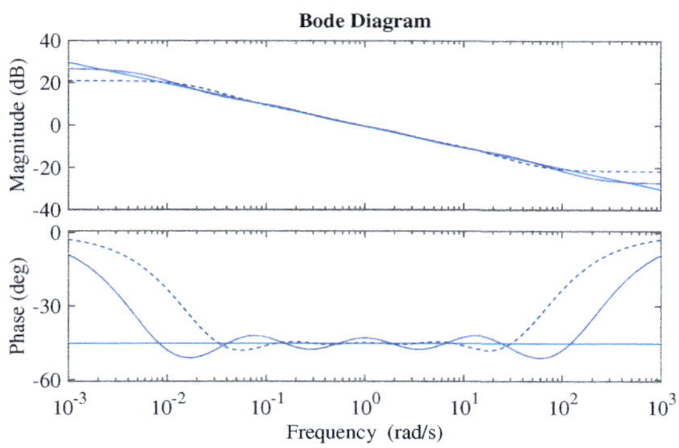

Fig. 5.6 Bode diagram comparison

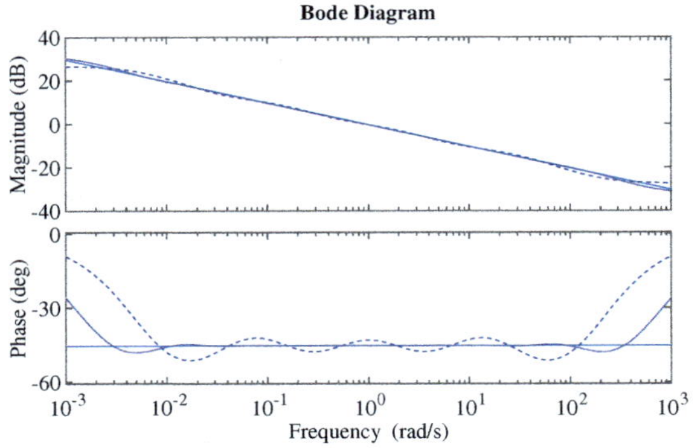

Fig. 5.7 Bode diagram comparison

```
>> w=logspace(-3,3); G1=matsuda_fod(-0.5,9,1e-2,1e2);
   G2=matsuda_fod(-0.5,17,1e-2,1e2), H=bode(s^-0.5);
   bode(H,G1,'--',G2,':',{1e-3,1e3})
```

With the filter, it is possible to feed a known signal into that filter, derive the 0.5th-order integral of that signal, and compare it to the exact solution calculated numerically, yielding the comparative results shown in Fig. 5.8. It can be seen that the G_2 filter almost coincides with the curve obtained from the numerical calculation, while the G_1 filter is a poorer fit (dashed line).

```
>> t=0:0.01:100; y=exp(-t).*sin(3*t+1);
   y0=glfdiff9(y,t,-0.5,3); y1=lsim(G1,y,t); y2=lsim(G2,y,t);
   plot(t,y0,'-',t,y1,'--',t,y2,':')
```

◀

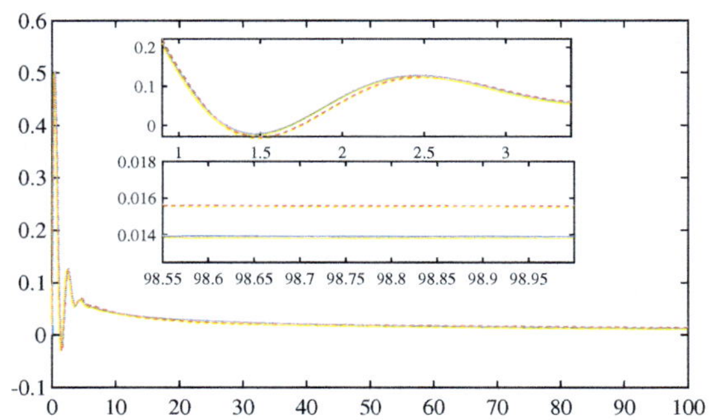

Fig. 5.8 Time domain response fitting effect

5.2.4 The Relationship between Fitting Quality and Filter Parameters

Various filter design methods were described previously, and some qualitative conclusions were drawn by comparing the effect of filter order and frequency band of interest on the approximation effect in terms of the time domain response. Here the effect of these filter parameters on the time response fitting will be explained more precisely in conjunction with the nature of the Laplace transform.

Theorem 5.1 Initial and final value theorems
If $F(s) = \mathscr{L}[f(t)]$, the initial and final values are

$$\lim_{s\to\infty} sF(s) = f(0), \quad \lim_{s\to 0} sF(s) = f(\infty). \tag{5.8}$$

Since the relationship between frequency and s is $s = \mathrm{j}\omega$, it can be seen from Theorem 5.1 that if we want to get a good fit at the initial moment (i.e., when t is small), we should improve the fit in the high-frequency band, i.e., increase the upper bound ω_h in the frequency band of interest. If a better time response accuracy when t is large is expected, the lower bound ω_b in the frequency band of interest should be reduced. If we want to get a better overall time domain response accuracy, we should make the frequency response curve in the middle frequency band better approximate the theoretical value of the slope. Increasing the order of the filter will of course increase the computational effort, but the increased computational effort here is not a great burden for the current computer hardware and software systems.

5.3 Oustaloup Filter Approximations

As demonstrated by the previous examples, the application of those filters is greatly compromised by the fact that the filter design method based on continued fraction expansion is not very good at fitting the frequency response data, especially since the continued fraction type of methods do not allow the user to freely choose the appropriate frequency band to be fitted. This section will introduce the more practical Oustaloup filter.

5.3.1 Oustaloup Filter

Assuming that the frequency band of interest is $(\omega_\mathrm{b}, \omega_\mathrm{h})$, the linear properties of the magnitude–frequency response of the fractional-order operator can be approximated by the set of poly-lines given in Fig. 5.9. The French scholar Prof. Alain Oustaloup and colleagues proposed a method for filter design based on such an idea [6], which is referred to as the Oustaloup filter in this book. All these poly-lines are generated from the integer-order zeros and poles, so that the slopes of the asymptotic lines of

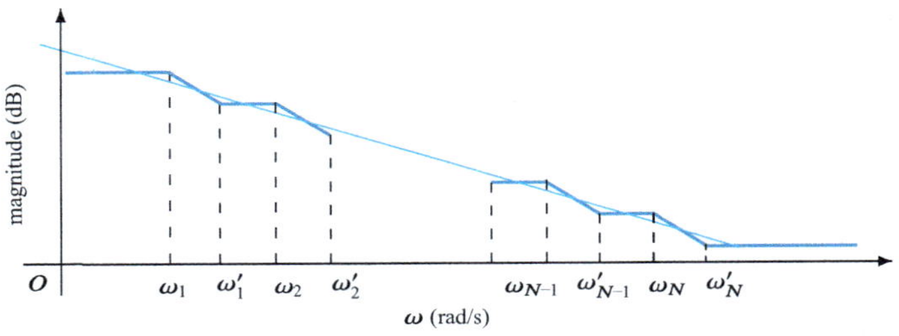

Fig. 5.9 Poly-line approximation of the Oustaloup filter

5

the Bode magnitude-frequency characteristics are alternating between 0 dB/dec and 20 dB/dec. The frequency response itself will be well approximated by a straight line.

Definition 5.4 Oustaloup filter

The standard form of Oustaloup filter is [6]

$$G(s) = K \prod_{k=1}^{N} \frac{s + \omega'_k}{s + \omega_k}, \tag{5.9}$$

where $k = 1, 2, \ldots, N$. The zeros, poles, and gain can be evaluated from

$$\omega'_k = \omega_{\rm b}\omega_{\rm u}^{(2k-1-\gamma)/N}, \quad \omega_k = \omega_{\rm u}^{2\gamma/N}\omega'_k, \quad K = \omega_{\rm h}^{\gamma}, \tag{5.10}$$

where

$$\omega_{\rm u} = \sqrt{\omega_{\rm h}/\omega_{\rm b}}. \tag{5.11}$$

The above design formulas are modified versions, which differ in the appearance from the ones in Reference [6]. Based on the above equations, the algorithm for designing the Oustaloup filter can be summarized.

Algorithm 5.3 Standard Oustaloup filter design

(1) Choose the fractional order γ, and select filter parameters $\omega_{\rm b}$, $\omega_{\rm h}$ and N.
(2) Compute $\omega_{\rm u}$ from (5.11), then with vectorized form to compute ω_k, ω'_k and K from (5.10).
(3) Construct Oustaloup filter $G(s)$ from (5.9).

The MATLAB function for the Oustaloup filter design can be written directly based on the above algorithm as follows

```
function G=ousta_fod(gam,N,wb,wh)
   arguments
      gam(1,1) double, N(1,1) {mustBePositiveInteger}=9
      wb(1,1)  double=1e-4;
      wh(1,1)  double {mustBeGreaterThan(wh,wb)}=1e4;
   end
   if round(gam)==gam, G=tf('s')^gam;
   else
      k=1:N; wu=sqrt(wh/wb);
      wkp=wb*wu.^((2*k-1-gam)/N); wk=wu^(2*gam/N)*wkp;
      G=zpk(-wkp,-wk,wh^gam);
end, end
```

The syntax of the function is G=ousta_fod(γ, N, ω_b, ω_h), where γ is the fractional order, N is the filter order, ω_b and ω_h are the lower and upper bounds, respectively, of the user-selected frequency band of interest. In general, the Bode diagram of the designed filter G works well in the specified frequency band, but not well outside that frequency band.

Remarks Oustaloup filter

(1) The filter implementation given here avoids the $\omega_\text{b}\omega_\text{h} = 1$ restriction and both sides can be chosen independently.

(2) If γ is an integer, it is straightforward to construct the integer-order operator $G(s) = s^\gamma$.

(3) Here γ can be either positive or negative, which means derivative and integral, respectively.

(4) The absolute value of γ is allowed to be greater than 1, for example, $\gamma = 3.7$ can still be fitted directly. However, in practical applications, it is recommended to restrict the order range to $-1 < \gamma < 1$, and the rest is expressed by integer-order operators. For example, $s^{3.7} = s^3 s^{0.7}$, or $s^{3.7} = s^4 s^{-0.3}$.

(5) Although the approximate magnitude–frequency characteristics look like varying on ◘ Fig. 5.9, these variations are actually asymptotes of the Bode diagram. The actual Bode diagram itself should be very close to the straight line if the parameters are chosen properly. □

Feeding the $y(t)$ signal into the Oustaloup filter, the output signal of the filter is ${}_0^{\text{RL}}\mathscr{D}_t^\gamma y(t)$, i.e., the Riemann–Liouville fractional-order derivative or integral of the input signal.

▶ Example 5.5

Suppose the frequency band of interest is $\omega_\text{b} = 0.01$ and $\omega_\text{h} = 1000$ rad/s. The Oustaloup filter can be designed by choosing different orders. If the input function is $f(t) = \text{e}^{-t}\sin(3t+1)$, design the filter with 0.5th-order integral and assess the approximation quality of the filter.

Solutions The fifth-order Oustaloup filter $G_1(s)$ and Matsuda–Fujii filter $G_2(s)$ can be designed directly using the following statements.

```
>> G1=ousta_fod(-0.5,5,0.01,1000), G1a=zpk(G1)
   G2=matsuda_fod(-0.5,11,0.01,1000), G2a=zpk(G2)
   s=fotf('s'); G=s^-0.5; w=logspace(-4,5);
   H=bode(G,w); bode(H,'-',G1,'--',G2,':')
```

The zero–pole forms of the two designed filters are

$$G_{1a}(s) = \frac{0.031623(s+562.3)(s+56.23)(s+5.623)(s+0.5623)(s+0.05623)}{(s+177.8)(s+17.78)(s+1.778)(s+0.1778)(s+0.01778)}$$

5

$$G_{2a}(s) = \frac{0.013865(s+1792)(s+71.46)(s+5.81)(s+0.503)(s+0.03427)}{(s+291.8)(s+19.88)(s+1.721)(s+0.1399)(s+0.00558)}.$$

Bode diagrams for these filters can be obtained, as shown in ◘ Fig. 5.10. Meanwhile with the Bode diagrams for the fractional-order operator $s^{-0.5}$, which is defined by the `fotf()` function to be discussed in detail in ▶ Chap. 6, it can be seen that for this example, the Matsuda–Fujii filter has a wider fitting bandwidth.

It is also possible to derive the numerical solution of the filtering results of the two filters with the Grünwald–Letnikov 0.5th-order integral as shown in ◘ Fig. 5.11. It can be seen that the resulting fractional-order integrals are not very accurate, and the approximation of the two filters is similar. If improvements are expected on the approximations, it is recommended to increase the frequency band of interest and increase the order.

```
>> t=0:0.001:pi; y=exp(-t).*sin(3*t+1);
   y1=lsim(G1,y,t); y2=lsim(G2,y,t);
   y0=glfdiff9(y,t,-0.5,3); plot(t,y0,t,y1,'--',t,y2,':')
```

◄

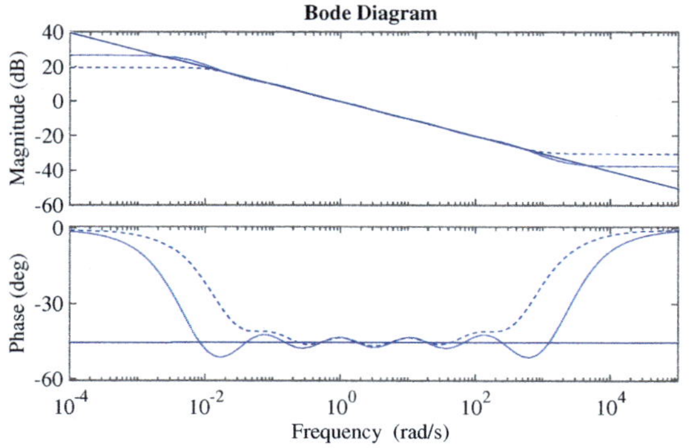

◘ **Fig. 5.10** Bode diagram fit of two filters

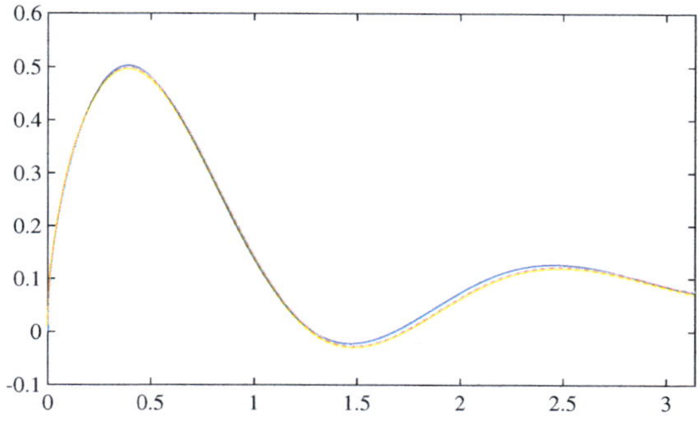

Fig. 5.11 Approximate effect of fractional-order integral

▶ Example 5.6

Now consider the design of an Oustaloup filter with 0.5th-order derivatives for the Heaviside function. Assume a large time range, and observe the impact of the Oustaloup filter parameters on the accuracy of the derivative evaluation.

Solutions The Heaviside function is with zero frequency. Now choosing the frequency band of interest (0.01, 1000) rad/s. Assuming that the time interval for the simulation is (0,20) s. The Oustaloup filter can be designed using the following statements. In this way, the fractional-order derivative curves are derived using both methods, as shown in Fig. 5.12. It can be seen that the results with the filter are very inaccurate when $t = 2$ or t is very large.

```
>> t=0:0.01:20;
   y=ones(size(t)); G=ousta_fod(0.5,5,0.01,1000),
   y1=lsim(G,y,t); y2=glfdiff9(y,t,0.5,3);
   plot(t,y1,t,y2,'--'), ylim([0,1])
```

Consider increasing the order of the Oustaloup filter, the fractional-order derivatives can be recalculated by the following statements. Although there is an improvement in the local fit, the error is still large at large t. This means that changing only the order of the filter does not improve the approximation for large time intervals.

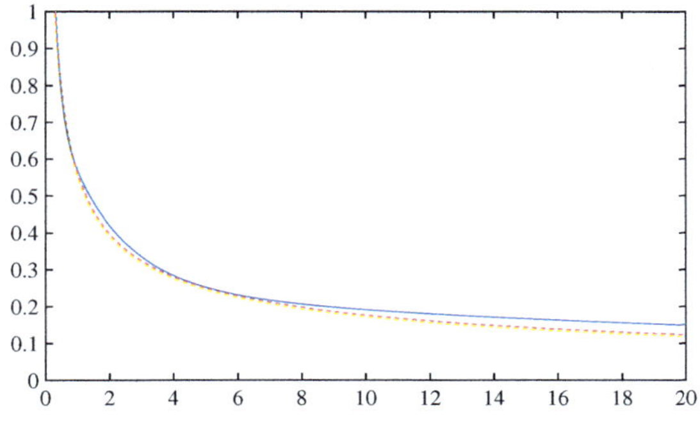

Fig. 5.12 Fractional-order derivatives of the Heaviside function

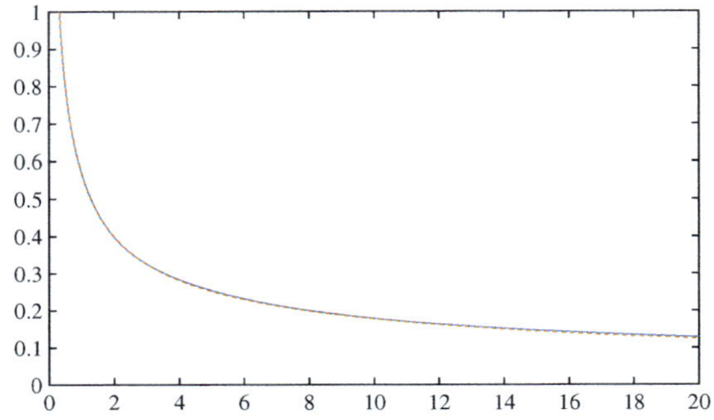

Fig. 5.13 Improved calculation results

```
>> G=ousta_fod(0.5,9,0.01,1000); y1=lsim(G,y,t);
   plot(t,y1,t,y2,'--'), ylim([0,1])
```

Since there is a large error at a large time t, which implies a poor fit of the Oustaloup filter in the low-frequency band. The value of ω_b should be reduced. For example, by choosing $\omega_b = 0.001$ rad/s. The fractional-order derivatives are evaluated under the new filter as shown in Fig. 5.13. It can be seen that a good computational accuracy has been obtained.

```
>> G=ousta_fod(0.5,9,0.001,1000); y1=lsim(G,y,t);
   plot(t,y1,t,y2,'--'), ylim([0,1])
```

◀

If one wants to further improve the computational accuracy, one can consider choosing a smaller lower frequency bound ω_b and a higher filter order.

▶ Example 5.7

From the previous example, it is seen that if we want to improve the accuracy of the calculation, we can enlarge the frequency band of interest. How do we choose the order of the filter?

Solutions As demonstrated in Example 5.5, if the frequency band of interest is $(10^{-2}, 10^3)$ rad/s, then a fifth-order Oustaloup filter works well enough. In general, if the frequency band is increased by 1 decade, the order should be increased by at least 1 to 2. For example, if the frequency band of interest is chosen as $(10^{-4}, 10^4)$ rad/s, i.e., the frequency band is increased by 3 decades, then an eleventh-order filter should be designed. In general, it is advisable to choose a higher order if computer resources allow. The Bode diagrams of different filter orders are shown in Fig. 5.14.

```
>> s=fotf('s'); G1=s^0.5;
   w=logspace(-5,5); H=bode(G1,w);
   G=ousta_fod(0.5,5,1e-4,1e4);    G1=ousta_fod(0.5,7,1e-4,1e4);
   G2=ousta_fod(0.5,9,1e-4,1e4);  G3=ousta_fod(0.5,11,1e-4,1e4);
   bode(H,G,'--',G1,':',G2,'-.')
```

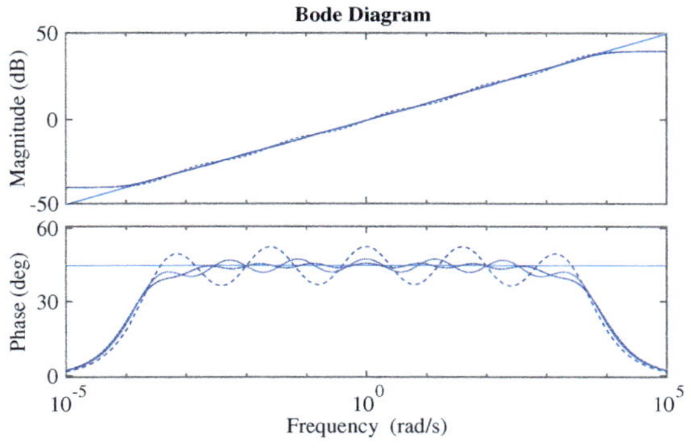

Fig. 5.14 Filters at different orders

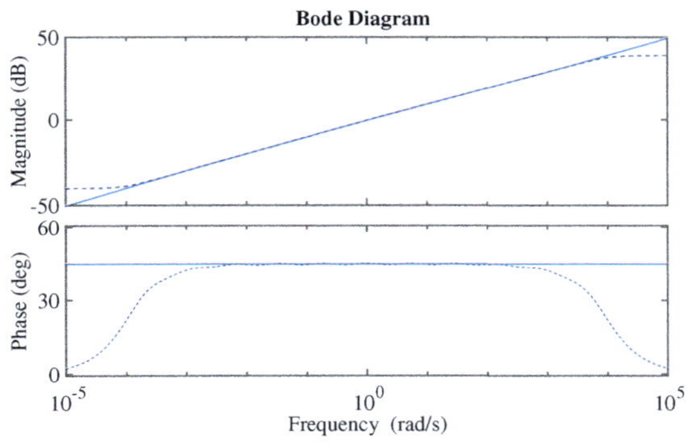

Fig. 5.15 Approximate frequency response of the eleventh-order oustaloup filter

In order to better demonstrate the eleventh-order Oustaloup filter, the fitting effect is given as shown in Fig. 5.15, and it can be seen that the frequency response curve is almost everywhere a straight line instead of a poly-line. ◀

Remarks Suggestions for selecting filter design parameters

(1) If the filtering approximation is poor at a larger time t, the lower frequency bound ω_b should be reduced accordingly, while the upper frequency bound ω_h should be increased if the initial moment approximation is poor.

(2) Do not worry about using very high filter orders and very wide frequency bands, e.g., $(10^{-6}, 10^{6})$ and $N = 30$. It does not significantly increase the computing time. □

5.3.2 An Improved Oustaloup Filter

From the previous examples, it can be seen that the approximation of the Oustaloup filter is less satisfactory at the chosen frequency band boundaries ω_h and ω_b. In addition, the order of the filter numerator is the same as the order of the denominator and the transfer function is not a strictly proper system, so an improved filter model can be considered.

Definition 5.5 Improved Oustaloup filter

The mathematical form of the improved Oustaloup filter is [7]

$$s^\gamma \approx \left(\frac{d}{b}\right)^\gamma \left(\frac{ds^2 + b\omega_\mathrm{h}s}{d(1-\gamma)s^2 + b\omega_\mathrm{h}s + d\gamma}\right) K \prod_{k=1}^{N} \frac{s+\omega_k'}{s+\omega_k}, \tag{5.12}$$

where

$$\omega_k' = \omega_\mathrm{b}\omega_\mathrm{u}^{(2k-1-\gamma)/N}, \ \omega_k = \omega_\mathrm{u}^{2\gamma/N}\omega_k', \ K = \omega_\mathrm{h}^\gamma, \ \omega_\mathrm{u} = \sqrt{\omega_\mathrm{h}/\omega_\mathrm{b}}. \tag{5.13}$$

In this way, the order γ of this filter must satisfy $\gamma \in (0, 1)$. In the normal case, the weighting parameters can be chosen as $b = 10$ and $d = 9$. A slight variation of Algorithm 5.3 can be used to design the improved Oustaloup filter. The following MATLAB functions can be directly written to improve the design of the improved Oustaloup filter.

```
function G=new_fod(r,N,wb,wh,b,d)
  arguments
     r(1,1) double, N(1,1){mustBePositiveInteger}=9
     wb(1,1)double=1e-4;
     wh(1,1){mustBeGreaterThan(wh,wb)}=1e4;
     b(1,1) double=10; d(1,1) double=9;
  end
  G=(d/b)^r*tf([d,b*wh,0],[d*(1-r),b*wh,d*r]);
  G=G*ousta_fod(r,N,wb,wh);
end
```

The syntax of the function is G=new_fod($\gamma, N, \omega_\mathrm{b}, \omega_\mathrm{h}, b, d$), where arguments b and d can be omitted, such that the default values can be adopted.

▶ Example 5.8

Consider the problem in Example 5.5, choosing $\omega_\mathrm{b} = 0.01$ rad/s and $\omega_\mathrm{h} = 1000$ rad/s. Observe the approximation results of the new filter and compare the approximation effect of the fractional-order derivatives.

Solutions The two filters can be designed from the following statements and the approximation of the Bode diagram is derived, as shown in Fig. 5.16. It can be seen that the improved algorithm has a wider fitting bandwidth.

```
>> G1=ousta_fod(0.5,5,0.01,1000);
   G2=new_fod(0.5,5,0.01,1000); zpk(G2)
   s=fotf('s'); G0=s^0.5; w=logspace(-5,5,100);
   H=bode(G0,w); bode(G1,'-',G2,'--',H,':')
```

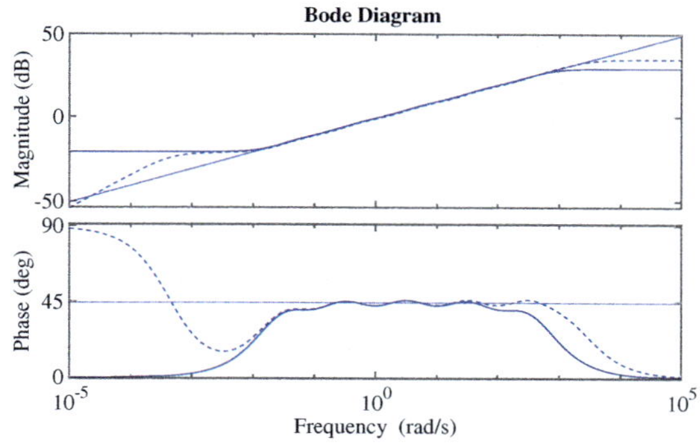

Fig. 5.16 Bode diagram comparisons

The zero–pole expression for the filter $G_2(s)$ is obtained as

$$G_2(s)=\frac{60s(s+0.01778)(s+0.1778)(s+1.778)(s+17.78)(s+177.8)(s+1111)}{(s+0.056)(s+0.56)(s+5.623)(s+56.23)(s+562.3)(s+2222)(s+0.00045)}.$$

The calculated fractional-order derivative curves are shown in Fig. 5.17. It can be seen that the improved filter yields slightly improved fitting results.

```
>> t=0:0.001:pi; y=exp(-t).*sin(3*t+1);
   y1=lsim(G1,y,t); y2=lsim(G2,y,t);
   y0=glfdiff9(y,t,0.5,3);
   plot(t,y1,t,y2,'--',t,y0,':'), axis([0,pi,-2 8])
```

◀

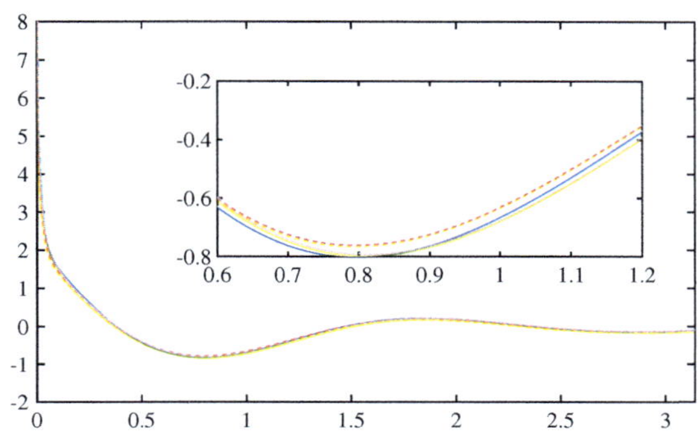

Fig. 5.17 Calculation of fractional-order derivatives under different filters

▶ Example 5.9

Consider the following FOTF model.

$$G(s) = \frac{s+1}{10s^{3.2} + 185s^{2.5} + 288s^{0.7} + 1}.$$

Compare the fitting effect of the two filters on the FOTF model.

Solutions If a 0.2th-order derivative filter is designed, the resulting Bode diagrams of the two filters are shown in Fig. 5.18. The Bode diagram of the FOTF can be drawn directly by the `bode()` overload function, provided in the FOTF Toolbox. The Bode diagrams of the two filters can also be drawn by the following commands, as shown in Fig. 5.19. It can be seen that the improved filter yields significantly better fitting results than the Oustaloup filter.

```
>> b=[1 1]; a=[10,185,288,1]; nb=[1 0]; na=[3.2,2.5,0.7,0];
   G0=fotf(a,na,b,nb); w=logspace(-4,4,200); H=bode(G0,w);
```

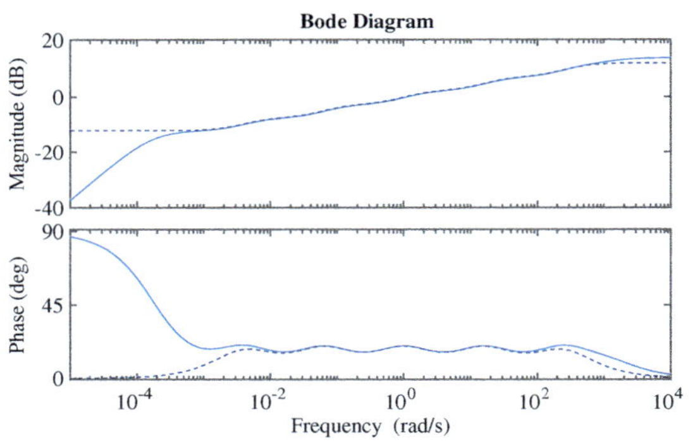

Fig. 5.18 Fitting of $s^{0.2}$ filter

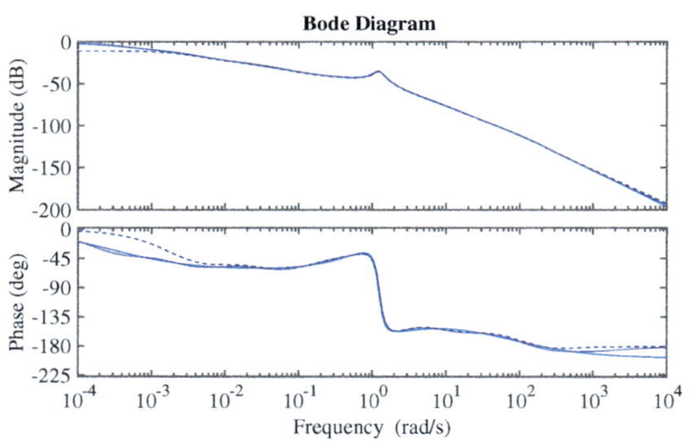

Fig. 5.19 Comparison of Bode diagrams for FOTFs

```
s=zpk('s'); N=5; w1=1e-3; w2=1e3; b=10; d=9;
g1=ousta_fod(0.2,N,w1,w2); a1=g1;
g2=ousta_fod(0.5,N,w1,w2); g3=ousta_fod(0.7,N,w1,w2);
G1=(s+1)/(10*s^3*g1+185*s^2*g2+288*g3+1);
g1=new_fod(0.2,N,w1,w2); g2=new_fod(0.5,N,w1,w2);
g3=new_fod(0.7,N,w1,w2); bode(g1,a1,'--');
figure, G2=(s+1)/(10*s^3*g1+185*s^2*g2+288*g3+1);
bode(H,G1,'--',G2,':',w)
```

◀

5.4 Integer-Order Approximation of FOTFs

5.4.1 High-Order Approximation of FOTFs

If an FOTF model given in Example 5.9 is encountered, a high-order approximation model using the method given in that example can be derived. However, it is tedious to build the high-order model with the low-level commands every time. A MATLAB function should be written to perform such an automatic conversion and derive the high-order transfer function.

Algorithm 5.4 High-order approximation of FOTFs

(1) Write a subfunction to handle the pseudo-polynomials by decomposing each s^γ factor into $s^m s^\alpha$, where $m = \lfloor\gamma\rfloor$, and $\alpha = \gamma - m$. Replacing s^α with the filter model.
(2) Process the numerator and denominator separately to derive the high-order representation from the pseudo-polynomials.
(3) Perform minimum realization to the resulting high-order transfer function.

Based on the above algorithm, a MATLAB function is written to perform the desired conversion, where many functions of the FOTF Toolbox are used. The function should be placed in the `@fotf` folder, as a part of the FOTF Toolbox.

```
function Ga=high_order(G0,filter,wb,wh,N)
   arguments, G0
      filter{mustBeMember(filter,...
         {'ousta_fod','new_fod','matsuda_fod'})}='ousta_fod'
      wb(1,1) double{mustBePositive}=1e-3
      wh(1,1) double {mustBeGreaterThan(wh,wb)}=1e3
      N(1,1) {mustBePositiveInteger}=5
   end
   [n,m]=size(G0); F=filter;
   for i=1:n, for j=1:m
      if G0(i,j)==fotf(0), Ga(i,j)=tf(0);
     else, G=simplify(G0(i,j)); [a na b nb]=fotfdata(G);
        G1=pseudo_poly(b,nb,F,wb,wh,N)/...
```

```
                pseudo_poly(a,na,F,wb,wh,N);
            Ga(i,j)=minreal(G1);
end, end, end, end
% pseudo-polynomial approximation
function G1=pseudo_poly(a,na,filter,wb,wh,N)
   G1=0; s=tf('s');
   for i=1:length(a), na0=na(i); n1=floor(na0);
      if na0>n1, g1=eval([filter '(na0-n1,N,wb,wh)']);
      else, g1=1; end
      G1=G1+a(i)*s^n1*g1;
end, end
```


The syntax of the function is G_a=high_order(G_0, filter, ω_b, ω_h, N), where G_0 is a FOTF matrix object that can directly handle multivariable problems; the input variable `filter` can be 'ousta_fod', 'new_fod' or 'matsuda_fod'; ω_b and ω_h are used to describe the frequency band of interest; N is the order of the filter for each fractional-order operator. It can be seen that the structure of the function is open, and if other filter design functions are available, they can be directly embedded following such a structure.

► Example 5.10

Consider the FOTF model in Example 5.9. Find its high-order transfer function approximation.

Solutions The following statements can be used to input the FOTF model directly and find the approximate model of the high-order transfer function, after replacing the fractional-order operators with the Oustaloup filters or the modified Oustaloup filters. Their frequency response fittings are exactly the same as the ones in Fig. 5.19.

```
>> b=[1 1]; a=[10,185,288,1]; nb=[1 0]; na=[3.2,2.5,0.7,0];
   G0=fotf(a,na,b,nb); N=4; w1=1e-3; w2=1e3; b=10; d=9;
   G10=high_order(G0,'ousta_fod',w1,w2,N)
   G20=high_order(G0,'new_fod',w1,w2,N)
   w=logspace(-3,3); bode(G0), hold on; bode(G10,'--',G20,':')
```

The approximate models thus derived are, respectively,

$$G_{10}(s)=\frac{\begin{array}{c}0.025119(s+595.7)(s+421.7)(s+251.2)(s+18.84)\times\\(s+13.34)(s+7.943)(s+1)(s+0.5957)(s+0.4217)\times\\(s+0.2512)(s+0.01884)(s+0.01334)\end{array}}{\begin{array}{c}(s+595.7)(s+507)(s+154.7)(s+40.89)(s+18.78)(s+7.377)\times\\(s+2.041)(s+0.4418)(s+0.2511)(s+0.05445)(s+0.01334)\times\\(s+0.002349)(s^2+0.3824s+1.613)\end{array}}$$

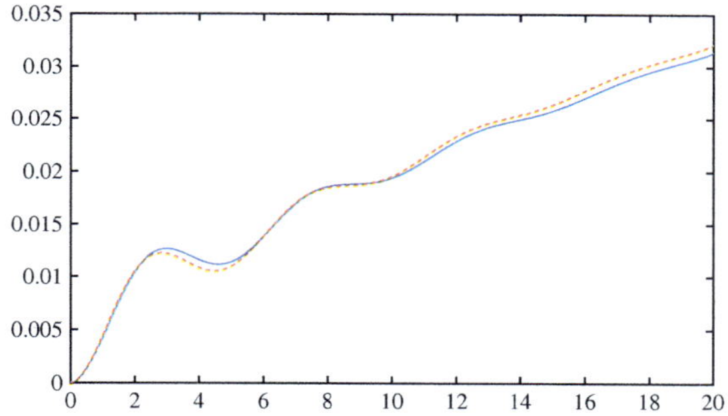

Fig. 5.20 Step response comparison

$$G_{20}(s)=\frac{\begin{array}{c}0.020523(s+1)(s+0.5957)(s+0.4217)(s+0.2512)(s+7.943)\times\\(s+13.34)(s+18.84)(s+251.2)(s+421.7)(s+595.7)(s+1389)\times\\(s+2222)(s+3704)(s+0.01884)(s+0.01334)(s+0.007943)\times\\(s+0.00063)(s+0.00045)(s+0.00018)\end{array}}{\begin{array}{c}(s+3704)(s+2325)(s+1111)(s+595.7)(s+488.2)(s+152.6)\times\\(s+40.07)(s+18.78)(s+7.358)(s+2.041)(s+0.4422)(s+0.2511)\times\\(s+0.05453)(s+0.01334)(s+0.007943)(s+0.002197)(s+0.00045)\times\\(s+0.0002201)(s+0.00018)(s^2+0.3762s+1.574)\end{array}}.$$

The step response curves derived from these two filter models can be obtained from the following statements, as shown in Fig. 5.20. It can be seen that the approximation of both filters is not satisfactory, and the reason is still due to the poor fitting in the low-frequency bound.

```
>> t=0:0.01:20; y=step(G0,t); y1=step(G10,t);
   plot(t,y,t,y1,'--')
```

For such a problem, it is still necessary to reduce the value of ω_b, for example, to 10^{-4} rad/s, and to increase the order of the filter appropriately, so that the step response curve can be obtained as shown in Fig. 5.21. It can be seen that even if a larger time interval is tested, the computational accuracy is still satisfactory.

```
>> N=7; w1=1e-4; w2=1e3; G10=high_order(G0,'ousta_fod',w1,w2,N)
   t=0:0.01:100; y=step(G0,t); y1=step(G10,t);
   plot(t,y,t,y1,'--')
```

◀

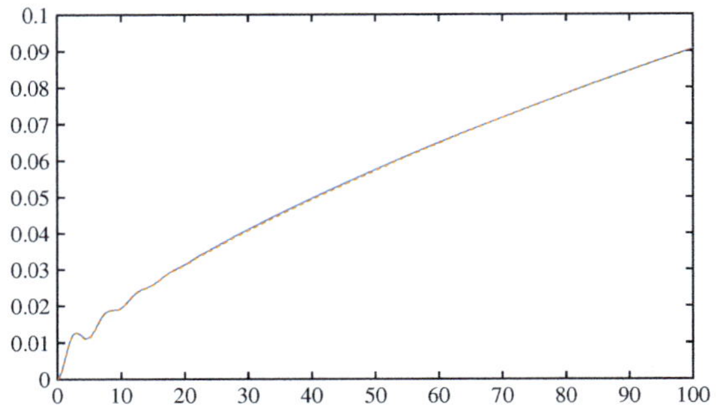

Fig. 5.21 Comparison of step response curves for a longer time period

5

5.4.2 Reduction of Fractional-Order Models

As the name implies, model reduction is to approximate a high-order model with a low-order one. Since the use of filters to approximate fractional-order operators often results in transfer functions of particularly high order, the behavior of approximating a fractional-order system with a low-order transfer function model is studied here.

Definition 5.6 Modelr reduction

The general form of reduced-order models is

$$G_{r/m,\tau}(s) = \frac{\beta_1 s^r + \beta_2 s^{r-1} + \cdots + \beta_{r+1}}{s^m + \alpha_1 s^{m-1} + \cdots + \alpha_{m-1}s + \alpha_m} e^{-\tau s}. \tag{5.14}$$

If the best reduced-order model is expected, then an optimization criterion should be defined. For example, if the original model and the reduced-order model are excited with the same signal, the error signal can be defined, from the two output signals. An objective function is then defined by this error signal. Based on such an objective function, the required problem is converted into a numerical optimization problem.

Definition 5.7 Optimization model for model reduction

Optimal model reduction problem can be converted into the following numerical optimization problem.

$$J = \min_{\boldsymbol{\theta}} \left\| \widehat{G}(s) - G_{r/m,\tau}(s) \right\|_2, \tag{5.15}$$

where $\boldsymbol{\theta}$ is the decision variable vector of the reduced-order model, i.e.

$$\boldsymbol{\theta} = \left[\beta_1, \beta_2, \ldots, \beta_r, \beta_{r+1}, \alpha_1, \alpha_2, \ldots, \alpha_m, \tau\right]^{\mathrm{T}}. \tag{5.16}$$

Since there is a delayed term in (5.15), Padé approximation can be considered to approximate the delayed term. In this way, the objective function can be turned into the following norm form and the optimization problem is transformed into

$$J = \min_{\boldsymbol{\theta}} \parallel \widehat{G}(s) - \widehat{G}_{r/m}(s) \parallel_2. \tag{5.17}$$

The above problem is without analytical solution. The original problem can only be solved by numerical optimization techniques. The idea given in Reference [8] leads to the optimal model reduction algorithm.

Algorithm 5.5 Optimal model reduction algorithm

(1) Input the high-order model $\widehat{G}(s)$.

(2) Select expected orders r and m of numerator and denominator of the reduced-order model.

(3) Write a MATLAB function to describe the objective function in (5.15).

(4) Use the `fminsearch()` solver to find the optimal reduced-order model $G_{r/m,\tau}(s)$.

It can be seen that the integer-order approximation model obtained from the original FOTF model is of very high order, so that the optimal reduced-order model method can be used to derive a low-order model with satisfactory results [8, 9]. The MATLAB function for optimal model reduction is written as

```
function Gr=opt_app(G,r,k,key,G0)
   arguments
      G, r(1,1) {mustBePositiveInteger}
      k(1,1) {mustBePositiveInteger}
      key(1,1){mustBeMember(key,[0,1])}=0, G0=0
   end
   GS=tf(G); Td=totaldelay(GS); GS.ioDelay=0;
   s=tf('s'); GS.InputDelay=0; GS.OutputDelay=0;
   if nargin<5, G0=(s+1)^r/(s+1)^k; end
   beta=G0.num{1}(k+1-r:k+1); alph=G0.den{1};
   Tau=1.5*Td; x=[beta(1:r),alph(2:k+1)];
   if abs(Tau)<1e-5, Tau=0.5; end
   dc=dcgain(GS); if key==1, x=[x,Tau]; end
   y=opt_fun(x,GS,key,r,k,dc);
   x=fminsearch(@opt_fun,x,[],GS,key,r,k,dc);
   alph=[1,x(r+1:r+k)];
   beta=x(1:r+1); beta(r+1)=alph(end)*dc;
   if key==0, Td=0; end
   if key==1, Tau=x(end)+Td; else, Tau=0; end
   Gr=tf(beta,alph,'ioDelay',Tau);
end
```

```
%object function evaluation
function y=opt_fun(x,G,key,r,k,dc)
   ff0=1e10; a=[1,x(r+1:r+k)]; b=x(1:r+1); b(end)=a(end)*dc;
   if key==1, tau=x(end);
      if tau<=0, tau=eps; end, [n,d]=pade(tau,3); gP=tf(n,d);
   else, gP=1; end
   Ge=G-tf(b,a)*gP;
   Ge.num{1}=[0,Ge.num{1}(1:end-1)]; [y,ierr]=geth2(Ge);
   if ierr==1, y=10*ff0; else, ff0=y; end
end
%compute the H2 norm
function [v,iE]=geth2(G)
   G=tf(G); num=G.num{1}; den=G.den{1};
   iE=0; v=0; n=length(den);
   if abs(num(1))>eps
      disp('System not strictly proper'); iE=1; return
   else, a1=den; b1=num(2:length(num)); end
   for k=1:n-1
      if (a1(k+1)<=eps), ierr=1; return
      else
         aa=a1(k)/a1(k+1); bb=b1(k)/a1(k+1);
         v=v+bb*bb/aa; k1=k+2;
         for i=k1:2:n-1
            a1(i)=a1(i)-aa*a1(i+1); b1(i)=b1(i)-bb*a1(i+1);
   end, end, end, v=sqrt(0.5*v);
end
```

The syntax of the function is G_r=opt_app(G, r, m, key, G_0), where r and m are the expected orders of the numerator denominator of the reduced-order model; key is the identifier indicating whether a delay term τ is needed in the reduced-order model. If necessary, a user's initial model G_0 can be supplied.

▶ Example 5.11

Consider the following FOTF model

$$G(s) = \frac{-2s^{0.63} - 4}{2s^{3.501} + 3.8s^{2.42} + 2.6s^{1.798} + 2.5s^{1.31} + 1.5}.$$

Find its optimal reduced-order model, and evaluate how well the reduced-order model matches the original model.

Solutions The high-order approximation model can be derived directly using the following statements.

```
>> b=[-2 -4]; nb=[0.63 0]; a=[2 3.8 2.6 2.5 1.5];
   na=[3.501 2.42 1.798 1.31 0]; G=fotf(a,na,b,nb);
   G1=high_order(G,'ousta_fod',1e-3,1e3,5); order(G1)
```

It can be seen that the model thus derived is an integer-order model of order 45. Now consider finding a reduced-order model, e.g., by choosing $r = 4$ and $m = 5$, then the reduced-order model can be derived directly and the Bode diagrams of the high-order and reduced-order models can be derived, as shown in ◻ Fig. 5.22. The phase diagrams here appear to be quite different, but the phase diagrams are parallel with a difference of 360°, so they can be assumed to be consistent.

```
>> G2=opt_app(G1,4,5,0), zpk(G2)
   w=logspace(-4,4); H=bode(G,w); bode(H,G1,'--',G2,':')
```

The reduced-order model thus obtained is

$$G_2(s) = \frac{-4.7911(s+0.2452)(s+0.01657)(s^2-8.983s+31.65)}{(s+140.9)(s+0.3799)(s+0.0173)(s^2+0.2303s+0.2479)}.$$

As can be seen, the frequency response of the resulting reduced-order model is still very close to that of the original model. The high-frequency part of the fit seems to have a large error with the original model. However, there is no need to worry too much because the unit used here for the Bode magnitude characteristics is dB (decibel). When the gain is less than -20 dB, the actual multiplier is less than -20 dB, i.e., 10^{-1}, and -50 dB is about $10^{-2.5}$, so both are very small gains and not too significantly different, at least not as much as ◻ Fig. 5.22. The difference looks so big in the several curves in the phase-frequency characteristic fit basically differ by 360° and 720°. The hand-modified phase-frequency curves are given here. It can be seen that the frequency domain response curves of the actual fractional-order system are almost identical to the high-order fitted model.

The step response of each model can also be found, as shown in ◻ Fig. 5.23. It can be seen that the reduced-order model and the original model are still very close.

```
>> t=0:0.01:10; y=step(G,t); y1=step(G1,t); y2=step(G2,t);
   plot(t,y,t,y1,'--',t,y2,':')
```

◀

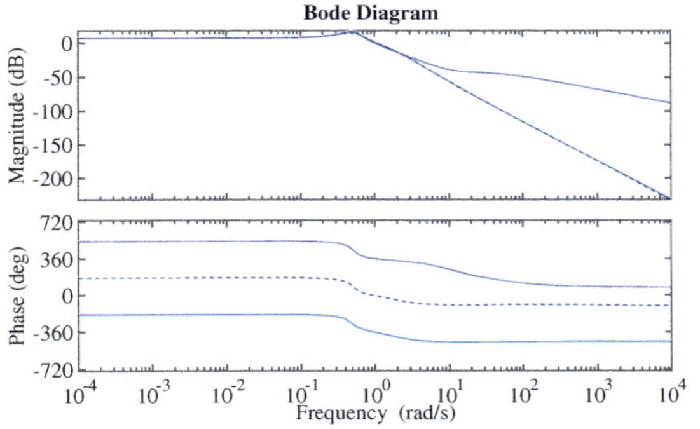

◻ **Fig. 5.22** Bode diagrams comparison

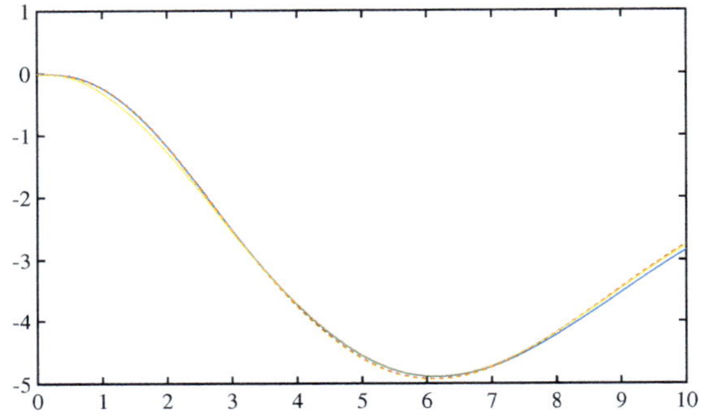

Fig. 5.23 Step responses comparison

▶ Example 5.12

Consider the following FOTF model. Find a reduced-order model and compare its effect.

$$G(s) = \frac{5}{s^{2.3} + 1.3s^{0.9} + 1.25}.$$

Solutions The high-order approximation of the original model is first derived. Then various reduced-order models are obtained. A comparison of the step responses is made as shown in Fig. 5.24. The step response curves here are automatically drawn by the `step()` function.

```
>> w1=1e-3; w2=1e3; N=9;
   s=fotf('s'); G=5/(s^2.3+1.3*s^0.9+1.25);
   G0=high_order(G,'ousta_fod',w1,w2,N); zpk(G0)
   G1=opt_app(G0,1,2,0), G2=opt_app(G0,2,3,0)
   G3=opt_app(G0,3,4,0), step(G0,G1,'--',G2,':',G3,'-.')
```

The above statement first yields a high-order approximate model of order 19, whose mathematical form is omitted, and also yield the optimal reduced-order models

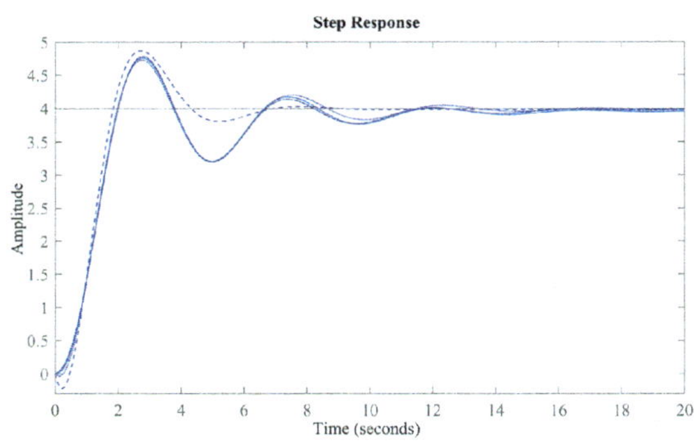

Fig. 5.24 Comparison of step responses

$$G_1(s) = \frac{-2.045s + 7.654}{s^2 + 1.159s + 1.917}, \quad G_2(s) = \frac{-0.5414s^2 + 4.061s + 2.945}{s^3 + 0.9677s^2 + 1.989s + 0.7378}$$

$$G_3(s) = \frac{-0.2592s^3 + 3.365s^2 + 4.9s + 0.3911}{s^4 + 1.264s^3 + 2.25s^2 + 1.379s + 0.09797}.$$

It can be seen that the second-order model $G_1(s)$ and the third-order model $G_2(s)$ do not fit very well. The responses of the fourth- and fifth-order approximate models are very close to those of the fractional-order model.

In some specific cases, such as the problem in Example 5.9, if the order of the reduced-order model is chosen too low, the approximation will be very poor, so the reduced-order model should be tested before it is actually used. ◄

5.5 Approximation of Irrational Fractional-Order Transfer Functions

In fractional-order control systems, there are times when a component of the system cannot be well described directly by a rational FOTF. For example, if the transfer function has a term of the form $\left[(as+b)/(cs+d)\right]^\alpha$, and α is not an integer, then the term can only be expanded into an infinite series of terms. It is even more difficult to handle if there are terms of the form $\left(as^\beta + b\right)^\alpha$. The approximation methods for such irrational transfer functions are discussed in this section.

5.5.1 Approximation of Implicit Irrational Models

Consider a simple irrational transfer function

$$G(s) = \frac{1}{(\tau s + 1)^\nu} = \tau^{-\nu} \frac{1}{(s + 1/\tau)^\nu}. \tag{5.18}$$

Theorem 5.2 Time domain representation of an irrational model
If the transfer function of the system is given by (5.18), then the relationship between input signal $u(t)$ and output signal $y(t)$ satisfies [10]

$$y(t) = \tau^{-\nu} \mathrm{e}^{-t/\tau} \mathscr{D}^{-\nu}\left[\mathrm{e}^{t/\tau} u(t)\right]. \tag{5.19}$$

Proof From the transfer function model, we have $Y(s) = G(s)U(s) = G(s + 1/\tau)U(s)$. Then $Y(s-1/\tau) = G(s-1/\tau)U(s-1/\tau) = \tau^{-\nu}U(s-1/\tau)/s^\nu$. Taking the inverse Laplace transform, then using the translation property $\mathscr{L}[\mathrm{e}^{-at}f(t)] = F(s+a)$ of the Laplace transform, and taking $a = -1/\tau$, yields

$$\mathscr{L}[Y(s-1/\tau)] = \tau^{-\nu}\mathscr{D}^{-\nu}\left[\mathrm{e}^{t/\tau}u(t)\right]. \tag{5.20}$$

The translation property is used again to yield the following equation, such that the theorem is proved.

$$e^{t/\tau}y(t) = \tau^{-\nu}\mathscr{D}^{-\nu}\left[e^{t/\tau}u(t)\right]. \tag{5.21}$$

□

Theorem 5.3 Implicit irrational differential equation
Choosing the state variable $x(t) = \mathscr{D}^{-\nu}\left[e^{t/\tau}u(t)\right]$, the corresponding fractional-order state space equation of the irrational system (5.18) is as follows [10]. This model is also known as the implicit fractional-order model.

$$\begin{cases} \mathscr{D}^{\nu}x(t) = e^{t/\tau}u(t) \\ y(t) = \tau^{-\nu}e^{-t/\tau}x(t). \end{cases} \tag{5.22}$$

5

Remarks Implicit irrational differential equation

(1) An approximation method is given here, but the method cannot yield an approximate model of the transfer function and cannot compare the frequency response fitting effect.

(2) Using the method presented here, the approximation method can be described by the block diagram shown in Fig. 5.25a. Based on the block diagram given here, a Simulink model can be built directly, as shown in Fig. 5.25b. Among them, the $1/s^{\nu}$ block can be implemented by the Oustaloup filter.

(3) Since there is a $t^{-\nu}$ term in the Simulink model, if $\nu > 0$, the initial value of t should not be set to 0 during the simulation. It can be set to a small value, such as 0.0001. □

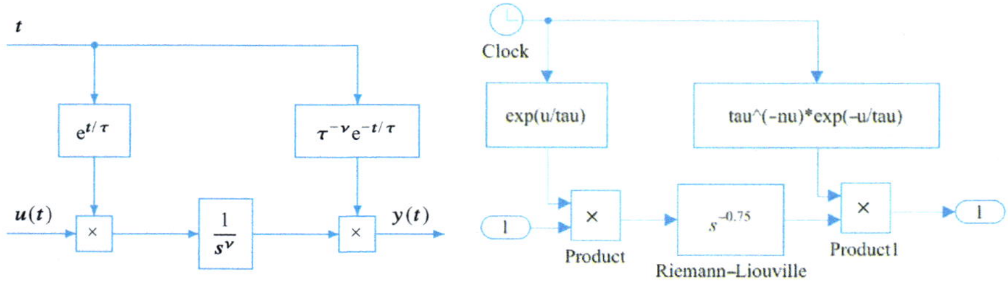

(a) block diagram (b) Simulink implementation (c5mimp.slx)

Fig. 5.25 Implicit irrational modeling

5.5.2 Frequency Response Fitting Methods

Of course, not all the irrational-order models can be expressed in the simple form of (5.18), so approximation methods for general irrational transfer function models need to be explored.

If the data of some frequency response points of a model are measured or calculated, an integer-order approximation model can be obtained by the `invfreqs()` function provided by MATLAB. In general, such a model can approximate the given frequency domain response data very well. The function is called in the syntax G=`invfreqs`(H, w, m, n), where H is the frequency response data, w is the frequency vector, and m and n are the expected numerator and denominator orders, respectively.

Algorithm 5.6 Rational fitting of irrational transfer functions
(1) A set of frequency response data is generated from the irrational transfer function model.
(2) Select the appropriation orders of the numerator and denominator of the approximate model.
(3) Call the `invfreqs()` function to obtain the fitted model from the frequency response data directly.
(4) Validate the obtained model. If the fitting effect is not satisfactory, increase the order of the expected model and then go to (2) to fit again, or regenerate a set of frequency response data to (1), until a satisfactory model is obtained.

▶ Example 5.13

Consider the fractional-order quantitative feedback theory (QFT) controller model given in Reference [11]

$$G_c(s) = 1.8393\left(\frac{s+0.011}{s}\right)^{0.96}\left(\frac{8.8\times 10^{-5}s+1}{8.096\times 10^{-5}s+1}\right)^{1.76}\frac{1}{(1+s/0.29)^2},$$

find the good rational approximation model.

Solutions The structure of such a controller is complicated. It can only be approximated with the help of a good rational transfer function model. Since the `frd()` function in the MATLAB Control System Toolbox can only handle integer-order models, this function returns a structure variable whose member `ResponseData` can be processed to obtain its non-integer powers, which eventually yields the exact frequency response data of the irrational transfer function. With the frequency response data, the `invfreqs()` function can be called to obtain the fitted model. For this example, one can first select the frequency band of interest $\omega \in (10^{-4}, 10^{2})$ rad/s, and then enter the following MATLAB commands.

```
>> w=logspace(-4,2); G1=tf([1 0.011],[1 0]); F1=frd(G1,w);
   G2=tf([8.8e-5 1],[8.096e-5 1]); F2=frd(G2,w);
   s=tf('s'); G3=1/(1+s/0.29)^2; F3=frd(G3,w); F=F1;
   h1=F1.ResponseData; h2=F2.ResponseData; h3=F3.ResponseData;
   h=1.8393*h1.^0.96.*h2.^1.76.*h3; F.ResponseData=h;
   [n,d]=invfreqs(h(:),w,4,4); H1=zpk(tf(n,d))
```

This yields an approximate controller of

$$H_1(s)=\frac{-1.5\times10^{-10}(s-3.913\times10^4)(s+2.635\times10^4)(s+0.01071)(s+0.0006019)}{(s+0.2922)(s+0.2878)(s+0.0007664)(s+1.284\times10^{-7})}.$$

From the results, it can be seen that there are stable zero–pole pairs that are close to each other, so they can be canceled by the minimum realization technique to obtain a lower order model.

```
>> H1a=minreal(H1,1e-3), H1b=minreal(H1,1e-2)
```

The lower order approximate models are

$$H_{1a}(s)=\frac{-1.5002\times10^{-10}(s-3.913\times10^4)(s+2.635\times10^4)(s+0.01071)}{(s+0.2922)(s+0.2878)(s+1.284\times10^{-7})}$$

$$H_{1b}(s)=\frac{-1.5002\times10^{-10}(s-3.913\times10^4)(s+2.635\times10^4)}{(s+0.2922)(s+0.2878)}.$$

For the problem of fitting irrational systems, one can also try using Matsuda–Fujii filters. The following statements are used to design such a filter directly, but the filter $H_2(s)$ is an unstable model, indicating that the Matsuda–Fujii filter design method does not guarantee the stability of the filter and should be used with care.

```
>> w=logspace(-4,2,11); F1=frd(G1,w); F2=frd(G2,w); F3=frd(G3,w);
   h1=F1.ResponseData; h2=F2.ResponseData; h3=F3.ResponseData;
   h=1.8393*h1.^0.96.*h2.^1.76.*h3; H2=zpk(matsuda_fod(h,w))
```

By choosing a larger frequency band of $(10^{-6}, 10^4)$ rad/s to validate the resulting model, the Bode diagram of the original controller with each lower order approximation can be derived, as shown in Fig. 5.26. The theoretical values are shown as dotted lines, and it can be seen that each magnitude–frequency approximation of the properties is ideal, and the phase frequency characteristics differ by 360°, which is also more desirable.

```
>> w=logspace(-6,4,200); F1=frd(G1,w); h1=F1.ResponseData;
   F2=frd(G2,w); h2=F2.ResponseData; F3=frd(G3,w);
   h3=F3.ResponseData; h=1.8393*h1.^0.96.*h2.^1.76.*h3;
   F=F1; F.ResponseData=h; bode(H1a,'--',H1b,F,':',H1,'-.',w)
```

From the comparison results, the minimum realized model $H_{1a}(s)$ with the fourth-order controller $H_1(s)$ is even close to the original irrational transfer function model, while the second-order approximate model $H_{1b}(s)$ is very poor, which implies that the original irrational transfer function model requires at least a third-order transfer function approximation. The $H_{1a}(s)$ derived above is derived from a minimal realization, which should actually be derived more directly from the `invfreqs()` function as follows.

```
>> w=logspace(-4,2); F1=frd(G1,w); F2=frd(G2,w); F3=frd(G3,w);
   h1=F1.ResponseData; h2=F2.ResponseData; h3=F3.ResponseData;
   h=1.8393*h1.^0.96.*h2.^1.76.*h3; [n,d]=invfreqs(h(:),w,3,3);
   H3=zpk(tf(n,d))
```

The directly derived third-order model is

$$H_3(s)=\frac{-1.4886\times10^{-10}(s-3.93\times10^4)(s+2.644\times10^4)(s+0.01009)}{(s+0.3028)(s+0.2768)(s+1.103\times10^{-5})}.$$

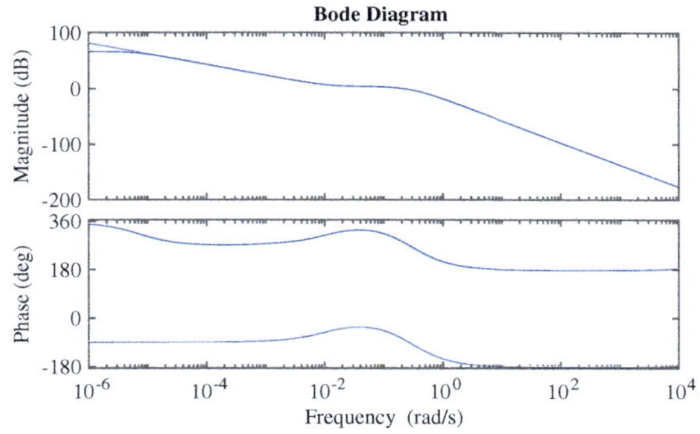

Fig. 5.26 Frequency response comparison

◀

▶ Example 5.14

Approximate the QFT controller given in Example 5.13 with implicit blocks.

Solutions The QFT controller transfer function is reorganized and can be written as

$$G_c(s) = \frac{1.8393/0.011^{0.96}}{s^{0.95}\,(1+s/0.29)^2}\left(\frac{s}{0.011}+1\right)^{0.96}(8.8\times10^{-5}s+1)^{1.76}(8.096\times10^{-5}s+1)^{-1.76}.$$

From this mathematical form, the Simulink simulation model shown in Fig. 5.27 can be constructed directly, whose integer-order part is given by the transfer function model, where the numerator edit box num fills `1.8393/0.011^0.96` and the denominator edit box den should fill `conv([1/0.29 1],[1/0.29 1])`. ◀

5.5.3 Charef Approximation

The Charef approximation technique is an effective approximation to the following irrational transfer function model [4, 12].

$$H(s) = \frac{1}{(1+s/p_{\mathrm{T}})^{\alpha}}. \tag{5.23}$$

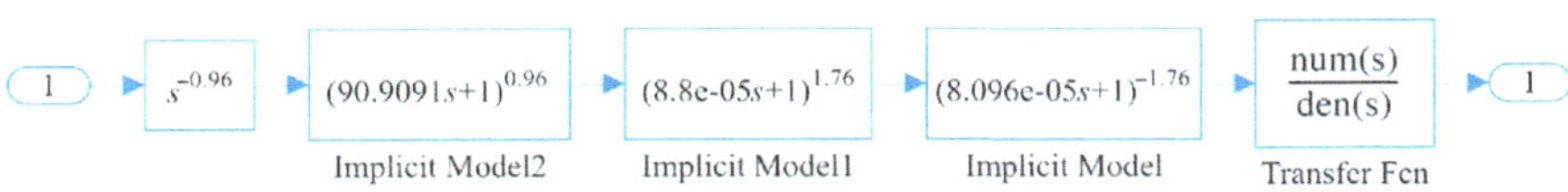

Fig. 5.27 Simulink model of QFT controller (`c5mqft.slx`)

Definition 5.8 Charef filter

The Charef filter technique is similar to the Oustaloup filter design method, which uses a poly-line asymptotic approximation of the irrational transfer function in the $(0, \omega_M)$ rad/s frequency band, yielding

$$H_1(s) = \frac{(1+s/z_0)(1+s/z_1)\cdots(1+s/z_{n-1})}{(1+s/p_0)(1+s/p_1)\cdots(1+s/p_n)}, \tag{5.24}$$

where, the following quantities are obtained

$$p_0 = p_T 10^{\delta/(20\alpha)},\quad a = 10^{\delta/[10(1-\alpha)]},\quad b = 10^{\delta/(10\alpha)}, \tag{5.25}$$

and the following formulas can be used to find the poles and zeros

$$p_{i+1} = p_0(ab)^{i+1},\quad z_i = ap_i,\quad i = 0, 1, \cdots, n-1. \tag{5.26}$$

The number of zero–pole pairs is determined directly with

$$n = \left\lceil \frac{\ln(\omega_M/p_0)}{\ln(ab)} \right\rceil, \tag{5.27}$$

where, δ is the error tolerance of frequency response fitting, usually taken as $\delta = 1$ dB. The error tolerance here is posed for asymptotes. The real fitting error is smaller than this quantity.

The following algorithm for designing Charef filters is presented.

Algorithm 5.7 Charef filter design

(1) Input the parameters p_T and α of the irrational transfer function model.
(2) Select the specifications δ and ω_M to design the filter.
(3) Calculate the quantities a, b, p_0 and n from (5.25) and (5.27).
(4) Calculate the zeros, poles and gain of the filter from (5.26), so that the Charef filter can be constructed.

The MATLAB function for the Charef filter design can be written based on the above algorithm.

```
function G=charef_fod(alpha,pT,delta,wM)
   arguments
      alpha(1,1) double, pT(1,1) double
      delta(1,1) double, wM(1,1) double
   end
   p0=pT*10^(delta/20/alpha);
   a=10^(delta/10/(1-alpha)); b=10^(delta/10/alpha);
   n=ceil(log(wM/p0)/log(a*b)); ii=1:n; p=p0*(a*b).^ii;
   p=[p0 p]; z=a*p(1:n); K=prod(p)/prod(z); G=zpk(-z,-p,K);
end
```

The syntax of the function is G=charef_fod(α, p_T, δ, ω_M), which has the same input arguments as the corresponding mathematical equations and the return variables as the designed Charef filter.

▶ Example 5.15

Design a Charef filter with irrational transfer function $G(s) = 1/(1 + 0.2s)^{0.5}$.

Solutions The error tolerance $\delta = 1$ dB and the upper bound of the frequency band $\omega_M = 1000$ rad/s can be chosen so that the Charef filter can be designed directly from the following statement.

```
>> G1=charef_fod(0.5,1/0.2,1,1000)
```

The designed Charef filter is

$$G_1(s) = \frac{99.763(s+9.976)(s+25.06)(s+62.95)(s+158.1)(s+397.2)(s+997.6)}{(s+6.295)(s+15.81)(s+39.72)(s+99.76)(s+250.6)(s+629.5)(s+1581)}.$$

In addition, an integer-order approximation to the original model can also be obtained from the frequency response fitting method described earlier.

```
>> w=logspace(-2,3); G0=tf(1,[0.2 1]);
   H=frd(G0,w); h=H.ResponseData; h=h.^0.5;
   [n,d]=invfreqs(h(:),w,5,5); G2=zpk(tf(n,d))
```

with the fitting model

$$G_2(s) = \frac{0.01541(s+8899)(s+1121)(s+340.2)(s+91.78)(s+17.16)}{(s+2412)(s+609.2)(s+183.6)(s+41.34)(s+7.41)}.$$

In fact, the Carlson filter described earlier is also suitable for the approximate irrational transfer function model, which can be designed directly by the following statements.

```
>> s=tf('s'); G=1/(1+0.2*s); H1=carlson_fod(0.5,G,2); zpk(H1)
```

The designed filter is

$$H_1(s) = \frac{0.11111(s+165.8)(s+20)(s+8.52)(s+6.667)(s+6.666)(s+5.662)(s+5)^6}{\begin{array}{c}(s+42.74)(s+12.1)(s+6.678)(s+4.887)(s^2+13.32s+44.37)\times \\ (s^2+10.33s+26.67)(s^2+9.859s+24.31)(s^2+10.08s+25.42)\end{array}}.$$

A comparison of the Bode diagrams of the irrational transfer function and the Charef filter is shown in ◘ Fig. 5.28. It can be seen that the frequency response of the Charef filter is also satisfactory.

```
>> w=logspace(-3,5); M=tf(1,[0.2 1]); H=frd(M,w);
   h=H.ResponseData; H.ResponseData=h.^0.5;
   bode(H,G1,'--',G2,':',H1,'-.'), xlim([1e-3 1e5])
```

The step response of the irrational transfer function $G(s)$ can be found by means of the numerical inverse Laplace transform, which is compared with the step response of the Charef filter and the frequency response fitting model as shown in ◘ Fig. 5.29, where the function `INVLAP_new()` is the solver based on the numerical Laplace transform. This solver will be introduced in detail in ▶ Chap. 6.

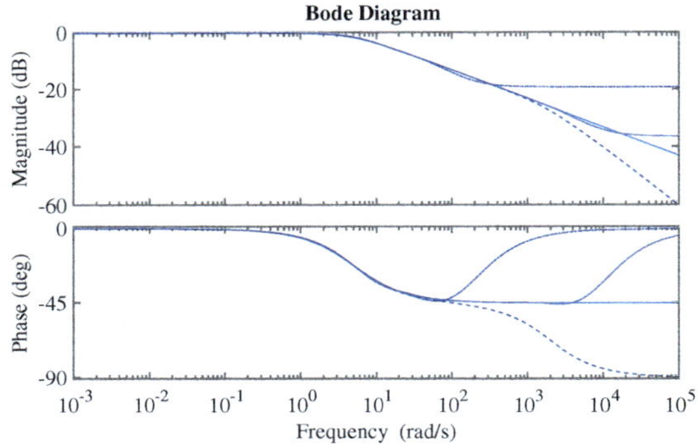

Fig. 5.28 Frequency response comparison

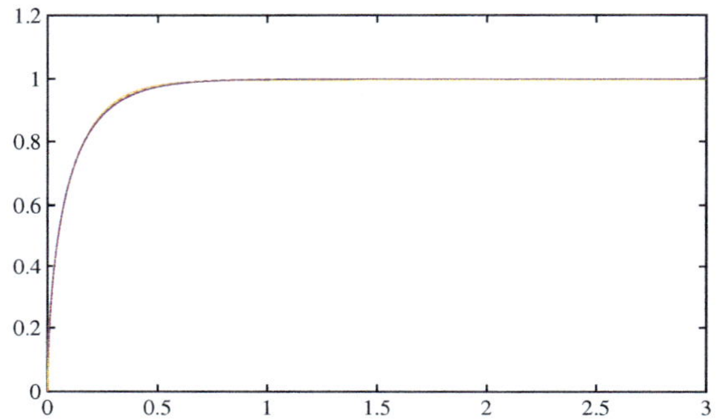

Fig. 5.29 Step response comparison

```
>> [t,y]=INVLAP_new('1/(1+0.2*s)^0.5',0,3,1001,0,'1/s');
   y1=step(G1,t); y2=step(G2,t); y3=step(H1,t);
   plot(t,y,t,y1,'--',t,y2,':',t,y3,'-.')
```

◀

▶ Example 5.16

Consider the following model with two irrational factors. Design a good Charef filter.

$$G(s) = \frac{1}{(1+s/1.6)^{0.6}\,(1+s/6.2)^{0.3}}.$$

Solutions The original model can be considered as the product of two irrational transfer functions, which yields the Charef filter.

```
>> G11=charef_fod(0.6,1.6,1,1000);
   G12=charef_fod(0.3,6.2,1,1000); G1=G11*G12; zpk(G1)
```

The obtained approximate model is

$$G_1(s)=\frac{\begin{array}{c}12018(s+3.447)(s+8.997)(s+12.64)(s+23.48)\times\\(s+37.85)(s+61.3)(s+113.3)(s+160)(s+339.2)\times\\(s+417.6)(s+1015)(s+1090)\end{array}}{\begin{array}{c}(s+1.938)(s+5.06)(s+9.1)(s+13.21)(s+27.24)\times\\(s+34.47)(s+81.55)(s+89.97)(s+234.8)(s+244.1)\times\\(s+613)(s+730.8)(s+1600)(s+2188)\end{array}}.$$

Choosing the frequency band of interest as $\omega \in (10^{-2}, 10^3)$ rad/s, it is also possible to derive the frequency response approximation model.

```
>> G01=tf(1,[1/1.6 1]); G02=tf(1,[1/6.2 1]);
   w=logspace(-2,3); H1=frd(G01,w); h1=H1.ResponseData;
   H2=frd(G02,w); h2=H2.ResponseData; H=H1;
   h=h1.^0.6.*h2.^0.3; [n,d]=invfreqs(h(:),w,5,5);
   H.ResponseData=h; G2=zpk(tf(n,d))
```

The model thus obtained is

$$G_2(s)=\frac{0.00011629(s+4.456\times10^4)(s+1348)(s+358.2)(s+81.78)(s+7.508)}{(s+1579)(s+407.5)(s+97.05)(s+10.91)(s+2.266)}.$$

If the original model is approximated by two frequency response fitting models in series, an approximate model of the 3rd integer-order transfer function can be derived as follows

```
>> h1=h1.^0.6; h2=h2.^0.3; [n,d]=invfreqs(h1(:),w,5,5);
   g=tf(n,d); [n,d]=invfreqs(h2(:),w,5,5); G3=zpk(tf(n,d)*g)
```

The model thus derived is

$$G_3(s)=\frac{\begin{array}{c}0.00029552(s+1.066\times10^4)(s+6244)(s+1122)(s+1016)\times\\(s+321.9)(s+319.3)(s+87.85)(s+76.86)(s+17.36)(s+9.993)\end{array}}{\begin{array}{c}(s+2975)(s+2079)(s+709.9)(s+526.2)(s+223.3)(s+143.8)\times\\(s+55.69)(s+24.43)(s+10.66)(s+2.536)\end{array}}.$$

A comparison of the Bode diagrams of these three models with the original irrational transfer function model is shown in ◘ Fig. 5.30. It can be seen that the piecewise fitting model $G_1(s)$ has the worst results.

```
>> w=logspace(-4,5); H1=frd(G01,w); h1=H1.ResponseData;
   H2=frd(G02,w); h2=H2.ResponseData; H=H1;
   h=h1.^0.6.*h2.^0.3; H.ResponseData=h;
   bode(H,G1,'--',G2,':',G3,'-.',w)
```

The step response curves of each model are shown in ◘ Fig. 5.31. It can be seen that all the three approximate models work relatively well. Of course, the Oustaloup filter method works the best and there is room for further improvement.

```
>> sys='1/(1+s/1.6)^0.6/(1+s/6.2)^0.3';
   [t,y]=INVLAP_new(sys,0,3,1001,0,'1/s');
   y1=step(G1,t); y2=step(G2,t); y3=step(G3,t);
   plot(t,y,t,y1,'--',t,y2,':',t,y3,'-.')
```

◀

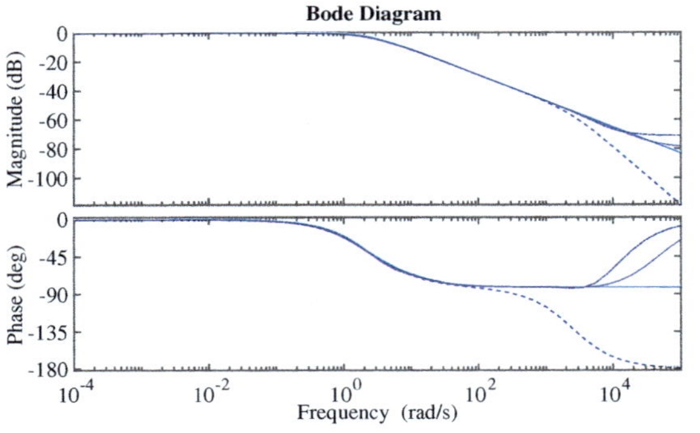

Fig. 5.30 Comparison of frequency response fitting

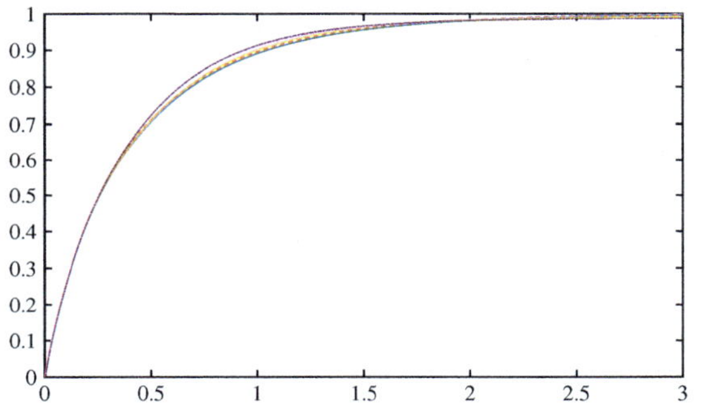

Fig. 5.31 Step response comparison

► Example 5.17

Consider again the irrational model in Example 5.16. Approximate the model using Simulink.

Solutions It is clear that the original model can be built by connecting two implicit models in series, and thus the Simulink model shown in Fig. 5.32 can be constructed. The step response curves shown in Fig. 5.33 can be derived from this model, and no difference can be observed between the curve and the numerical Laplace transform result.

```
>> sys='1/(1+s/1.6)^0.6/(1+s/6.2)^0.3';
   [t,y]=INVLAP_new(sys,0,3,1001,0,'1/s'); N=15;
   ww=[1e-4 1e4]; [t1,~,y1]=sim('c4msim1',3);
   plot(t,y,t1,y1,'--')
```

◄

Fig. 5.32 Simulink model (c5msim1.slx)

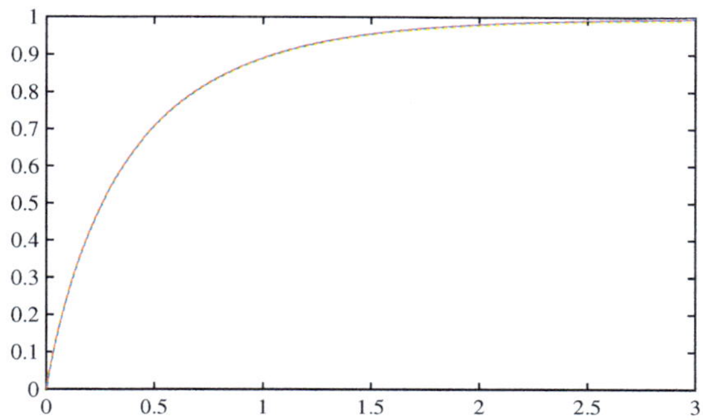

Fig. 5.33 Step response comparison

5.5.4 Optimum Charef Filter Design for Complicated Irrational Models

Now recall the Charef filter in Definition 5.8, where the ratios in this filter are $z_i/p_i = a$ and $p_{i+1}/z_i = b$, $i = 0, 1, \ldots, n-1$, where a and b are fixed values. If the ratios are no longer fixed, the concept of optimal Charef filter [13] can be introduced by finding the best filter under some performance indices.

Definition 5.9 Charef filter format

Complicated irrational transfer function $G(s)$ can be approximated by a Charef-type filter in the frequency band of interest $(\omega_{\rm m}, \omega_{\rm M})$ rad/s. The parameters in the model can be found by optimization method.

$$H(s) = K\frac{(1+s/z_0)\cdots(1+s/z_{n-1})}{(1+s/p_0)(1+s/p_1)\cdots(1+s/p_n)}. \tag{5.28}$$

For the frequency band of interest $(\omega_{\rm m}, \omega_{\rm M})$, the centers ω_{u_i} of a series of zeros and poles pairs (z_i, p_i) are defined such that

$$\frac{\omega_{u_{i+1}}}{\omega_{u_i}} = c, \quad \omega_{u_{i+1}} = c\omega_{u_i}, \quad i = 0, 1, 2, \ldots, n. \tag{5.29}$$

The poles and zeros of the filter can be found directly

$$p_i = 10^{y_i\omega_{u_i}}, i = 0, 1, \ldots, n; \; z_i = 10^{\omega_{u_i}/y_i}, \; i = 0, 1, \ldots, n-1. \tag{5.30}$$

where y_i are the parameters for optimization.

A set of frequency points $\omega_1, \omega_2, \ldots, \omega_N$ is chosen and the frequency response data $\boldsymbol{h}_a$ of the irrational transfer function is computed. The objective of optimization is to obtain the values of a set of decision variables $\boldsymbol{x} = \left[c, w_{u_0}, k, y_0, y_1, \ldots, y_n\right]$ such that the selected objective function is minimized. Here the optimized objective function is

$$f(\boldsymbol{x}) = w_1||\Delta\hat{\boldsymbol{h}}(\mathrm{j}\omega_i)||_\infty + \frac{w_2}{N}\sum_{i=1}^{N}\Delta\hat{\boldsymbol{h}}(\mathrm{j}\omega_i) + w_3||\Delta\angle\hat{\boldsymbol{h}}(\mathrm{j}\omega_i)||_\infty + \frac{w_4}{N}\sum_{i=1}^{N}\Delta\angle\hat{\boldsymbol{h}}(\mathrm{j}\omega_i). \tag{5.31}$$

.

where w_i are weighting coefficients; $\boldsymbol{h}(\mathrm{j}\omega)$ is the frequency response data of the filter model, and

$$\Delta\hat{\boldsymbol{h}}(\mathrm{j}\omega_i) = \big||\boldsymbol{h}(\mathrm{j}\omega_i)| - |\boldsymbol{h}_a(\mathrm{j}\omega_i)|\big| \tag{5.32}$$

$$\Delta\angle\hat{\boldsymbol{h}}(\mathrm{j}\omega_i) = \big|\angle h(\mathrm{j}\omega_i) - \angle h_a(\mathrm{j}\omega_i)\big|. \tag{5.33}$$

If the infinite norm is chosen, it means that the maximum of the absolute values of all elements of the vector is found.

Definition 5.10 Optimization model

The following constrained optimization model for Cheref filter design is proposed in Reference [13]

$$\min_{\boldsymbol{x} \text{ s.t.} \begin{cases} c>1 \\ \omega_{u_0}>0 \\ 0<y_i<1, i=0,1,\ldots,n \end{cases}} f(\boldsymbol{x}) \tag{5.34}$$

Algorithm 5.8 Optimal Charef filter design

(1) Selecting frequency vector $\boldsymbol{\omega}_\mathrm{r}$ and compute frequency response data of the irrational transfer function.
(2) Write the objective function for the problem in (5.31).
(3) Generate an initial search point $\boldsymbol{x}_0$ and solve the nonlinear programming problem.
(4) Construct optimal Charef filter from the optimization result.

Based on the above algorithm, the MATLAB function for the optimal Charef filter design can be written as follows.

```
function Ga=charef_opt(wr,n,G,wt,a,wc)
   arguments, wr(:,1), n(1,1){mustBePositiveInteger}
      G(:,1), w(1,4)=[1 1 1 1], a(1,1)=0.5, wc(1,1)=1
   end
   e=1; x0(1)=10^(e/(10*a*(1-a)));
   x0(2)=10^(e/(20*(1-a)))*wc*10^(e/(20*a));
   x0(3:n+3)=ones(1,n+1)/10^(e/(20*(1-a)));
   ff=optimset; ff.Display='iter';
   A=[];B=[];Aeq=[];Beq=[];CF=[];
   xm=[1 0 zeros(1,n+1)]; xM=[3 10 ones(1,n+1)];
   x1=fmincon(@charef_obj,x0,A,B,Aeq,Beq,xm,xM,CF,ff,wr,G,wt);
   c=x1(1); wu0=x1(2); wu=wu0*c.^[0:n]; p=wu.*x1(3:end);
   z=wu.^2./p; k=prod(p)/prod(z); Ga=zpk(-z',-p',k);
end
% subfunction to compute the objective function
function f=charef_obj(x,wr,G,wt)
   n=length(x)-2;
   c=x(1); wu0=x(2); wu=wu0*c.^(0:(n-1)); p=wu.*x(3:end);
   z=wu.^2./p; k=prod(p)/prod(z); Ga=zpk(-z',-p',k);
   Ga1=frd(Ga,wr); h=Ga1.ResponseData; Ga_fr=h(:);
   f=wt(1)*norm(abs(angle(Ga_fr)-angle(G)),inf)+...
      wt(2)*sum(abs(angle(Ga_fr)-angle(G)))/length(wr)+...
      wt(3)*norm(abs(abs(Ga_fr)-abs(G)),inf)+...
      wt(4)*sum(abs(abs(Ga_fr)-abs(G)))/length(wr);
end
```

where such an objective function can be described by a MATLAB subfunction with $\boldsymbol{x}$ represents the decision variables. All variables except $\boldsymbol{x}$ are additional variables and should be passed to the solution function `fmincon()` in the form of additional arguments. The additional arguments here include: a is the initial value of the order; ω_c is the initial value of the first-order system time constant ω_T; $\boldsymbol{\omega}_r$ is the vector of frequency points; $\boldsymbol{G}$ is the frequency response data of the original irrational transfer function at the frequency point ω_r; $\boldsymbol{w}_t$ is the vector of weighting coefficients w_1, w_2, w_3 and w_4, and the default value of these weighting values is 1. It can be seen that the values of ω_c and a are not very important in the actual optimization process. They are the initial values of the numerical optimization, and the default values can be used. This function calls the MATLAB function `fmincon()` to find the numerical solution of the optimization problem.

The syntax of the function is G_a=`charef_opt`$(\boldsymbol{\omega}_r, n, \boldsymbol{G}, \boldsymbol{w}_t, a, \omega_c)$, where $\boldsymbol{\omega}_r$ and $\boldsymbol{G}$ provide information on the frequency response of the irrational transfer function and n is the expected order of the optimal Charef filter.

Remarks Optimal Charef filter

(1) The frequency band of interest is given by the frequency vector $\boldsymbol{\omega}_r$, and the algorithm takes into account the fitting of magnitude and phase data.

(2) The choice of frequency ω_m does not have much impact on the final result, and the actual lower bound of the frequency is determined by the turning point of the plant model.

(3) The static gain of the original model must be 1, otherwise it should be extracted beforehand.

(4) In practice, it is better to set the upper bound of c to 3, so that the zero–pole pairs (z_i, p_i) may not cross over.

(5) Suggestions for choosing the order n: first, choose an initial integer n_0 to derive the optimal Charef filter and show the fitting error, then choose the order $n_0 + 1$. Further, choose a higher order if the fitting error is significantly lower, otherwise accept n_0. □

5

▶ Example 5.18

Design an optimal Charef filter for the model of Example 5.16.

Solutions By choosing a frequency band of interest $\omega = (10^{-2}, 10^3)$ rad/s, the frequency response data of the irrational transfer function can be calculated. Then choose the expected order of the optimal Charef filter to be 6. The optimal Charef filter can be designed by the following statements.

```
>> G01=tf(1,[1/1.6 1]); G02=tf(1,[1/6.2 1]);
   w=logspace(-2,3); H1=frd(G01,w); h1=H1.ResponseData;
   H2=frd(G02,w); h2=H2.ResponseData; H=H1;
   h=h1.^0.6.*h2.^0.3; h=h(:); G1=charef_opt(w,6,h)
```

The resulting optimal Charef filter model is

$$G_1(s) = \frac{0.00096744(s+6.692)(s+14.16)(s+46.48)(s+126.5)\times (s+299.6)(s+929)(s+3537)}{(s+2.274)(s+7.892)(s+17.67)(s+47.71)(s+148)(s+350.5)(s+676.4)}.$$

A comparison of the Bode diagrams of the two models is shown in Fig. 5.34. It can be seen that the approximation effect of the sixth-order filter model is similar to that of the seventeenth-order filter obtained in Example 5.16.

```
>> H.ResponseData=h; bode(H,G1,'--',w)
```

A larger frequency band of $(10^{-2}, 10^6)$ rad/s can also be specified to design an optimal Charef filter of order 16, which gives a perfect fit for the specified range, as shown in Fig. 5.35, with almost no difference visible in the fitting results.

```
>> w=logspace(-2,7,100); H1=frd(G01,w); h1=H1.ResponseData;
   H2=frd(G02,w); h2=H2.ResponseData; H=H1;
   h=h1.^0.6.*h2.^0.3; h=h(:); H1.ResponseData=h;
   G2=charef_opt(w,15,h), bode(H1,G2,'--',w)
```

The step response curves of the original irrational transfer function and the filter model are shown in Fig. 5.36. The responses of the two filters are close, but slightly different from the original model, obtained with numerical inverse Laplace transform.

```
>> f='1/(1+s/1.6)^0.6/(1+s/6.2)^0.3/s';
   [t,y]=INVLAP_new(f,0,3,1001);
   step(G1,G2,'--',t); hold on, plot(t,y,':'), hold off
```

◀

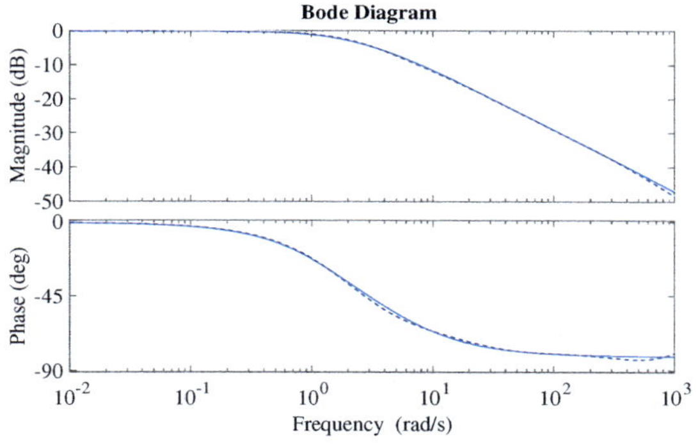

Fig. 5.34 Frequency response comparison

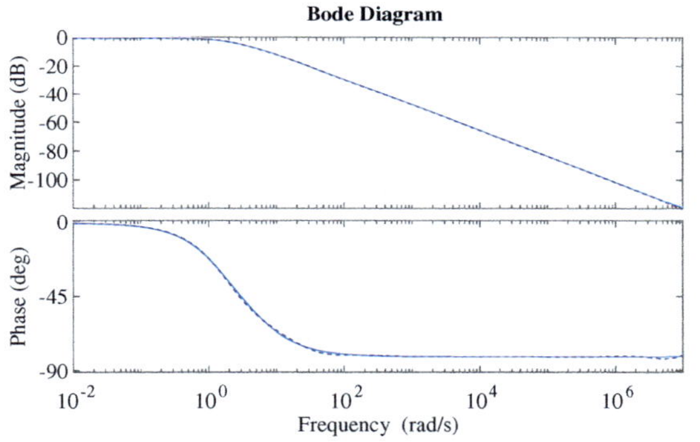

Fig. 5.35 Comparison of frequency response fits for larger intervals

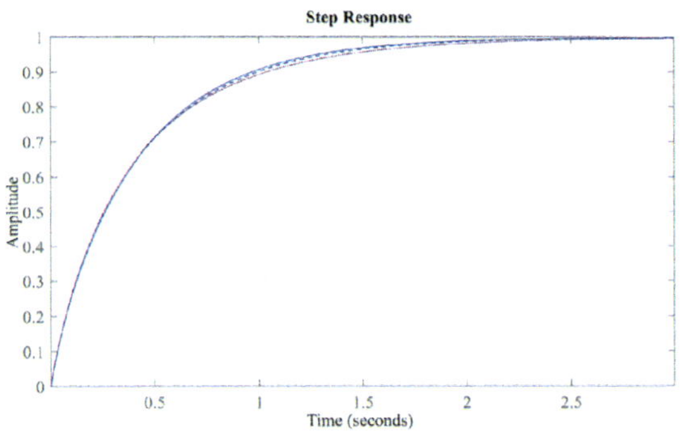

Fig. 5.36 Step response comparison

▶ Example 5.19

Now consider a more complicated model of an irrational transfer function.

$$G(s) = \frac{(1+s/6.2)^{0.3}}{(1+s/1.6)^{0.6}\,(s^{0.7}/3+1)^{0.2}}.$$

Design an optimal Charef filter and analyze its effect.

Solutions Generate a vector in the frequency band $(10^{-4}, 10^5)$ rad/s, calculate the frequency response of the original irrational transfer function, and then design the optimal Charef filter.


```
>> G01=tf(1,[1/1.6 1]); G02=tf([1/6.2 1],1);
   w=logspace(-4,5,200); H1=frd(G01,w); h1=H1.ResponseData;
   H2=frd(G02,w); h2=H2.ResponseData; H=H1;
   G3=fotf([1/3 1],[0.7 0],1,0); H3=bode(G3,w);
   h3=H3.ResponseData; h=h1.^0.6.*h2.^0.3.*h3.^0.2;
   H1.ResponseData=h; G1=charef_opt(w,11,h(:));
   bode(H1,G1,{1e-4,1e5})
```

The Bode diagram of the original irrational transfer function with the integer-order fitted model is shown in Fig. 5.37. It can be seen that the fitting effect is still satisfactory.

The step response curve of the irrational system can be derived using the numerical inverse Laplace transform technique, and the step response curve of the approximate model $G_1(s)$ can also be obtained, as shown in Fig. 5.38. It can be seen that the responses of the two systems are very close.

```
>> f='(1+s/6.2)^0.3/(1+s/1.6)^0.6/(s^0.7/3+1)^0.2';
   [t,y]=INVLAP_new(f,0,3,1000,0,'1/s'); step(G1,t); line(t,y)
```

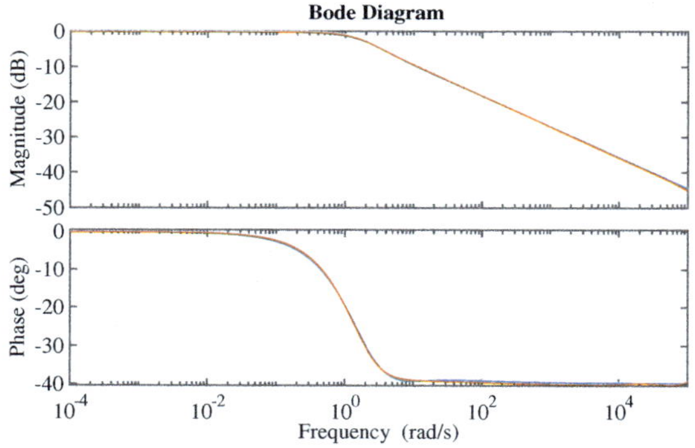

Fig. 5.37 Frequency response comparison

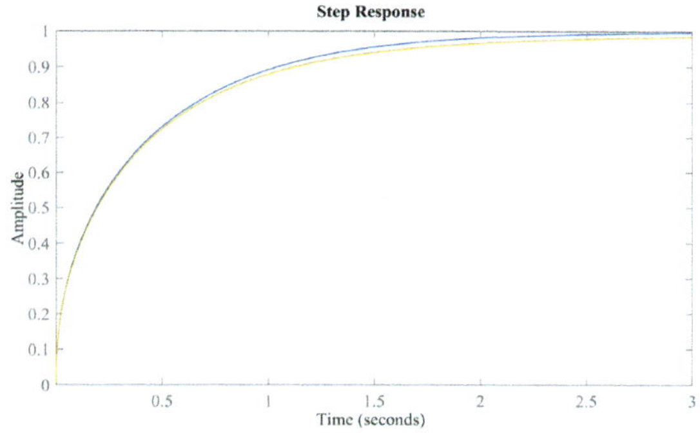

Fig. 5.38 Comparison of step responses

The designed twelfth-order optimal Charef filter is

$$G_1(s) = \frac{\begin{array}{c} 0.0027751(s+3.591)(s+8.44)(s+23.97)(s+64.13)\times \\ (s+179.1)(s+488.1)(s+1351)(s+3681)(s+1.019\times 10^4)\times \\ (s+2.772\times 10^4)(s+7.561\times 10^4)(s+2.515\times 10^5) \end{array}}{\begin{array}{c} (s+1.757)(s+5.638)(s+14.97)(s+42.22)(s+114)\times \\ (s+315.6)(s+860.2)(s+2380)(s+6487)\times \\ (s+1.799\times 10^4)(s+4.973\times 10^4)(s+1.128\times 10^5) \end{array}}.$$

◀

5.6 Discrete Filter Approximations

The previous section describes the continuous filter approximation for fractional-order and irrational systems. In this section, discrete filter approximation methods for these systems are presented.

Definition 5.11 Discrete filter

The mathematical form of the discrete filter

$$H(z) = \frac{b_1 + b_2 z^{-1} + b_3 z^{-2} + \cdots + b_{n+1} z^{-n}}{1 + a_1 z^{-1} + a_2 z^{-2} + \cdots + a_m z^{-m}}. \tag{5.35}$$

Assuming that the input signal is $u(n)$, the output signal after the filter can be expressed by the difference equation as

$$\begin{aligned} y(k) = -a_1 y(k-1) - a_2 y(k-1) - \cdots - a_m y(k-m) + \\ b_1 u(k) + b_2 u(k-1) + \cdots + b_{n+1} u(k-n). \end{aligned} \tag{5.36}$$

In practical applications, the following two special forms of discrete filters are often used.

Definition 5.12 FIR filter

The finite impulse response (FIR) filter needs to assign $m = 0$ in (5.35), where a is a scalar. It is also known as moving average (MA) model. Vector $\boldsymbol{b}$ is used to describe the filter.

Definition 5.13 IIR filter

All-pole infinite impulse response (IIR) is also known as auto-regressive (AR) model, where $n = 0$, that is b is a scalar. Vector $\boldsymbol{a}$ can be used to represent the filter.

Compared with all-pole IIR filter, the order of the FIR filter to achieve the same quality is usually higher. The advantage is that stable filters can always be designed [14].

5.6.1 FIR Filter Approximation

For discrete filters that can approximate fractional-order systems, the introduction of FIR filters [15] or IIR filters [16] in the form of a filter can also be considered for their approximation. The FIR filter design function for fractional-order operators [17] was developed by Prof. Ivo Petráš using the `filt()` function in the MATLAB Filter Design Toolbox. The list of its core parts, rewritten in a style consistent with this book, is as follows.

```
function H=dfod2(n,T,r)      %modifications are made here
   arguments
      n(1,1) {mustBePositiveInteger}
      T(1,1) {mustBePositive}, r(1,1) double
   end
   if r>0
      bc=cumprod([1,1-((r+1)./[1:n])]); H=filt(bc,T^r,T);
   elseif r<0
      bc=cumprod([1,1-((-r+1)./[1:n])]); H=filt(T^(-r),bc,T);
end, end
```

where n is the expected filter order, T is the sample time, r is the order of the derivative. Filter H can be designed directly with the function. If a signal is feed to filter H, the rth-order derivative is computed. The output signal can be evaluated from function `lsim()`.

▶ Example 5.20

Differentiate the signal given in Example 5.5 by fractional order using a discrete FIR filter.

Solutions The filters can be designed directly with the following statements. The Bode diagrams are shown in Fig. 5.39. From the resulting frequency response, $n = 100$ is clearly better than the filter at $n = 50$, but obviously much worse than the continuous filter designed with the Oustaloup algorithm presented earlier.

```
>> t=0:0.001:pi;
   y=exp(-t).*sin(3*t+1); y3=glfdiff(y,t,0.5);
   G1=dfod2(50,0.001,0.5); G2=dfod2(100,0.001,0.5);
   bode(G1,G2,'--')
```

The fractional-order derivative of a given signal can also be derived under these filters, as shown in Fig. 5.40. However, the computational accuracy is not ideal and cannot be really used.

```
>> y1=lsim(G1,y,t); y2=lsim(G2,y,t);
   plot(t,y1,t,y2,'--',t,y3,':'), ylim([-1 8])
```

◀

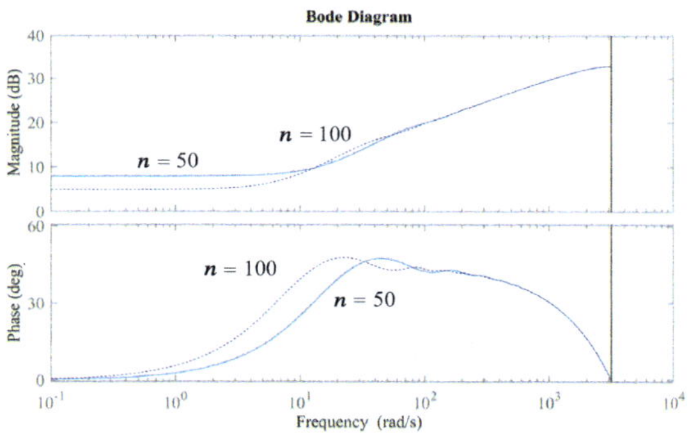

Fig. 5.39 Comparison of frequency responses of numerical differentiators for different (n, a) combinations

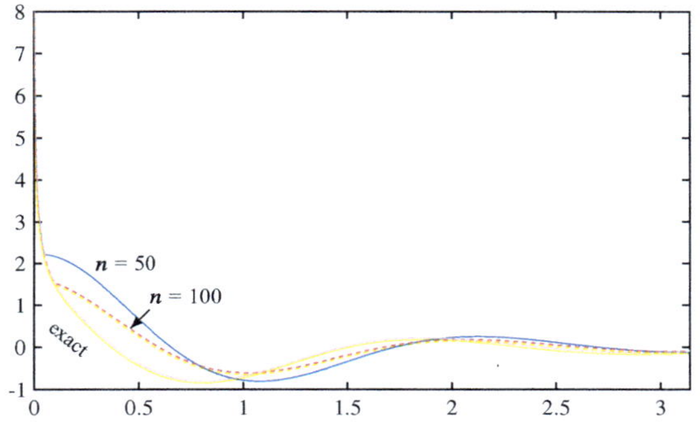

Fig. 5.40 Fractional-order derivatives

5.6.2 IIR Filter Approximation

The FIR filter has a higher order. The IIR filter can be considered to generate a fractional-order derivatives signal. From the continuous system point of view, the fractional-order operators can be written as $s^{\pm\gamma}$ using the Laplace transform. By discretizing it, the transform function $s = w\left(z^{-1}\right)$ can be introduced to approximate the fractional-order derivative or integral actions.

A new method for selecting the transform function is presented in Reference [16] that takes into account the Simpson integral algorithm and the trapezoidal integral algorithm, which can implement the discrete fractional-order operator relatively well.

5

Algorithm 5.9 IIR filter design

Introducing weighting coefficient a, the filter can be expressed as

$$H(z) = aH_S(z) + (1-a)H_T(s), \quad a \in [0, 1], \tag{5.37}$$

so that the IIR filter can be formulated

$$G\left(z^{-1}\right) = k_0^{\pm\gamma}\left[\frac{1-z^{-2}}{(1+bz^{-1})^2}\right]^{\pm\gamma}, \tag{5.38}$$

where $b = (3+a-2\sqrt{3a})/(3-a)$, $k_0 = 6b/[T(3-a)]$, and $\gamma \in (0, 1)$. T is the sample time.

For higher order derivatives, integer-order derivatives can be performed first, and the results are then subjected to fractional-order operators using various filters. Reference [16] also gives a filter design method and a solution function based on the continuous fractional form. However, the function is calculated based on Maple's continued fraction function, which is no longer available under the current versions of MATLAB. MuPAD under the latest MATLAB version does not support the continued fraction form properly, so the core part is replaced by Padé approximation here. The modified function is as follows.

```
function H=iir_pade(r,a,n,T)
   arguments, r(1,1) double, a(1,1){mustBeInRange(a,0,1)}=0
      n(1,1){mustBePositiveInteger}=3
      T(1,1){mustBePositive}=0.01
   end
   syms x, b=(3+a-2*sqrt(3*a))/(3-a);
   k0=6*b/T/(3-a); f=((1-x^2)/(1+b*x)^2)^r;
   c=taylor(f,x,'Order',2*n); c=sym2poly(c);
   c=c(end:-1:1); [N,D]=padefcn(c,n-1,n);
   H=k0^r*tf(N(end:-1:1),D(end:-1:1),...
      'Variable','z^-1','Ts',T);
end
```

The syntax of the function is $H = $ `iir_pade`(r, a, n, T), where r is the order, $a \in [0, 1]$ is the weighting coefficient, n is the filter order, T is the sample time. The new discrete filter H can be designed directly with the function. The following example demonstrates the design of the IIR filter and the fractional-order operator approximation.

▶ Example 5.21

construct an $s^{0.5}$ filter with different combinations of n and a and compare the frequency responses.

Solutions The following filter can be designed immediately using the procedure given earlier. The Bode diagrams of the filters are drawn as shown in Fig. 5.41. From the resulting frequency response curves, if $a = 0.5$ is chosen, the phase curve is not constant in the given region. So the fitting effect will not be ideal, but the assignment at high frequencies will have a decreasing trend, which will have a suppression effect on the noises.

```
>> H1=iir_pade(0.5,0,3,0.01); H2=iir_pade(0.5,0.5,3,0.01);
   H3=iir_pade(0.5,0,4,0.01); H4=iir_pade(0.5,0,5,0.01);
   bode(H1,'-',H2,'--',H3,':',H4,'-.')
```

The filters are denoted by $H_{4,a}\left(z^{-1}\right)$, and the various types of filters designed can be derived. For example

$$H_{4,0}\left(z^{-1}\right) = \frac{113.1 - 113.1z^{-1} - 56.57z^{-2} + 56.57z^{-3}}{8 - 8z^{-2} + z^{-4}}.$$

If the order of the filter is increased, e.g., by choosing $n = 10$ and $a = 0$ and 0.5, respectively, the discrete filters can be designed and the resulting time response fitting results are shown in Fig. 5.42. H_1 is not plotted since its amplification to the noise is too big at high frequencies. It can be seen that even though the selected $a = 0.5$ has suppressed the high frequency gain, the overall time response curve is still not very satisfactory.

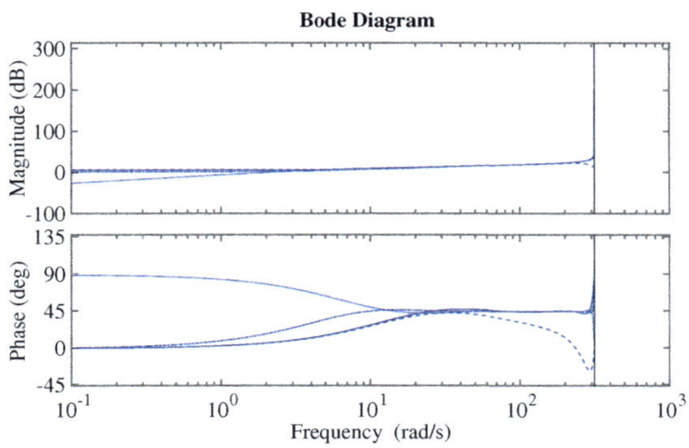

Fig. 5.41 Frequency response fitting for different (n, a) combinations

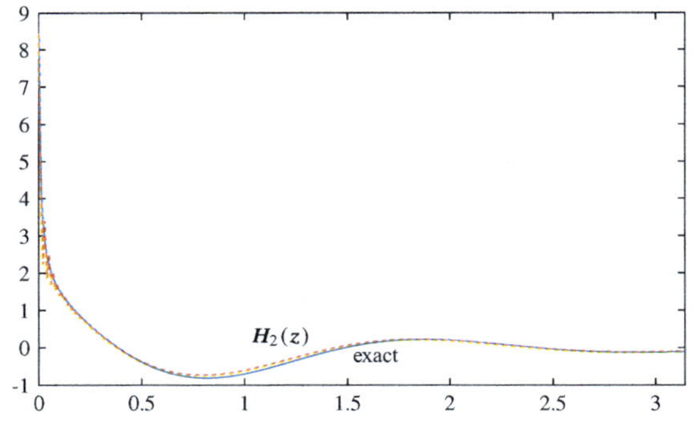

Fig. 5.42 Time domain response comparison

```
>> H1=iir_pade(0.5,0,10,0.01); H2=iir_pade(0.5,0.5,10,0.01);
   t=0:0.01:pi; y=exp(-t).*sin(3*t+1); y0=glfdiff(y,t,0.5);
   y1=lsim(H1,y,t); y2=lsim(H2,y,t); plot(t,y0,t,y2,'--')
```

◀

Compared with the continuous filters introduced earlier, especially the Oustaloup filter, the frequency band fitted by the discrete filter is much narrower than that of the continuous filter. It is often accompanied by ultra-high gain at high frequencies, which is not suitable for finding the fractional-order derivatives of the signal and is not recommended.

5.6.3 Discrete Filters for Step and Impulse Response Invariants

A set of MATLAB functions to design discrete filters for fractional-order operators and even fractional-order irrational transfer functions is listed in Table 5.1. These filters are designed on the basis of step or impulse response invariant [18]. The FOTF Toolbox also provides these three functions that can be called directly.

Table 5.1 Discrete filters for fractional-order operators

Function names	Function syntaxes	Function descriptions
`irid_fod()`	`G=irid_fod(`α, T, N`)`	impulse response invariant for s^α filter design
`srid_fod()`	`G=srid_fod(`α, T, N`)`	s^α filter designed with step response invariant
`irid_folpf()`	`G=irid_folpf(`τ, α, T, N`)`	$(\tau s+1)^{-\alpha}$ filter designed with impulse response invariant

▶ Example 5.22

Choosing a sample time $T = 0.01$ s and making the filter order 5, so that the first two functions are used to design discrete filters with 0.5th-order integrators and compare their effects.

Solutions With the following commands, the two filters can be designed directly. The Bode diagrams of the filters are shown in Fig. 5.43.

```
>> G1=irid_fod(-0.5,0.01,5); G2=srid_fod(-0.5,0.01,5);
   s=fotf('s'); H=bode(s^(-0.5),logspace(-3,3));
   bode(H), hold on, bode(G1,'--',G2,':'), hold off
```

The designed discrete filters are

$$G_1(z) = \frac{0.09354z^5 - 0.2395z^4 + 0.2094z^3 - 0.06764z^2 + 0.003523z + 0.0008224}{z^5 - 3.163z^4 + 3.72z^3 - 1.966z^2 + 0.4369z - 0.02738}$$

$$G_2(z) = \frac{2.38\times 10^{-6}z^5 + 0.1128z^4 - 0.367z^3 + 0.4387z^2 - 0.2269z + 0.04241}{z^5 - 3.671z^4 + 5.107z^3 - 3.259z^2 + 0.882z - 0.05885}.$$

The known $y(t) = \mathrm{e}^{-t}\sin(3t+1)$ signal can be fed into these two filters, and the integral curves of the function can be derived, as shown in Fig. 5.44. From the derived results, the $G_1(z)$ filter is not effective and the $G_2(z)$ filter (i.e., a filter designed based on step response invariant) has a better approximation of the integral behavior, if the order of the filter is increased. Better filters can be expected if the order is increased.

```
>> t=0:0.01:pi; y=exp(-t).*sin(3*t+1);
   y0=glfdiff9(y,t,-0.5,5);
   y1=lsim(G1,y,t); y2=lsim(G2,y,t); plot(t,y0,t,y1,'--',t,y2,':')
```

◀

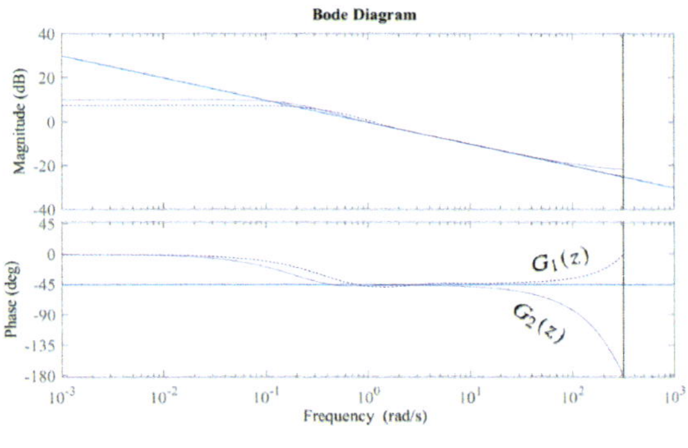

Fig. 5.43 Bode diagram comparison of filters

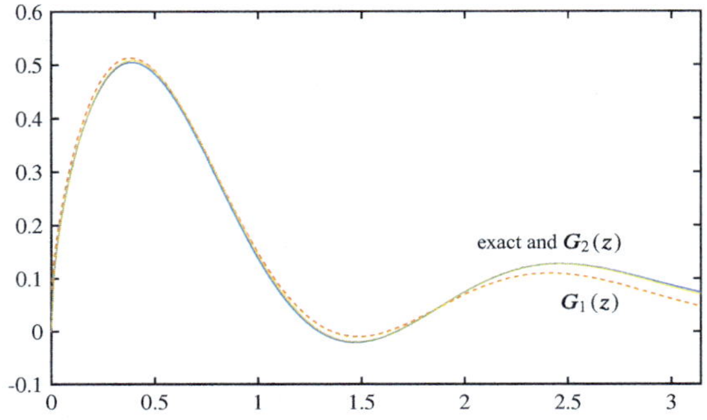

Fig. 5.44 Comparison of the integral effect of filters

▶ Example 5.23

Now consider the benchmark problem of Example 4.9. Design a discrete filter for an integral operator and compare it with the Oustaloup filter to evaluate its approximate accuracy.

Solutions Two discrete filters can be designed by choosing a sample time $T = 0.005$ s. The comparison of the fractional-order integral effect obtained under these filters with the theoretical value is shown in Fig. 5.45. The figure also gives the calculation results derived from the `caputo9()` function, which has a maximum error of 0.0358, resulting in a curve represented by a dotted line with a relatively large error. The maximum error derived from the step response invariant discrete filter is 0.0092, with the dashed line in the figure being closer to the exact value. The maximum error of the impulse response invariant discrete filter is 0.0496, with the dotted line in the figure. Also the maximum error of the Oustaloup filter is 0.0010 and cannot be distinguished in the curve, which coincides exactly with the exact value. It can be seen that the effect of the discrete filter is much worse than that of the Oustaloup filter.

```
>> f=@(t)1/gamma(1.3)*t.^0.3-2/gamma(2.3)*(t-1).^1.3.*(t>1);
   y=@(t)t+1-(t-1).^2.*(t>1); T=0.005; t=[0:T:2]'; y0=f(t);
   N=5; G=srid_fod(-0.7,T,N); ya=y(t); y1=lsim(G,y0,t)+1;
   G=irid_fod(-0.7,T,N); ya=y(t); y4=lsim(G,y0,t)+1;
   G1=ousta_fod(-0.7,15,1e-4,1e3); y3=lsim(G1,y0,t)+1;
```

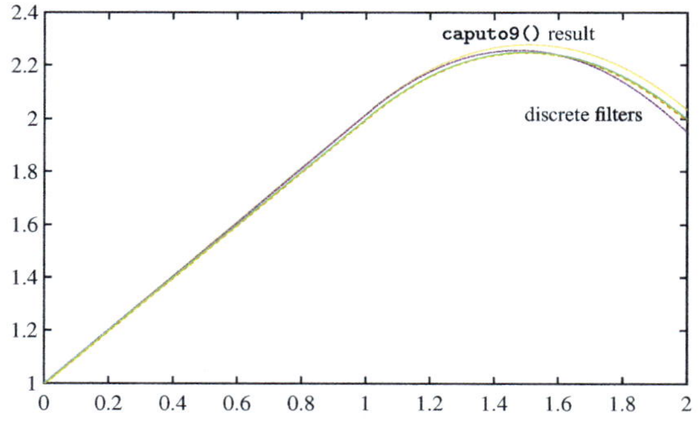

Fig. 5.45 Comparison of solutions for the benchmark problem

```
t1=0:0.02:2; y01=f(t1); ya1=y(t1); y2=caputo9(y01,t1,-0.7,6)+1;
plot(t,ya,t,y1,'--',t1,y2,':',t,y4,'-.',t,y3)
max(abs(y1-ya)), max(abs(y2-ya1(:)))
max(abs(y3-ya)), max(abs(y4-ya))
```

◀

It is important to note that when designing discrete filters and controllers, if the order is obtained too high, it may yield unstable results.

5.7 Exercises

(1) Approximate the fractional-order integral $1/s^{0.8}$ by the function `contfrac()` and apply the Bode diagram to assess the approximation effect.

(2) Use the function `carlson_fod ()` to approximate fractional-order integral $1/s^{0.51}$, with the transfer function and the previously introduced function `glfdiff9()` to calculate ${}_0^{\mathrm{RL}}\mathscr{D}_t^{-0.51} \sin t^2$. Check whether the results obtained are consistent.

(3) Try the function `matsuda_fod()` to approximate the fractional-order integral $1/s^{0.6}$. Choose different parameters to obtain different transfer functions. Use these transfer functions to evaluate ${}_0^{\mathrm{RL}}\mathscr{D}_t^{-0.6}\mathrm{e}^{-t}$, and validate the result with `ml_func()` function.

(4) Approximate the fractional-order integral $1/s^{0.7}$ by the function `ousta_fod()` and calculate ${}_0^{\mathrm{RL}}\mathscr{D}_t^{-0.7}\sqrt{\sin t + 3}$. Design a numerical algorithm to check whether the previous results are correct.

(5) Use the function `ousta_fod()` to approximate the fractional-order derivative $s^{0.3}$. Choose different parameters can get different transfer functions. Use these designed transfer functions to calculate ${}_0^{\mathrm{RL}}\mathscr{D}_t^{0.3}\sqrt{\cosh t}$, and validate the results.

(6) Approximate the fractional-order derivative $s^{0.2}$ with the function `new_fod()` and calculate ${}_0^{\mathrm{RL}}\mathscr{D}_t^{0.2}\left(1/\Gamma(t)\right)$. Check whether the previous calculation is correct.

(7) In this chapter, various continuous filter design methods are introduced, including the continued fraction method, Matsuda–Fujii method, Carlson method, Oustaloup method, and Charef method. Again, $\alpha = 1/2$ and $\alpha = -1/2$ are chosen to find the best possible fit that can be obtained in the frequency band of $\omega \in [10^{-4}, 10^4]$ by adjusting the design parameters. Also, the accuracy of the fractional-order derivative and integral calculation is evaluated using the designed filter.

(8) Considering the problem of Example 5.6. Find the appropriate frequency band and order so that the Oustaloup filter can be used in $t \in [0, 5000]$ s interval fits well the behavior of the $s^{-0.5}$ fractional-order operator.

(9) Input the FOTF into the MATLAB workspace [8].

$$G(s) = \frac{\left(s^{0.3}+3\right)^2}{\left(s^{0.2}+3\right)\left(s^{0.4}+4\right)\left(s^{0.4}+3\right)}.$$

(10) Assume that the models of a typical unity negative feedback control system are

$$G(s) = \frac{0.8s^{1.2}+2}{1.1s^{1.8}+0.8s^{1.3}+1.9s^{0.5}+0.4},\quad G_c(s) = \frac{1.2s^{0.72}+1.5s^{0.33}}{3s^{0.8}}.$$

input this model into MATLAB workspace [8].

(11) Input the following FOTF model into the MATLAB workspace, and use the function `ousta_fod()` and `new_fod()` are fitted to this model.

$$G(s) = \frac{0.5s^{0.9}+1}{s^3+23s^{2.5}+19s^{0.7}+1}.$$

(12) Consider the following irrational fractional-order model and apply the frequency response approximation method to find its rational approximation model.

$$G(s) = \frac{1}{\sqrt{s^2+s+1}}.$$

(13) The multivariable FOTF matrix is

$$\boldsymbol{G}(s) = \begin{bmatrix} \dfrac{1}{1.35s^{1.2}+2.3s^{0.9}+1} & \dfrac{2}{4.13s^{0.7}+1} \\ \dfrac{1}{0.52s^{1.5}+2.03s^{0.7}+1} & -\dfrac{1}{3.8s^{0.8}+1} \end{bmatrix}.$$

Input it into MATLAB workspace.

(14) Consider the irrational fractional-order model $G(s) = 1/\sqrt{s+1}$. Use the function `charef_fod()` to find its rational approximation.

(15) Consider the irrational model $G(s) = \mathrm{e}^{-\sqrt{s}}/(\sqrt{s}+1)$. Draw the step response curve. The step response of the system is shown below. Evaluate the accuracy of the computational step response at different step sizes.

$$y(t) = \mathrm{erfc}\left(\frac{1}{2\sqrt{t}}\right) - \mathrm{e}^{t+1}\mathrm{erfc}\left(\sqrt{t}+\frac{1}{2\sqrt{t}}\right).$$

(16) Consider the implicit irrational system model $G(s) = 1/(5s+1)^{0.8}$. The analytical solution of its step response can be derived as $y(t) = (t/5)^{0.8}\,\mathrm{E}_{1,1.8}^{0.8}(-t/5)$. Compare the numerical inverse Laplace transform, continuous and discrete filters, and evaluate the effect of the parameters on the accuracy of the solutions.

(17) The following is a model for the identification of ion-exchange polymerized metal materials [19].

$$G(s) = \frac{340}{s^{0.756}(s^2+3.85s+5880)^{1.15}}.$$

find its rational approximation models using various methods provided in this chapter and compare the frequency domain properties of these rational approximation models.

(18) Consider the following irrational fractional-order model:

$$G(s) = \frac{1}{(1+2s)^{0.4}(1+4s)^{0.2}}.$$

Build a Simulink block diagram using the Implicit model block. Let the input signal be $u(t) = \cos t/\Gamma(t)$, use the Simulink block diagram to calculate the output of the model, and try to verify the solution.

(19) Select the sample time $T = 0.01$ s, filter order 5. Use the functions `irid_fod()` and `srid_fod()` to design the discrete filters for the 0.7th-order integrator. Apply the obtained filter to calculate ${}_{0}^{\rm RL}\mathscr{D}_t^{-0.7}\big(\cosh t/\Gamma(t)\big)$, and use `glfdiff9()` function to validate the results.

(20) Choose a suitable integer-order transfer function to approximate the following fractional-order model and compare the frequency response fit [8].

$$G(s) = \frac{25}{(s^2+8.5s+25)^{0.2}}, \quad G(s) = \frac{562920(s+1.0118)^{0.6774}}{(s^2+54.7160s+590570)^{0.8387}}.$$

(21) Known fractional-order models [8]

$$G_1(s) = \frac{5}{s^{2.3}+1.3s^{0.9}+1.25}, \quad G_2(s) = \frac{5s^{0.6}+2}{s^{3.3}+3.1s^{2.6}+2.89s^{1.9}+2.5s^{1.4}+1.2}.$$

Find the integer-order model that can fit the original model well. Discuss which order combination can give better results. Compare the reduced-order model of the system in terms of frequency response and step response.

References

1. La OA (1995) dérivation non entière: Théorie, synthèse et applications[M]. Hermès, Paris
2. Petráš I (2011) Fractional-order nonlinear systems: modelling, analysis and simulation[M]. Higher Education Press, Beijing
3. Chen YQ, Vinagre BM (2003) A new IIR-type digital fractional order differentiator[J]. Signal Process 83(11):2359~2365
4. Petráš I, Podlubny I, O'Leary P (2002) Analogue realization of fractional order controllers[R]. Fakulta BERG, TU Košice
5. Matsuda K, Fujii H (1993) $\mathcal{H}_\infty$-optimized wave-absorbing control: analytical and experimental results[J]. J Guid Control Dyn 16(6):1146~1153
6. Oustaloup A, Levron F, Nanot F et al (2000) Frequency band complex non integer differentiator: characterization and synthesis[J]. IEEE Trans Circuits Syst I: Fund Theory Appl 47(1):25~40
7. Xue DY, Zhao CN, Chen YQ (2006) A modified approximation method of fractional order system[C]. In: Proceedings of IEEE conference on mechatronics and automation, Luoyang, China, pp 1043~1048
8. Xue DY (2022) Computer-aided design of control systems with MATLAB (in Chinese)[M], 4th edn. Tsinghua University Press, Beijing
9. Xue DY, Chen YQ (2007) Suboptimum $\mathcal{H}_2$ pseudo-rational approximations to fractional-order linear time invariant systems. // Sabatier J, Agrawal OP, Machado JAT (2007) Advances in fractional calculus—theoretical developments and applications in physics and engineering[M]. Springer, Dordrecht

10. Sabatier J, Farges C (2015) Analysis of fractional models physical consistency[J]. J Vib Control 23(6):895~908
11. Monje CA, Chen YQ, Vinagre BM et al (2010) Fractional-order systems and controls: fundamentals and applications[M]. Springer, London
12. Charef A, Sun HH, Tsao YY et al (1992) Fractal system as represented by singularity function[J]. IEEE Trans Autom Control 37(9):1465~1470
13. Meng L, Xue DY (2012) A new approximation algorithm of fractional order system models based optimization[J]. J Dyn Syst Meas Control 134(4):504~511
14. MathWorks (2007) Signal processing toolbox user's guide [Z]
15. Tseng C-C, Pei S-C, Hsia S-C (2000) Computation of fractional derivatives using Fourier transform and digital FIR differentiator[J]. Signal Process 80(1):151~159
16. Chen YQ, Vinagre BM (2003) A new IIR-type digital fractional order differentiator[J]. Signal Process 83:2359~2365
17. Petráš I (2022) Digital fractional order differentiator/integrator—FIR type[OL]. [Dec 8, 2022] ► https://www.mathworks.com/matlabcentral/fileexchange/3672-digital-fractional-order-differentiator-integrator-iir-type
18. Chen YQ, Impulse response invariant discretization of fractional order integrators/differentiators [OL]. ► http://www.mathworks.cn/matlabcentral/fileexchange/21342-impulse-response-invariant-discretization-of-fractional-order-integrators-differentiators
19. Caponetto R, Dongola G, Fortuna L et al (2010) Fractional order systems: modeling and control applications[M]. World Scientific, Singapore

5

Analytical and Numerical Solutions of Linear Fractional-Order Differential Equations

Contents

D. Y. Xue and L. Bai, *Fractional Calculus*, https://doi.org/10.1007/978-981-99-2070-9_6

In the past, the familiar dynamical systems are described by integer-order differential equations. Accordingly, the fractional-order dynamical systems are described by fractional-order differential equations (FODEs). Therefore, the solution of FODEs is a very crucial part in the analysis and design of fractional-order systems.

In this chapter, the solution problems of linear FODEs are explored. A general form of linear FODEs is presented in ▸ Sect. 6.1, then some common Laplace transform formulas are listed. It is shown that the description of Caputo equations is more suitable for nonzero initial value problems. An important formula for Laplace transform is given from which the analytical solutions of many FODEs can be obtained. The analytical solutions of some linear FODEs are discussed in ▸ Sect. 6.2. In ▸ Sect. 6.3, the analytical solutions of linear commensurate-order differential equations are introduced, based on partial fraction expansions and Laplace transform. The numerical solutions of differential equations are explored for which the analytical solutions do not exist. In ▸ Sect. 6.4, a closed-form algorithm is proposed for linear FODEs with zero initial values. A high-precision numerical algorithm is proposed, and the matrix method is further explored. The numerical solution algorithm is presented in ▸ Sect. 6.5 for the Caputo equation with nonzero initial values, and introduces the equivalent initial condition estimation method The high-precision closed-form solution method is proposed for linear Caputo equations. In ▸ Sect. 6.6, the state space equation description method is introduced for linear fractional-order systems, the solution method for the state transition matrix is studied. The numerical solution of the irrational transfer function can be found directly by the numerical Laplace transform and inverse Laplace transform tools, which are discussed in ▸ Sect. 6.7, giving the time response of systems that are difficult to describe by differential equations. The stability assessment methods are introduced in ▸ Sect. 6.8 for various types of linear fractional-order systems. The stability criterion for commensurate-order systems is introduced first, and then the stability assessment methods for non-commensurate-order systems and even irrational systems are explored.

The command-driven methods for the numerical solutions of nonlinear FODEs are described in detail in ▸ Chap. 7 of this book. The block diagram-based approach to solving FODEs is presented in ▸ Chap. 8, while numerical solutions of other FODEs are explored in ▸ Chap. 9.

6.1 Introduction to Linear Fractional-Order Differential Equations

The standard form of linear FODEs is given first in this section. The FOTF models and FODEs with nonzero initial values are then discussed, and the Laplace transform methods under different fractional calculus definitions are explored.

6.1.1 The General Form of Linear Fractional-Order Differential Equations

Definition 6.1 Linear time-invariant FODE

A class of linear time-invariant (LTI) FODEs [1] is given by

$$\begin{aligned} &a_1\mathscr{D}_t^{\eta_1}y(t)+a_2\mathscr{D}_t^{\eta_2}y(t)+\cdots+a_{n-1}\mathscr{D}_t^{\eta_{n-1}}y(t)+a_n\mathscr{D}_t^{\eta_n}y(t)\\ &=b_1\mathscr{D}_t^{\gamma_1}u(t)+b_2\mathscr{D}_t^{\gamma_2}u(t)+\cdots+b_m\mathscr{D}_t^{\gamma_m}u(t), \end{aligned} \tag{6.1}$$

where b_i and a_i are real coefficients; γ_i and η_i are orders. Signal $u(t)$ can be regarded as the input to the system described by the FODE, while $y(t)$ is the output signal.

6

If the initial values of the function $y(t)$ and its derivatives of all orders are zero, then the FODE is referred to as the FODE with zero initial values. In this case, the various fractional calculus definitions are equivalent.

Just as in the case of linear integer-order systems, transfer function can be defined as the ratio of the Laplace transforms of the output signal to the input signal, a fractional-order transfer function (FOTF) model can be defined from the FODEs. In fact, the FOTF model is the gain in the s-domain of the system described by the linear FODE.

Definition 6.2 Fractional-order transfer function

The general form of a FOTF is

$$G(s)=\frac{b_1s^{\gamma_1}+b_2s^{\gamma_2}+\cdots+b_ms^{\gamma_m}}{a_1s^{\eta_1}+a_2s^{\eta_2}+\cdots+a_{n-1}s^{\eta_{n-1}}+a_ns^{\eta_n}}. \tag{6.2}$$

Definition 6.3 Pseudo-polynomials

Pseudo-polynomial can be described by two vectors ($\boldsymbol{c}$ and $\boldsymbol{\alpha}$).

$$p(s)=c_1s^{\alpha_1}+c_2s^{\alpha_2}+\cdots+c_{n-1}s^{\alpha_{n-1}}+c_ns^{\alpha_n}. \tag{6.3}$$

A FOTF model is the ratio of two pseudo-polynomials. The FOTF model is not only an effective method for describing linear FODEs but also an effective tool for the analysis and design of linear fractional-order systems. The FOTF Toolbox accompanying this book is developed on the basis of the FOTF models.

6.1.2 Initial Value Problems of Fractional-Order Derivatives Under Different Definitions

If the initial values of a function $f(t)$ and its derivatives of all orders are not zero, then the inverse Laplace transform of the integer-order derivative of the function can be found as follows:

$$\mathscr{L}\left[\frac{\mathrm{d}^n}{\mathrm{d}t^n}f(t)\right] = s^n F(s) - \sum_{k=0}^{n-1} s^{n-k-1} f^{(k)}(t_0). \tag{6.4}$$

Now we will discuss how to compute the Laplace transform of the fractional-order derivative and integral of a function under the Riemann–Liouville definition. Let $F(s) = \mathscr{L}[f(t)]$, then the Laplace transform of the fractional-order derivatives and integrals of a function $f(t)$ satisfies the following theorem.

Theorem 6.1 Laplace transform of Riemann–Liouville derivatives
If $n = \lceil \gamma \rceil$, the Laplace transform of the Riemann–Liouville fractional-order derivative and integral of function $f(t)$ satisfies [1]

$$\mathscr{L}\left[{}_{t_0}^{\mathrm{RL}}\mathscr{D}_t^{-\beta} f(t)\right] = s^{-\beta} F(s) \tag{6.5}$$

$$\mathscr{L}\left[{}_{t_0}^{\mathrm{RL}}\mathscr{D}_t^{\gamma} f(t)\right] = s^{\gamma} F(s) - \sum_{k=0}^{n-1} s^k \, {}_{t_0}^{\mathrm{RL}}\mathscr{D}_t^{\gamma-k-1} f(t)\Big|_{t=t_0}. \tag{6.6}$$

Proof Let us first look at the integral formula. The Laplace transform of the convolution function can be written as

$$\mathscr{L}[f(t) * g(t)] = F(s)G(s).$$

Let $g(t) = t^{\beta-1}$, it follows that

$${}_{t_0}^{\mathrm{RL}}\mathscr{D}_t^{-\beta} f(t) = \frac{1}{\Gamma(\beta)} \int_0^t \frac{f(\tau)}{(t-\tau)^{1-\beta}} \mathrm{d}\tau = t^{\beta-1} * f(t).$$

It can be shown that $G(s) = \Gamma(\beta) s^{-\beta}$, such that the integral formula is proved.

$$\mathscr{L}\left[{}_{t_0}^{\mathrm{RL}}\mathscr{D}_t^{-\beta} f(t)\right] = s^{-\beta} F(s). \tag{6.7}$$

Now let us explore the Laplace transform formula for the fractional-order derivatives. Let ${}_{t_0}^{\mathrm{RL}}\mathscr{D}_t^{\gamma} f(t) = g^{(n)}(t)$, where

$$g(t) = \frac{1}{\Gamma(n-\gamma)} {}_{t_0}^{\mathrm{RL}}\mathscr{D}_t^{-(n-\gamma)} f(t) = \int_0^t \frac{f(\tau)}{(t-\tau)^{\gamma+1-n}} \mathrm{d}\tau.$$

It is straightforward to see from (6.7) that

$$g^{(n-k-1)}(t) = \frac{\mathrm{d}^{n-k-1}}{\mathrm{d}t^{n-k-1}} {}_{t_0}^{\mathrm{RL}}\mathscr{D}_t^{-(n-\gamma)} f(t) = {}_{t_0}^{\mathrm{RL}}\mathscr{D}_t^{\gamma-k-1} f(t). \tag{6.8}$$

Substituting (6.7) and (6.8) into (6.4), it is found that (6.6) is satisfied, thus proving the derivative property. □

Similar to linear integer-order differential equations, the Laplace transform is an effective tool for dealing with linear FODEs. A very useful Laplace transform formula is given directly below.

6

Theorem 6.2 Laplace transform of Caputo derivatives
If $n = \lceil \gamma \rceil$, the Laplace transform of Caputo derivative satisfies [2]

$$\mathscr{L}\left[{}_{t_0}^{\mathrm{C}}\mathscr{D}_t^{\gamma} f(t)\right] = s^{\gamma} F(s) - \sum_{k=0}^{n-1} s^{\gamma-k-1} f^{(k)}(t_0). \tag{6.9}$$

It can be seen that the initial values of the function $f(t)$ and its integer-order derivatives are involved. Therefore, Caputo equations are more suitable for describing FODEs or fractional-order systems with nonzero initial values.

From (6.1), it can be seen that if $q = \lceil \max(\eta_i) \rceil$, then the condition for the FODE to have a unique solution is that q independent initial values are known.

Theorem 6.3 Laplace transform of fractional-order integrals
The Laplace transforms of Grünwald–Letnikov and Caputo fractional-order integrals both satisfy

$$\mathscr{L}\left[{}_{t_0}^{\mathrm{GL}}\mathscr{D}_t^{\gamma} f(t)\right] = \mathscr{L}\left[{}_{t_0}^{\mathrm{C}}\mathscr{D}_t^{\gamma} f(t)\right] = s^{\gamma} F(s), \quad \gamma \leqslant 0. \tag{6.10}$$

Since the definition of two fractional-order integrals is equivalent, this theorem can be proved directly by Theorem 6.1. In addition, since this book is not a strictly mathematical work, the proofs of some theorems are not given and the interested reader can refer to other works, such as Ref. [1].

The Caputo equation is suitable for describing practical systems with nonzero initial values, while the Grünwald–Letnikov and Riemann–Liouville equations are more suitable for describing fractional-order systems that are initially at rest.

6.1.3 An Important Laplace Transform Formula

Similar to linear integer-order differential equations, the Laplace transform is an effective tool for dealing with linear FODEs. A very useful Laplace transform formula is given directly below.

Lemma 6.1 If $\mathbb{R}(\gamma) > 0$, $\mathbb{R}(\beta) > 0$ and $|z| < 1$, then there exists [3]

$$\frac{1}{(1+z)^\gamma} = \int_0^\infty \mathrm{e}^{-x} x^{\beta-1} \mathrm{E}_{\alpha,\beta}^{\gamma}\left(-x^\alpha z\right) \mathrm{d}x. \tag{6.11}$$

Theorem 6.4 A very important Laplace transform formula
A very important Laplace transform formula is given below [3].

$$\mathscr{L}^{-1}\left[\frac{s^{\alpha\gamma-\beta}}{(s^\alpha+a)^\gamma}\right] = t^{\beta-1}\mathrm{E}_{\alpha,\beta}^{\gamma}\left(-at^\alpha\right). \tag{6.12}$$

Proof It can be seen from Lemma 6.1 that this theorem can be proved if $z = as^{-\alpha}$ and $x = st$ are substituted into the integral expression of (6.11).

$$\begin{aligned}\frac{1}{(1+as^{-\alpha})^\gamma} &= \frac{s^{\alpha\gamma}}{(s^\alpha+a)^\gamma} = \int_0^\infty \mathrm{e}^{-st}(st)^{\beta-1}\mathrm{E}_{\alpha,\beta}^{\gamma}\left(-(st)^\alpha as^{-\alpha}\right) s\,\mathrm{d}t \\ &= s^\beta \mathscr{L}\Big[t^{\beta-1}\mathrm{E}_{\alpha,\beta}^{\gamma}\left(-at^\alpha\right)\Big].\end{aligned} \tag{6.13}$$

□

6.2 Analytical Solutions of Some Linear FODEs

A typical linear n-term FODE refers to a differential equation with the weighted sum of n output signals and their fractional-order derivatives in the equation, and (6.1) is an example of such an equation. In this section, the analytical solution methods for one-, two-, three- and general n-term FODEs are discussed.

6.2.1 One-Term FODEs

Definition 6.4 Linear one-term FODE

The general form of a linear one-term FODE is

$$a\,\mathscr{D}_t^{\eta} y(t) = u(t). \tag{6.14}$$

It can be seen that the solution of the FODE with the zero initial values can be written directly as

$$y(t) = \frac{1}{a}\,\mathscr{D}_t^{-\eta} u(t). \tag{6.15}$$

That is, the solution of this differential equation is in fact the fractional-order integral of the input signal $u(t)$. The numerical solution of this FODE can be evaluated with the high-precision numerical algorithm described in ▶ Chap. 4.

6.2.2 Two-Term FODEs

Definition 6.5 Linear two-term FODE

The mathematical form of linear two-term FODE is

$$a\,\mathscr{D}_t^{\eta} y(t) + by(t) = u(t). \tag{6.16}$$

The Laplace transform is taken at both sides of the equation, and the output signal of this FODE can be directly derived as

$$Y(s) = \frac{1}{as^{\eta} + b}\mathscr{L}\big[u(t)\big].$$

Theorem 6.5 Step and impulse responses of linear two-term FODEs

For signal input signals such as $u(t) = \delta(t)$ and $u(t) = H(t)$, i.e., when the input is step or impulsive signals, whose Laplace transforms are respective $\mathscr{L}\big[\delta(t)\big] = 1$ or $\mathscr{L}\big[H(t)\big] = 1/s$, the analytical solution of the FODE are

$$y_{\delta}(t) = \frac{t^{\eta-1}}{a}\mathrm{E}_{\eta,\eta}\left(-\frac{b}{a}t^{\eta}\right) \tag{6.17}$$

$$y_{\mathrm{H}}(t) = \frac{t^{\eta}}{a}\mathrm{E}_{\eta,\eta+1}\left(-\frac{b}{a}t^{\eta}\right). \tag{6.18}$$

Proof Assuming that the input signal is a unit impulsive signal with $u(t) = \delta(t)$, the Laplace transform expression $Y(s)$ for the output signal can be derived directly as

$$Y(s) = \frac{1}{as^\eta + b} = \frac{1}{a}\frac{1}{s^\eta + b/a}.$$

Recalling the Laplace transform equation in (6.12), if $\gamma = 1, \alpha = \eta$ and $\beta = \eta$ are chosen, the output signal in (6.17) can be derived.

If the input signal is a Heaviside function with $u(t) = H(t)$, the Laplace transform of the output signal is

$$Y(s) = \frac{1}{as^\eta + b}\frac{1}{s} = \frac{1}{a}\frac{1}{s(s^\eta + b/a)}.$$

Therefore, if we choose $\gamma = 1$, $\alpha = \eta$ and $\beta = \eta + 1$, the result in (6.18) can be derived, and this theorem is proved. □

Since such a FODE is too simple, it is only a special case of what will be introduced in ▸ Sect. 6.3, so its solution will not be discussed in depth here for now.

6.2.3 Three-Term FODEs

Definition 6.6 Linear three-term FODE

The general form of the linear three-term FODE is

$$a_1 \mathscr{D}_t^{\eta_1} y(t) + a_2 \mathscr{D}_t^{\eta_2} y(t) + a_3 y(t) = u(t), \tag{6.19}$$

where the initial values of $y(t)$ and its derivatives are zero.

Theorem 6.6 Step response of linear three-term FODE

If input signal $u(t)$ is a unit step signal, the analytical solution of the FODE can be written as [4]

$$y(t) = \frac{1}{a_1}\sum_{k=0}^{\infty} \frac{(-1)^k \hat{a}_3^k t^{-\hat{a}_2+(k+1)\eta_1}}{k!} \frac{\mathrm{d}^k}{\mathrm{d}t^k} \mathrm{E}_{\eta_1-\eta_2,\eta_1+\eta_2 k+1}\big(-\hat{a}_2 t^{\eta_1-\eta_2}\big), \tag{6.20}$$

where $\hat{a}_3 = a_3/a_1$ and $\hat{a}_2 = a_2/a_1$.

If the numerical implementation is considered, the output signal $y(t)$ can be evaluated by the truncation algorithm, while the Mittag-Leffler function and its derivatives are directly evaluated by the `ml_func()` function, so that the `ml_step3()` function can be written to find the numerical solution of the step response of the linear three-term FODE as follows.

```
function y=ml_step3(a,b,t,eps0)
   arguments, a(1,3), b(1,2), t(:,1), eps0(1,1)=eps; end
   y=0; k=0; ya=1;
   a1=a(1); a2=a(2)/a1; a3=a(3)/a1; b1=b(1); b2=b(2);
   while max(abs(ya))>=eps0
      ya=(-1)^k/gamma(k+1)*a3^k*t.^((k+1)*b1).*...
          ml_func([b1-b2,b1+b2*k+1],-a2*t.^(b1-b2),k,eps0);
      y=y+ya; k=k+1;
   end, y=y/a1;
end
```

The syntax of the function is y = ml_step3(a, b, t, ϵ_0), where $\boldsymbol{a} = [a_1, a_2, a_3]$, $\boldsymbol{b} = [\eta_1, \eta_2]$, ϵ_0 is the error tolerance, and it can be omitted, implying that its default value eps is adopted.

▶ Example 6.1

Find the analytical and numerical solutions of the step response of a linear three-term FODE.

$$\mathscr{D}_t^{0.8}y(t) + 0.75\mathscr{D}_t^{0.4}y(t) + 0.9y(t) = 5u(t).$$

Solutions It is obvious that $a_1 = 1$, $a_2 = 0.75$, $a_3 = 0.9$, $\eta_1 = 0.8$, and $\eta_2 = 0.4$, so it is straightforward to derive the analytical solution of the step response of the differential equation as

$$y(t) = \sum_{k=0}^{\infty} \frac{(-1)^k 0.9^k t^{-0.75+0.8(k+1)}}{k!} \frac{\mathrm{d}^k}{\mathrm{d}t^k} \mathrm{E}_{0.4,1.8+0.4k}\left(-0.75t^{0.4}\right). \tag{6.21}$$

The numerical solution of this equation can be found directly from the following statements, and the plot obtained is as shown in ◻ Fig. 6.1.

```
>> t=0:0.001:5;
   y=ml_step3([1,0.75,0.9],[0.8,0.4],t); plot(t,5*y)
```

◀

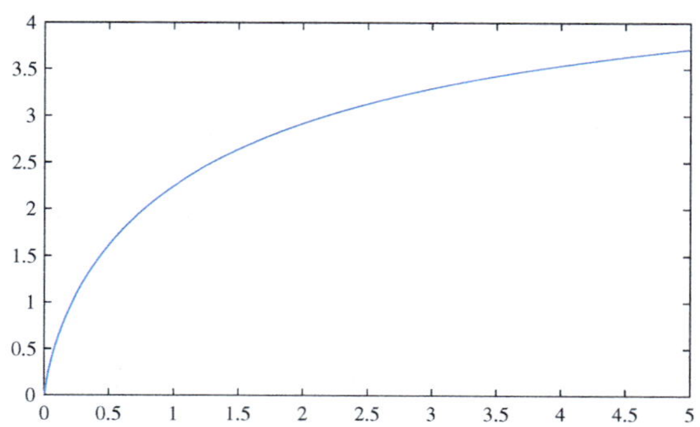

◻ **Fig. 6.1** Numerical solution of the step response of the three-term equation

6.2.4 General *n*-Term FODEs

Definition 6.7 Linear n-term FODE

The mathematical form of the linear n-term FODE is

$$a_1 \mathscr{D}_t^{\eta_1} y(t) + a_2 \mathscr{D}_t^{\eta_2} y(t) + \cdots + a_{n-1} \mathscr{D}_t^{\eta_{n-1}} y(t) + a_n \mathscr{D}_t^{\eta_n} y(t) = u(t). \tag{6.22}$$

Definition 6.8 Multinomial coefficients

The definition of multinomial coefficients is

$$(m; k_2, k_3, \ldots, k_n) = \frac{m!}{k_2! k_3! \ldots k_n!}. \tag{6.23}$$

Theorem 6.7 Analytical solutions of linear n-term FODE

If the input $u(t)$ is a unit step signal, the analytical solution of linear n-term FODE is [1]

$$\begin{aligned} y(t) = \frac{1}{a_1} \sum_{m=0}^{\infty} \frac{(-1)^m}{m!} \sum_{\substack{k_2+\cdots+k_n=m \\ k_2\geqslant 0,\ \ldots,\ k_n\geqslant 0}} (m; k_2, k_3, \ldots, k_n) \times \\ \prod_{i=2}^{n} \left(\frac{a_i}{a_1}\right)^{k_i} t^{(\eta_1-\eta_2)m+\eta_1+\sum\limits_{j=2}^{n}(\eta_2-\eta_j)k_j-1} \times \\ \frac{\mathrm{d}^m}{\mathrm{d}t^m} \mathrm{E}_{\eta_1-\eta_2,\,-\eta_2+\sum\limits_{j=2}^{n}(\eta_2-\eta_j)k_j} \left(-\frac{a_2}{a_1} t^{\eta_1-\eta_2}\right). \end{aligned} \tag{6.24}$$

where $(m; k_2, k_3, \ldots, k_n)$ are multinomial coefficients.

As can be seen from the previous analytical solution expressions, it is extremely difficult to actually write an analytical solution for a general n-term FODE, and the solution is limited to a step-response solution, so the solution is not of much practical value. This method can be used to solve at most three-term differential equations with step response analytically. Thus, better algorithms and tools are needed to solve linear FODEs with an arbitrary number of terms and arbitrary input signals.

6.3 Analytical Solutions of Linear Commensurate-Order FODEs

A particularly important property of the Laplace transform is given in (6.12), and we would like to use this property to derive the analytical solutions of the step response and impulse response of the linear commensurate-order differential equations.

6.3.1 The General Form of Linear Commensurate-Order Differential Equations

Consider the order of the FODE of (6.1). If all the orders are rational fractions, then these order terms can be written as

$$(\eta_1, \eta_2, \ldots, \eta_n, \gamma_1, \gamma_2, \ldots, \gamma_m) = (c_1/d_1, c_2/d_2, \ldots, c_M/d_M). \tag{6.25}$$

where $M \leqslant n+m$ is the number of terms after eliminating duplicate terms. Assuming that the pairs (c_i, d_i) are coprime integers, the least common multiple (LCM) of d_i and the greatest common multiple (GCD) of c_i can be found, respectively,

$$d = \mathrm{lcm}(d_1, d_2, \ldots, d_M), \quad c = \gcd(c_1, c_2, \ldots, c_M). \tag{6.26}$$

The resulting order $\alpha = c/d$ is referred to as the base order of the FODE. Letting $\lambda = s^\alpha$, the original FODE can be mapped into an integer-order transfer function with respect to λ. If it is used only for the analysis of system stability, it is sufficient to extract the base order for the order of the denominator.

Definition 6.9 Commensurate-order differential equation

If the base order α is selected, the commensurate-order differential equation can be written as

$$\begin{aligned} &a_1 \mathscr{D}_t^{n\alpha} y(t) + a_2 \mathscr{D}_t^{(n-1)\alpha} y(t) + \cdots + a_n \mathscr{D}_t^{\alpha} y(t) + a_{n+1} y(t) \\ &= b_1 \mathscr{D}_t^{m\alpha} u(t) + b_2 \mathscr{D}_t^{(m-1)\alpha} u(t) + \cdots + b_m \mathscr{D}_t^{\alpha} u(t) + b_{m+1} u(t). \end{aligned} \tag{6.27}$$

Note that the coefficients a_i and b_i are different from those in (6.1).

Definition 6.10 Commensurate-order transfer functions

The integer-order transfer function $\lambda = s^\alpha$ and base order α can be used to express directly the commensurate-order transfer functions.

$$G(\lambda) = \frac{b_1 \lambda^m + b_2 \lambda^{m-1} + \cdots + b_m \lambda + b_{m+1}}{a_1 \lambda^n + a_2 \lambda^{n-1} + \cdots + a_n \lambda + a_{n+1}}. \tag{6.28}$$

Theorem 6.8 Special partial fraction expansion
For commensurate-order system of λ, if there is a set of distinct poles $-p_i$, the partial fraction expansion of the integer-order transfer function can be written as

$$G(\lambda) = \sum_{i=1}^{n} \frac{r_i}{\lambda + p_i} = \sum_{i=1}^{n} \frac{r_i}{s^\alpha + p_i}. \tag{6.29}$$

If there exists a repeated pole $-p_i$ and its multiplicity is k, then the corresponding terms in the partial fraction expansion become

$$\frac{r_{i1}}{s^\alpha + p_i} + \frac{r_{i2}}{(s^\alpha + p_i)^2} + \cdots + \frac{r_{ik}}{(s^\alpha + p_i)^k} = \sum_{j=1}^{k} \frac{r_{ij}}{(s^\alpha + p_i)^j}. \tag{6.30}$$

Definition 6.11 Partial fraction expansion of commensurate-order system

For the commensurate-order system with base order α, the general form of the partial fraction expansion is

$$G(s) = \sum_{i=1}^{N} \sum_{j=1}^{m_i} \frac{r_{ij}}{(s^\alpha + p_i)^j}, \tag{6.31}$$

where p_i and r_{ij} are complex numbers. The multiplicity m_i of the ith pole p_i is a positive integer, and $m_1 + m_2 + \cdots + m_N = n$.

The partial fraction expansion of the commensurate-order system can be obtained directly by using MATLAB's `residue()` function or, if symbolic computing is used, by using the `partfrac()` function.

▶ Example 6.2

Perform a partial fraction expansion of the following fractional-order transfer function with numerical and analytical methods, respectively:

$$G(s) = \frac{1}{s^{2.4} + 10s^2 + 40s^{1.6} + 82s^{1.2} + 91s^{0.8} + 52s^{0.4} + 12}.$$

Solutions From the existing orders, it can be seen that the base order is $\alpha = 0.4$. Denote $\lambda = s^{0.4}$, the transfer function can be entered into the MATLAB workspace, where $s^{0.4}$ is substituted by λ. To avoid unnecessary trouble, it is necessary to set s and λ as positive numbers temporarily. Take partial fraction expansion of the λ expression and replace t with $s^{0.4}$ to get the desired partial fraction expression.

```
>> syms s lam positive
   G=1/(s^2.4+10*s^2+40*s^1.6+82*s^1.2+91*s^0.8+52*s^0.4+12);
   G1=subs(G,s^0.4,lam), G2=subs(partfrac(G1),lam,s^0.4)
```

the partial fraction expansion can be found.

$$G_2(s)=\frac{17}{8(s^{0.4}+1)}-\frac{5}{4(s^{0.4}+1)^2}-\frac{2}{s^{0.4}+2}+\frac{1}{2(s^{0.4}+1)^3}-\frac{1}{(s^{0.4}+2)^2}-\frac{1}{8(s^{0.4}+3)}.$$

Meanwhile, an expression of integer order for λ can be derived.

$$G_1(\lambda)=\frac{1}{\lambda^6+10\lambda^5+40\lambda^4+82\lambda^3+91\lambda^2+52\lambda+12}.$$

By entering the numerator and denominator polynomial coefficients of this system into the MATLAB workspace, the information on the partial fraction expansions $\boldsymbol{r}=[-0.125, -2, -1, 2.125, -1.25, 0.5]$ and $\boldsymbol{p}=[-3, -2, -2, -1, -1, -1]$ are also found numerically, which are in agreement with the $G_2(s)$ derived above. It must be noted that due to the numerical approach and the fact that the denominator polynomial has repeated roots, so the numerical solution is not very precise and it is recommended to use the analytical derivation method.

6

```
>> num=1; den=[1 10 40 82 91 52 12]; [r,p,K]=residue(num,den)
```

◀

6.3.2 Some Commonly Used Laplace Transform Formulas for Linear FODEs

Recall the important inverse Laplace transform property in (6.12), rewritten as follows:

$$\mathscr{L}^{-1}\left[\frac{s^{\alpha\gamma-\beta}}{(s^\alpha+a)^\gamma}\right]=t^{\beta-1}\mathrm{E}_{\alpha,\beta}^{\gamma}\big(-at^\alpha\big). \tag{6.32}$$

For different combinations of parameters, the following meaningful derivation forms can be deduced, see the individual corollaries below.

Corollary 6.1 If $\gamma=1$ and $\alpha\gamma=\beta$, i.e., $\beta=\alpha$, (6.32) can be written as

$$\mathscr{L}^{-1}\left[\frac{1}{s^\alpha+a}\right]=t^{\alpha-1}\mathrm{E}_{\alpha,\alpha}\big(-at^\alpha\big). \tag{6.33}$$

From the control system point of view, since the Laplace transform of the impulsive signal is 1, the above equation can be interpreted as the analytical solution of the time response of the FOTF $1/(s^\alpha+a)$ under the excitation of the unit impulsive signal. It can be seen that the Mittag-Leffler function in a fractional-order system is similar to the exponential function $\mathrm{e}^{-\lambda t}$ in an integer-order system. With this equation, the analytical solution of the impulse response of a class of linear fractional-order systems can be derived by combining the partial fraction expansion technique.

Corollary 6.2 If $\gamma=1$ and $\alpha\gamma-\beta=-1$, i.e., $\beta=\alpha+1$, (6.32) can be written as

$$\mathscr{L}^{-1}\left[\frac{1}{s(s^\alpha+a)}\right]=t^{\alpha}\mathrm{E}_{\alpha,\alpha+1}\big(-at^\alpha\big). \tag{6.34}$$

This formula can be interpreted as the analytical solution of the unit step response of the FOTF model $1/(s^\alpha + a)$.

Theorem 6.9 Analytical solution of step response
The step response of model $1/(s^\alpha + a)$ can alternatively be written as

$$\mathscr{L}^{-1}\left[\frac{1}{s(s^\alpha + a)}\right] = \frac{1}{a}\left[1 - \mathrm{E}_\alpha(-at^\alpha)\right]. \tag{6.35}$$

Proof The theorem can be easily proved from the definition of the Mittag-Leffler function.

$$\begin{aligned} at^\alpha \mathrm{E}_{\alpha,\alpha+1}(-at^\alpha) &= -\sum_{k=0}^{\infty} \frac{(-at^\alpha)^{k+1}}{\Gamma(\alpha k + \alpha + 1)} \\ &= 1 - \sum_{(k+1)=0}^{\infty} \frac{(-at^\alpha)^{k+1}}{\Gamma(\alpha(k+1)+1)} = 1 - \mathrm{E}_\alpha(-at^\alpha). \end{aligned}$$

□

Corollary 6.3 If $\gamma = k$ is an integer, and $\alpha\gamma = \beta$, i.e., $\beta = \alpha k$, (6.32) can be written as

$$\mathscr{L}^{-1}\left[\frac{1}{(s^\alpha + a)^k}\right] = t^{\alpha k - 1}\mathrm{E}_{\alpha,\alpha k}^{k}\left(-at^\alpha\right). \tag{6.36}$$

This can be viewed as the analytical solution of the impulse response of the FOTF $1/(s^\alpha + a)^k$.

Corollary 6.4 If $\gamma = k$ is an integer, and $\alpha\gamma - \beta = -1$, i.e., $\beta = \alpha k + 1$, (6.32) can be written as

$$\mathscr{L}^{-1}\left[\frac{1}{s(s^\alpha + a)^k}\right] = t^{\alpha k}\mathrm{E}_{\alpha,\alpha k+1}^{k}(-at^\alpha). \tag{6.37}$$

This can be viewed as the analytical solution of the step response of the FOTF $1/(s^\alpha + a)^k$.

6.3.3 Analytical Solutions of Linear Commensurate-Order Differential Equations

As introduced earlier, the linear commensurate-order transfer function with a base order α can be written in the form of the partial fraction expansion given in (6.29). If there are no repeated poles, the Laplace transforms in (6.33) and (6.34) lead to the analytical solutions of the impulse response and step response of the system, respectively.

$$\mathscr{L}^{-1}\left[\sum_{i=1}^{n}\frac{r_i}{s^\alpha+p_i}\right]=\sum_{i=1}^{n}r_i t^{\alpha-1}\mathrm{E}_{\alpha,\alpha}\left(-p_i t^\alpha\right) \tag{6.38}$$

$$\mathscr{L}^{-1}\left[\sum_{i=1}^{n}\frac{r_i}{s(s^\alpha+p_i)}\right]=\sum_{i=1}^{n}r_i t^{\alpha}\mathrm{E}_{\alpha,\alpha+1}\left(-p_i t^\alpha\right). \tag{6.39}$$

The latter one can also be equivalently written as

$$\mathscr{L}^{-1}\left[\sum_{i=1}^{n}\frac{r_i}{s(s^\alpha+p_i)}\right]=\sum_{i=1}^{n}\frac{r_i}{p_i}\left[1-\mathrm{E}_{\alpha}\left(-p_i t^\alpha\right)\right]. \tag{6.40}$$

6

Theorem 6.10 Analytical solutions of commensurate-order equations
More generally, since linear commensurate-order transfer functions can be written in partial fraction expansion form in Definition 6.11, the analytical solution of the impulse response can be written as

$$y_\delta(t)=\mathscr{L}^{-1}\left[\sum_{i=1}^{N}\sum_{j=1}^{m_i}\frac{r_{ij}}{(s^\alpha+p_i)^j}\right]=\sum_{i=1}^{N}\sum_{j=1}^{m_i}r_{ij}\,t^{\alpha j-1}\mathrm{E}_{\alpha,\alpha j}^{j}\left(-p_i t^\alpha\right), \tag{6.41}$$

while the analytical solution of step response can be written as

$$y_u(t)=\mathscr{L}^{-1}\left[\sum_{i=1}^{N}\sum_{j=1}^{m_i}\frac{r_{ij}}{s(s^\alpha+p_i)^j}\right]=\sum_{i=1}^{N}\sum_{j=1}^{m_i}r_{ij}\,t^{\alpha j}\mathrm{E}_{\alpha,\alpha j+1}^{j}\left(-p_i t^\alpha\right). \tag{6.42}$$

► Example 6.3

Derive the analytical and numerical solutions of the step response of the system in Example 6.1.

$$\mathscr{D}_t^{0.8}y(t)+0.75\,\mathscr{D}_t^{0.4}y(t)+0.9y(t)=5u(t).$$

Solutions It can be seen that the base order of the system is $\alpha=0.4$, and denote that $\lambda=s^{0.4}$, the commensurate-order transfer function model corresponding to the original equation can be derived, and the original system is written as an integer-order one, with respect to λ.

$$G(s)=\frac{5}{s^{0.8}+0.75s^{0.4}+0.9}\quad\Rightarrow\quad G(\lambda)=\frac{5}{\lambda^2+0.75\lambda+0.9}.$$

With the command

```
>> [r,p,k]=residue(5,[1 0.75 0.9])
```

The partial fraction expansion model of the system $G(\lambda)$ can be derived as

$$G(s) = \frac{-2.8689\mathrm{j}}{s^{0.4}+0.3750-0.8714\mathrm{j}} + \frac{2.8689\mathrm{j}}{s^{0.4}+0.3750+0.8714\mathrm{j}}.$$

This immediately shows that the analytical solution of the step response of the original system can be written as

$$y(t) = -2.87\mathrm{j}\, t^{0.4}\mathrm{E}_{0.4,1.4}\left((-0.38+0.87\mathrm{j})t^{0.4}\right) + 2.87\mathrm{j}\, t^{0.4}\mathrm{E}_{0.4,1.4}\left((-0.38-0.87\mathrm{j})t^{0.4}\right).$$

The following statement also yields the numerical solution of the original differential equation, which is identical to the curve of Example 6.1.

```
>> t=0:0.001:5; t1=t.^0.4;
   y=r(1)*t1.*ml_func([0.4,1.4],p(1)*t1)+...
     r(2)*t1.*ml_func([0.4,1.4],p(2)*t1);
   y1=ml_step3([1,0.75,0.9],[0.8,0.4],t); plot(t,y,t,5*y1,'--')
```

◀

▶ Example 6.4

Find the analytical and numerical solutions for the impulse response of the following fractional-order system:

$$G(s) = \frac{s^{1.2}+3s^{0.4}+5}{s^{1.6}+10s^{1.2}+35s^{0.8}+50s^{0.4}+24}.$$

Solutions Let $\lambda = s^{0.4}$, the original system can be expressed as

$$G(\lambda) = \frac{\lambda^3+3\lambda+5}{\lambda^4+10\lambda^3+35\lambda^2+50\lambda+24}.$$

Call MATLAB function `residue()`

```
>> n=[1 0 3 5]; d=[1 10 35 50 24]; [r,p,K]=residue(n,d)
```

The partial fraction expansion of the system can be derived as

$$G(s) = \frac{71/6}{s^{0.4}+4} + \frac{-31/2}{s^{0.4}+3} + \frac{9/2}{s^{0.4}+2} + \frac{1/6}{s^{0.4}+1}.$$

Using the properties in (6.33), the analytical solution of the impulse response of the original system can be derived as

$$\begin{aligned} y(t) = {} & 71t^{-0.6}\mathrm{E}_{0.4,0.4}(-4t^{0.4})/6 - 31t^{-0.6}\mathrm{E}_{0.4,0.4}(-3t^{0.4})/2+ \\ & 9t^{-0.6}\mathrm{E}_{0.4,0.4}(-2t^{0.4})/2 + t^{-0.6}\mathrm{E}_{0.4,0.4}(-t^{0.4})/6. \end{aligned}$$

Based on the above analytical solution, the numerical solution of the impulse response of the system can be plotted by the following MATLAB command, as shown in ◘ Fig. 6.2.

```
>> t=[0:0.001:0.2]'; t1=t.^(0.4);
   y=71/6*t.^(-0.6).*ml_func([0.4,0.4],-4*t1)-...
     31/2*t.^(-0.6).*ml_func([0.4,0.4],-3*t1)+...
```

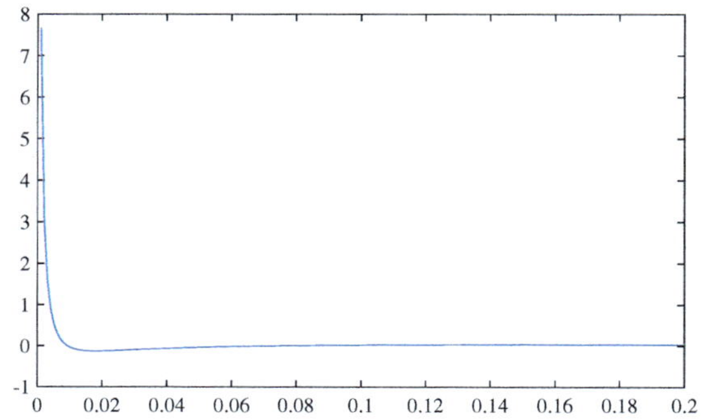

Fig. 6.2 Numerical solution of the impulse response of the fractional-order system


```
  9/2*t.^(-0.6).*ml_func([0.4,0.4],-2*t1)+...
  1/6*t.^(-0.6).*ml_func([0.4,0.4],-t1);
plot(t,y)
```

◄

► Example 6.5

Solve the following FODE with zero initial values:

$$\mathscr{D}_t^{1.2}y(t) + 5\mathscr{D}_t^{0.9}y(t) + 9\mathscr{D}_t^{0.6}y(t) + 7\mathscr{D}_t^{0.3}y(t) + 2y(t) = u(t),$$

where $u(t)$ is the unit step input or unit impulsive input.

Solutions The base order of the system is chosen to be 0.3, then let $\lambda = s^{0.3}$, and the integer-order transfer function with respect to λ can be derived.

$$G(\lambda) = \frac{1}{\lambda^4 + 5\lambda^3 + 9\lambda^2 + 7\lambda + 2}.$$

The partial fraction expansion is found by the following MATLAB command.

```
>> num=1; den=[1 5 9 7 2]; [r,p]=residue(num,den)
```

The resulting expansion equation is

$$G(\lambda) = -\frac{1}{\lambda+2} + \frac{1}{\lambda+1} - \frac{1}{(\lambda+1)^2} + \frac{1}{(\lambda+1)^3}.$$

If the input signal $u(t)$ is a unit impulsive signal, the Laplace transform expression for the output signal is

$$Y(s) = G(s) = -\frac{1}{s^{0.3}+2} + \frac{1}{s^{0.3}+1} - \frac{1}{(s^{0.3}+1)^2} + \frac{1}{(s^{0.3}+1)^3}.$$

From (6.33) and (6.36), the analytical solution of the impulse response of the system is

$$y_1(t) = -t^{-0.7}\mathrm{E}_{0.3,0.3}\left(-2t^{0.3}\right) + t^{-0.7}\mathrm{E}_{0.3,0.3}\left(-t^{0.3}\right) - t^{-0.4}\mathrm{E}_{0.3,0.6}^{2}\left(-t^{0.3}\right) + t^{-0.1}\mathrm{E}_{0.3,0.9}^{3}\left(-t^{0.3}\right).$$

If the input $u(t)$ is a unit step signal, the Laplace transform of the output signal is

$$Y(s) = \frac{1}{s}G(s) = -\frac{1}{s(s^{0.3}+2)} + \frac{1}{s(s^{0.3}+1)} - \frac{1}{s(s^{0.3}+1)^2} + \frac{1}{s(s^{0.3}+1)^3}.$$

The analytical solution of the step response of the system is thus

$$y_2(t) = -t^{0.3}\mathrm{E}_{0.3,1.3}\left(-2t^{0.3}\right) + t^{0.3}\mathrm{E}_{0.3,1.3}\left(-t^{0.3}\right) - t^{0.6}\mathrm{E}_{0.3,1.6}^{2}\left(-t^{0.3}\right) + t^{0.9}\mathrm{E}_{0.3,1.9}^{3}\left(-t^{0.3}\right).$$

In addition, an alternative form of the analytical solution of the step response can be written from (6.35).

$$y_3(t) = \frac{1}{2} + \frac{1}{2}\mathrm{E}_{0.3}(-2t^{0.3}) - \mathrm{E}_{0.3}\left(-t^{0.3}\right) - t^{0.6}\mathrm{E}_{0.3,1.6}^{2}\left(-t^{0.3}\right) + t^{0.9}\mathrm{E}_{0.3,1.9}^{3}\left(-t^{0.3}\right).$$

The numerical solution curves of the step response and impulse response of the system can be plotted directly by the following command, as shown in Fig. 6.3. It can be seen that the step response curves $y_2(t)$ and $y_3(t)$ derived from the two methods are identical.

```
>> t=[0:0.002:0.5]'; T=t.^0.3; T1=t.^0.6; T2=t.^0.9;
   y1=-t.^-0.7.*ml_func([0.3,0.3],-2*T)...
      +t.^-0.7.*ml_func([0.3,0.3],-T)...
      -t.^-0.4.*ml_func([0.3,0.6,2],-T)...
      +t.^-0.1.*ml_func([0.3,0.9,3],-T);
   y2=-T.*ml_func([0.3,1.3],-2*T)+T.*ml_func([0.3,1.3],-T)...
      -T1.*ml_func([0.3,1.6,2],-T)+T2.*ml_func([0.3,1.9,3],-T);
   y3=1/2+0.5*ml_func(0.3,-2*T)-ml_func(0.3,-T)...
      -T1.*ml_func([0.3,1.6,2],-T)+T2.*ml_func([0.3,1.9,3],-T);
   plot(t,y1,t,y2,'--',t,y3,':')
```

◄

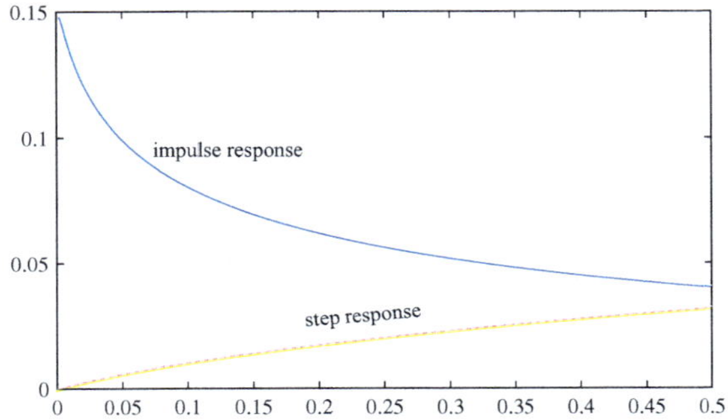

Fig. 6.3 Step response and impulse response curves

It must be noted that the numerical solution here is based on the Mittag-Leffler functions. Due to the limitations of the `ml_func()` function itself, it does not handle the problem of taking larger values of t. If $t = 50$, the truncation algorithm of this function will fail, and the embedded `mlf()` function cannot solve such a problem. Thus, other efficient numerical methods are needed.

6.4 A Closed-Form Algorithm for Linear FODEs with Zero Initial Conditions

It can be seen that the linear FODEs discussed so far have significant limitations, such as requiring the system to be commensurate-order, while the input to the system is an impulsive or step signal. Therefore, this section will explore the numerical solution methods for general linear FODEs with zero initial values, subjected to arbitrary inputs. A closed-form algorithm is proposed, and high-precision algorithm and matrix algorithm are explored.

6.4.1 A Closed-Form Algorithm

If the input signal $u(t)$ and the output signal $y(t)$ of a linear FODE and their derivatives of all orders are zero at $t = 0$, the excitation at the right side of the differential equation can be equated to $\hat{u}(t)$, the original fractional-order differential equation can be equivalently simplified to

$$a_1 \mathscr{D}_t^{\gamma_1} y(t) + a_2 \mathscr{D}_t^{\gamma_2} y(t) + \cdots + a_{n-1} \mathscr{D}_t^{\gamma_{n-1}} y(t) + a_n \mathscr{D}_t^{\gamma_n} y(t) = \hat{u}(t), \tag{6.43}$$

where $\hat{u}(t)$ is a linear combination of the input signal $u(t)$ and its fractional-order derivatives, which can be evaluated independently as

$$\hat{u}(t) = b_1 \mathscr{D}_t^{\eta_1} u(t) + b_2 \mathscr{D}_t^{\eta_2} u(t) + \cdots + b_m \mathscr{D}_t^{\eta_m} u(t). \tag{6.44}$$

For simplicity, it is safe to assume that $\gamma_1 > \gamma_2 > \cdots > \gamma_{n-1} > \gamma_n > 0$. If the following two special cases arise, the transformation should be executed first and then solved.

Remarks Special cases

(1) If the orders of the equations do not satisfy the above expressions, they should be sorted first.

(2) If there exists negative order γ_i, fractional-order integro-differential equation is involved, which is not easy to solve. A new variable $z(t) = \mathscr{D}_t^{\gamma_n} y(t)$ can be introduced to convert the original equation into the FODE of $z(t)$. Then numerical integral is performed from the solution $z(t)$ to derive the output signal $y(t)$. □

Theorem 6.11 Closed-form solution of linear FODE
The closed-form solution of linear FODE is [5]

$$y_t = \frac{1}{\displaystyle\sum_{i=1}^{n} \frac{a_i}{h^{\gamma_i}}} \left[\hat{u}_t - \sum_{i=1}^{n} \frac{a_i}{h^{\gamma_i}} \sum_{j=1}^{[(t-t_0)/h]} w_j^{(\gamma_i)} y_{t-jh} \right]. \tag{6.45}$$

Proof Consider the Grünwald–Letnikov definition in (3.19), whose improved discrete form is

$$\mathscr{D}_t^{\gamma_i} y(t) \approx \frac{1}{h^{\gamma_i}} \sum_{j=0}^{[(t-t_0)/h]} w_j^{(\gamma_i)} y_{t-jh} = \frac{1}{h^{\gamma_i}} \left[y_t + \sum_{j=1}^{[(t-t_0)/h]} w_j^{(\gamma_i)} y_{t-jh} \right], \tag{6.46}$$

where $w_j^{(\gamma_i)}$ can still be derived recursively using the following formula:

$$w_0^{(\gamma_i)} = 1, \quad w_j^{(\gamma_i)} = \left(1 - \frac{\gamma_i + 1}{j}\right) w_{j-1}^{(\gamma_i)}, \; j = 1, 2, \ldots \tag{6.47}$$

Substituting the above equation into (6.43), it is proved that the closed-form solution of the original equation is (6.45). □

Now consider the general form of the linear FODE given in (6.1). The algorithm presented here looks like it should compute the signal $\hat{u}(t)$ on the right-hand side of the equal sign first and transform the differential equation into the form of (6.43) before solving it, and this idea is shown in Fig. 6.4a. However, such a method is sometimes not feasible. For example, if the input signal is a step one and the right-hand side of the equal sign happens to have an integer-order derivative, the contribution from this term is completely ignored, and wrong results may be obtained. Therefore, a new idea should be conceived.

Different ideas can be considered in practical programming. For example, the original linear problem can be equated as the output signal $\hat{y}(t)$ by first excitation of the equation on the left-hand side of the equal sign with $u(t)$, and then the weighted sum of fractional-order derivatives of $\hat{y}(t)$ by the right-hand side of the equal sign to derive the closed-form solution of the time domain response of the original system,

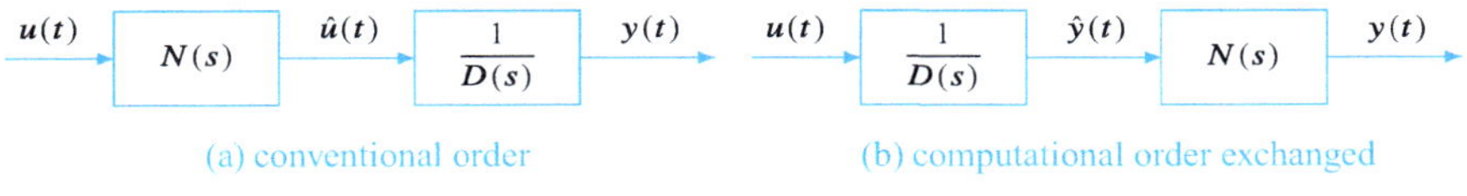

Fig. 6.4 Illustration of computational order exchange

which is shown in Fig. 6.4b. Since the original system is linear, it can be decomposed into two parts, $N(s)$ and $1/D(s)$, where both $N(s)$ and $D(s)$ are pseudo-polynomials. In the framework of a linear system, their orders are interchangeable and the resulting $y(t)$ is consistent. In this way, the following new algorithm can be proposed.

Algorithm 6.1 Closed-form algorithm of linear FODE

(1) Compute first the coefficients w_j of different orders recursively from (6.47).
(2) Let $\hat{u}(t)=u(t)$, compute the solution $y(t)$ from (6.45), and denote it as $\hat{y}(t)$.
(3) Numerical integral is used to compute the actual FODE solutions $y(t)$.

6

Based on the above algorithm, the following MATLAB function `fode_sol()` can be written to find the numerical solution of the linear FODE with zero initial values. In this function, the variable $\boldsymbol{W}$ is a matrix whose jth row is used to store the $\boldsymbol{w}^{(\gamma_j)}$ row vector of the jth-order γ_j.

```
function y=fode_sol(a,na,b,nb,u,t)
   arguments
      a(1,:),na(1,:), b(1,:),nb(1,:), u(:,1), t(:,1)
   end
   h=t(2)-t(1); D=sum(a./h.^na); nT=length(t);
   D1=b(:)./h.^nb(:); nA=length(a); vec=[na nb];
   y1=zeros(nT,1); W=ones(nT,length(vec));
   for j=2:nT, W(j,:)=W(j-1,:).*(1-(vec+1)/(j-1)); end
   for i=2:nT
      A=y1(i-1:-1:1)'*W(2:i,1:nA);
      y1(i,1)=(u(i)-sum(A.*a./h.^na))/D;
   end
   for i=2:nT, y(i,1)=(W(1:i,nA+1:end)*D1)'*y1(i:-1:1); end
end
```

The syntax of the function is $\boldsymbol{y}=$ `fode_sol`$(\boldsymbol{a},\boldsymbol{n}_a,\boldsymbol{b},\boldsymbol{n}_b,\boldsymbol{u},\boldsymbol{t})$, where the time and input vectors are provided in vectors $\boldsymbol{t}$ and $\boldsymbol{u}$.

► Example 6.6

Find the numerical solution of the step response of Example 6.5.

$$\mathscr{D}_t^{1.2}y(t)+5\mathscr{D}_t^{0.9}y(t)+9\mathscr{D}_t^{0.6}y(t)+7\mathscr{D}_t^{0.3}y(t)+2y(t)=u(t).$$

Solutions The numerical solution of the step response of the original system can be solved directly by using the following statement, and the resulting solution is exactly the same as the one obtained in Example 6.5.

```
>> a=[1 5 9 7 2]; na=1.2:-0.3:0;
   t=[0:0.002:0.5]'; u=ones(size(t));
   y2=-t.^0.3.*ml_func([0.3,1.3],-2*t.^0.3)...
      +t.^0.3.*ml_func([0.3,1.3],-t.^0.3)...
      -t.^0.6.*ml_func([0.3,1.6,2],-t.^0.3)...
```

```
   +t.^0.9.*ml_func([0.3,1.9,3],-t.^0.3);
y=fode_sol(a,na,1,0,u,t); plot(t,y,t,y2,'--')
```

With such a numerical algorithm even if a large time range $t \in (0, 50)$ is chosen, when the analytical solution is extremely time-consuming or even unsolvable, the numerical method here yields an immediate solution to the original problem. To check the correctness of the resulting solution, two step sizes $h = 0.01$ and $h = 0.002$ can be chosen, using which exactly the same results can be obtained, as shown in Fig. 6.5.

```
>> t1=0:0.01:50; u=ones(size(t1)); y1=fode_sol(a,na,1,0,u,t1);
   t2=0:0.002:50; u=ones(size(t2));
   y2=fode_sol(a,na,1,0,u,t2); plot(t1,y1,t2,y2,'--')
```

◀

Under normal circumstances, a numerical solution to a FODE is required, and a verification session of the solution is needed. The simplest verification method is to choose two different step sizes and observe whether the resulting results are consistent. If not then the step size should be reduced and further verified until finally a consistent and correct result is obtained.

6.4.2 Impulse Responses of Linear FODEs

Using the `fode_sol()` function does not directly solve for the impulse response of a FODE, because the impulsive input signal cannot be described by an input vector. The reader can use the following two alternative methods to find the impulse response of a given system.

(1) The step response $y_1(t)$ of the system is found first, and then the numerical solution of the first-order derivative of $y_1(t)$ is obtained to derive the numerical solution of the impulse response of the original equation.

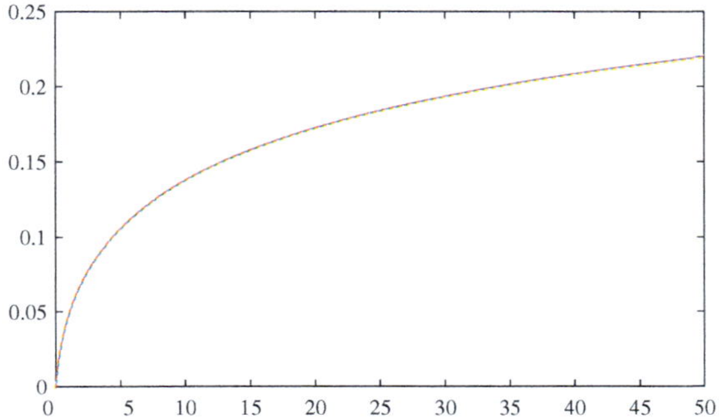

Fig. 6.5 Numerical solution of step response for large time range

(2) Add 1 to all the orders on the right-hand side of the FODE, and then find the step response of the new differential equation, which is the numerical solution of the impulse response of the original differential equation.

► Example 6.7

Find the impulse response of the differential equation in Example 6.5.

Solutions For this specific problem, the impulse response curves shown in Fig. 6.6 are obtained by the two approaches described above. It can be seen that the results obtained by the two approaches are almost identical.

```
>> t=0:0.01:50; u=ones(size(t));
   y1=fode_sol([1 5 9 7 2],1.2:-0.3:0,1,0,u,t);
   y2=fode_sol([1 5 9 7 2],1.2:-0.3:0,1,1,u,t);
   y3=glfdiff9(y1,t,1,4); plot(t,y2,t,y3,'--')
```

6

For convenience, the latter method is recommended, but of course, it works because the idea of Fig. 6.4b is implemented in the function `fode_sol()`, and the result is zero if the idea of Fig. 6.4a is used. ◄

► Example 6.8

If the input signal is $u(t) = \sin t^2$, find the numerical solution of the FODE.

$$\mathscr{D}_t^{3.5}y(t)+8\mathscr{D}_t^{3.1}y(t)+26\mathscr{D}_t^{2.3}y(t)+73\mathscr{D}_t^{1.2}y(t)+90\mathscr{D}_t^{0.5}y(t) = 30u'(t)+90\mathscr{D}_t^{0.3}u(t).$$

Solutions Such a FODE cannot be solved directly by the method described in the previous section, because the input signal here is neither step nor impulsive, so the numerical algorithm described in this section should be considered for its direct solution.
From the given equations, the vectors $\boldsymbol{a}$, $\boldsymbol{n}_a$, $\boldsymbol{b}$ and $\boldsymbol{n}_b$ can be immediately extracted, the equallyspaced time vectors are created and the vectors of the input signals are calculated so that the `fode_sol()` function can be called to solve the original equations directly. The evaluated results are shown in Fig. 6.7. Here, two step sizes of 0.002 and 0.001 are used to verify the results, and it can be found that the two results are in perfect agreement, indicating that the resulting solutions are correct.

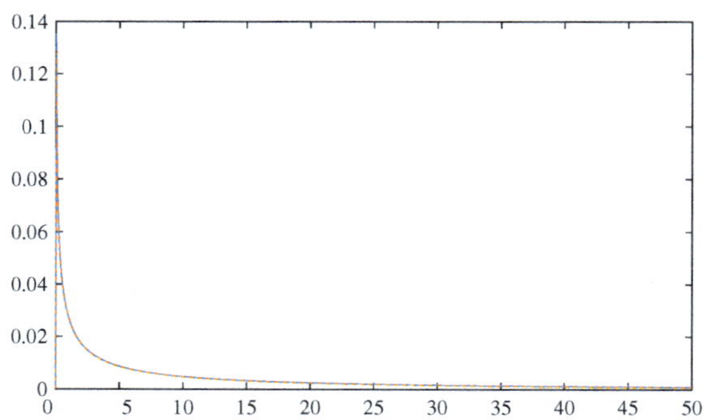

Fig. 6.6 Impulse response of the FODE

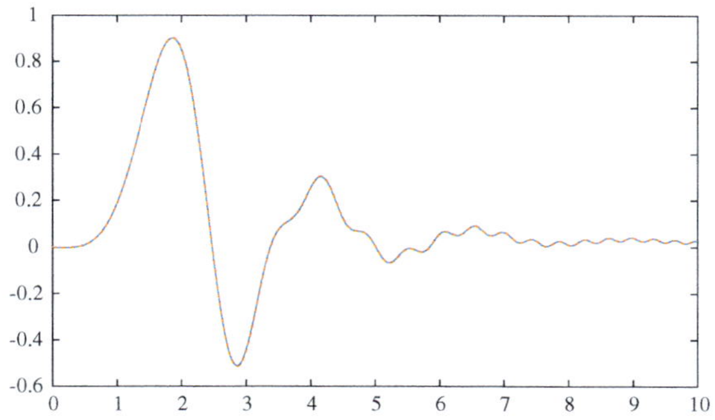

Fig. 6.7 Numerical solutions of the time responses of the system for different step sizes

```
>> a=[1,8,26,73,90]; na=[3.5,3.1,2.3,1.2,0.5];
   b=[30,90]; nb=[1,0.3]; uf=@(t)sin(t.^2);
   t=[0:0.002:10]'; u=uf(t); y=fode_sol(a,na,b,nb,u,t);
   t1=[0:0.001:10]'; u1=uf(t1); y1=fode_sol(a,na,b,nb,u1,t1);
   plot(t,y,t1,y1,'--')
```

◀

6.4.3 Validating Numerical FODE Solutions

A closed-form solution method for linear FODE was given earlier, and it was verified that consistent results were indeed obtained by choosing different step sizes. In fact, there is a more exact test for the solution of the differential equation, which is to substitute the numerical solution back to the original equation and observe whether the equation is satisfied. An example of the verification will be given here.

▶ Example 6.9

Verify that the numerical solution in Example 6.8 satisfies the original differential equation by back substitution.

Solutions By choosing the step size $h = 0.002$, the differential equation can be solved again by the following statements. The result is then verified by substituting back into each side of the original equation and then finding the resulting difference and observing the magnitude of the difference. For this example, the maximum error observed is 0.0065, which shows that the solution basically satisfies the original equation. It is also possible to simultaneously plot the function curves on both sides of the equation, as shown in Fig. 6.8. As can be seen, even though the magnitude of the resulting error is large, the resulting result is still quite close and completely indistinguishable on the curves, considering the magnitude of the function itself.

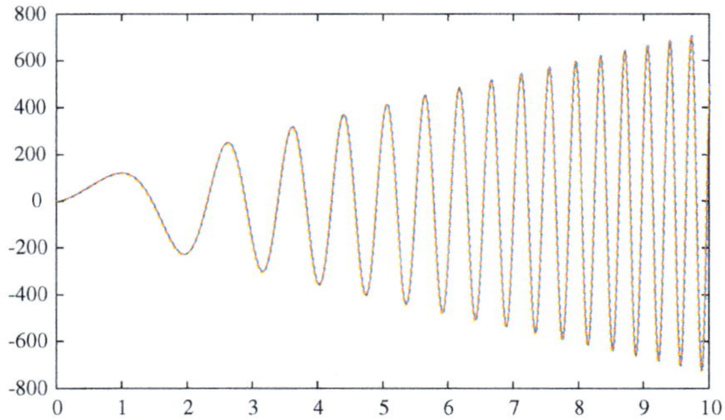

Fig. 6.8 Comparison of curves at both sides of the equation


```
>> a=[1,8,26,73,90]; na=[3.5,3.1,2.3,1.2,0.5];
   b=[30,90]; nb=[1,0.3]; uf=@(t)sin(t.^2);
   t=[0:0.001:10]'; u=uf(t); y=fode_sol(a,na,b,nb,u,t);
   yy=0; for i=1:length(a), yy=yy+a(i)*glfdiff(y,t,na(i)); end
   uu=0; for i=1:length(b), uu=uu+b(i)*glfdiff(u,t,nb(i)); end
   max(diff(yy-uu)), plot(t,yy,t,uu,'--')
```

◄

Remarks Validating the FODE solutions

(1) Both the solution and the test here use the algorithm with $O(h)$ accuracy, which may reduce the value of the error if the accuracy of the algorithm is improved.

(2) This test method can also be extended to nonlinear FODEs. □

6.4.4 A Matrix-Based Algorithm

The closed-form solution algorithm introduced in (6.43) can also be implemented by matrix methods. In the matrix algorithm a coefficient matrix $\boldsymbol{W}^{(\gamma_i)}$ can be constructed for each fractional-order operator $\mathscr{D}_t^{\gamma_i}$. Let $N = (t - t_0)/h$, then

$$\boldsymbol{W}^{(\gamma_i)} = \frac{1}{h^{\gamma_i}} \begin{bmatrix} w_0^{(\gamma_i)} & 0 & 0 & 0 & \cdots & 0 \\ w_1^{(\gamma_i)} & w_0^{(\gamma_i)} & 0 & 0 & \cdots & 0 \\ w_2^{(\gamma_i)} & w_1^{(\gamma_i)} & \omega_0^{(\gamma_i)} & 0 & \cdots & 0 \\ \vdots & \vdots & \vdots & \vdots & \ddots & \vdots \\ w_{N-1}^{(\gamma_i)} & \cdots & \omega_2^{(w_i)} & w_1^{(\gamma_i)} & w_0^{(\gamma_i)} & 0 \\ w_N^{(\gamma_i)} & w_{N-1}^{(\gamma_i)} & \cdots & w_2^{(\gamma_i)} & w_1^{(\gamma_i)} & w_0^{(\gamma_i)} \end{bmatrix}. \tag{6.48}$$

It can be seen that such a matrix is the Hankel matrix after rotation.

Two column vectors $\hat{\boldsymbol{y}}$ and $\boldsymbol{u}$ can also be created separately to represent the sample points $\hat{\boldsymbol{y}} = \left[\hat{y}(0), \hat{y}(h), \ldots, \hat{y}(Nh)\right]^{\mathrm{T}}$, $\boldsymbol{u} = \left[u(0), u(h), \ldots, u(Nh)\right]^{\mathrm{T}}$ of the output signal $\hat{y}(t)$ and the input signal $u(t)$. Using the idea in Fig. 6.4b, it is seen that

$$\sum_{i=0}^{n} a_i \boldsymbol{W}^{(\gamma_i)} \hat{\boldsymbol{y}} = \boldsymbol{u}. \tag{6.49}$$

Therefore, the output signal can finally be derived from the following formula:

$$\boldsymbol{y} = \boldsymbol{B}\boldsymbol{A}^{-1}\boldsymbol{u} = \left[\sum_{i=0}^{m} b_i \boldsymbol{W}^{(\eta_i)}\right]\left[\sum_{i=0}^{n} a_i \boldsymbol{W}^{(\gamma_i)}\right]^{-1} \boldsymbol{u}, \tag{6.50}$$

where

$$\boldsymbol{A} = \sum_{i=0}^{n} a_i \boldsymbol{W}^{(\gamma_i)},\ \ \boldsymbol{B} = \sum_{i=0}^{m} b_i \boldsymbol{W}^{(\eta_i)}. \tag{6.51}$$

The matrix algorithm for finding the closed-form solution of the linear FODE with zero initial values in (6.43) is given below.

Algorithm 6.2 Matrix implementation closed-form algorithm

(1) Establish vectors $\boldsymbol{w}^{(\gamma_i)}$ and $\boldsymbol{w}^{(\eta_i)}$ from (6.47).

(2) Configure matrices $\boldsymbol{A}$ and $\boldsymbol{B}$ by rotating Hankel ones.

(3) Compute the closed-form solution $y(t)$ from (6.50).

The MATLAB function `fode_solm()` can be written based on the algorithm given here, which can be used to find the numerical solution of a linear FODE in exactly the same format as `fode_sol()`.

```
function y=fode_solm(a,na,b,nb,u,t)
   arguments
      a(1,:),na(1,:), b(1,:),nb(1,:), u(:,1), t(:,1)
   end
   h=t(2)-t(1); u=u(:); A=0; B=0; g=double(genfunc(1));
   nt=length(t); n=length(a); m=length(b);
   for i=1:n, A=A+get_vecw(na(i),nt,g)*a(i)/(h^na(i)); end
   for i=1:m, B=B+get_vecw(nb(i),nt,g)*b(i)/(h^nb(i)); end
   A=rot90(hankel(A(end:-1:1)));
   B=rot90(hankel(B(end:-1:1))); y=B*(A\u);
end
```

▶ Example 6.10

Solve again the linear FODE in Example 6.8 using the matrix algorithm.

Solutions Choosing a step size of $h = 0.002$, the numerical solution of the original FODE can be found by the closed-form algorithm and the matrix algorithm, respectively, and both methods yield identical results.

```
>> a=[1,8,26,73,90]; b=[30,90];
   na=[3.5,3.1,2.3,1.2,0.5]; nb=[1,0.3]; t=0:0.002:10;
   u=sin(t.^2); tic, y=fode_sol(a,na,b,nb,u,t); toc
   tic, y1=fode_solm(a,na,b,nb,u,t); toc, plot(t,y,t,y1)
```

However, comparing the time elapse of the two functions, it is found that the closed-form solution algorithm takes 0.18 s, while the matrix algorithm takes 0.56 s. This is due to the fact that the matrix algorithm is not efficient as it needs to handle operations on 5001×5001 large-scale matrices under the matrix algorithm. ◀

6

It must be noted that if N is very large, the matrix storage space requirement is relatively large, which can make the computation significantly slower or even impossible to store. Therefore, the matrix algorithm is not recommended.

6.4.5 A High-Precision Closed-Form Algorithm

The closed-form algorithm for linear FODEs with zero initial values was presented earlier, and its application is wide, although the accuracy of its algorithm is only $O(h)$, which is not suitable for operations with larger step sizes. If the $O(h^p)$ high-precision numerical method introduced in ▶ Chap. 4 is considered, a high-precision numerical solution of the FODE may be obtained if it can be applied to solve FODEs.

If the FODE in (6.43) has zero initial conditions, the fractional-order operator in the equation can be reconstructed from the definition of Grünwald–Letnikov fractional calculus to produce a closed-form solution of the differential equation similar to (6.45). If the coefficients w_j in (6.45) are replaced by the high-precision algorithm of ▶ Chap. 4, the high-precision closed-form solution of the FODE can be obtained using the following algorithm.

Algorithm 6.3 High-precision closed-form algorithm

(1) The fractional-order definitions in (6.43) are replaced by the Grünwald–Letnikov operator.

(2) For each order, high-precision algorithm is used to compute w_j recursively.

(3) Compute the numerical solutions y_k with (6.45).

Based on such an algorithm, the MATLAB function `fode_sol9()` can be written to find the analytical solution of a linear FODE with zero initial values. The syntax of the function is close to that of `fode_sol()`, except that it has an additional order parameter p.

```
function y=fode_sol9(a,na,b,nb,u,t,p)
   arguments
      a(1,:), na(1,:), b(1,:), nb(1,:), u(:,1), t(:,1)
      p(1,1) {mustBePositiveInteger}=5
   end
   h=t(2)-t(1); n=length(t); vec=[na nb];
   g=double(genfunc(p)); W=[];
   for i=1:length(vec), W=[W; get_vecw(vec(i),n,g)]; end
   D1=b./h.^nb; nA=length(a); y1=zeros(n,1);
   W=W.'; D=sum((a.*W(1,1:nA))./h.^na);
   for i=2:n, A=y1(i-1:-1:1)'*W(2:i,1:nA);
      y1(i)=(u(i)-sum(A.*a./h.^na))/D;
   end
   for i=2:n, y(i,1)=(W(1:i,nA+1:end)*D1)'*y1(i:-1:1); end
end
```

A necessary condition for the successful use of the pth-order algorithm in Algorithm 6.4 is that the first p samples of $y(t)$ have values of zero or very close to zero to avoid the influence of some missing values of w_j in Table 4.4 on the calculation results. If a particular FODE does not satisfy these special conditions, the high-precision algorithm can be derived directly by the later algorithm.

▶ Example 6.11

Find the numerical solution of the following FODEs with zero initial values differential equation:

$$y'''(t) + {}_0^{\mathrm{C}}\mathscr{D}_t^{2.5}y(t) + y(t) = -1 + t - t^2/2 - t^{0.5}\mathrm{E}_{1,1.5}(-t).$$

The exact solution of this FODE is known to be $y(t) = -1 + t - t^2/2 + \mathrm{e}^{-t}$. Solve the FODE under different values of p and assess the error.

Solutions Since this FODE is with zero initial value, the algorithm described earlier can be used to solve it directly. Choosing a larger step size $h = 0.1$, the following statement can be used to find the numerical solution of the FODE at different values of p. The resulting numerical solution error is not very large.

```
>> a=[1 1 1]; na=[3 2.5 0]; b=1; nb=0;
   t=[0:0.1:1]'; y=-1+t-t.^2/2+exp(-t);
   u=-1+t-t.^2/2-t.^0.5.*ml_func([1,1.5],-t);
   y1=fode_sol9(a,na,b,nb,u,t,1);
   y2=fode_sol9(a,na,b,nb,u,t,2);
   y3=fode_sol9(a,na,b,nb,u,t,3);
   y4=fode_sol9(a,na,b,nb,u,t,4);
   e1=y-y1; e2=y-y2; e3=y-y3; e4=y-y4;
   plot(t,e1,t,e2,'--',t,e3,':',t,e4,'-.')
```

The error curves for these four different p values can be derived from the above statements, as shown in Fig. 6.9. It can be seen that the accuracy of the result for $p = 2$ is significantly higher than that of the result for $p = 1$, which is the result of the `fode_sol()` function, and then increases to $p = 3$ and $p = 4$, where the accuracy is decreasing. If the first four samples of $y(t)$ are observed, they are 0, −0.0002, −0.0013, and −0.0042, it is

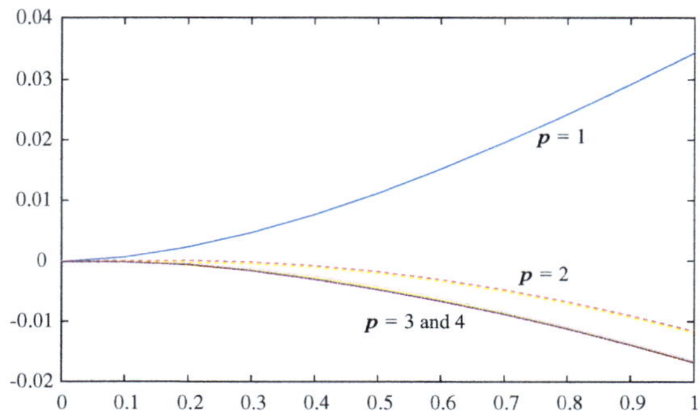

Fig. 6.9 Errors for different values of p

seen that the first value is zero, the second value is close to zero, and the rest of the values are relatively large, which affects the final approximation. ◀

It is also possible to build up a matrix version of the algorithm function, although it is not recommended for practical applications because the matrix algorithm itself is too time-consuming.

6.5 Numerical Solutions of Caputo Equations with Nonzero Initial Conditions

6.5.1 Mathematical Descriptions of Caputo Equations

The numerical solution of linear FODEs with zero initial values is described in the previous section. For nonzero initial value problems, it is necessary to establish and solve the Caputo equations. In this section, the numerical algorithm and implementation of the linear Caputo equations are discussed.

Definition 6.12 Linear Caputo equations

The general form of linear Caputo equations is

$$a_1 \, {}_{t_0}^{\mathrm{C}}\mathscr{D}_t^{\gamma_1} y(t) + a_2 \, {}_{t_0}^{\mathrm{C}}\mathscr{D}_t^{\gamma_2} y(t) + \cdots + a_n \, {}_{t_0}^{\mathrm{C}}\mathscr{D}_t^{\gamma_n} y(t) = \hat{u}(t), \tag{6.52}$$

and it is safe to assume that $\gamma_1 > \gamma_2 > \cdots > \gamma_n$.

Definition 6.13 Caputo equations with nonzero initial conditions

The initial conditions of the Caputo equation are $y(t_0) = c_0, y'(0) = c_1, \ldots, y^{(q-1)}(0) = c_{q-1}$. If $c_i = 0, 0 \leqslant i \leqslant q-1$, the FODE is changed into zero initial value problems;

otherwise, it is referred to as nonzero initial condition problem. To ensure the FODE has a unique solution, let $q = \lceil \gamma_1 \rceil$.

If the dynamic expression $\hat{u}(t)$ exists on the right-hand side of the equation, the equivalent signal $\hat{u}(t)$ needs to be evaluated first. This signal can be computed directly by calling the high-precision function `caputo9()`.

$$\hat{u}(t) = b_1 \mathscr{D}_t^{\eta_1} u(t) + b_2 \mathscr{D}_t^{\eta_2} u(t) + \cdots + b_m \mathscr{D}_t^{\eta_m} u(t). \tag{6.53}$$

It must be noted that since the initial conditions in the Caputo definition is the initial value of the $y(t)$ signal and its integer-order derivatives, this type of definition is more suitable to requirements in practical applications. In-depth discussion on the numerical solution of the Caputo equation is needed.

6.5.2 Taylor Axillary Functions

If (6.52) contains nonzero initial conditions, it is necessary to introduce an auxiliary function $T(t)$, which decomposes the original FODE into a FODE with zero initial values [6].

Definition 6.14 Taylor axillary function

A Taylor axillary function is created, such that

$$T(t) = \sum_{k=0}^{q-1} \frac{y^{(k)}(t_0)}{k!}(t - t_0)^k, \tag{6.54}$$

and

$$y(t) = z(t) + T(t), \tag{6.55}$$

where the initial conditions of the axillary signal $T(t)$ are identical to those of signal $y(t)$. Function $z(t)$ is the one with zero initial conditions.

Substituting $y(t)$ in (6.52) by $z(t) + T(t)$, the differential equation for $y(t)$ can be replaced with $z(t)$, which is the equation with zero initial conditions.

$$a_1 {}_{t_0}^{\rm C}\mathscr{D}_t^{\gamma_1} z(t) + a_2 {}_{t_0}^{\rm C}\mathscr{D}_t^{\gamma_2} z(t) + \cdots + a_n {}_{t_0}^{\rm C}\mathscr{D}_t^{\gamma_n} z(t) = \hat{u}(t) - P(t), \tag{6.56}$$

where $P(t)$ can be expressed by

$$P(t) = \left(a_1 {}_{t_0}^{\rm C}\mathscr{D}_t^{\gamma_1} + a_2 {}_{t_0}^{\rm C}\mathscr{D}_t^{\gamma_2} + \cdots + a_n {}_{t_0}^{\rm C}\mathscr{D}_t^{\gamma_n}\right) T(t). \tag{6.57}$$

Since $z(t)$ is the signal with zero initial conditions, ${}_{t_0}^{\rm C}\mathscr{D}^{\alpha} z(t) = {}_{t_0}^{\rm GL}\mathscr{D}^{\alpha} z(t)$. In this way, the equivalent input signal $\hat{u}(t) - P(t)$ on the right-hand side of the equation is found, and then the solution z_m to the zero initial condition problem can be solved with the method similar to (6.45).

Theorem 6.12 Closed-form solution of Caputo equations
The closed-form solution of the linear FODE with nonzero initial conditions is

$$y_m = \frac{1}{\sum_{i=1}^{n} \frac{a_i}{h^{\gamma_i}}} \left(\hat{u}_m - P_m - \sum_{i=1}^{n} \frac{a_i}{h^{\gamma_i}} \sum_{j=1}^{m} w_j y_{m-j} \right) + T_m. \tag{6.58}$$

▶ Example 6.12

Solve the Bagley–Torwik FODE [7].

$$Ay''(t) + B{}_0^{\rm C}\mathscr{D}_t^{3/2} y(t) + Cy(t) = C(t+1), \text{ and } y(0) = y'(0) = 1.$$

Show that the solutions are independent of the constants A, B, and C.

Solutions The Taylor auxiliary function $T(t) = t + 1$ can be written from the initial conditions, so that the original output signal can be decomposed into $y(t) = z(t) + t + 1$, and $z(t)$ is a signal with zero initial conditions. With such a transformation, the Caputo equation for $z(t)$, i.e., the Grünwald–Letnikov equation with zero initial values, can be established.

Recall the definition of Caputo fractional-order derivatives, since the αth-order Caputo derivative of $y(t)$ requires the second-order derivative of $y(t)$ followed by the integral operation, this means that if $\alpha \in (1, 2)$, the compensation term $t + 1$ is eliminated, so that the original equation eventually changes to

$$Az''(t) + B\mathscr{D}_t^{3/2} z(t) + Cz(t) + \cancel{C(t+1)} = \cancel{C(t+1)}.$$

The $C(t+1)$ terms at both sides of the equal sign can be canceled and the original differential equation degenerates to

$$Az''(t) + B\mathscr{D}_t^{3/2} z(t) + Cz(t) = 0.$$

It can be seen that since $z(t)$ is with zero initial conditions, and there is no external excitation in the FODE, implying that $z(t) \equiv 0$. The solution of the original equation is $y(t) = t + 1$, i.e., the solution of the equation is independent of the choice of the constants A, B, and C. ◀

The auxiliary function $T(t)$ is the Taylor series expansion of the output signal $y(t)$ at the initial moment. The difference between the signal $T(t)$ and $y(t)$ at the initial moment should be small, however, since $y(t)$ is a bounded function, $|y(t) - T(t)|$ may increase as t increases, and when the value of t is large enough, $|T(t)|$ will become an increasing function and $|z(t)| = |y(t) - T(t)|$ is also an increasing function. Since these two functions are evaluated separately, it is possible that the two increasing functions do not cancel perfectly when t is large, which makes the final calculation error large. Obviously, it is difficult to guarantee the accuracy of the result by such a direct method. Such a phenomenon will be demonstrated later by an example.

Algorithm 6.4 Caputo equation algorithm

(1) Compute T_m from the given initial conditions using (6.54).
(2) Compute the equivalent signal $\hat{u}$ from (6.53).
(3) Compute the new equivalent signal $\tilde{u} = \hat{u} - P$ from (6.57).
(4) Find the Grünwald–Letnikov numerical solution z from Algorithm 6.1.
(5) Adding back T_m such that the solution $y_m = z_m + T_m$ is found.

Based on the previous algorithm, the following MATLAB function can be written.

```
function y=fode_caputo0(a,na,b,nb,y0,u,t)
   arguments
      a(1,:),na(1,:), b(1,:),nb(1,:), y0(:,1),u(:,1),t(:,1)
   end
   h=t(2)-t(1); T=0; P=0; U=0;
   for i=1:length(y0), T=T+y0(i)*t.^(i-1)/gamma(i); end
   for i=1:length(na), P=P+a(i)*caputo9(T,t,na(i),5); end
   for i=1:length(nb), U=U+b(i)*caputo9(u,t,nb(i),5); end
   z=fode_sol(a,na,1,0,U-P,t); y=z+T;
end
```

The syntax of the function is $\boldsymbol{y}$ = fode_caputo0($\boldsymbol{a}, \boldsymbol{n}_a, \boldsymbol{b}, \boldsymbol{n}_b, \boldsymbol{y}_0, \boldsymbol{u}, \boldsymbol{t}$), where the initial condition vector is $\boldsymbol{y}_0 = [c_0, c_1, \ldots, c_{q-1}]$. In this function, the default $O(h^5)$ solver is used to evaluate fractional-order derivatives.

▶ Example 6.13

Solve the following Caputo equation:

$$y'''(t) + \frac{1}{16}{}_0^{\mathrm{C}}\mathscr{D}_t^{2.5}y(t) + \frac{4}{5}y''(t) + \frac{3}{2}y'(t) + \frac{1}{25}{}_0^{\mathrm{C}}\mathscr{D}_t^{0.5}y(t) + \frac{6}{5}y(t) = \frac{172}{125}\cos\frac{4t}{5},$$

where the initial conditions are $y(0) = 1$, $y'(0) = 4/5$, $y''(0) = -16/25$, and $0 \leqslant t \leqslant 30$. The exact solution is $y(t) = \sqrt{2}\sin(4t/5 + \pi/4)$.

Solutions Introducing the Taylor auxiliary function, $T(t) = 1 + 4t/5 - 8t^2/25$. Letting $y(t) = z(t) + T(t)$, the numerical solution of (6.58) can be obtained. The numerical solutions are shown in Fig. 6.10, when the step sizes $h = 0.01$ and $h = 0.1$ are chosen, respectively.

```
>> a=[1 1/16 4/5 3/2 1/25 6/5]; na=[3 2.5 2 1 0.5 0];
   b=1; nb=0; t=[0:0.01:30]; u=172/125*cos(4*t/5);
   y0=[1 4/5 -16/25]; y1=fode_caputo0(a,na,b,nb,y0,u,t);
   t1=[0:0.1:30]; u2=172/125*cos(4*t1/5);
   y2=fode_caputo0(a,na,b,nb,y0,u2,t1);
   y=sqrt(2)*sin(4*t/5+pi/4);
   plot(t,y,t,y1,'--',t1,y2,':')
```

It can be seen that the numerical and exact solutions are close to each other at the step size of $h = 0.01$, and the error becomes larger if the step size is increased. ◀

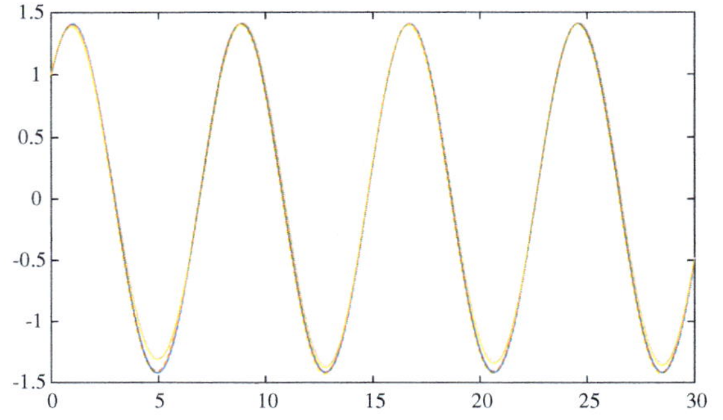

Fig. 6.10 Numerical solution and the exact one

It must be noted that in earlier literature, because of the $O(h)$ algorithm for solving the equation, it may lead to large errors or even divergence when h is large, making the Taylor auxiliary function method ineffective and forcing the introduction of other convergence functions, such as exponential auxiliary functions [6]. Nowadays, high-precision numerical derivative algorithms are available, so good solutions can be obtained by using Taylor auxiliary functions in high-precision Caputo equation-solving algorithms.

6.5.3 High-Precision Algorithm for Caputo Equations

As pointed out earlier, if a pth-order algorithm is required, the necessary condition is that the first p values of the signal $z(t) = y(t) - T(t)$ are all zero or very close to zero. Unfortunately, in the actual Caputo equation, such a condition is not always satisfied. The phenomenon that has been demonstrated in Example 6.11.

Since the order p can be chosen independently, there are only two possible relationships between p and the highest order q of the actual Caputo equation: one possibility, if $p \leqslant q$, this necessary condition can be guaranteed to be satisfied so that a high-precision numerical solution can be derived directly from Algorithm 6.3; the other possibility, if $p > q$, requires the reconstruction of p equivalent initial conditions such that Algorithm 6.3 can be used to derive a high-precision solution. In the latter case, a two-step solution algorithm is required.

1. Equivalent Initial Condition Problem

Whatever the value of q in Caputo equation, it is necessary to construct a Taylor auxiliary function

$$T(t) = \sum_{k=0}^{p-1} c_k \frac{t^k}{k!}. \tag{6.59}$$

If $p \leqslant q$, then when $k = 0, 1, \ldots, p-1$, let $c_k = y^{(k)}(0)$, this step is complete; if $p > q$, then when $k = 0, 1, \ldots, q-1$, let $c_k = y^{(k)}(0)$, and the remaining of $p - q$ initial values of c_k need to be recalculated, in which case it is necessary to let $T(t)$

have the same initial values as $y(t)$ with undetermined values in $q \leqslant k \leqslant p-1$ still denoted as c_k. In this way, the $z(t)$ signal is the signal with zero initial conditions.

For ease of description, $T(t)$ is redescribed as

$$T(t) = \sum_{k=0}^{p-1} c_k T_k, \tag{6.60}$$

where $T_k = t^k/k!$. Since $z(t)$ is a function with zero initial conditions, the first p values of its interpolating polynomials are all zero, and the first p values of the auxiliary signal $T(t)$ are consistent with the $y(t)$ function. In other words, $T(t)$ satisfies the original Caputo equation at the first p values. Substituting (6.60) into the original equation, we can obtain

$$\sum_{k=0}^{p} c_k x_k(t) = \hat{u}(t), \tag{6.61}$$

where

$$x_k = \left(a_1 {}_{t_0}^{\rm C}\mathscr{D}_t^{\eta_1} + a_2 {}_{t_0}^{\rm C}\mathscr{D}_t^{\eta_2} + \cdots + a_n {}_{t_0}^{\rm C}\mathscr{D}_t^{\eta_n}\right) T_k(t). \tag{6.62}$$

It can be seen that the value of x_k at the first p points can be found directly. In (6.61), let $t = h, 2h, \ldots, Kh$, where $K = p-q$, then the following equation can be established:

$$\begin{bmatrix} x_0(h) & x_1(h) & \cdots & x_p(h) \\ x_0(2h) & x_1(2h) & \cdots & x_p(2h) \\ \vdots & \vdots & \ddots & \vdots \\ x_0(Kh) & x_1(Kh) & \cdots & x_p(Kh) \end{bmatrix} \begin{bmatrix} c_0 \\ c_1 \\ \vdots \\ c_{p-1} \end{bmatrix} = \begin{bmatrix} \hat{u}(h) \\ \hat{u}(2h) \\ \vdots \\ \hat{u}(Kh) \end{bmatrix}. \tag{6.63}$$

From this equation it follows that $c_k, 0 \leqslant k \leqslant q-1$ is equal to the initial condition of the original equation, and the number of unknown variables is K. These unknown variables can be solved directly from the linear algebraic equation in (6.63). The coefficients thus derived c_i, $0 \leqslant i \leqslant p-1$, can be viewed as new equivalent initial conditions. With these initial conditions, the Taylor auxiliary function $T(t)$ can be established from (6.60). The first p values of the output signal $y(t)$ can be calculated directly from the signal $T(t)$. The above idea will be summarized in the algorithm given below.

Algorithm 6.5 Constructing equivalent initial conditions
(1) Construct $T_k = t^k/k!$ for $0 \leqslant k \leqslant p-1$, and compute x_k from (6.62).
(2) Let $K = p-q$, the coefficient matrix in (6.63) can be established with the obtained x_k.
(3) Compute $\hat{u}(h), \hat{u}(2h), \ldots, \hat{u}(Kh)$, which are the equivalent input on the right of (6.52).

(4) The first q values of c_k equal the initial conditions, and the remaining ones can be obtained by solving the linear equation in (6.63), such that the equivalent initial conditions are found.

Based on the above algorithm, the following MATLAB function caputo_ics() can be written for the high-precision Caputo equation solution process of the equation, to set up p equivalent initial conditions.

```
function [c,y]=caputo_ics(a,na,b,nb,y0,u,t)
   arguments
      a(1,:),na(1,:), b(1,:),nb(1,:), y0(:,1),u(:,1),t(:,1)
   end
   na1=ceil(na); q=max(na1); K=length(t);
   p=K+q-1; d1=y0./gamma(1:q)';
   I1=1:q; I2=(q+1):p; X=zeros(K,p);
   for i=1:p, for k=1:length(a)
      if i>na1(k)
        X(:,i)=X(:,i)+...
            a(k)*t.^(i-1-na(k))*gamma(i)/gamma(i-na(k));
   end, end, end
   u1=0;
   for i=1:length(b), u1=u1+b(i)*caputo9(u,t,nb(i),K-1); end
   X(1,:)=[]; u2=u1(2:end)-X(:,I1)*d1; d2=X(:,I2)\u2;
   c=[d1;d2]; y=0; for i=1:p, y=y+c(i)*t.^(i-1); end
end
```

The syntax of the function is $[\boldsymbol{c}, \boldsymbol{y}] =$ caputo_ics($\boldsymbol{a}, \boldsymbol{n}_a, \boldsymbol{b}, \boldsymbol{n}_b, \boldsymbol{y}_0, \boldsymbol{u}, \boldsymbol{t}$), where the vectors $\boldsymbol{a}$, $\boldsymbol{n}_a$, $\boldsymbol{b}$ and $\boldsymbol{n}_b$ stores, respectively, the coefficients and orders on the two sides of the equation, $\boldsymbol{y}_0$ is the initial value vector, whose length is q; $\boldsymbol{u}$ and $\boldsymbol{t}$ are the input vector and the time vector, whose length is $p+1$, thus, giving the function an input p value. The returned variable $\boldsymbol{c}$ is the equivalent initial conditions, while y returns the first p values of the output signal.

Remarks Equivalent initial conditions

(1) If the accuracy requirement is not very high, i.e., $p \leqslant q$, then the given initial condition $\boldsymbol{y}_0$ is sufficient.

(2) If high accuracy is required, i.e., $p > q$, the given vector $\boldsymbol{y}_0$ is not sufficient, so that the resulting equivalent initial condition is in fact the higher order term of the Taylor series expansion of the output signal $y(t)$.

(3) In practice, since the $y(t)$ signal is unknown, the higher order terms of the Taylor series can only be reconstructed by solving the linear algebraic equations in the algorithm. □

▶ Example 6.14

Consider the Caputo equation from Example 6.13. The exact solution is known to be $y(t) = \sqrt{2}\sin(4t/5 + \pi/4)$. Assess the accuracy of the higher order terms of the reconstructed Taylor series is tested for different step sizes h.

Solutions A theoretical Taylor series expansion of the original exact solution function can be performed by the following statement.

```
>> syms t; y=sqrt(2)*sin(4*t/5+pi/4),
   F=taylor(y,t,'Order',7), y0a=sym2poly(F)
```

It can be seen that the first 7 terms of its Taylor series expansion are

$$y(t) = 1 + \frac{4}{5}t - \frac{8}{25}t^2 - \frac{32}{375}t^3 + \frac{32}{1875}t^4 + \frac{128}{46875}t^5 - \frac{256}{703125}t^6 + o(h^6).$$

If the first few terms of the Taylor series expansion are reconstructed using the functions of Algorithm 6.5, equivalent initial conditions can be computed for different p and step sizes h. The errors are given in Table 6.1.

```
>> a=[1 1/16 4/5 3/2 1/25 6/5]; na=[3 2.5 2 1 0.5 0];
   y0=[1 4/5 -16/25]; b=1; nb=0;
   h=[0.1,0.05,0.02,0.01,0.005,0.002,0.001];
   for i=1:7, for p=1:6
      t=[0:h(i):p*h(i)]; u=172/125*cos(4/5*t);
      y=sqrt(2)*sin(4*t/5+pi/4);
      [ee,yy]=caputo_ics(a,na,b,nb,y0,u,t); err(p,i)=norm(yy-y');
   end, end
```

◀

It can be seen that the errors derived by this algorithm are small. If the equivalent initial conditions are used, the necessary conditions of the previously mentioned high-precision algorithm are satisfied and the high-precision numerical solution of the Caputo equation can be derived directly.

Table 6.1 Errors at different orders and step sizes

p	$h=0.1$	$h=0.05$	$h=0.02$	$h=0.01$	$h=0.005$	$h=0.002$	$h=0.001$
1	5.0×10^{-6}	3.2×10^{-7}	8.2×10^{-9}	5.1×10^{-10}	3.2×10^{-11}	8.2×10^{-13}	5.1×10^{-14}
2	1.6×10^{-6}	5.7×10^{-8}	6.1×10^{-10}	2.0×10^{-11}	6.2×10^{-13}	6.0×10^{-15}	2.0×10^{-16}
3	3.7×10^{-7}	5.9×10^{-9}	2.4×10^{-11}	3.8×10^{-13}	6.0×10^{-15}	3.0×10^{-16}	2.0×10^{-16}
4	4.0×10^{-8}	3.8×10^{-10}	6.8×10^{-13}	5.3×10^{-15}	3.0×10^{-16}	2.0×10^{-16}	3.0×10^{-16}
5	7.2×10^{-9}	2.9×10^{-11}	1.9×10^{-14}	2.2×10^{-16}	3.8×10^{-16}	3.1×10^{-16}	3.8×10^{-16}
6	5.2×10^{-10}	1.4×10^{-12}	5.0×10^{-16}	5.4×10^{-16}	4.4×10^{-16}	3.1×10^{-16}	3.8×10^{-16}

2. High-Precision Algorithm for the FODEs

Since the equivalent initial conditions and the numerical solution $y(t)$ of the first p values can be evaluated using Algorithm 6.5, the Taylor auxiliary function $T(t)$ can be built up directly so that the method introduced earlier can directly lead to a high-precision solution of the Caputo equations. By summarizing the above ideas, the following solution algorithm is proposed.

Algorithm 6.6 High-Precision Algorithm for Caputo Equations

(1) Compute the equivalent initial conditions with Algorithm 6.5.
(2) Construct the auxiliary function $T(t)$ from the equivalent initial conditions, and decompose $y(t)$ into $T(t)+z(t)$.
(3) Find the high-precision solution $z(t)$ of the FODE with zero initial conditions using Algorithm 6.3.
(4) Find the high-precision solution of the original FODE with $y(t)=T(t)+z(t)$.

Based on the above algorithm, the following function `fode_caputo9()` can be written, which embeds the two functions introduced earlier, namely, functions `fode_sol9()` and `caputo_ics()`.

```
function y=fode_caputo9(a,na,b,nb,y0,u,t,p)
   arguments
      a(1,:),na(1,:), b(1,:),nb(1,:), y0(:,1),u(:,1),t(:,1)
      p(1,1) {mustBePositiveInteger}=5
   end
   T=0; dT=0;
   if p>length(y0)
      yy0=caputo_ics(a,na,b,nb,y0,u(1:p),t(1:p));
      y0=yy0(1:p).*gamma(1:p)';
   elseif p==length(y0)
      yy0=caputo_ics(a,na,b,nb,y0,u(1:p+1),t(1:p+1));
      y0=yy0(1:p+1).*gamma(1:p+1)';
   end
   for i=1:length(y0), T=T+y0(i)/gamma(i)*t.^(i-1); end
   for i=1:length(na), dT=dT+a(i)*caputo9(T,t,na(i),p); end
   u=u-dT; y=fode_sol9(a,na,b,nb,u,t,p)+T;
end
```

The syntax of the function is $y=$ `fode_caputo9`$(\boldsymbol{a},\boldsymbol{n}_a,\boldsymbol{b},\boldsymbol{n}_b,\boldsymbol{y}_0,\boldsymbol{u},\boldsymbol{t},p)$. The syntax is very close to the `fode_caputo0()` function presented earlier, with one additional parameter p. This function can be used directly to obtain a high-precision numerical solution to a linear Caputo equation with nonzero initial conditions.

▶ Example 6.15

Solve again the equation in Example 6.13 using a high-precision algorithm and evaluate the accuracy of the numerical solution for different choices of step sizes and orders.

Solutions The numerical solution of the original equation can be obtained by choosing a large step size $h = 0.1$ and different orders p. The computational errors at different time instances compared to the known exact solution are presented in Table 6.2. It can be seen that the resulting error decreases significantly when the accuracy requirement p increases, with $p = 6$ a counterexample. The error also increases when t is large, although it is still significantly better than the traditional $O(h)$ algorithm. For this example, $p = 5$ is a good choice.

```
>> h=0.1; t=[0:h:30]'; y=sqrt(2)*sin(4*t/5+pi/4);
   u=172/125*cos(4/5*t); ii=31:30:301; y=y(ii);
   a=[1 1/16 4/5 3/2 1/25 6/5]; na=[3 2.5 2 1 0.5 0];
   b=1; nb=0; y0=[1 4/5 -16/25]; T=[];
   for p=1:6, y1=fode_caputo9(a,na,b,nb,y0,u,t,p);
      err=y-y1(ii); T=[T abs(err)];
   end
```

If a smaller step size of $h = 0.01$ is chosen, the same error can be obtained for each choice of p, as given in Table 6.3. From the results, the accuracy is not as good as that for $h = 0.1$, because as the step size decreases, the number of calculation points increases and the cumulative error increases. $p = 3$ is a good choice for this step size.

Table 6.2 Errors for the step size $h = 0.1$

t	$p=1$	$p=2$	$p=3$	$p=4$	$p=5$	$p=6$
3	0.00103	0.01469	0.002614	0.000146	5.150×10^{-7}	1.911×10^{-7}
6	0.00035	0.00974	0.000599	4.576×10^{-5}	3.034×10^{-7}	2.101×10^{-7}
9	0.02348	0.00347	0.001267	4.883×10^{-5}	2.303×10^{-9}	5.381×10^{-7}
12	0.04216	0.00468	0.000448	0.000137	4.837×10^{-7}	3.045×10^{-6}
15	0.03881	0.01212	0.000228	0.000193	5.032×10^{-7}	1.281×10^{-5}
18	0.00876	0.01661	0.001434	0.000202	2.723×10^{-6}	3.677×10^{-5}
21	0.03357	0.01653	0.000876	0.000162	8.828×10^{-7}	9.081×10^{-5}
24	0.06635	0.01189	0.001411	8.124×10^{-5}	7.378×10^{-6}	0.000180
27	0.06989	0.00409	0.000479	1.812×10^{-5}	8.030×10^{-6}	0.000357
30	0.04048	0.00433	0.000626	0.000112	1.618×10^{-5}	0.000614

Table 6.3 Errors for step size $h = 0.01$

t	$p=1$	$p=2$	$p=3$	$p=4$	$p=5$	$p=6$
3	0.0002661	0.0008371	0.0002531	9.896×10^{-7}	6.242×10^{-9}	1.622×10^{-7}
6	0.0004601	8.888×10^{-5}	4.260×10^{-5}	2.256×10^{-6}	9.681×10^{-8}	6.299×10^{-5}
9	0.0003474	0.0005399	0.0001029	1.245×10^{-5}	1.145×10^{-7}	0.00125
12	0.0005890	0.0010615	5.969×10^{-5}	2.683×10^{-5}	1.155×10^{-5}	0.00756
15	0.0007764	0.0013316	5.806×10^{-5}	6.094×10^{-5}	4.532×10^{-5}	0.02636
18	0.0038395	0.0012920	0.0001429	0.00010372	0.0001179	0.06931
21	0.0071045	0.0009483	0.0001485	0.00016832	0.0001865	0.15579
24	0.0086094	0.0003837	9.830×10^{-5}	0.00026625	0.0003858	0.30641
27	0.0067604	0.0002648	0.0001268	0.00034755	0.0009031	0.55351
30	0.0016280	0.0008419	7.436×10^{-5}	0.00052638	0.0019239	0.94305

6

```
>> h=0.01; t=[0:h:30]'; y=sqrt(2)*sin(4*t/5+pi/4);
   u=172/125*cos(4/5*t); ii=301:300:3001; y=y(ii); T=[];
   for p=1:6, y1=fode_caputo9(a,na,b,nb,y0,u,t,p);
      err=y-y1(ii); T=[T abs(err)];
   end
```

◀

As can be seen, in order to avoid possible cumulative errors, the step size cannot be selected too small, and in general the total number of calculation points should not exceed 1500, otherwise, the cumulative errors may affect the final results. Therefore, the algorithm given here is not suitable for a large time range, for instance, $0 \leq t \leq 5000$, for solving linear FODEs.

▶ Example 6.16

Consider the zero initial condition problem in Example 6.11. In the previous example, the direct use of the high-precision algorithm did not work well, so now try to solve the original problem again using the high-precision Caputo equation function.

Solutions It was pointed out in Example 6.11 that, the reason for the failure of the high-precision algorithm is that the necessary conditions for the high-precision solution are not satisfied. Since the new algorithm reconstructs the equivalent initial conditions, it is possible to find the problem directly using the new algorithm. The resulting error curve is shown in Fig. 6.11. It can be seen that if $p = 4$ is chosen, the error is not visible from Fig. 6.11.

```
>> a=[1 1 1]; na=[3 2.5 0]; b=1; nb=0;
   t=[0:0.1:1]'; y=-1+t-t.^2/2+exp(-t); y0=0;
   u=-1+t-t.^2/2-t.^0.5.*ml_func([1,1.5],-t);
```

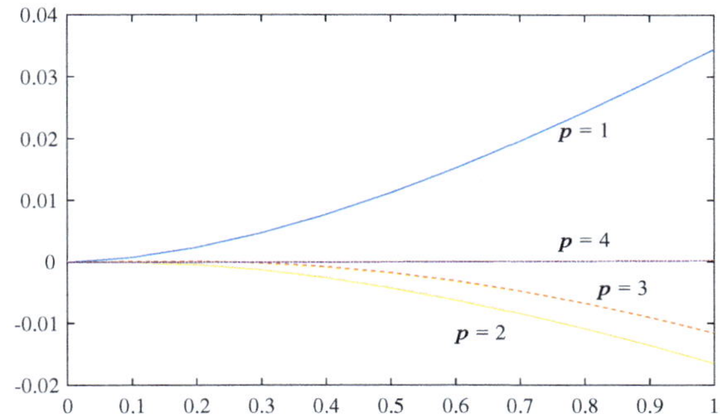

Fig. 6.11 Comparison of calculation errors under different p choices

```
y1=fode_caputo9(a,na,b,nb,y0,u,t,1); e1=y-y1;
y2=fode_caputo9(a,na,b,nb,y0,u,t,2); e2=y-y2;
y3=fode_caputo9(a,na,b,nb,y0,u,t,3); e3=y-y3;
y4=fode_caputo9(a,na,b,nb,y0,u,t,4); e4=y-y4;
plot(t,e1,t,e2,'--',t,e3,':',t,e4,'-.')
```

◀

Therefore, it is recommended here to use `fode_caputo9()` function to solve FODEs, both zero and nonzero initial values. However, note that it is not advisable to calculate too many points.

6.6 Solutions of Linear Fractional-Order State Space Models

As with integer-order systems, the concept of fractional-order state space (FOSS) equation model is introduced in addition to the transfer function description. This section first introduces the commensurate-order state space model and state transition matrix, and then gives the extended state space model for general non-commensurate-order differential equations.

6.6.1 State Space Descriptions of Linear FODEs

Consider the transfer function model for a commensurate-order system described in Definition 6.10. Similarly to the integer-order systems, if the same vector of state variables, $\boldsymbol{y}(t)$, is chosen, the equation of state space model for the fractional-order system can be written.

Definition 6.15 Commensurate-order state space models

The general form of a linear time invariant commensurate-order state space model is

$$\begin{cases} \mathscr{D}_t^{\alpha} \boldsymbol{x}(t) = \boldsymbol{A}\boldsymbol{x}(t) + \boldsymbol{B}\boldsymbol{u}(t) \\ \boldsymbol{y}(t) = \boldsymbol{C}\boldsymbol{x}(t) + \boldsymbol{D}\boldsymbol{u}(t), \end{cases} \tag{6.64}$$

where $\boldsymbol{x}(t) \in \mathbb{R}^{n\times 1}$ is the state variable vector, n is the order of the system, $\boldsymbol{A} \in \mathbb{R}^{n\times n}$, $\boldsymbol{B} \in \mathbb{R}^{n\times p}$, $\boldsymbol{C} \in \mathbb{R}^{q\times n}$ and $\boldsymbol{D} \in \mathbb{R}^{q\times p}$ are constant matrices.

Remarks Commensurate-order state space models

(1) The term "commensurate" is emphasized here; otherwise, α in the model is no longer a scalar. This case is not considered for the time being.

6

(2) The state space model here is suitable for fractional-order derivatives under any definition. □

▶ Example 6.17

Consider again the commensurate-order differential equation from Example 6.5. Write out its FOSS model.

$$\mathscr{D}_t^{1.2} y(t) + 5\mathscr{D}_t^{0.9} y(t) + 9\mathscr{D}_t^{0.6} y(t) + 7\mathscr{D}_t^{0.3} y(t) + 2y(t) = u(t).$$

Solutions It is seen from the orders of the equation that, the base order is $\alpha = 0.3$. If the states $x_1(t) = y(t)$, $x_2(t) = \mathscr{D}_t^{0.3} y(t)$, $x_3(t) = \mathscr{D}_t^{0.6} y(t)$ and $x_4(t) = \mathscr{D}_t^{0.9} y(t)$ are chosen, the state space model can be written directly in the form

$$\mathscr{D}_t^{0.3} \boldsymbol{x}(t) = \begin{bmatrix} 0 & 1 & 0 & 0 \\ 0 & 0 & 1 & 0 \\ 0 & 0 & 0 & 1 \\ -2 & -7 & -9 & -5 \end{bmatrix} \boldsymbol{x}(t) + \begin{bmatrix} 0 \\ 0 \\ 0 \\ 1 \end{bmatrix} u(t), \ y(t) = [1, 0, 0, 0]\boldsymbol{x}(t).$$

◀

6.6.2 State Transition Matrix

The problem of analytical solution of the Caputo fractional-order state space models will be presented here, since the Caputo definition is significantly better than the other definitions in describing nonzero initial value problems. Here the form of the analytical solution of the state space model will be analyzed, the definition of the state transition matrix will be given and its computational problems will be solved.

Recall the integer-order state space models.

$$\boldsymbol{x}'(t) = \boldsymbol{A}\boldsymbol{x}(t) + \boldsymbol{B}\boldsymbol{u}(t). \tag{6.65}$$

For convenience let $t_0 = 0$. If the initial condition is known to be $\boldsymbol{x}(0)$, the analytical solution is

$$x(t) = e^{At}x(0) + \int_0^t e^{A(t-\tau)}Bu(\tau)d\tau, \tag{6.66}$$

where e^{At} is referred to as the state transition matrix, denoted as $\Phi(t)$.

Definition 6.16 Analytical solution of autonomous state space model

For autonomous Caputo state space model

$${}_0^C\mathscr{D}_t^\alpha x(t) = Ax(t), \tag{6.67}$$

with initial states $x(0)$, the analytical solution can be written as

$$x(t) = \Phi(t)x(0), \tag{6.68}$$

where $\Phi(t)$ is referred to as the state transition matrix of the fractional-order system.

Theorem 6.13 State transition matrix of the fractional-order system

The state transition matrix $\Phi(t)$ in Definition 6.16 is

$$\Phi(t) = E_\alpha(At^\alpha). \tag{6.69}$$

Proof Assume that $u(t) \equiv 0$, the analytical solution of the FOSS equation is

$$x(t) = A_0 + A_1t^\alpha + A_2t^{2\alpha} + \cdots + A_mt^{m\alpha} + \cdots \tag{6.70}$$

Taking αth-order Caputo derivative to both sides, and recall Theorem 3.7.

$${}_0^C\mathscr{D}_t^\alpha t^\lambda = \frac{\Gamma(\lambda+1)}{\Gamma(\lambda+1-\alpha)}t^{\lambda-\alpha},$$

it is found that

$$\begin{aligned}{}_0^C\mathscr{D}_t^\alpha x(t) = \mathbf{0} + A_1\Gamma(1+\alpha) + \frac{A_2\Gamma(1+2\alpha)}{\Gamma(1+\alpha)}t^\alpha + \frac{A_3\Gamma(1+3\alpha)}{\Gamma(1+2\alpha)}t^{2\alpha} + \cdots +\\ \frac{A_k\Gamma(1+k\alpha)}{\Gamma(1+(k-1)\alpha)}t^{(k-1)\alpha} + \cdots = Ax(t).\end{aligned}$$

Comparing the coefficient matrix above with the one in (6.70), it is seen that

$$A_1 = \frac{AA_0}{\Gamma(1+\alpha)}, \quad A_k = \frac{\Gamma(1+(k-1)\alpha)}{\Gamma(1+k\alpha)}AA_{k-1}, \tag{6.71}$$

and $\boldsymbol{A}_0 = \boldsymbol{x}(0)$. It is easily seen that

$$\begin{cases} \boldsymbol{A}_1 = \dfrac{\boldsymbol{A}}{\Gamma(1+\alpha)}\boldsymbol{x}(0) \\ \boldsymbol{A}_2 = \dfrac{\Gamma(1+\alpha)}{\Gamma(1+2\alpha)}\boldsymbol{A}\boldsymbol{A}_1 = \dfrac{\boldsymbol{A}^2}{\Gamma(1+2\alpha)}\boldsymbol{x}(0) \\ \vdots \\ \boldsymbol{A}_k = \dfrac{\Gamma\left(1+(k-1)\alpha\right)}{\Gamma\left(1+k\alpha\right)}\boldsymbol{A}\boldsymbol{A}_{k-1} = \dfrac{\boldsymbol{A}^k}{\Gamma(1+k\alpha)}\boldsymbol{x}(0). \end{cases} \tag{6.72}$$

The analytical solution of (6.70) can be derived as follows which completes the proof.

$${}_0^{\mathrm{C}}\mathscr{D}_t^{\alpha}\boldsymbol{x}(t) = \sum_{k=0}^{\infty} \frac{\left(\boldsymbol{A}t^{\alpha}\right)^k}{\Gamma(1+k\alpha)}\boldsymbol{x}(0) = \mathrm{E}_{\alpha}\left(\boldsymbol{A}t^{\alpha}\right)\boldsymbol{x}(0). \tag{6.73}$$

□

Theorem 6.14 Analytical solution of FOSS equation
Caputo state space equation is

$${}_0^{\mathrm{C}}\mathscr{D}_t^{\alpha}\boldsymbol{x}(t) = \boldsymbol{A}\boldsymbol{x}(t) + \boldsymbol{B}\boldsymbol{u}(t). \tag{6.74}$$

Under initial state $\boldsymbol{x}(0)$, the analytical solution is written as

$$\boldsymbol{x}(t) = \boldsymbol{\Phi}(t)\boldsymbol{x}(0) + \int_0^t \widehat{\boldsymbol{\Phi}}(t-\tau)\boldsymbol{B}\boldsymbol{u}(\tau)\mathrm{d}\tau, \tag{6.75}$$

where

$$\boldsymbol{\Phi}(t) = \widehat{\boldsymbol{\Phi}}(t) * \chi_{\alpha-1}(t) \text{ and } \chi_{\alpha-1}(t) = \begin{cases} t^{-\alpha}/\Gamma(1-\alpha), & \alpha < 1 \\ \delta(t), & \alpha = 1, \end{cases} \tag{6.76}$$

where $\delta(t)$ is a Dirac function, and [8]

$$\widehat{\boldsymbol{\Phi}}(t) = \mathscr{L}^{-1}\left[(s^{\alpha}\boldsymbol{I}-\boldsymbol{A})^{-1}\right], \ \boldsymbol{\Phi}(t) = \mathscr{L}^{-1}\left[(s^{\alpha}\boldsymbol{I}-\boldsymbol{A})^{-1}s^{\alpha-1}\right]. \tag{6.77}$$

▶ Example 6.18
The FOSS is given below. Find its state transition matrix $\boldsymbol{\Phi}(t)$.

$${}_0^{\mathrm{C}}\mathscr{D}_t^{\alpha}\boldsymbol{x}(t) = \begin{bmatrix} -2 & 0 & -1 & 0 \\ -1 & -3 & 1 & 0 \\ 2 & 1 & 1 & 1 \\ 0 & 1 & -2 & -2 \end{bmatrix}\boldsymbol{x}(t).$$

Solutions The MATLAB function `funmsym()` given in the Ref. [9] can be used directly to find the state transition matrix for a given state space equation. Denoting the Mittag-Leffler function $\mathrm{E}_\alpha(x)$ by the symbolic function $E(x)$, the state transition matrix can be computed directly using the following statements.

```
>> syms t x a E(x)
   A=[-2,0,-1,0; -1,-3,1,0; 2,1,1,1; 0,1,-2,-2];
   Phi=funmsym(A,E(x*t^a),x)
```

The mathematical form of the state transition matrix is obtained as follows (to facilitate understanding of the results obtained, the following interpretation is made: the first group below shows the first column of the matrix, the second group shows the second column of the matrix, and the third group shows the third and fourth columns of the matrix)

$$\boldsymbol{\Phi}(t)=\left[\begin{array}{c} \mathrm{E}_\alpha\left(-t^\alpha\right)-t^{2\alpha}\,\mathrm{E}_\alpha^{(2)}\left(-t^\alpha\right)/2-t^\alpha\,\mathrm{E}_\alpha^{(1)}\left(-t^\alpha\right) \\ \mathrm{E}_\alpha\left(-3t^\alpha\right)-\mathrm{E}_\alpha\left(-t^\alpha\right)+t^{2\alpha}\mathrm{E}_\alpha^{(2)}\left(-t^\alpha\right)/2+t^\alpha\mathrm{E}_\alpha^{(1)}\left(-t^\alpha\right) \\ t^{2\alpha}\mathrm{E}_\alpha^{(2)}\left(-t^\alpha\right)/2+2t^\alpha\mathrm{E}_\alpha^{(1)}\left(-t^\alpha\right) \\ \mathrm{E}_\alpha\left(-t^\alpha\right)-\mathrm{E}_\alpha\left(-3t^\alpha\right)-t^{2\alpha}\mathrm{E}_\alpha^{(2)}\left(-t^\alpha\right)/2-2t^\alpha\mathrm{E}_\alpha^{(1)}\left(-t^\alpha\right) \end{array}\right.$$

$$\begin{array}{c} -t^{2\alpha}\,\mathrm{E}_\alpha^{(2)}\left(-t^\alpha\right)/2 \\ \mathrm{E}_\alpha\left(-3t^\alpha\right)+t^{2\alpha}\mathrm{E}_\alpha^{(2)}\left(-t^\alpha\right)/2 \\ t^{2\alpha}\mathrm{E}_\alpha^{(2)}\left(-t^\alpha\right)/2+t^\alpha\mathrm{E}_\alpha^{(1)}\left(-t^\alpha\right) \\ \mathrm{E}_\alpha\left(-t^\alpha\right)-\mathrm{E}_\alpha\left(-3t^\alpha\right)-t^{2\alpha}\mathrm{E}_\alpha^{(2)}\left(-t^\alpha\right)/2-t^\alpha\mathrm{E}_\alpha^{(1)}\left(-t^\alpha\right) \end{array}$$

$$\left.\begin{array}{c|c} -t^{2\alpha}\mathrm{E}_\alpha^{(2)}\left(-t^\alpha\right)/2-t^\alpha\mathrm{E}_\alpha^{(1)}\left(-t^\alpha\right) & -t^{2\alpha}\mathrm{E}_\alpha^{(2)}\left(-t^\alpha\right)/2 \\ t^{2\alpha}\mathrm{E}_\alpha^{(2)}\left(-t^\alpha\right)/2+t^\alpha\mathrm{E}_\alpha^{(1)}\left(-t^\alpha\right) & t^{2\alpha}\mathrm{E}_\alpha^{(2)}\left(-t^\alpha\right)/2 \\ \mathrm{E}_\alpha\left(-t^\alpha\right)+t^{2\alpha}\mathrm{E}_\alpha^{(2)}\left(-t^\alpha\right)/2+2t^\alpha\mathrm{E}_\alpha^{(1)}\left(-t^\alpha\right) & t^{2\alpha}\mathrm{E}_\alpha^{(2)}\left(-t^\alpha\right)/2+t^\alpha\mathrm{E}_\alpha^{(1)}\left(-t^\alpha\right) \\ -t^{2\alpha}\mathrm{E}_\alpha^{(2)}\left(-t^\alpha\right)/2-2t^\alpha\mathrm{E}_\alpha^{(1)}\left(-t^\alpha\right) & \mathrm{E}_\alpha\left(-t^\alpha\right)-t^{2\alpha}\mathrm{E}_\alpha^{(2)}\left(-t^\alpha\right)/2-t^\alpha\mathrm{E}_\alpha^{(1)}\left(-t^\alpha\right) \end{array}\right],$$

where $\mathrm{E}_\alpha^{(k)}(\cdot)$ is the kth-order derivative of the one-parameter Mittag-Leffler function with respect to t, $k\in\mathbb{N}$. ◀

With the state transition matrix, the `ml_func()` function can be called directly to obtain the numerical solution of the autonomous linear commensurate-order differential equations.

Remarks States and state transition matrix [8]

(1) As already mentioned, the states in fractional calculus are pseudo-states because their values are also related to the historical information between t_0 and the current moment t. Therefore, in order to determine the future behavior of the fractional-order system, all the information of $[t_0, t]$ is needed.

(2) Unlike the integer-order systems, the matrix $\boldsymbol{\Phi}(t)$ is not a state transition matrix in the traditional sense. It is only a state transition matrix $\boldsymbol{\Phi}(t, t_0)$ from t_0 to t time. In the time period $t_0<\tau<t$, the state transition matrix property $\boldsymbol{\Phi}(t, t_0)=\boldsymbol{\Phi}(t,\tau)\boldsymbol{\Phi}(\tau, t_0)$ does not hold. Therefore, it is better to refer such matrices as the pseudo-state transition matrices. □

6.6.3 Commensurate-Order State Space Models

If the linear FOSS equation is not a commensurate-order one, it is difficult to describe such a system by the ordinary state space equation, and the concept of extended FOSS equation needs to be introduced.

Definition 6.17 Extended FOSS equation

If the order α is no longer a scalar base order, but an order vector, the extended FOSS equation is defined

$$\begin{cases} \mathscr{D}_t^{\boldsymbol{\alpha}} \boldsymbol{x}(t) = \boldsymbol{A}\boldsymbol{x}(t) + \boldsymbol{B}\boldsymbol{u}(t) \\ \boldsymbol{y}(t) = \boldsymbol{C}\boldsymbol{x}(t) + \boldsymbol{D}\boldsymbol{u}(t). \end{cases} \tag{6.78}$$

6

In contrast to the FOSS equation given in Definition 6.15, the sizes of matrix $\boldsymbol{A}$ are significantly reduced in many examples.

► Example 6.19

Consider the FOTF model in Example 5.11. Write its extended FOSS equation.

$$G(s) = \frac{-2s^{0.63} - 4}{2s^{3.501} + 3.8s^{2.42} + 2.6s^{1.798} + 2.5s^{1.31} + 1.5}.$$

Solutions The FODE corresponding to the original transfer function can be written as

$$\mathscr{D}_t^{3.501} y(t) + 1.9\mathscr{D}_t^{2.42} y(t) + 1.3\mathscr{D}_t^{1.798} y(t) + 1.25\mathscr{D}_t^{1.31} y(t) + 0.75y(t) = \hat{u}(t).$$

If the base order is chosen, it is seen that $\alpha = 0.001$, and the state space equation thus created will have 3501 states, which will be extremely cumbersome to handle. Now reselect the state variables $x_1 = y(t)$, $x_2(t) = \mathscr{D}_t^{1.31} y(t)$, $x_3(t) = \mathscr{D}_t^{1.798} y(t)$, $x_4(t) = \mathscr{D}_t^{2.42} y(t)$, the extended state space model of the system can be written as

$$\begin{cases} \mathscr{D}_t^{1.31} x_1(t) = x_2(t) \\ \mathscr{D}_t^{0.488} x_2(t) = x_3(t) \\ \mathscr{D}_t^{0.612} x_3(t) = x_4(t) \\ \mathscr{D}_t^{1.081} x_4(t) = -0.75x_1(t) - 1.25x_2(t) - 1.3x_3(t) - 1.9x_4(t) + \hat{u}(t). \end{cases}$$

The related matrices in the extended state space model are

$$\boldsymbol{A} = \begin{bmatrix} 0 & 1 & 0 & 0 \\ 0 & 0 & 1 & 0 \\ 0 & 0 & 0 & 1 \\ -0.75 & -1.25 & -1.3 & -1.9 \end{bmatrix}, \quad \boldsymbol{B} = \begin{bmatrix} 0 \\ 0 \\ 0 \\ 1 \end{bmatrix}, \quad \boldsymbol{\alpha} = \begin{bmatrix} 1.31 \\ 0.488 \\ 0.612 \\ 1.081 \end{bmatrix}.$$

If the numerator of the original fractional-order transfer function also contains a pseudo-polynomial in s, the states of the original system need to be chosen again, since the order of the numerator pseudo-polynomial should be considered equally. In this case, the new state variables can be chosen $x_1(t) = y(t)$, $x_2(t) = \mathscr{D}_t^{0.63} y(t)$, $x_3(t) = \mathscr{D}_t^{1.31} y(t)$, $x_4(t) = \mathscr{D}_t^{1.798} y(t)$ and $x_5(t) = \mathscr{D}_t^{2.42} y(t)$, then the state space equation can be written as

$$\begin{cases} \mathscr{D}_t^{0.63} x_1(t) = x_2(t) \\ \mathscr{D}_t^{0.68} x_2(t) = x_3(t) \\ \mathscr{D}_t^{0.488} x_3(t) = x_4(t) \\ \mathscr{D}_t^{0.612} x_4(t) = x_5(t) \\ \mathscr{D}_t^{1.081} x_5(t) = -0.75x_1(t) - 1.25x_3(t) - 1.3x_4(t) - 1.9x_5(t) + u(t), \end{cases}$$

and $y(t) = \begin{bmatrix} -2, & -1, & 0, & 0, & 0 \end{bmatrix} \boldsymbol{x}(t)$.
The related matrices are then

$$\boldsymbol{A} = \begin{bmatrix} 0 & 1 & 0 & 0 & 0 \\ 0 & 0 & 1 & 0 & 0 \\ 0 & 0 & 0 & 1 & 0 \\ 0 & 0 & 0 & 0 & 1 \\ -0.75 & 0 & -1.25 & -1.3 & -1.9 \end{bmatrix}, \quad \boldsymbol{B} = \begin{bmatrix} 0 \\ 0 \\ 0 \\ 0 \\ 1 \end{bmatrix}, \quad \boldsymbol{\alpha} = \begin{bmatrix} 0.63 \\ 0.68 \\ 0.488 \\ 0.612 \\ 1.081 \end{bmatrix}.$$

If dealing with nonzero initial value problems, the state space equation should also be written by choosing all integer-order derivatives of $y(t)$ as state variables. ◀

6.7 Numerical Solutions of Irrational Differential Equations

6.7.1 Descriptions of Irrational Transfer Functions

The standard rational FOTF in (6.2) has been used in the previous analysis. In practice, there are times when the model with a $p^{\gamma}(s)$ term, where γ is not an integer, then such a term can only be approximated by an infinite series, and the transfer function cannot be represented by the rational transfer function given in (6.2). In this case, a class of irrational FOTF models should be considered for representation and analysis. More generally, any function containing non-integer powers of s and nonlinear operations can be considered as an irrational transfer function model.

Definition 6.18 A class of irrational transfer function models

A class of commonly used irrational transfer function models is

$$G(s) = \frac{\prod_{i=1}^{m} \left(b_1 s^{\eta_1} + b_2 s^{\eta_2} + \cdots + b_{r_i} s^{\eta_{r_i}}\right)^{\beta_i}}{\prod_{i=1}^{n} \left(a_1 s^{\gamma_1} + a_2 s^{\gamma_2} + \cdots + a_{k_i} s^{\gamma_{k_i}}\right)^{\alpha_i}}. \tag{6.79}$$

6.7.2 Solutions Based on Numerical Laplace Inverse Transform

The analytical solution of the Laplace transform and its inverse of a given function can be attempted directly by calling the functions `laplace()` or `ilaplace()`. In fact, the scope of these two functions is very narrow, and the vast majority of functions in the field of fractional calculus do not have analytical solutions of Laplace and its inverse, in which case, the numerical solution methods should be considered.

Many kinds of MATLAB functions for numerical inverse Laplace transform can be found from the Internet, and the `INVLAP()` function developed by Juraj Valsa presented here is one of the more reliable ones [10, 11]. To better utilize this tool for solving typical feedback control system simulation problems with irrational models, this book modifies and extends the function to make it more widely applicable. The new function is renamed as `INVLAP_new()` [12].


```
function [t,y]=INVLAP_new(G,t0,tn,N,H,tx,ux)
   arguments
      G, t0(1,1) double, tn(1,1) {mustBeGreaterThan(tn,t0)}
      N(1,1){mustBePositiveInteger}=100;
      H(1,1)=0; tx='1'; ux double=0
   end
   G=add_dots(G); if ischar(H), H=add_dots(H); end
   if ischar(tx), tx=add_dots(tx); end
   a=6; ns=20; nd=19; t=linspace(t0,tn,N);
   if t0==0, t=[1e-6 t(2:N)]; end
   n=1:ns+1+nd; alfa=a+(n-1)*pi*1j; bet=-exp(a)*(-1).^n;
   n=1:nd; bet(1)=bet(1)/2;
   bdif=fliplr(cumsum(gamma(nd+1)./...
      gamma(nd+2-n)./gamma(n)))./2^nd;
   bet(ns+2:ns+1+nd)=bet(ns+2:ns+1+nd).*bdif;
   if isnumeric(H), H=num2str(H); end
   for i=1:N
      tt=t(i); s=alfa/tt; bt=bet/tt; sG=eval(G); sH=eval(H);
      if ischar(tx), sU=eval(tx);
      else
         if isnumeric(tx)
            f=@(x)interp1(tx,ux,x,'spline').*exp(-s.*x);
         else, f=@(x)tx(x).*exp(-s.*x); end
         sU=integral(f,t0,tn,'ArrayValued',true);
      end
      btF=bt.*sG./(1+sG.*sH).*sU; y(i,1)=sum(real(btF));
   end
   % remove the dot operations, and then add back the dots uniformly
   function F=add_dots(F)
      F=strrep(strrep(strrep(F,'.*','*'),'./','/'),'.^','^');
      F=strrep(strrep(strrep(F,'*','.*'),'/','./'),'^','.^');
end end
```

For a general numerical inverse Laplace transform of a given function, the function can be called in the following syntax.

```
[t,y]=INVLAP_new(F,t0,tn,N)
```

Remarks `INVLAP_new()` function arguments

(1) The original `INVLAP()` function uses the string description F and requires that dot operations are used in the string, which may lead to errors if some dot operation representations are forgotten somewhere in the actual use of the string description. In this new version, the Laplace transform string F does not need to use dot operations intentionally, because the internal mechanism of the function will automatically convert the required operations into dot operations uniformly.

(2) Input variables (t_0, t_n) describe the time interval of interest, N is expected to calculate the number of points, the restrictions that the original function cannot have $t_0 = 0$ have been lifted, $t_0 = 0$ is automatically replaced with an offset of 10^{-6}.

(3) Different N can be selected to verify the simulation results.

(4) The other syntaxes of this function will be introduced in the subsequent contents. □

▶ Example 6.20

Find the numerical solution of the inverse Laplace transform of the function

$$G(s) = \frac{-17s^5 - 7s^4 + 2s^3 + s^2 - s + 1}{s^6 + 11s^5 + 48s^4 + 106s^3 + 125s^2 + 75s + 17}.$$

Solutions The analytical solution of this inverse Laplace transform in the strict sense does not exist, and its numerical solution can be obtained with high accuracy by using the MATLAB Symbolic Math Toolbox. For the same function, the numerical solution of the inverse Laplace transform can also be found by representing it as a string. By choosing different number of computation points N, the time consumed and the relative error can be measured as shown in Table 6.4. Note that a 2-norm is used to represent the relative error, and for the sake of fair comparison, the 2-norm obtained is divided by the number of points and included in Table 6.4. From the measured results, the computational effort increases with the increase of N, but the change in the error index is not significant. The accuracy of the numerical solutions is relatively high.

```
>> syms s t;                          % declare symbolic variable and function
   G=(-17*s^5-7*s^4+2*s^3+s^2-s+1)...
       /(s^6+11*s^5+48*s^4+106*s^3+125*s^2+75*s+17);
   f=ilaplace(G,s,t); fun=char(G); N=200; tic,
   [t1,y1]=INVLAP_new(fun,0,5,N); toc  % numerical inverse Laplace transform
   y0=subs(f,t,t1); y0=double(y0); % theoretical computing, very time consuming
   err=norm((y1-y0)./y0)/N           % compute relative error
```

◀

Table 6.4 The impact of different calculation points on the results

Points N	50	100	300	500	800	1000	2000
Error	2.5×10^{-7}	1.8×10^{-6}	4.6×10^{-7}	3.1×10^{-7}	2.8×10^{-7}	3.7×10^{-7}	2.8×10^{-7}
Time elapse	0.0033	0.0062	0.0197	0.0335	0.0528	0.0683	0.1220

▶ **Example 6.21**

Consider the following irrational transfer function model:

$$G(s)=\frac{(s^{0.4}+0.4s^{0.2}+0.5)}{\sqrt{s}(s^{0.2}+0.02s^{0.1}+0.6)^{0.4}(s^{0.3}+0.5)^{0.6}}.$$

Draw the impulse response curve of this system in time interval $t\in(0,1)$.

Solutions Unlike the previous example, this function cannot be treated with `ilaplace()` to obtain its inverse Laplace transform. The numerical solution is the only option. By choosing $N=200$, the numerical solution of the inverse Laplace transform can be obtained directly by the following commands, as shown in Fig. 6.12. If N is increased to $N=5000$, an identical curve is also obtained, indicating that the resulting result is correct.

```
>> f=['(s^0.4+0.4*s^0.2+0.5)/sqrt(s)/',...
      '(s^0.2+0.02*s^0.1+0.6)^0.4/(s^0.3+0.5)^0.6'];
   [t,y]=INVLAP_new(f,0,1,200); plot(t,y), ylim([0 20])
```

◀

It can be seen from this example that if there is a $p^{\gamma}(s)$ term in the equation, it corresponds to an infinite series, so it is not possible to derive an analytical solution of the inverse Laplace transform. The numerical solution function here is still computationally efficient in practical applications.

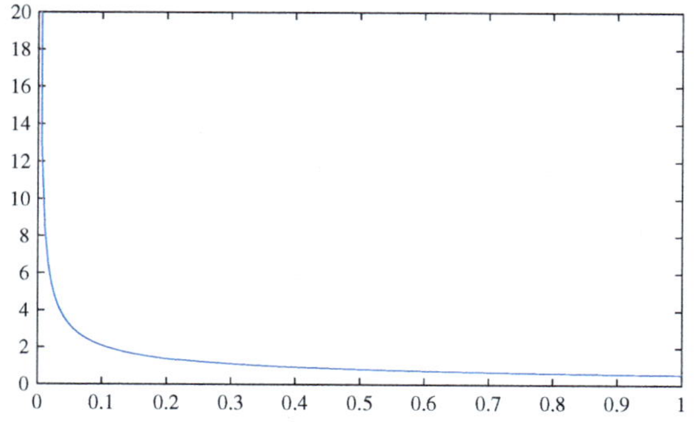

Fig. 6.12 Results from numerical inverse Laplace transform

6.7.3 Time Response Computing of Closed-Loop Irrational Systems

In practice, if the transfer function $G(s)$ of a complicated system is known and the Laplace transform $R(s)$ of the input signal is also known, the Laplace transform of the output signal can be obtained from $Y(s) = G(s)U(s)$. If the analytical solution of $Y(s)$ does not exist, the system can be simulated using the numerical inverse Laplace transform technique.

If the original LAPINV() function is used, the entire closed-loop system needs to be represented as a complete string, before solving numerically the inverse Laplace transform problem. Such a string would be extremely lengthy, so the INVLAP_new() function is extended with the syntax $[\boldsymbol{t}, \boldsymbol{y}]=$ INVLAP_new$(G, t_0, t_\mathrm{n}, N, H, U)$.

Remarks Arguments in the new function syntaxes

(1) The forward path transfer function model can be represented by the string G, the feedback path transfer function by the string H. These strings should be written as strings of s, but it is not necessary to represent in the string dot operations.

(2) The unity negative feedback structure can be assigned by denoting $H = 1$, and the unit positive feedback is set to $H = -1$.

(3) The input argument U is written as a Laplace transform string of the actual input signal. For example, the unit step input should be represented as the string '1/s' and the impulsive input can be omitted.

(4) The input arguments (t_0, t_n) is the time interval and N is the number of calculation points.

(5) The returned arguments $\boldsymbol{t}$ and $\boldsymbol{y}$ are the result of time and output. □

▶ Example 6.22

Suppose the complicated irrational open-loop model is as follows [13]:

$$G(s) = \left[\frac{\sinh(w\sqrt{s})}{w\sqrt{s}}\right]^2 \frac{1}{\sqrt{s}\sinh(\sqrt{s})}.$$

Draw the step response curve of a unity negative feedback system, where $w = 0.1$.

Solutions The open-loop irrational transfer function can be represented by a string so that the closed-loop step response curve of the system is directly obtained by the following statements, as shown in Fig. 6.13.

```
>> G='(sinh(0.1*sqrt(s))/0.1/sqrt(s))^2/sqrt(s)/sinh(sqrt(s))';
   [t,y]=INVLAP_new(G,0,10,1000,1,'1/s'); plot(t,y)
```

◀

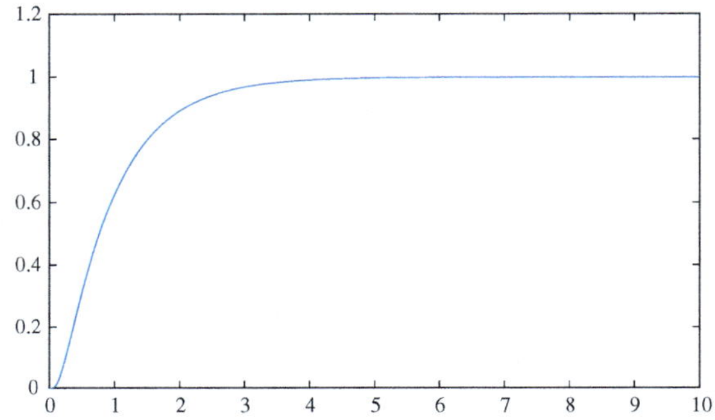

Fig. 6.13 Step response curve of closed-loop system

▶ Example 6.23

The plant model of the system is $G(s) = 3.13\,\mathrm{e}^{-5s}/(433.33s + 1)$. Consider the controller model based on quantitative feedback theory design given in the Ref. [8] as follows:

$$G_c(s) = 1.8393\left(\frac{s+0.011}{s}\right)^{0.96}\left(\frac{8.8\times 10^{-5}s+1}{8.096\times 10^{-5}s+1}\right)^{1.76}\frac{1}{(1+s/0.29)^2}.$$

Draw the step response curve of the closed-loop system.

Solutions The plant model given here is an integer-order transfer function model. A good approximation to the controller is found in Example 5.13. From the following statements, the closed-loop system model can be approximated with the integer-order transfer function to derive the step response curve. Meanwhile, the step response of the system can be obtained using the numerical inverse Laplace transform-based method. The results of the two methods are shown in Fig. 6.14. As can be seen, the curves obtained by the two methods are very close to each other.

```
>> w=logspace(-6,4,200);        % select the frequency fitting range
   s=tf('s'); G=3.13*exp(-5*s)/(433.33*s+1);
```

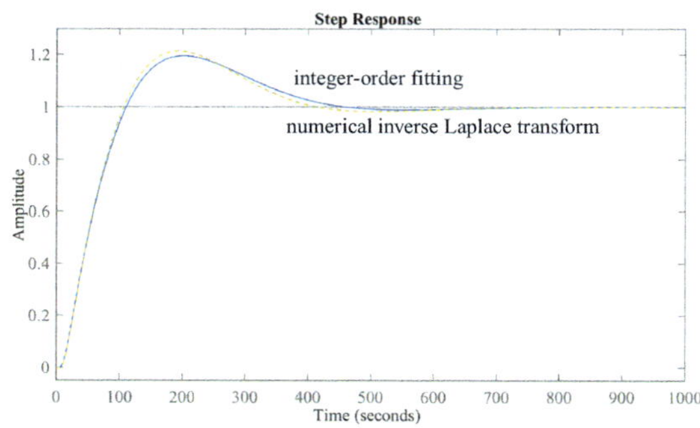

Fig. 6.14 Step response curves of the closed-loop system derived from the two methods

```
G1=(s+0.011)/s; F1=frd(G1,w);
G2=(8.8e-5*s+1)/(8.096e-5*s+1); F2=frd(G2,w);
G3=1/(1+s/0.29)^2; F3=frd(G3,w);
h1=F1.ResponseData; h2=F2.ResponseData;
h3=F3.ResponseData; h=1.8393*h1.^0.96.*h2.^1.76.*h3;
[n,d]=invfreqs(h(:),w,4,4); Gc=zpk(tf(n,d)); %integer-order fitting
step(feedback(G*Gc,1),1000);                  %closed-loop response
G2='((s+0.011)/s)^0.96*((8.8e-5*s+1)/(8.096e-5*s+1))^1.76';
G1='1.8393*3.13*exp(-5*s)/(433.33*s+1)/(1+s/0.29)^2';
F=[G2,'*',G1]; [t,y]=INVLAP_new(F,0,1000,1000,1,'1/s');
hold on; plot(t,y,'--'), hold off %numerical inverse Laplace transform
```

◀

6.7.4 Time Responses Under Arbitrary Input Signals

If the analytical solution of the Laplace transform of the input signal $u(t)$ cannot be derived, the time domain response of the irrational system can still be obtained numerically. As seen from the internal code of the `INVLAP_new()` function, the main function is implemented by a loop structure, and within each loop step, a vector $\boldsymbol{s}$ is generated, which can be integrated numerically to yield numerical solution of the Laplace transform.

$$\mathscr{L}\big[u(t)\big] = \int_0^\infty u(t)\mathrm{e}^{-st}\mathrm{d}t = U(s). \tag{6.80}$$

Because of the existence of the e^{-st} term in the integrand, in practice, if the interval $\big(0, t_\mathrm{n}\big)$ is large enough, the integrand will converge to zero. Finite time integral can be used to approximate the infinite integral in the definition of Laplace transform, so that the numerical solution of the Laplace transform can be achieved directly. If the mathematical expression of the input function $u(t)$ is unknown, while only the vectors $\boldsymbol{x}_0$ and $\boldsymbol{u}_0$ of its samples are known, the spline interpolation method can be considered to reconstruct the input signal so as to find the numerical solution of its Laplace transform.

The Laplace transform of the output signal can be obtained if the result of the dot product of the Laplace transforms of the input signal and the transfer function $G(s)$ can be derived. In this way, the numerical solution of the output signal can be obtained by the MATLAB function `integral()`. The `INVLAP_new()` function given earlier implemented such a calculation method, and its syntaxes are

[$\boldsymbol{t}$,$\boldsymbol{y}$]=INVLAP_new(G,t_0,t_n,N,H,f)
[$\boldsymbol{t}$,$\boldsymbol{y}$]=INVLAP_new(G,t_0,t_n,N,H,$\boldsymbol{t}_x$,$\boldsymbol{u}_x$)

where f is a handle to a given input signal, which can be described by an anonymous function or an M-function, or by the time vector. In the latter case, $\boldsymbol{u}_x$ is the vector of input samples.

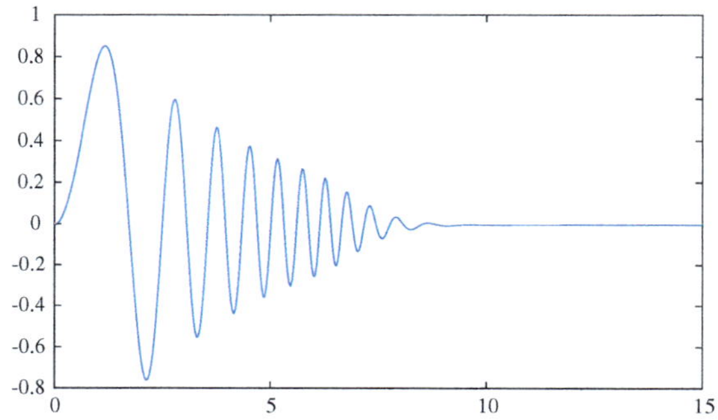

Fig. 6.15 Time domain response of the complicated irrational system

6

▶ Example 6.24

Assuming that the Laplace transform expression $G(s)$ for the irrational transfer function model is given by Example 6.21. Draw the time response curve of the system under the excitation of the signal $u(t) = \mathrm{e}^{-0.3t} \sin t^2$.

Solutions The transfer function model can still be entered using the format in Example 6.21. Since the mathematical expression of the input signal is known, it can be expressed as an anonymous function so that the time domain response curve of the system can be derived from the following statements, as shown in Fig. 6.15. The numerical integral operation embedded in this function is time-consuming, and the time elapse of these commands is about 5.62 s.

```
>> f=@(t)exp(-0.3*t).*sin(t.^2);      %if input function is known
   G=['(s^0.4+ 0.4*s^0.2+0.5)/(s^0.2+0.02*s^0.1+0.6)',...
      '^0.4/(s^0.3+0.5)^0.6'];
   tic, [t,y]=INVLAP_new(G,0,15,400,0,f); toc, plot(t,y)
```

Now, assuming that the mathematical expression of the input signal is unknown and only a set of sample points in $t \in (0, 15)$ are known, one can consider reconstructing the input signal by means of spline interpolation and finally deriving the numerical solution of the time domain response of the system. It can be seen that the following function yields the same curve as the previous one, but this method is more time-consuming, taking about 28.04 s when the number of computed points is only half of the previous one.

```
>> x0=0:0.1:15; u0=exp(-0.3*x0).*sin(x0.^2);
   tic, [t,y]=INVLAP_new(G,0,15,200,0,x0,u0); toc, plot(t,y)
```

◀

6.8 Stability Assessment of Linear Fractional-Order Systems

The stability of a system is the most important property of an engineering system. If a system is unstable, and it is not controlled, all other properties are meaningless. So far, there is hardly any method for judging the stability of other fractional-order or irrational systems in the literatures, except for a special method for determining the stability of linear commensurate-order systems. This section first introduces the stability determination methods for linear commensurate-order systems, and then discusses the stability analysis methods for linear non-commensurate-order systems and irrational systems.

6.8.1 Stability Assessment of Linear Commensurate-Order Systems

Consider the commensurate-order transfer function system given in (6.28). For the convenience of the description, its transfer function model is given here again.

$$G(\lambda) = \frac{b_1\lambda^m + b_2\lambda^{m-1} + \cdots + b_m\lambda + b_{m+1}}{a_1\lambda^n + a_2\lambda^{n-1} + \cdots + a_n\lambda + a_{n+1}}, \tag{6.81}$$

where $\lambda = s^\alpha$, and α is the base order of the system.

Theorem 6.15 Stability assessment theorem
If all the poles p_i of the system regarding λ can be found, the system $G(s)$ is stable, if and only if [14]

$$\left|\arg(p_i)\right| > \alpha\frac{\pi}{2}, \tag{6.82}$$

where $\arg(z)$ is the argument of the complex variable z.

A graphical depiction of the stability region of the system is shown in Fig. 6.16. An informal proof of the theorem is given below.

Proof Since the original fractional-order system is an integer-order system with $\lambda = s^\alpha$ and the stable argument boundary for s is $\pm\pi/2$, the stable argument boundary for λ should be $\pm\alpha\pi/2$. □

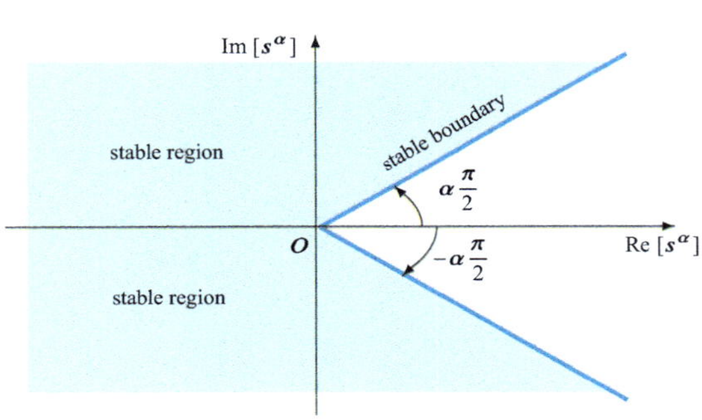

Fig. 6.16 Illustration of the stability region of the commensurate-order system

For a given commensurate-order transfer function model,+ if roots of the polynomial equation can be found

$$a_1\lambda^n + a_2\lambda^{n-1} + \cdots + a_n\lambda + a_{n+1} = 0, \tag{6.83}$$

we can check whether each of these roots lies in the shaded region in Fig. 6.16, and then using Theorem 6.15 to determine the stability of the system. The following example will demonstrate the problems that may be faced in finding the roots of a pseudo-polynomial.

Corollary 6.5 If the commensurate-order state space equation is given by Definition 6.15, it is also possible to find the eigenvalues p_i of the $\boldsymbol{A}$ matrix, so that the stability of the system can be judged in conjunction with Theorem 6.15.

6

▶ Example 6.25

Assess the stability of a system by the pseudo-polynomial equation [9]

$$s^{2.3}+5s^{1.6}+6s^{1.3}-5s^{0.4}+7=0.$$

Solutions From the given orders, it is clear that the base order of this pseudo-polynomial is $\alpha = 0.1$. Let $\lambda = s^{0.1}$, a polynomial equation $\lambda^{23} + 5\lambda^{16} + 6\lambda^{13} - 5\lambda^4 + 7 = 0$ can be found. The distribution of characteristic roots with respect to λ can be obtained directly from the following statements and the rays with angles of $\pm\alpha\pi/2$ and $\pm\alpha\pi$, respectively, as shown in Fig. 6.17.

```
>> syms z; f1=z^23+5*z^16+6*z^13-5*z^4+7;
   p=sym2poly(f1); r=roots(p); alpha=0.1;
   xm=xlim; xm(1)=0; a1=tan(alpha*pi/2)*xm; a2=tan(alpha*pi)*xm;
   hold on, A=1.5; x0=[0,A,A,-A,-A,A,A,0];
   y0=[0,-a1(2),-A,-A,A,A,a1(2),0]; fill(x0,y0,'c'),
   plot(xm,[a1; -a1; a2; -a2],real(r),imag(r),'x'), hold off
```

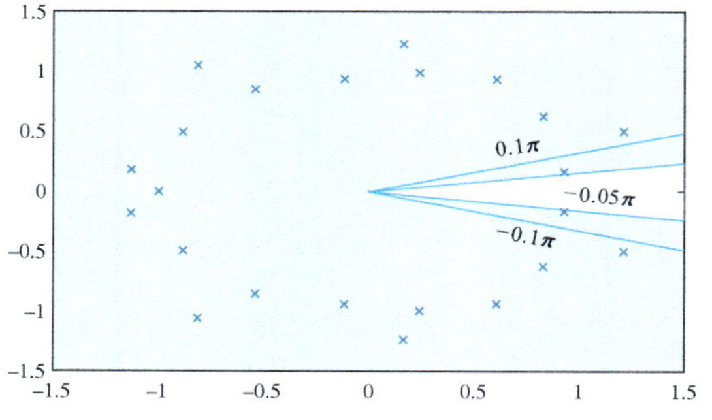

Fig. 6.17 Characteristic root distribution and stability boundary

It can be seen that all the 23 roots of the equation can be found using the above commands. From the distribution of the roots, all of them are located in the stability region limited by the stability boundary $\pm 0.05\pi$ angles, so it can be concluded that the system is stable.

With these 23 roots, one can then use $s = r^{10}$ to try to find all the roots that satisfy the original pseudo-polynomial equation for s. However, substituting the resulting roots back into the pseudo-polynomial equation reveals that only one pair of roots (the last two roots $s_{1,2} = -0.1076 \pm 0.5562\mathrm{j}$) truly satisfy the pseudo-polynomial equation and the rest are extraneous roots. These two roots are located in the left half of the s-plane, so the system is stable. In fact, the roots corresponding to the truly satisfying pseudo-polynomial equation are located on the first Riemann sheet, i.e., the angle is in the region $\pm 0.1\pi$.

```
>> syms s; f=s^2.3+5*s^1.6+6*s^1.3-5*s^0.4+7;
   r1=r.^10, double(subs(f,s,r1)), r1([22,23])
```

◀

Remarks About stability

(1) Note that the stability boundary mentioned here is for $\lambda = s^\alpha$, and not for s. The stability boundary for s is still the imaginary axis.

(2) When the base order $\alpha = 1$, the system is an integer-order system, and its stability boundary coincides exactly with that of the integer-order system.

(3) If $\alpha \geqslant 2$, then the system is unstable.

(4) As demonstrated in Example 6.25, many of the resulting roots with respect to λ are extraneous roots that do not satisfy the pseudo-polynomial equation, and one might question the correctness of the theorem. In fact, both theoretically and through practical examples, the roots with the smallest angle satisfy the pseudo-polynomial equation. Therefore, the theorem is correct. □

Conjecture 6.1 About the extraneous roots
All the roots satisfying the pseudo-polynomial equation satisfy $|\arg(p_i)| < \alpha\pi$. These roots are located on the first Riemann sheet. The other roots are extraneous roots, while the pseudo-polynomial equation is not satisfied.

In other words, it is sufficient to find the characteristic roots in the region between $\pm\alpha\pi/2$ and $\pm\alpha\pi$ angles, and it is not necessary to seek to obtain all of them.

6.8.2 Stability Assessment of Non-Commensurate-Order Systems

From the example given earlier, if the original system is a commensurate-order transfer function model, it is always possible to find the base order α. Let $\lambda = s^\alpha$, the characteristic equation on s can be mapped into an integer-order polynomial equation of λ. With the help of the solution function of integer-order polynomials, all the solutions of λ are derived. Then the stability of the system is determined by combining the conditions given in Theorem 6.15.

This naturally brings us a new question: how to judge the stability of a non-commensurate-order transfer function? For example, if the pseudo-polynomial equation is $s^{\sqrt{5}}+3s+6=0$, there is no base order at all, and it is impossible to convert the equation into an integer-order polynomial equation in λ. How can we find the roots and judge the stability for such a system? The existing tools provided by MATLAB are unable to solve such problems.

In Ref. [9], a solver is provided for finding all the roots of an arbitrary system of nonlinear algebraic equations. Further improvements and expansions have been made in Ref. [15].

6

Algorithm 6.7 Finding all the roots

(1) For the equation $\boldsymbol{Y}=\boldsymbol{F}(\boldsymbol{X})$, choose an initial solution set $\boldsymbol{X}$, solution range $\boldsymbol{A}$ and error tolerance ϵ.
(2) Use loop structure. Inside the loop, an initial search point $\boldsymbol{x}_0$ in the interested region is generated randomly, and find a solution $\boldsymbol{x}$ with `fsolve()` function.
(3) If $\boldsymbol{x}$ does not exist in set $\boldsymbol{X}$, record the new root $\boldsymbol{x}$ in $\boldsymbol{X}$ and return to (2).
(4) If no new solution is found in a certain time $t_{\rm lim}$, complete the solver.

A MATLAB function `more_sols()` [15] can be written based on such an algorithm, and the user can interrupt the solution process at any time by pressing the Ctrl+C keys, or by waiting for the natural end of the function.

```
function more_sols(f,X0,A,tol,tlim,ff)
   arguments, f(1,1)
      X0, A=1000, tol(1,1)=eps, tlim(1,1)=20, ff=optimset;
   end
   X=X0; nA=length(A); E=1e-4; [n,m,i]=size(X0);
   if nA==1, a=-0.5*A; b=0.5*A; else, a=A(1); b=A(2); end
   ar=real(a); br=real(b); ai=imag(a); bi=imag(b);
   ff.Display='off'; ff.TolX=tol; ff.TolFun=1e-20; tic
   if i==0, X0=zeros(n,m);   %judge whether 0 is an isolated solution
      if norm(f(X0))<tol, i=1; X(:,:,i)=X0; end
   end
   while (1)     %infinite loop, press Ctrl+C to terminate or wait
      x0=ar+(br-ar)*rand(n,m);          %generate a random matrix
      if ~isreal(A), x0=x0+(ai+(bi-ai)*rand(n,m))*1i; end
      try, [x,~,key]=fsolve(f,x0,ff); catch, continue; end
      t=toc; if t>tlim, break; end   %no new one after long wait
      if key>0, N=size(X,3);            %read the number of solutions
         for j=1:N
            if norm(X(:,:,j)-x)<E; key=0; break; end
         end
         if key==0                      %more accurate solution found
            if norm(f(x))<norm(f(X(:,:,j))), X(:,:,j)=x; end
         elseif key>0, X(:,:,i+1)=x; %record the new solution
```

```
            if norm(imag(x))>1e-8,i=i+1;X(:,:,i+1)=conj(x);
            end, assignin('base','X',X); i=i+1, tic  %update
         end, assignin('base','X',X);
end, end, end
```

The function is with the syntax more_sols(f, X_0, A, ϵ, $t_{\rm lim}$). The explanation of its individual variables is given in the following remarks.

Remarks Finding roots of nonlinear systems of equations

(1) The argument A is used to describe the solution interval of interest, which can be a scalar or an interval $[a, b]$.

(2) If A is a scalar, the solution interval is $(-A/2, A/2)$. If the default value of $A = 1000$ is used, the equation can be solved in a wide range.

(3) If A contains a nonzero imaginary part, then it is necessary to derive the complex roots; otherwise, only the real roots are required.

(4) The default value of the error tolerance ϵ is eps.

(5) The default parameter $t_{\rm lim}$ is 20, which means that if no new equation solution is found within 20 s, the solution process is completed.

(6) The user can press Ctrl+C keys at any time to terminate the solution process. The resulting ith root is stored in X(:,:,i).

(7) If needed, the high-precision solution function more_vpasols() introduced in Ref. [9] can be used, which is also provided in the corresponding FOTF Toolbox of this book. □

It must be noted that more_sols() given in this book is a general-purpose function that can be used to solve nonlinear matrix equations with multiple solutions. The scalar pseudo-polynomial equation $D(s) = 0$ studied here is only a special case of it, which can be considered to be solved directly with the command more_sols(D, zeros(1,1,0),100+100i).

Conjecture 6.2 Stability assessment

Assume that the irrational transfer function is written in the ratio form of s

$$G(s) = \frac{N(s)}{D(s)}, \tag{6.84}$$

where $D(s)$ is a pseudo-polynomial. If there is no positive real part in the roots of the characteristic equation $D(s) = 0$, the system is stable.

Remarks Stability conjecture I

(1) The function $N(s)$ is not restricted to the pseudo-polynomial form, it can be any nonlinear function.

(2) The stability of the system can be judged as long as all roots of $D(s) = 0$ with respect to s can be found.

(3) This conjecture lacks a rigorous mathematical proof, and the authors have tried many examples but found no counterexamples. □

▶ Example 6.26

Use the `more_sols()` function to solve again the problem in Example 6.25, to obtain all roots with respect to s.

Solutions Anonymous function can be used to describe the pseudo-polynomial equation. All the roots of the pseudo-polynomial equation can be found by the `more_sols()` function . It can be seen that the equation has only two roots, $s = -0.1076 \pm \mathrm{j}0.5562$, which are exactly the same as Example 6.25.

```
>> f=@(x)x.^2.3+5*x.^1.6+6*x.^1.3-5*x.^0.4+7;
   more_sols(f,zeros(1,1,0),100+100i), s=X(:)
```

◀

▶ Example 6.27

Judge the stability of the following system for $K = 3$:

$$G(s) = \frac{1}{s^{\sqrt{5}} + 25s^{\sqrt{3}} + 16s^{\sqrt{2}} - Ks^{0.4} + 7}.$$

In addition, find the critical gain K value that makes $G(s)$ start to be unstable.

Solutions Since the orders of the equation contain irrational numbers, it is not possible to convert the equation into a commensurate-order one. The search method introduced here has to be used, and it seems to be the only choice. The characteristic equation of the system can be directly represented by the following anonymous function, so that all the roots of the equation can be found by calling the `more_sols()` function.

```
>> K=3; f=@(s)s^sqrt(5)+25*s^sqrt(3)+16*s^sqrt(2)-K*s^0.4+7;
   more_sols(f,zeros(1,1,0),100+100i); s=X(:)
```

The equation has only one pair of complex roots $s = -0.0812 \pm 0.2880\mathrm{j}$, and both of them lie in the left-half of the s-plane, i.e., both have negative real parts, so the system $G(s)$ is stable.

Letting $K = 10$ and running the above code again reveals that the resulting pair of roots is $0.0338 \pm 0.2238\mathrm{j}$, with positive real parts, so the system is unstable. This leads to the conclusion that there exists at least one point within $K \in (3, 10)$ that lies on the stability boundary. This boundary can be found by considering the bisection method as follows.

```
>> a=3; b=10;
   while (b-a)>0.001, K=0.5*(a+b);
      f=@(s)s^sqrt(5)+25*s^sqrt(3)+16*s^sqrt(2)-K*s^0.4+7;
      more_sols(f,zeros(1,1,0),100+100i,eps,1);
      if real(X(1))>0, b=K; else, a=K; end
   end, K0=K
```

This yields a critical gain of $K_0 = 7.8492$. It can be seen that if $K < K_0$, the system is stable, and increasing the gain further will turn the system into an unstable one. To verify such a result, the step response curve of the system at the critical gain K_0 can be obtained, as shown in Fig. 6.18. It can be seen that the system response curve is basically an equal-amplitude oscillation, so the resulting critical gain is correct.

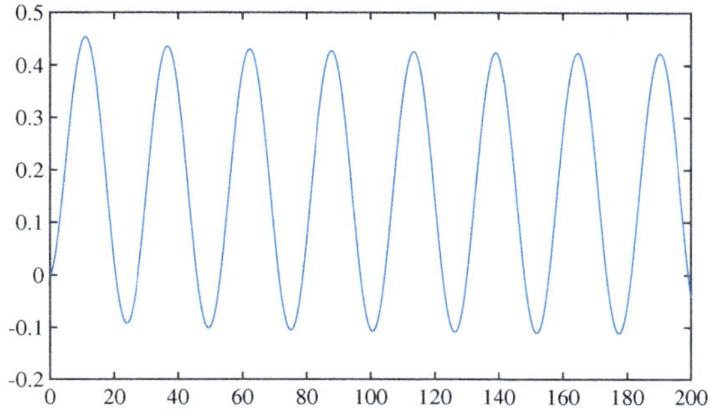

Fig. 6.18 Step response curve at critical gain

```
>> na=[sqrt(5),sqrt(3),sqrt(2),0.4,0]; a=[1 25 16 -K0 7];
   t=0:0.1:200; y=fode_caputo9(a,na,1,0,0,ones(size(t)),t,4);
   plot(t,y)
```

◀

6.8.3 Stability Assessment of Irrational Systems

In fact, if the irrational transfer function model $G(s)$ is known, and can be expressed in the form $G(s) = N(s)/D(s)$, where $D(s) = 0$ is a pseudo-polynomial equation, its stability can be judged by the method given in Conjecture 6.2.

If the irrational transfer function cannot be expressed in the above form, or if the pseudo-polynomial $D(s)$ cannot be derived, then it is not possible to judge the stability of the system with Conjecture 6.2, by finding the roots of the characteristic equation.

Theorem 6.16 Transfer function is always linear
If a system can be described by a transfer function $G(s)$, no matter how complicated mathematical form of $G(s)$ is, the system is linear.

Proof If a system satisfies the superposition principle, meaning that if the signals $u_1(t)$ and $u_2(t)$ excite the system alone, yielding output signals $y_1(t)$ and $y_2(t)$, the system output is $y(t) = ay_1(t) + by_2(t)$ when the system is excited by $u(t) = au_1(t) + bu_2(t)$, regardless of how a and b are taken. Obviously, from the Laplace transform representation

$$Y(s) = aG(s)U_1(s) + bG(s)U_2(s) = G(s)\left[aU_1(s) + bU_2(s)\right],$$

the superposition principle is satisfied, so the system $G(s)$ is linear. □

For complicated systems, especially for irrational closed-loop systems like Example 6.22, it is not possible to judge the stability using the closed-loop characteristic roots method. Therefore, the simulation method becomes the only option to judge the stability. The good thing is that for a linear system, the system is stable if it is stable for a certain nonzero signal. Therefore, one can try to find the numerical inverse Laplace transform of $G(s)$ or $G(s)/s$ directly. If the curve does not diverge, the system is stable.

▶ Example 6.28

Consider the irrational plant model given in Example 6.22. Assuming that the plant model also has 0.05 s of transport delay. If a PID controller $G_c(s) = 8 + 0.4/s + 0.2s$ is designed for the plant, and assuming a unity negative feedback structure, judge the stability of the closed-loop system.

6

Solutions For such a complicated system, there is no other known way to judge the stability of the closed-loop system. The simulation method is the only feasible way. The PID controller can also be written in the form of a string, which can be concatenated with the plant and the time delay $e^{-0.05s}$, to form a new forward path model string. In this way, the closed-loop step response curve of the system can be directly obtained by the `INVLAP_new()` function, as shown in Fig. 6.19. From the results, it is clear that such a closed-loop system is stable.

```
>> G='(sinh(0.1*sqrt(s))/0.1/sqrt(s))^2/sqrt(s)/sinh(sqrt(s))';
   Gc='(8+0.4/s+0.2*s)'; GG=[Gc '*' G '*exp(-0.05*s)'];
   [t,y]=INVLAP_new(GG,0,5,200,1,'1/s'); plot(t,y)
```

◀

▶ Example 6.29

Consider the open-loop model and controller given in Example 6.28.

$$G(s) = \left[\frac{\sinh(w\sqrt{s})}{w\sqrt{s}}\right]^2 \frac{1}{\sqrt{s}\sinh(\sqrt{s})} e^{-Ts}, \quad G_c(s) = 8 + 0.4\frac{1}{s} + 0.2s, \quad w = 0.1.$$

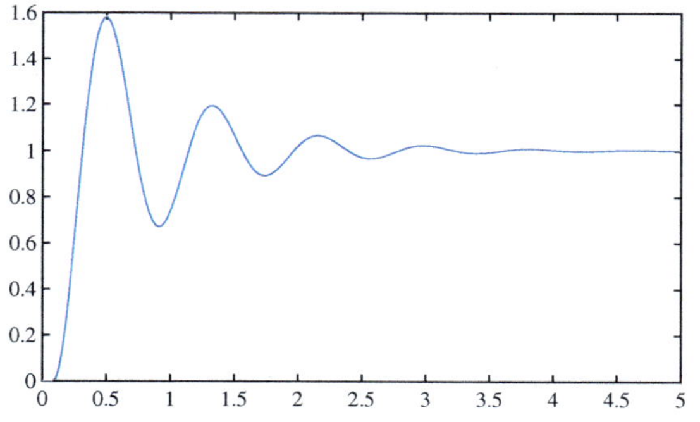

Fig. 6.19 Step response curve of closed-loop system

The closed-loop system is stable for a delay constant $T = 0.05$. If a delay constant $T = 0.2$ is tried, the closed-loop system is unstable. Is it possible to find the critical delay constant T, such that the closed-loop system is stable?

Solutions The characteristic equation of the closed-loop system can be written from the plant, the time delay and the controller.

$$1 + G(s)G_c(s)e^{-Ts} = 0.$$

The `more_sols()` function can be used to find all of its characteristic roots. Bisection method can be used again to find the smallest T value such that the real part of the root is greater than 0. These can be implemented by the following statements, which yield a critical time constant of $T = 0.1103$. If the value of T is greater than the critical value, the roots with positive real parts appear.

```
>> G=@(s)(sinh(0.1*sqrt(s))/0.1/sqrt(s))^2/sqrt(s)/sinh(sqrt(s));
   Gc=@(s)(8+0.4/s+0.2*s); a=0.05; b=0.2;
   while (b-a)>1e-6, T=(b+a)/2; f=@(s)(1+Gc(s)*G(s)*exp(-T*s));
       more_sols(f,zeros(1,1,0),100+100i,1e-10,3);
       if any(real(X(:))>0), b=T; else, a=T; end
   end, T, X
```

Let $T = 0.1103$, the step response of the closed-loop system can be obtained using the following commands. The response curve is then a sinusoidal curve of equal-amplitude oscillation (graph omitted). This confirms that the obtained T value is the critical time constant value for the stability of the closed-loop system.

```
>> G='(sinh(0.1*sqrt(s))/0.1/sqrt(s))^2/sqrt(s)/sinh(sqrt(s))';
   Gc='(8+0.4/s+0.2*s)'; GG=[Gc '*' G '*exp(-0.1103*s)'];
   [t,y]=INVLAP_new(GG,0,5,200,1,'1/s'); plot(t,y)
```

◄

Conjecture 6.3 Stability of linear irrational feedback control systems
For linear irrational feedback control systems, if the forward path transfer function is $G(s)$, while the feedback one is $H(s)$, the characteristic equation $1 + G(s)H(s) = 0$ can be constructed. If all the roots are located on the left-half of the s-plane, the closed-loop system is stable.

Remarks Stability conjecture II

(1) There is no other known solution to the problem given in Example 6.29 or any other problems alike.

(2) The problem can be solved directly from this conjecture, and the correctness and feasibility of this method can be verified by numerical experiments.

(3) No matter how complicated the mathematical form of the characteristic equation is, it is always possible to use the `more_sols()` function to obtain all of its roots and to judge the stability of the irrational system with the aid of simulation methods.

□

6.9 Exercises

(1) Derive the ramp response of the partial fraction expansion of the transfer function model from Theorem 6.4. The mathematical form of the ramp function and its Laplace transform are, respectively, $u(t) = t$ and $\mathscr{L}[u(t)] = 1/s^2$.

(2) Derive the analytical solution of the step response of the implicit differential equation $G(s) = 1/(\tau s + 1)^\gamma$ from Theorem 6.4.

(3) The transfer function of a linear fractional-order system is known as follows. Find the analytical solution of the step response. Solve the problem again by numerical methods and verify the correctness of the results obtained.

$$G(s) = \frac{7s^{1.8} + 42s^{1.5} + 106s^{1.2} + 147s^{0.9} + 120s^{0.6} + 55s^{0.3} + 11}{s^{2.1} + 8s^{1.8} + 27s^{1.5} + 50s^{1.2} + 55s^{0.9} + 36s^{0.6} + 13s^{0.3} + 2}.$$

Convert the FOTF model with zero initial values into a FODE model. Find the numerical solution of the step response. Compare with the exact solution and assess the resulting error.

(4) Solve the following linear Riemann–Liouville equation with zero initial values:

$${}_0^{\rm RL}\mathscr{D}_t^{\pi} y(t) + 4\,{}_0^{\rm RL}\mathscr{D}_t^{1.5} y(t) + 2y(t) = \sqrt{4.2}\,{\rm e}^{-\sqrt{t}} \sin\sqrt{5}t^2.$$

Use different methods to assess the correctness of the resulting solution: choose different step sizes and compare the results; use the back substitution method to check the correctness of the solution by substituting the resulting numerical solution back into the equation.

(5) Find the numerical solution of the following linear FODE and verify that the analytical solution is $y(t) = t + 1$.

$$y'(t) + {}_0^{\rm C}\mathscr{D}_t^{0.5} y(t) + y(t) = t + 2\sqrt{\frac{t}{\pi}} + 2,$$

where the initial value is $y(0) = 1$.

(6) Find the numerical solution of the following linear FODE:

$${}_0^{\rm C}\mathscr{D}_t^{0.7} y(t) = {}_0^{\rm RL}\mathscr{D}_t^{-0.3} \left(\frac{1}{\Gamma(t)}\right),$$

with initial value $y(0) = 0$, and the exact solution is $y(t) = \displaystyle\int_0^t \frac{1}{\Gamma(\tau)} {\rm d}\tau$.

(7) Find the numerical solution of the following linear FODE:

$${}_0^{\rm C}\mathscr{D}_t^{0.5} y(t) = 2\,{}_0^{\rm RL}\mathscr{D}_t^{-0.5} t \sin t^2.$$

It is known that the initial value condition of the equation is $y(0) = 0$, and the exact solution is $y(t) = 1 - \cos t^2$.

(8) Find the numerical solution of the following linear FODE:

$$ {}_0^{\mathrm{C}}\mathscr{D}_t^{1.7} y(t) = \frac{t^{0.3}}{2}\big(\mathrm{E}_{1,1.3}(t) - \mathrm{E}_{1,1.3}(-t)\big), $$

where the initial conditions are $y(0) = 0$ and $y'(0) = 1$. The exact solution is $y(t) = \sinh t$.

(9) Solve the following linear Caputo equation and verify the result by back substitution method:

$$ {}_0^{\mathrm{C}}\mathscr{D}_t^{\pi} y(t) + 4\,{}_0^{\mathrm{C}}\mathscr{D}_t^{1.5} y(t) + 2y(t) = \sqrt{4.2}\,\mathrm{e}^{-\sqrt{t}} \sin\sqrt{5}t^2 + 2, $$

where $y(0) = 1$, $y'(0) = 0$ and $y''(0) = 0$. Compare the solutions with the one in Exercise (4).

(10) If the denominator pseudo-polynomial of the system is $s^{\sqrt{11}} + 4s^{\pi} + 2s^{\sqrt{\pi}} + 5 = 0$, judge the stability of the system.

(11) Consider the nonlinear equation $(18s^{1.1} + 9)\mathrm{e}^{-s} + 3 + 29s^{1.2} + 46s^{\sqrt{3}} + 20s^{\pi} = 0$. Is it possible to find all the roots of this equation? If e^{-s} is replaced by $\mathrm{e}^{-\sqrt{s}}$, how many roots does the equation have?

(12) If the commensurate-order state space equation is given by Example 6.17, find the analytical expression of the state transition matrix.

(13) Consider the transfer function model of Example 6.19. If it is necessary to select each integer-order derivative of $y(t)$ as a state variable as well, try to reselect the state variables and write the extended FOSS equation.

(14) The analytical solutions of the impulsive and step responses of the linear FODE in Example 6.5 are given in Fig. 6.3. Superimpose the response curves derived from the numerical inverse Laplace transform and the `fode_sol()` function on this figure and assess their correctness.

(15) Consider the irrational open-loop transfer function model in Example 6.22. Select two different sets of input signals and verify that the model is linear by simulation.

(16) Consider the open-loop transfer function model in Ref. [13] as follows. If the model has a time delay of 0.1 s, find the critical w value that makes the closed-loop transfer function unstable under a unity negative feedback structure.

$$ G(s) = \left[\frac{\sinh(w\sqrt{s})}{w\sqrt{s}}\right]^2 \frac{1}{\sqrt{s}\sinh(\sqrt{s})}. $$

(17) Consider the irrational open-loop plant model in Exercise (16). If $w = 1$, there is no time delay, and the PD controller is known to be $G_{\mathrm{c}}(s) = 10.1351 + 1.0686s^{1.0728}$ [12]. Draw the step response of the closed-loop system.

(18) Several conjectures about stability are made in this chapter. Can you find out examples where the conjectures are incorrect?

References

1. Podlubny I (1999) Fractional differential equations[M]. Academic, San Diego
2. Caputo M (1967) Linear models of dissipation whose Q is almost frequency independent II[J]. Geophys J Int 13(5):529–539
3. Chamati H, Tonchev NS (2006) Generalized Mittag-Leffler functions in the theory of finite-size scaling for systems with strong anisotropy and/or long-range interaction[J]. J Phys A: Math General 39(3):469–478
4. Podlubny I (1999) Fractional-order systems and $\mathrm{PI}^{\lambda}\mathrm{D}^{\mu}$-controllers[J]. IEEE Trans Autom Control 44(1):208–214
5. Zhao CN, Xue DY (2008) Closed-form solutions to fractional-order linear differential equations[J]. Front Electr Electr Eng China 3(2):214–217
6. Xue DY, Bai L (2017) Numerical algorithms for Caputo fractional-order differential equations[J]. Int J Control 90(6):1201–1211
7. Diethelm K (2010) The analysis of fractional differential equations: an application-oriented exposition using differential operators of Caputo type[M]. Springer, New York
8. Monje CA, Chen YQ, Vinagre BM et al (2010) Fractional-order systems and controls: fundamentals and applications[M]. Springer, London
9. Xue DY (2023) MATLAB solutions of advanced applied mathematical problems (in Chiness)[M], 5th edn. Tsinghua University Press, Beijing
10. Valsa J, Brančik L (1998) Approximate formulae for numerical inversion of Laplace transforms[J]. Int J Numer Modell: Electr Netw Devices Fields 11(3):153–166
11. Valsa J (2023) Numerical inversion of Laplace transforms in MATLAB[OL]. [2023/10/17 03:42:11]. ▶ https://ww2.mathworks.cn/matlabcentral/fileexchange/32824-numerical-inversion-of-laplace-transforms-in-matlab
12. Xue DY (2022) Computer-aided design of control systems with MATLAB (in Chinese)[M], 4th edn. Tsinghua University Press, Beijing
13. Callier FM, Winkin J (1993) Infinite dimensional system transfer functions. In: Curtain RF, Bensoussan A, Lions JL (eds) Analysis and optimization of systems: state and frequency domain approaches for infinite-dimensional systems[M]. Springer, Berlin
14. Matignon D (1998) Stability properties for generalized fractional differential equations[C]. ESAIM: Proc Fract Diff Syst: Models Methods Appl 5:145–158
15. Xue DY (2021) Optimization computing with MATLAB[M]. De Gruyter Press, Berlin

6

Numerical Solutions of Nonlinear FODEs

Contents

D. Y. Xue and L. Bai, *Fractional Calculus*, https://doi.org/10.1007/978-981-99-2070-9_7

Analytical solution methods for linear FODEs were given in ▶ Chap. 6. High-precision numerical algorithms for Riemann–Liouville and Caputo differential equations were given. In control system research and other fields of applications, nonlinear behaviors are unavoidable. This chapter will focus on the numerical solution methods for nonlinear FODE.

To better understand the solution of differential equations, several forms of nonlinear FODEs are listed in ▶ Sect. 7.1, including general explicit differential equations, commensurate-order state space equations, and extended state space equations.

The subsequent part of this chapter focuses on the numerical solution methods for nonlinear FODEs. There are not as many numerical algorithms in the field of solving nonlinear FODEs as there are for integer-order ordinary differential equations, and only a limited number of commonly used methods can be found in the literature that have been panned to the fractional-order domain. For example, the Adams–Bashforth–Moulton-based predictor–corrector (PECE) algorithm is well described in Ref. [1]. It will be presented in ▶ Sect. 7.2 to derive and implement a numerical solution method for the extended state space equation. The efficiency and adaptation range of this method is limited in practical applications, so an efficient and high-precision numerical solution algorithm is proposed and implemented in ▶ Sect. 7.3, which is far superior to the other methods presented here in several aspects, from solution accuracy to solution efficiency.

To better evaluate the various numerical solution methods, this chapter also compares the advantages and disadvantages of these algorithms through some benchmark problems, designed by the authors and gives suggestions to improve the efficiency of the solutions.

Of course, the algorithms that will be presented in this chapter have a number of limitations. First, the methods presented here are fixed-step algorithms, and second, the methods presented here are artificially given a step size to derive the numerical solution of the differential equation by iterative or recursive methods. This method is closer to open-loop control, where the result is derived directly after a given step size, without any monitoring mechanism in between. This is different from the adaptive variable-step numerical solution of integer-order differential equations, where there is a monitoring mechanism inside the variable-step algorithm that automatically adjusts the step size according to the magnitude of the error, to ensure the accuracy and efficiency of the solution. The latter methods are similar to closed-loop control.

These methods in the chapter are collectively referred to as command-driven solution methods in this book. Compared to the block diagram-based solution algorithms that will be presented in ▶ Chaps. 8 and ▶ 9, the methods presented in this chapter are very limited in the types of differential equations that can be solved.

7.1 Descriptions of FODEs

▶ Chapter 6 provides a more detailed introduction to linear FODEs and their solutions, and the remainder of the book explores numerical solutions to various types of nonlinear FODEs.

7.1.1 General form of FODEs

The most basic form of simple explicit FODEs is given here first. Then several different types of explicit FODEs are introduced.

Definition 7.1 Explicit FODE

The basic form of explicit FODEs is

$$ {}_{t_0}^{\mathrm{C}}\mathscr{D}^{\alpha}y(t) = f\big(t, y(t), {}_{t_0}^{\mathrm{C}}\mathscr{D}^{\alpha_1}y(t), {}_{t_0}^{\mathrm{C}}\mathscr{D}^{\alpha_2}y(t), \cdots, {}_{t_0}^{\mathrm{C}}\mathscr{D}^{\alpha_n}y(t)\big), \tag{7.1} $$

where it is safe to assume that $\alpha > \alpha_1 > \alpha_2 > \cdots > \alpha_n > 0$. The initial values of $y(t)$ and its integer-order derivatives are known

$$ y(t_0) = y_0,\ y'(t_0) = y_1,\ y''(t_0) = y_2, \cdots, y^{(q-1)}(t_0) = y_{q-1}, \tag{7.2} $$

with $q = \lceil \alpha \rceil$.

In addition to the explicit FODEs given here, FODEs may contain other forms, e.g., systems of explicit FODEs, implicit FODEs, fractional-order delay differential equations, and other forms, and may also contain boundary value problems for FODEs. These will be further explored in ▶ Chap. 9.

▶ Example 7.1

A test problem for nonlinear FODEs is presented in Ref. [1].

$$ {}_0^{\mathrm{C}}\mathscr{D}_t^{1.455}y(t) = -t^{0.1}\frac{\mathrm{E}_{1.545}(-t)}{\mathrm{E}_{1.445}(-t)}\,\mathrm{e}^t y(t)\,{}_0^{\mathrm{C}}\mathscr{D}_t^{0.555}y(t) + \mathrm{e}^{-2t} - \big[y'(t)\big]^2, \tag{7.3} $$

where $y(0) = 1$ and $y'(0) = -1$. It is claimed that the exact solution of the FODE is $y(t) = \mathrm{e}^{-t}$. As a matter of fact, e^{-t} is not the exact solution of (7.3).

The authors of this book found that the two-parameter Mittag-Leffler functions in the original equation was mistakenly written as one-parameter functions. They reconstructed the nonlinear FODE on the basis of the original equation so that its exact solution is $y(t) = \mathrm{e}^{-t}$.

$$ {}_0^{\mathrm{C}}\mathscr{D}_t^{1.455}y(t) = -t^{0.1}\frac{\mathrm{E}_{1,1.545}(-t)}{\mathrm{E}_{1,1.445}(-t)}\,\mathrm{e}^t y(t)\,{}_0^{\mathrm{C}}\mathscr{D}_t^{0.555}y(t) + \mathrm{e}^{-2t} - \big[y'(t)\big]^2. \tag{7.4} $$

◀

▶ Example 7.2

Show that $y(t) = \mathrm{e}^{-t}$ is the exact solution of the FODE (7.4).

Solutions To prove that a solution satisfies a differential equation is a relatively easy matter. Just substitute the solution back into the equation and observe whether the original differential equation is satisfied. Clearly, $y(0) = \mathrm{e}^{-0} = 1$ and $y'(0) = -\mathrm{e}^{-0} = -1$, indicating that the initial conditions of the equation are satisfied.

Using the important property given by Theorem 3.6: ${}_0^{\rm C}\mathscr{D}_t^{\alpha} {\rm e}^{\lambda t} = \lambda^m t^\gamma {\rm E}_{1,\gamma+1}(\lambda t)$, where $m = \lceil \alpha \rceil$ and $\gamma = m - \alpha$. It can be seen that if $y(t) = {\rm e}^{-t}$, then $m = 2$, $\gamma = 0.545$, and $\lambda = -1$ in ${}_0^{\rm C}\mathscr{D}_t^{1.455} y(t)$, it follows that ${}_0^{\rm C}\mathscr{D}_t^{1.455} y(t) = (-1)^2 t^{0.545} {\rm E}_{1,1.545}(-t)$. Similarly, in ${}_0^{\rm C}\mathscr{D}_t^{0.555} y(t)$, there are $m = 1$ and $\gamma = 0.445$, which yields ${}_0^{\rm C}\mathscr{D}_t^{0.555} y(t) = (-1)^1 t^{0.445} {\rm E}_{1,1.445}(-t)$. Substituting them into the right end of (7.3),

$$-t^{0.1} \frac{{\rm E}_{1,1.545}(-t)}{{\rm E}_{1,1.445}(-t)} {\rm e}^t {\rm e}^{-t} (-1)^1 t^{0.445} {\rm E}_{1,1.445}(-t) + {\rm e}^{-2t} - [{\rm e}^{-t}]^2 = t^{0.545} {\rm E}_{1,1.545}(-t).$$

Obviously, the right-hand side of the equation is also consistent with ${}_0^{\rm C}\mathscr{D}_t^{1.455} y(t)$, which proves that (7.4) is solved as $y(t) = {\rm e}^{-t}$. ◀

7.1.2 Commensurate-Order State Space Models

The commensurate-order differential equation model is a special case of the class of general explicit FODEs given by Definition 7.1. Its general mathematical form is given below.

Definition 7.2 Explicit commensurate-order Caputo equation

The mathematical form of a class of nonlinear commensurate-order Caputo equation is

$$ {}_0^{\rm C}\mathscr{D}_t^{n\alpha} y(t) = f\big(t, y(t), {}_0^{\rm C}\mathscr{D}_t^{\alpha} y(t), {}_0^{\rm C}\mathscr{D}_t^{2\alpha} y(t), \cdots, {}_0^{\rm C}\mathscr{D}_t^{(n-1)\alpha} y(t)\big), \tag{7.5}$$

with its initial conditions defined in (7.2).

Introducing a set of state variables

$$x_1(t) = y(t),\ x_2(t) = {}_0^{\rm C}\mathscr{D}_t^{\alpha} y(t), \cdots,\ x_n(t) = {}_0^{\rm C}\mathscr{D}_t^{(n-1)\alpha} y(t), \tag{7.6}$$

the corresponding state space model can be written as

$$\begin{cases} {}_0^{\rm C}\mathscr{D}_t^{\alpha} x_1(t) = x_2(t) \\ {}_0^{\rm C}\mathscr{D}_t^{\alpha} x_2(t) = x_3(t) \\ \quad \vdots \\ {}_0^{\rm C}\mathscr{D}_t^{\alpha} x_{n-1}(t) = x_n(t) \\ {}_0^{\rm C}\mathscr{D}_t^{\alpha} x_n(t) = f\big(t, x_1(t), x_2(t), \cdots, x_n(t)\big), \end{cases} \tag{7.7}$$

from which the commensurate-order state space equations can be obtained.

Definition 7.3 Commensurate-order state space equations

The mathematical form of commensurate-order state space equations is

$$ {}_0^{\mathrm{C}}\mathscr{D}_t^{\alpha}\boldsymbol{x}(t)=\boldsymbol{F}\big(t,\boldsymbol{x}(t)\big), \tag{7.8}$$

where $\boldsymbol{x}(t)$ is referred to as the state variable vector. The initial conditions of the state variables can be assigned with the following regulations [1]: if $(i-1)\alpha$ is an integer, then $x_i(0)=y_{(i-1)\alpha}$, otherwise, $x_i(0)=0$.

7.1.3 Extended State Space Models

This section first demonstrates the limitations of the commensurate-order equation of state space model introduced earlier by examples. Then the mathematical expressions are given for general non-commensurate-order extended state space equations.

▶ Example 7.3

Consider the correct nonlinear Caputo differential equation given in Example 7.1. Write the commensurate-order state space model.

Solutions To convert the original equation into a commensurate-order state space equation, we first need to find its base order. To find the base order, the following MATLAB commands can be used.

```
>> [n d]=rat([1.455, 1, 0.555]);     % find all the orders
   a=gcd(sym(n))/lcm(sym(d)), n=1.455/a, n1=0.555/a, n2=1/a
```

From the results, it can be seen that the base order of the equation should be chosen as $\alpha=0.005$. It can be seen that the original equation has $n=291$ states, and its key states can be chosen as $x_1(t)=y(t)$, $x_{112}(t)=\mathscr{D}^{0.555}y(t)$, $x_{201}(t)=y'(t)$, $x_{291}=\mathscr{D}^{1.450}y(t)$. The state space equation can be written as

$$\begin{cases} {}_0^{\mathrm{C}}\mathscr{D}_t^{\alpha}x_i(t)=x_{i+1}(t),\ i=1,2,\cdots,290\\ {}_0^{\mathrm{C}}\mathscr{D}_t^{\alpha}x_{291}(t)=-t^{0.1}\dfrac{\mathrm{E}_{1,1.545}(-t)}{\mathrm{E}_{1,1.445}(-t)}\mathrm{e}^t x_1(t)x_{112}(t)+\mathrm{e}^{-2t}-x_{201}^2(t). \end{cases}$$

According to the initial value rules given earlier, only the initial values of the state variables corresponding to $y(t)$ and its integer-order derivatives can be inherited from the known initial values of the equation, while the initial values of the rest of the state variables are zero. Therefore, we have $x1(0)=y(0)=1$, $x_{201}(0)=y'(0)=-1$. The initial values of the remaining state variables are zero. ◀

From the state space model derived here, the number of state variables is excessive, and therefore a more concise representation of the FOSS model needs to be considered for introduction.

▶ Example 7.4

Consider again the nonlinear Caputo differential equation model in Example 7.1. Write a more concise state space equation, by selecting again the state variables.

Solutions Selecting state variables $x_1(t) = y(t)$, $x_2(t) = {}_0^{\mathrm{C}}\mathscr{D}_t^{0.555}y(t)$ and $x_3(t) = y'(t)$, the following extended FOSS expression can be established.

$$\begin{cases} {}_0^{\mathrm{C}}\mathscr{D}_t^{0.555}x_1(t) = x_2(t) \\ {}_0^{\mathrm{C}}\mathscr{D}_t^{0.445}x_2(t) = x_3(t) \\ {}_0^{\mathrm{C}}\mathscr{D}_t^{0.455}x_3(t) = -t^{0.1}\dfrac{\mathrm{E}_{1,1.545}(-t)}{\mathrm{E}_{1,1.445}(-t)}\mathrm{e}^t x_1(t)x_2(t) + \mathrm{e}^{-2t} - x_3^2(t). \end{cases}$$

If a state is the output signal $y(t)$ or its integer-order derivatives, its initial value can be assigned directly from the given initial values of the Caputo equation, otherwise, the initial values of the states are zero. It can be seen from the example here that, the initial states are $x_1(0) = 1$, $x_2(0) = 0$ and $x_3(0) = -1$. ◀

7

This example gives a hint about the choice of initial values: even if there are no $y'(t)$ or other integer-order derivative terms in the original FODE, they should be set as state variables in the choice of state variables; otherwise, their initial values will not be reflected in the state space equation. The solution of the resulting differential equation will be problematic.

Summarizing the above presentation, a general nonlinear extended FOSS model can be derived, whose standard form is defined as follows.

Definition 7.4 Extended FOSS model

The general form of a nonlinear extended FOSS model is

$$ {}_0^{\mathrm{C}}\mathscr{D}_t^{\boldsymbol{\alpha}}\boldsymbol{x}(t) = \boldsymbol{f}(t, \boldsymbol{x}(t)), \tag{7.9}$$

where the initial state vector is $\boldsymbol{x}(0)$. The length of the order vector $\boldsymbol{\alpha}$ is the same as that of vector $\boldsymbol{x}(t)$.

Comparing the mathematical expressions in Definitions 7.3 and 7.4 that, the major difference between the two is that the state variable derivative order: α in Definition 7.3 is a scalar, while $\boldsymbol{\alpha}$ in Definition 7.4 is a vector.

7.2 Numerical Solutions of Nonlinear Caputo Equations

The solution of linear Caputo differential equations has been discussed extensively in depth in ▶ Chap. 6. MATLAB functions for solving general linear Caputo equations with high precision have been proposed and developed. This section will explore the general numerical solution of nonlinear Caputo differential equations.

In this section, we will first explore the method for solving the simplest form of the nonlinear Caputo differential equation, the one-term FODE, and extend the algorithm to the solution of the multi-term FODE. Then explore the numerical solution of the extended FOSS models.

7.2.1 Numerical Solutions of Scalar Commensurate-Order Equations

The commensurate-order state space equation given in Definition 7.3 is also known as the one-term FODE. For convenience of description, $\boldsymbol{x}(t)$ is rewritten as $\boldsymbol{y}(t)$, i.e., the prototype of the equation is the state space equation defined by Caputo equation as ${}_0^{\mathrm{C}}\mathscr{D}_t^{\alpha}\boldsymbol{y}(t)=\boldsymbol{f}(t,\boldsymbol{y}(t))$, which corresponds to the integer-order differential equation in the standard form $\boldsymbol{y}'(t)=\boldsymbol{f}\big(t,\boldsymbol{y}(t)\big)$

Theorem 7.1 Analytical of commensurate-order differential equations
The analytical solution of (7.8) can be written as [1]

$$\boldsymbol{y}(t)=\sum_{k=0}^{q-1}\frac{t^k}{k!}\boldsymbol{y}^{(k)}(0)+\frac{1}{\Gamma(\alpha)}\int_0^t\frac{\boldsymbol{f}\big(\tau,\boldsymbol{y}(\tau)\big)}{(t-\tau)^{1-\alpha}}\mathrm{d}\tau, \tag{7.10}$$

where $q=\lceil\alpha\rceil$.

For convenience, the following is a summary review of the classical Adams–Bashforth–Moulton predictor–corrector algorithm for integer-order differential equations. The mathematical form of the original explicit differential equation is

$$\boldsymbol{y}'(t)=\boldsymbol{f}\big(t,\boldsymbol{y}(t)\big), \tag{7.11}$$

with initial conditions $\boldsymbol{y}(0)=\boldsymbol{y}_0$. Assuming that the time interval of interest is $t\in(0,t_{\mathrm{n}})$ and a fixed step size h is chosen, one can set up equally spaced time instances at $0,h,2h,\cdots$. The interested interval is divided into several subintervals. Assume that the solution $\boldsymbol{y}_k=\boldsymbol{y}(kh)$ at subinterval k has been derived, the solution at the next point, $\boldsymbol{y}_{k+1}$, can be directly written as

$$\boldsymbol{y}_{k+1}=\boldsymbol{y}_k+\int_{t_k}^{t_{k+1}}\boldsymbol{f}\big(\tau,\boldsymbol{y}(\tau)\big)\mathrm{d}\tau. \tag{7.12}$$

If the step size h is very small, the integral term in the solution can be approximated by the two-point trapezoidal area formula:

$$\boldsymbol{y}_{k+1}=\boldsymbol{y}_k+\frac{h}{2}\big[\boldsymbol{f}\big(t_k,\boldsymbol{y}_k\big)+\boldsymbol{f}\big(t_{k+1},\boldsymbol{y}_{k+1}\big)\big]. \tag{7.13}$$

Since the $\boldsymbol{y}_{k+1}$ term occurs at both sides of the equation, if we want to avoid solving the nonlinear equations, we need to replace the $\boldsymbol{y}_{k+1}$ term in the right-hand side of the equation by a predictor solution $y_{k+1}^{\rm p}$. Therefore, with the substitution, the y_{k+1} term can be found in an iterative way. Equation (7.13) can be rewritten as

$$y_{k+1} = y_k + \frac{h}{2}\left[f(t_k, y_k) + f(t_{k+1}, y_{k+1}^{\rm p})\right]. \tag{7.14}$$

This equation is also known as the corrector formula.

The initial predicted value of $y_{k+1}^{\rm p}$ can be derived directly from the Euler formula:

$$y_{k+1}^{\rm p} = y_k + hf(t_k, y_k). \tag{7.15}$$

This equation is also known as the predictor equation.

Algorithm 7.1 Adams–Bashforth–Moulton predictor–corrector algorithm

(1) Set k to 0 and the value of the first term to y_0.
(2) For each k value, the initial predicted value $y_{k+1}^{\rm p}$ is calculated from (7.15).
(3) Calculate the corrected value y_{k+1} from (7.14).

Similar to the integer-order predictor–corrector algorithm, a fractional-order corrector equation can be derived for the Caputo equation described in (7.8) [1].

$$y_{k+1} = \sum_{i=0}^{q-1} \frac{t_{k+1}^i}{i!} y^{(i)}(0) + \frac{1}{\Gamma(\alpha)}\left[a_{k+1,k+1} f(t_{k+1}, y_{k+1}^{\rm p}) + \sum_{i=0}^{k} a_{i,k+1} f(t_i, y_i)\right], \tag{7.16}$$

where the coefficient a_{ij} can be derived recursively from the following equation, for a time axis grid with unequally spaced partitions.

$$a_{0,k+1} = \frac{(t_{k+1}-t_1)^{\alpha+1} + t_{k+1}^{\alpha}\left[(\alpha+1)t_1 - t_{k+1}\right]}{t_1\alpha(\alpha+1)} \tag{7.17}$$

$$\begin{aligned} a_{i,k+1} = {} & \frac{(t_{k+1}-t_{i-1})^{\alpha+1} + (t_{k+1}-t_i)^{\alpha}\left[\alpha(t_{i-1}-t_i) + t_{i-1} - t_{k+1}\right]}{(t_i - t_{i-1})\alpha(\alpha+1)} + \\ & \frac{(t_{k+1}-t_{i-1})^{\alpha+1} - (t_{k+1}-t_i)^{\alpha}\left[\alpha(t_i - t_{i+1}) - t_{i+1} + t_{k+1}\right]}{(t_{i+1} - t_i)\alpha(\alpha+1)}, \end{aligned} \tag{7.18}$$

where $1 \leqslant i \leqslant k$, and

$$a_{k+1,k+1} = \frac{(t_{k+1}-t_k)^{\alpha}}{\alpha(\alpha+1)}. \tag{7.19}$$

If equally spaced grids are assigned, the above equation can be simplified to

$$a_{i,k+1} = \begin{cases} \dfrac{h^\alpha}{\alpha(\alpha+1)}\left[k^{\alpha+1} - (k-\alpha)(k+1)^\alpha\right], & i=0 \\ \dfrac{h^\alpha}{\alpha(\alpha+1)}\left[(k-i+2)^{\alpha+1} + (k-i)^{\alpha+1} - 2(k-i+1)^{\alpha+1}\right], & 1 \leqslant i \leqslant k \\ \dfrac{h^\alpha}{\alpha(\alpha+1)}, & i=k+1. \end{cases} \tag{7.20}$$

The predictor equation for the fractional-order Adams–Bashforth–Moulton (ABM) algorithm can also be written.

$$\boldsymbol{y}_{k+1}^{\mathrm{p}} = \sum_{i=0}^{q-1} \frac{t_{k+1}^i}{i!} \boldsymbol{y}^{(i)}(0) + \frac{1}{\Gamma(\alpha)} \sum_{i=0}^{k} b_{i,k+1} \boldsymbol{f}(t_i, \boldsymbol{y}_i), \tag{7.21}$$

where

$$b_{i,k+1} = \frac{1}{\alpha}\left[(t_{k+1} - t_i)^\alpha - (t_{k+1} - t_{i+1})^\alpha\right]. \tag{7.22}$$

For equally spaced time axis time instances, the $b_{i,k+1}$ coefficient can be derived from the following equation:

$$b_{i,k+1} = \frac{h^\alpha}{\alpha}\left[(k+1-i)^\alpha - (k-i)^\alpha\right]. \tag{7.23}$$

It can be shown that if $\alpha \geqslant 1$, the accuracy of the algorithm is $O(h^2)$. If $\alpha < 1$, the accuracy of the algorithm is $O(h^{1+\alpha})$. Thus, the complete fractional-order ABM predictor–corrector algorithm can be summarized as follows.

Algorithm 7.2 Fractional-order ABM predictor–corrector algorithm

(1) For each k, calculate the predictor coefficient $b_{i,k+1}$ from (7.23).
(2) Calculate the corrector coefficient $a_{i,k+1}$ from (7.20).
(3) From (7.21), calculate the predictor solution $\boldsymbol{y}_{k+1}^{\mathrm{p}}$.
(4) Calculate the solution $\boldsymbol{y}_{k+1}$ of the equation from (7.16).
(5) If $y(t)$ is not a scalar function, use the iterative algorithm for each value of k until a convergent $\boldsymbol{y}_{k+1}$ is derived.

Remarks Predictor–corrector algorithm

(1) In practice, if the iterative process is enabled, the predictor equation is not very important because it only provides the initial value of the iterative process. If the predictor algorithm is skipped, the computational effort may be greatly reduced, often by 30%–50%.

(2) If the predictor algorithm is skipped, the predicted value $\boldsymbol{y}_1^{\mathrm{p}}$ in the first step can be chosen as $\boldsymbol{y}_0$ and the initial value of the subsequent points is chosen as $\boldsymbol{y}_{k+1}^{\mathrm{p}} = \boldsymbol{y}_k$, which gives a good prediction of the effect.

(3) In this algorithm, it is assumed that the time axis is equally spaced. If it is unequally spaced, the corrector coefficient $a_{i,k+1}$ can be calculated from (7.18), and the predictor is skipped using the above method.

(4) Although a_{ij} and b_{ij} look like matrices, in practice, it is sufficient to store them with vectors in each step k.

(5) If $\boldsymbol{y}(t)$ is a vector function, and the iterative process is not used, the result may be very inaccurate. □

Based on the above algorithm, the following predictor–corrector algorithm solver can be written in MATLAB language.

```
function [y,t]=pepc_nlfode(f,alpha,y0,h,tn,err,key)
   arguments
      f, alpha(1,1), y0, h(1,1), tn(1,1) {mustBePositive}
      err(1,1)=1e-6; key(1,1){mustBeMember(key,1:3)}=0
   end
   t=0; y1=y0; q=ceil(alpha); a1=alpha+1; m=round(tn/h);
   K=h^alpha/alpha/a1; Ga=1/gamma(alpha);
   y=[y1 zeros(size(y0,1),m)]; n=length(y1);
   for k=0:m-1, tk1=t(end)+h; t=[t,tk1]; ii=0:k; y01=0;
      for i=1:q,y01=y01+tk1^(i-1)/factorial(i-1)*y0(:,i);end
      if key==1, yp=y(:,k+1);
      else, yp=y01;
         b=h^alpha/alpha*((k+1-ii).^alpha-(k-ii).^alpha);
         for i=0:k, yp=yp+b(i+1)*f(t(i+1),y(:,i+1))*Ga; end
      end
      if key~=2
         ii=1:k; a2=(k-ii+2).^a1+(k-ii).^a1-2*(k-ii+1).^a1;
         a0=k^a1-(k-alpha)*(k+1)^alpha; a=[a0 a2 1]*K;
         while (1), y1=y01+a(k+2)*f(tk1,yp)*Ga;
            for i=0:k, y1=y1+a(i+1)*f(t(i+1),y(:,i+1))*Ga;end
            if (norm(y1-yp)<err||(key==0&&n==1)||key==3),break
            else, yp=y1; end
         end
      else, y1=yp; end, y(:,k+2)=y1;
   end, y=y'; t=t';
end
```

The syntax of the function is $[\boldsymbol{y}, \boldsymbol{t}] = \texttt{pepc\_nlfode}(f, \alpha, \boldsymbol{y}_0, h, t_n, \epsilon, \texttt{key})$, where f is the function handle for the right-hand side function $\boldsymbol{f}(t, \boldsymbol{y})$; α is the order; input argument $\boldsymbol{y}_0$ is the vector of initial values; h is the step size; t_n is the termination time; ϵ is the error tolerance for the corrector algorithm, with default value of 10^{-6}. If the input key is set to 1, the predictor algorithm is skipped; if key is 3, the predictor solution is found first, then find the solution $\boldsymbol{t}$ and $\boldsymbol{y}$ with iteration method; if key is set 2, the regular predictor–corrector algorithm is used, without iterations. In this case, the error tolerance err is not used.

▶ Example 7.5

Solve the following nonlinear multi-term Caputo equation [1]:

$$\mathscr{D}_t^{\alpha} y(t) = \frac{40320}{\Gamma(9-\alpha)} t^{8-\alpha} - 3\frac{\Gamma(5+\alpha/2)}{\Gamma(5-\alpha/2)} t^{4-\alpha/2} + \frac{9}{4}\Gamma(\alpha+1) + \left(\frac{3}{2}t^{\alpha/2} - t^4\right)^3 - y^{3/2}(t),$$

where the time interval is $t \in (0, 1)$ and the initial values are $y(0) = 0$ and $y'(0) = 0$. The exact solution of this equation is known to be $y(t) = t^8 - 3t^{4+\alpha/2} + 9t^{\alpha}/4$. If $\alpha = 1.25$, evaluate the accuracy and speed of the algorithm.

Solutions It can be seen that the equation has only one state and the $f\big(t, y(t)\big)$ function is a scalar one, which can be directly described by the following anonymous function.

```
>> alpha=1.25; y0=[0,0];
   f=@(t,y)40320/gamma(9-alpha)*t.^(8-alpha)-...
      3*gamma(5+alpha/2)/gamma(5-alpha/2)*t.^(4-alpha/2)+...
      9/4*gamma(alpha+1)+(3/2*t.^(alpha/2)-t.^4).^3-y.^(3/2);
```

If the step size is chosen to be $h = 0.01$, and no prediction is used, the equation can be solved directly from the following statements. The resulting solution is shown in Fig. 7.1. The solution derived using the predictor method alone is also given in Fig. 7.1 with the exact solution superimposed. It can be seen that there is a visible error in the predictor solution. In fact, the norm of the error vector under iterations is 5.2198×10^{-6}.

```
>> h=0.01; err=1e-10; key=1; tn=1;
   tic, [y,t]=pepc_nlfode(f,alpha,y0,h,tn,err,1);  toc
   tic, [y1,t]=pepc_nlfode(f,alpha,y0,h,tn,err,2); toc
   tic, [y2,t]=pepc_nlfode(f,alpha,y0,h,tn,err,3); toc
   ya1=t.^8-3*t.^(4+alpha/2)+9*t.^alpha/4;
   plot(t,y,t,ya1,'--',t,y1,':',t,y2,'-.')
   N=length(t); norm(y-ya1)/N
```

If different step sizes are chosen, and different `key` values are set, the computation accuracy and time elapse can be measured, as shown in Table 7.1. It can be seen that the corrector algorithm cannot be turned off in terms of computational accuracy, but the predictor solution can be skipped. In practical applications, it is recommended to set the value of `key` to 1. ◀

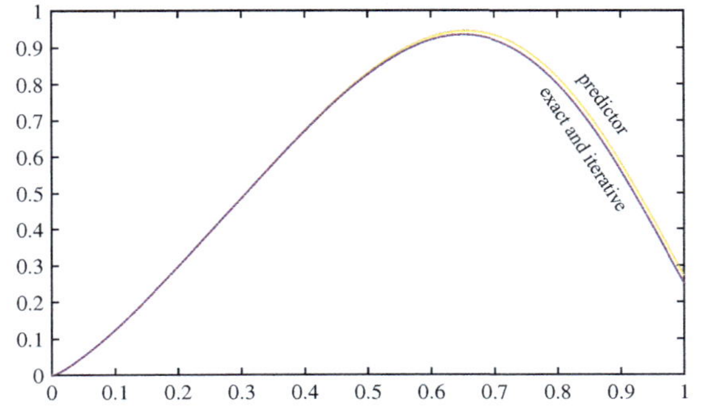

Fig. 7.1 Numerical and exact solutions of the FODE

Table 7.1 Comparison of algorithms with different step sizes

Step	Skip predictor (key= 1)		Use predictor initially (key= 3)		Only predictor (key= 2)	
Sizes h	Error norm	Time elapse	Error norm	Time elapse	Error norm	Time elapse
0.05	2.9228×10^{-4}	0.002	4.0080×10^{-4}	0.004	0.0130	0.017
0.01	5.2198×10^{-6}	0.069	6.5977×10^{-6}	0.101	0.0012	0.026
0.005	9.2358×10^{-7}	0.182	1.1287×10^{-6}	1.083	4.2120×10^{-4}	0.053
0.001	1.6536×10^{-8}	3.330	1.8968×10^{-8}	2.312	3.7732×10^{-5}	1.188
0.0005	2.9236×10^{-9}	13.02	3.2832×10^{-9}	9.012	1.3343×10^{-5}	4.793

7.2.2 Solutions of Commensurate-Order Caputo Equations

7

As seen from the previous introduction, the one-term Caputo equation seems to cover a narrow range, since the right-hand side of the explicit equation contains only the $y(t)$ signal but not the fractional-order derivatives of $y(t)$. This section will eventually explore the numerical solutions of a class of more general Caputo equations described in Definition 7.1.

In fact, the algorithm presented in Ref. [1] is difficult to implement. Only one of its special cases, the commensurate-order form, is considered here using the one-term equation solution method. It can be seen that the nonlinear commensurate-order differential equation can be converted into the form of a one-term FODE and solved directly. The process of solving these FODEs will be demonstrated by the following examples.

▶ Example 7.6

Solve the Bagley–Torwik differential equation given in Example 6.12.

$$Ay''(t) + B\,{}_0^{\mathrm{C}}\mathscr{D}_t^{3/2}y(t) + Cy(t) = C(t+1) \text{ and } y(0) = y'(0) = 1.$$

Solutions The base order should be chosen as $\alpha_0 = 1/2$, and then the state variables are

$$x_1(t) = y(t),\ x_2(t) = {}_0^{\mathrm{C}}\mathscr{D}_t^{1/2}y(t),\ x_3(t) = y'(t),\ x_4(t) = {}_0^{\mathrm{C}}\mathscr{D}_t^{3/2}y(t).$$

The corresponding state space equation can be written as

$${}_0^{\mathrm{C}}\mathscr{D}_t^{1/2}\boldsymbol{x}(t) = \begin{bmatrix} x_2(t) \\ x_3(t) \\ x_4(t) \\ C(t+1) - Cx_1(t)/A - Bx_4(t)/A \end{bmatrix},\quad \boldsymbol{x}_0 = \begin{bmatrix} 1 \\ 0 \\ 1 \\ 0 \end{bmatrix}.$$

The initial values of the state variables can be chosen as follows. Since the first and third states satisfy the condition that $i\alpha_0$ is an integer, they should be set to $y(0)$ and $y'(0)$, respectively, while the initial values of the other two state variables should be set to zero.

Assuming that the parameters $A = 1$, $B = 2$, and $C = 3$ are chosen, the time response curves of each state can be found, as shown in Fig. 7.2, where the first state coincides with the $t+1$ curve, with a maximum error of 0.0015, indicating that the solution function is accurate.

```
>> A=1; B=2; C=3; x0=[1; 0; 1; 0];
   f=@(t,x)[x(2); x(3); x(4); C*(t+1)-C*x(1)/A-B*x(4)/A];
   [x,t]=pepc_nlfode(f,1/2,x0,0.01,3,1e-10,1);
   plot(t,x,t,1+t,'--'), err=x(:,1)-t-1; max(abs(err))
```

In fact, as pointed out in Example 6.12, the analytical solution for the first state is always $x_1(t) = t + 1$, regardless of the choice of parameters A, B, and C. ◀

▶ Example 7.7

Find the numerical solution of the commensurate-order Caputo equation in Example 7.3 using the traditional predictor–corrector method.

Solutions The traditional predictor–corrector method is to use (7.21) to derive the predictor solution first, and then calculate the corrector solution directly from (7.16), i.e., the solution method in which the value of `key` is 3 when the `pepc_nlfode()` function is called.

From the method given in Example 7.3, choosing a base order of $\alpha = 0.005$, we can build a commensurate-order state space equation with 291 states.

$$\begin{cases} {}_0^{\mathrm{C}}\mathscr{D}_t^{\alpha} x_i(t) = x_{i+1}(t), \ i = 1, 2, \cdots, 290 \\ {}_0^{\mathrm{C}}\mathscr{D}_t^{\alpha} x_{291}(t) = -t^{0.1}\dfrac{\mathrm{E}_{1,1.545}(-t)}{\mathrm{E}_{1,1.445}(-t)}\mathrm{e}^t x_1(t)x_{112}(t) + \mathrm{e}^{-2t} - x_{201}^2(t), \end{cases}$$

and $x_1(0) = y(0) = 1$, $x_{201}(0) = y'(0) = -1$, whereas the initial values of remaining state variables are zero. The MATLAB function corresponding to this state space equation is as follows.

```
>> f=@(t,x)[x(2:end);
        -t^0.1*ml_func([1,1.545],-t)/ml_func([1,1.445],-t)...
          *exp(t)*x(1)*x(112)+exp(-2*t)-x(201)^2];
```

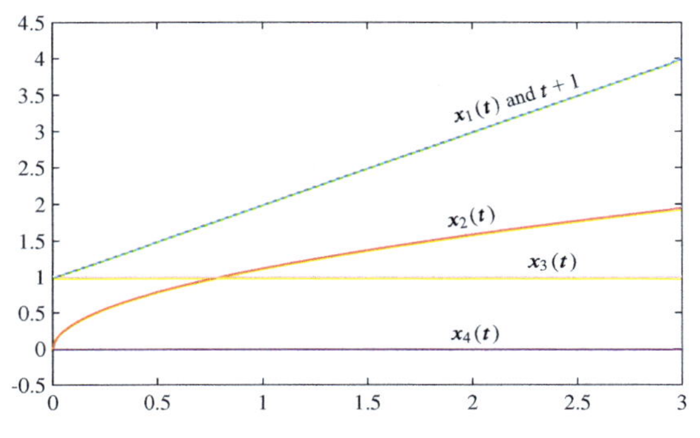

Fig. 7.2 Numerical solution of Bagley–Torwik equation

The initial values $y(0)$ and $y'(0)$ can be assigned to the state variables $x_1(t)$ and $x_{201}(t)$, respectively, and the initial values of the other state variables can be made zero. The step size $h = 0.001$ can be chosen. Considering first the traditional predictor–corrector algorithm, i.e., setting the key to 3, the following statement can be given.

```
>> y0=zeros(291,1); y0([1,201])=[1; -1];
   alpha=0.005; h=0.001; tn=1; key=3;    % find predictor-corrector solution
   tic, [y,t]=pepc_nlfode(f,alpha,y0,h,tn,1e-10,key); toc
   max(abs(y(:,1)-exp(-t)))              % find maximum error
   plot(t,y(:,1),t,exp(-t),'--')
```

After a wait of 27.829 s, the results are calculated with a maximum error of 0.0943. A comparison of the resulting results with the analytical solution is shown in Fig. 7.3.

The measured data in Table 7.2 can be obtained by repeating the above commands with different h values during the calculation. The predictor–corrector solutions are presented in Fig. 7.4. It is clear that even with very small step sizes, the resulting predictor–corrector solutions have a segment of horizontal solutions around the initial moment, which is the main source of the overall error. It can be seen that the time consumption for further decreasing the step size will increase sharply and the accuracy will not change radically. ◄

It can be seen that if $\boldsymbol{y}(t)$ is a vector function, the direct use of the predictor–-corrector algorithm yields incorrect results [1]. A modified approach must be used to solve the differential equation. The method of introducing an iterative process in each step of the solution, i.e., deriving the corrector solution and then updating the corrector solution from (7.16) with its predicted value, can be considered. Repeated use of update method and stop the iteration when the solution of two iterations is

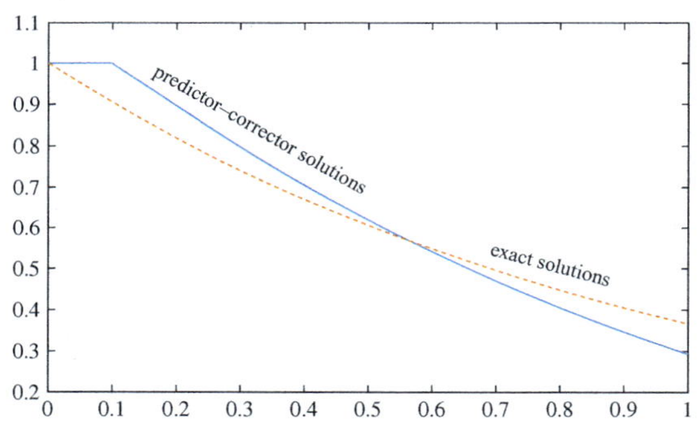

Fig. 7.3 Direct results of the predictor–corrector algorithm

Table 7.2 Comparison of algorithm performance at different step sizes h

Step size h	**0.005**	**0.002**	**0.001**	**0.0005**	**0.0002**
Time elapse	0.3904	8.2563	27.829	71.0745	582.88
Maximum error	1.0954	0.1796	0.0943	0.0483	0.0204

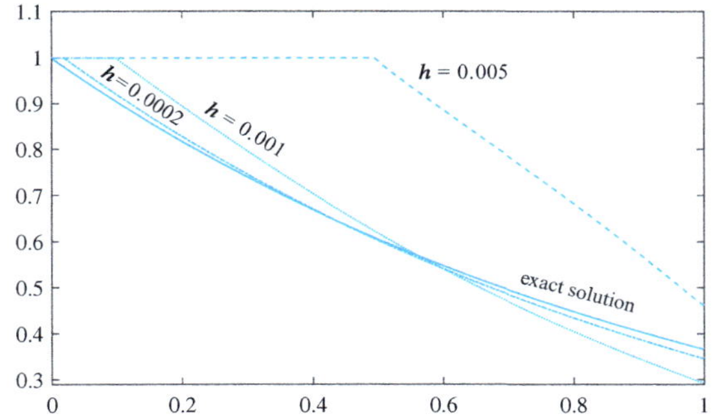

Fig. 7.4 Predictor–corrector solutions for different step sizes h

close enough to obtain the numerical solution of the current step. This idea is already implemented in `pepc_nlfode()`, where the `key` is assigned as 1 or 2 to achieve such iterative operations. If the `key` is 2, the predictor solution is obtained before iteration, while if the `key` is 1, the solution of the previous step can be used as the initial value to iterate directly without the predictor solution.

▶ Example 7.8

Solve again the commensurate-order Caputo equation in Example 7.3 using the iterative predictor–corrector algorithm.

Solutions Before solving the differential equation, the differential equation in state space form is described by an anonymous function.

```
>> f=@(t,x)[x(2:end);
        -t^0.1*ml_func([1,1.545],-t)/ml_func([1,1.445],-t)...
          *exp(t)*x(1)*x(112)+exp(-2*t)-x(201)^2];
```

Let the value of `key` be 2, the FODE can be solved directly by the following statements.

```
>> key=2; h=0.001; err=1e-10; alpha=0.005;
   tic, [y,t]=pepc_nlfode(f,alpha,y0,h,tn,err,key); toc
   max(abs(y(:,1)-exp(-t)))
```

Unfortunately, the efficiency of the algorithm after using iterations is extremely low, taking 4326.54 s. But the advantage is that the maximum error of the solution is 4.7712×10^{-4}, which is much more accurate than the maximum error of 0.06 obtained in Ref. [1] by choosing a smaller step size $h = 1/1600$. The time consumption and computational accuracy of the algorithm with different parameters can be observed by choosing different error tolerances and step sizes, which are listed in Table 7.3. This algorithm is time-consuming and not very accurate, so better numerical solution methods need to be explored. ◀

It can be seen that if the multi-term non-commensurate-order Caputo equation is solved in this way, it will be extremely time-consuming or even impossible to solve. More efficient solution algorithms and tools are needed for such problems.

Table 7.3 Comparison of algorithm performance under different parameters

Step size h	0.001	0.001	0.001	0.01	0.005
Error tolerance ϵ	$\epsilon=10^{-10}$	$\epsilon=10^{-6}$	$\epsilon=10^{-5}$	$\epsilon=10^{-10}$	$\epsilon=10^{-10}$
Time elapse	4326.54	645.22	402.27	76.367	261.41
Maximum error	4.7712×10^{-4}	4.7649×10^{-4}	4.9907×10^{-4}	0.0045	0.0023

7.2.3 Numerical Solutions of Extended FOSS Models

Consider first a simple one-term scalar Caputo equation with zero initial values.

$$ {}_0^{\rm C}\mathscr{D}_t^{\alpha} z(t) = f\big(t, z(t)\big). \tag{7.24}$$

7

From the Grünwald–Letnikov definition, it is straightforward to write

$$\frac{1}{h^\alpha}\sum_{j=0}^{m} w_j^{(\alpha)} z(t_{k-j}) = \frac{1}{h^\alpha}\left[z(t_k)+\sum_{j=1}^{m} w_j^{(\alpha)} z(t_{k-j})\right] = f\big(t_k, z(t_k)\big), \tag{7.25}$$

where $m=\lceil t_k/h\rceil+1$. It is seen that

$$z(t_k) = h^\alpha f\big(t_k, z(t_k)\big) - \sum_{j=1}^{m} w_j^{(\alpha)} z(t_{k-j}), \tag{7.26}$$

where z_k appears at both sides of the equation. Thus, if the step size h is small enough, the first term on the right-hand side of the equation can be approximately replaced by $h^\alpha f\big(t_k, z(t_{k-1})\big)$, so the original equation becomes an iterative equation.

For the system of extended nonlinear explicit FODEss given in Definition 7.4, each differential equation can be written in the following form:

$$ {}_0^{\rm C}\mathscr{D}_t^{\alpha_i} z_i(t) = f_i\big(t, z_1(t), z_2(t), \cdots, z_n(t)\big),\quad i=1,2,\cdots,n. \tag{7.27}$$

The original equation corresponding to the approximate iterative equation can be constructed as follows [2]:

$$\begin{cases} z_1(t_k) = h^{\alpha_1} f_1\big(t_k, z_1(t_{k-1}), z_2(t_{k-1}), \cdots, z_n(t_{k-1})\big) - \displaystyle\sum_{j=1}^{m} w_j^{(\alpha_1)} z_1(t_{k-j}) \\ z_2(t_k) = h^{\alpha_2} f_2\big(t_k, z_1(t_k), z_2(t_{k-1}), \cdots, z_n(t_{k-1})\big) - \displaystyle\sum_{j=1}^{m} w_j^{(\alpha_2)} z_2(t_{k-j}) \\ \quad\vdots \\ z_n(t_k) = h^{\alpha_n} f_n\big(t_k, z_1(t_k), z_2(t_k), \cdots, z_n(t_{k-1})\big) - \displaystyle\sum_{j=1}^{m} w_j^{(\alpha_n)} z_n(t_{k-j}). \end{cases} \tag{7.28}$$

Define a set of state variables $z_k(t) = x_k(t) - x_k(0)$ with zero initial values, then its crucial recursive formula can be changed to

$$\begin{cases} x_1(t_k) = h^{\alpha_1} f_1\big(t_k, x_1(t_{k-1}), x_2(t_{k-1}), \cdots, x_n(t_{k-1})\big) + x_1(0) - \psi_1 \\ x_2(t_k) = h^{\alpha_2} f_2\big(t_k, x_1(t_k), x_2(t_{k-1}), \cdots, x_n(t_{k-1})\big) + x_2(0) - \psi_2 \\ \quad\vdots \\ x_n(t_k) = h^{\alpha_n} f_n\big(t_k, x_1(t_k), x_2(t_k), \cdots, x_n(t_{k-1})\big) + x_n(0) - \psi_n, \end{cases} \tag{7.29}$$

where

$$\psi_i = \sum_{j=1}^{m} w_j^{(\alpha_i)} z_i(t_{k-j}).$$

If the number of simulation points to be computed is large, the summation term in the approximation equation can be considered using the short-time memory effect. Assuming that only the latest L_0 samples are kept, the summation term can be approximated as

$$\psi_i = \sum_{j=1}^{m} w_j^{(\alpha_i)} z_i(t_{k-j}) \approx \sum_{j=1}^{\min(L_0,k)} w_j^{(\alpha_i)} z_i(t_{k-j}). \tag{7.30}$$

Based on the above considerations, the following algorithm for the numerical solution of the nonlinear extended FOSS equation can be developed.

Algorithm 7.3 Solutions of extended FOSS equations
(1) Use a double loop structure to implement the equation-solving process.
(2) Solve (7.29) recursively and store it in $\boldsymbol{z}$, while $\boldsymbol{x}_1$ is used to store the latest $\boldsymbol{x}$ vector.
(3) Compute the numerical solution of the original equation from $\boldsymbol{x} = \boldsymbol{z} + \boldsymbol{x}_0$.

The MATLAB function to find the numerical solution of the nonlinear extended Caputo state space equation can be written based on the above algorithm and its program list is as follows.

```
function [x,t]=nlfode_vec(f,alpha,x0,h,tn,L0)
   arguments
      f, alpha(1,:), x0(:,1), h(1,1){mustBePositive}
      tn(1,1){mustBePositive}, L0(1,1){mustBePositive}=1e20;
   end
   n=length(x0); m=round(tn/h)+1; t=0; g=double(genfunc(1));
   ha=h.^alpha; z=zeros(n,m); x1=x0;
   for i=1:n, W(i,:)=get_vecw(alpha(i),min(m,L0+1),g); end
   for k=2:m, tk=(k-1)*h; L=min(k-1,L0);
      for i=1:n
         x1(i)=f(tk,x1,i)*ha(i)-...
               W(i,2:L+1)*z(i,k-1:-1:k-L).'+x0(i);
      end
      t=[t,tk]; z(:,k)=x1-x0;
   end
```

```
    x=(z+repmat(x0,[1,m])).'; t=t(:);
end
```

The syntax of the function is $[t, x] = $ `nlfode_vec(fun,`$\boldsymbol{\alpha}, \boldsymbol{x}_0, h, t_n, L_0$`)`, where `fun` is the function handle describing the right-hand side expression of the original explicit equation, the exact format of which will be demonstrated later by examples; $\boldsymbol{\alpha}$ is the order vector; $\boldsymbol{x}_0$ is the initial state vector; the input arguments h and t_n are the step size and the termination simulation time, respectively; the optional input argument L_0 is the memory duration of the short-time memory effect. If this parameter is not given, the short-time memory effect will not be used. The output variables $\boldsymbol{t}$ and $\boldsymbol{y}$ are the solutions of the equation, where the ith column vector of $\boldsymbol{y}$ is the numerical solution of the state variable $x_i(t)$, and $\boldsymbol{t}$ is the column vector.

7

► Example 7.9

Simulate the dynamical response of the fractional-order Chua equation [2].

$$\begin{cases} {}_0^{\mathrm{C}}\mathscr{D}_t^{\alpha_1} x_1(t) = \alpha\big[x_2(t) - x_1(t) + f\big(x_1(t)\big)\big] \\ {}_0^{\mathrm{C}}\mathscr{D}_t^{\alpha_2} x_2(t) = x_1(t) - x_2(t) + x_3(t) \\ {}_0^{\mathrm{C}}\mathscr{D}_t^{\alpha_3} x_3(t) = -\beta x_2(t) - \gamma x_3(t), \end{cases}$$

where the mathematical form of the nonlinear function $f(\cdot)$ is

$$f\big(x_1(t)\big) = m_1 x_1(t) + \frac{1}{2}(m_0 - m_1)\big(|x_1(t)+1| - |x_1(t)-1|\big),$$

where $\alpha = 10.725$, $\beta = 10.593$, $\gamma = 0.268$, $m_0 = -1.1726$, $m_1 = -0.7872$, $\alpha_1 = 0.93$, $\alpha_2 = 0.99$, and $\alpha_3 = 0.92$. The initial states are $x_1(0) = 0.2$, $x_2(0) = -0.1$, and $x_3(0) = 0.1$ [2].

Solutions The kth expression on the right-hand side of the original state space equation can be described by the following MATLAB function.

```
function y=c7mchua(~,x,k)
   a=10.725; b=10.593; c=0.268; m0=-1.1726; m1=-0.7872;
   switch k
      case 1
         f=m1*x(1)+(m0-m1)/2*(abs(x(1)+1)-abs(x(1)-1));
         y=a*(x(2)-x(1)-f);
      case 2, y=x(1)-x(2)+x(3);
      case 3, y=-b*x(2)-c*x(3);
end, end
```

Note that, even though the right-hand side of the original equation does not explicitly contain the time variable t, the placeholder t or $\sim$ should be given as the input argument in this function; otherwise, an error will occur.

By choosing the step size $h = 0.001$ and the termination time $t_n = 200$, the numerical solution of this chaotic equation can be solved by the following commands, whose phase plane trajectory is shown in ◘ Fig. 7.5. The time required to solve this equation is about 347.68 s.

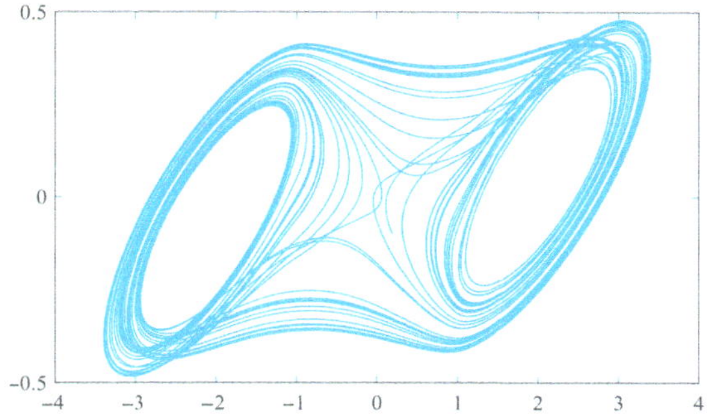

Fig. 7.5 Phase plane trajectory of fractional-order Chua equation

```
>> alpha=[0.93,0.99,0.92]; x0=[0.2; -0.1; 0.1];
   h=0.001; tn=200;
   tic, [x,t]=nlfode_vec(@c7mchua,alpha,x0,h,tn); toc
   plot(x(:,1),x(:,2))
```

As can be seen, the direct application of the above solution method is extremely time-consuming for this example, mainly because of the large number of computational points (200001 points in this case). If the short-time memory effect approximation is used, and the memory duration $L_0 = 10000$ is chosen, and the termination time is increased to $t_n = 500$, the total time required is only about 47.20 s. This shows that the efficiency of the simulation is greatly improved, and the resulting phase plane trajectory is shown in Fig. 7.6. Although the resulting trajectory trend is very similar to the previous one, some differences can be seen in the trajectory details from the initial moment, so the short-time memory approximation is not reliable.

```
>> alpha=[0.93,0.99,0.92]; x0=[0.2; -0.1; 0.1];
   h=0.001; N=10000; tn=500;
   tic, [x,t]=nlfode_vec(@c7mchua,alpha,x0,h,tn,N); toc
   plot(x(:,1),x(:,2))
```

◀

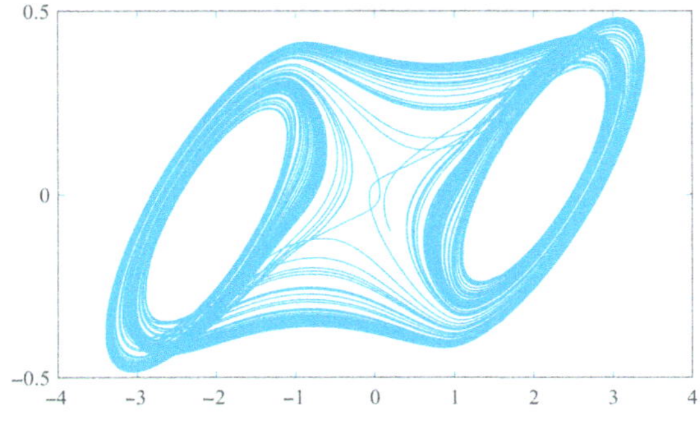

Fig. 7.6 Phase plane trajectory for termination time $t_n = 500$

The nonlinear FOSS equation introduced previously has no analytical solution and the accuracy of the algorithm cannot be evaluated conveniently. So the authors designed a set of benchmark problems [3], of which the fifth benchmark problem is designed for the nonlinear state space equation. The accuracy and efficiency of this algorithm for solving this state space equation are discussed below.

▶ Example 7.10

Consider the FOSS equation from Ref. [3], which has been partially modified here for convenience.

$$\begin{cases} {}_0^{\mathrm{C}}\mathscr{D}_t^{0.5}x(t) = \dfrac{1}{\sqrt{\pi}}\left(\sqrt[5]{(y(t)-0.5)(z(t)-0.3)}+\sqrt{t}\right) \\ {}_0^{\mathrm{C}}\mathscr{D}_t^{0.2}y(t) = \Gamma(2.2)\left[x(t)-1\right] \\ {}_0^{\mathrm{C}}\mathscr{D}_t^{0.1}z(t) = \dfrac{\Gamma(2.3)}{\Gamma(2.2)}\left[y(t)-0.5\right], \end{cases}$$

where $x(0) = 1$, $y(0) = 0.5$, and $z(0) = 0.3$. The analytical solution of this FOSS equation is known to be $x(t) = t + 1$, $y(t) = t^{1.2} + 0.5$ and $z(t) = t^{1.3} + 0.3$, and $0 \leqslant t \leqslant 10$. Test the accuracy and time consumption of the algorithm for different step sizes.

Solutions From the given state space equation, the following functions can be written to compute the value of the kth equation at different values of k, respectively.

```
function y=c7mbench5(t,x,k)
   switch k
      case 1
         y=(((x(2)-0.5)*(x(3)-0.3))^(1/5)+sqrt(t))/sqrt(pi);
      case 2, y=(x(1)-1)*gamma(2.2);
      case 3, y=(x(2)-0.5)*gamma(2.3)/gamma(2.2);
end, end
```

In this way, the following statements can be executed to solve the corresponding differential equations directly. The resulting time response curves for the three state variables are shown in Fig. 7.7. It can be seen that these obtained results coincide exactly with the theoretical values, indicating that the algorithm here is more accurate.

```
>> alpha=[0.5,0.2,0.1]; x0=[1,0.5,0.3];
   h=0.001; tn=10;
```

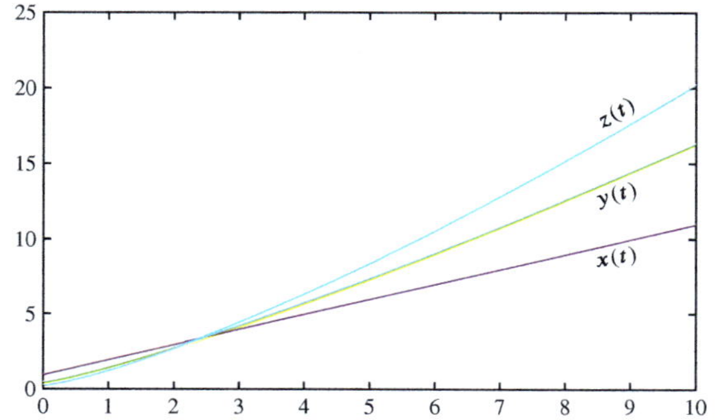

Fig. 7.7 Time domain response of state variables versus theoretical values

Table 7.4 Dependence of the algorithm on the step sizes

Step size h	0.01	0.005	0.001	0.0005	0.0001
Time elapse	0.033	0.058	0.6195	2.786	80.729
Maximum error	0.0063	0.0031	6.0513×10^{-4}	3.0051×10^{-4}	5.9398×10^{-5}

```
tic, [x,t]=nlfode_vec(@c7mbench5,alpha,x0,h,tn); toc
xa=[t+1, t.^1.2+0.5, t.^1.3+0.3];
err=x-xa; max(abs(err(:))), plot(t,x,'--',t,xa)
```

The time consumed and the maximum error measured by running the above code repeatedly for different step sizes are listed in Table 7.4. As can be seen, the smaller the step size is, the smaller the error is, but the time consumed will increase geometrically. ◀

▶ **Example 7.11**

Consider the extended FOSS equation given in Example 7.4 as follows. Use the new algorithm given here to solve again the Caputo equation.

$$\begin{cases} {}_0^{\rm C}\mathscr{D}^{0.555}x_1(t)=x_2(t) \\ {}_0^{\rm C}\mathscr{D}^{0.445}x_2(t)=x_3(t) \\ {}_0^{\rm C}\mathscr{D}^{0.455}x_3(t)=-t^{0.1}\dfrac{{\rm E}_{1,1.545}(-t)}{{\rm E}_{1,1.445}(-t)}{\rm e}^t x_1(t)x_2(t)+{\rm e}^{-2t}-x_3^2(t). \end{cases}$$

Solutions The initial states are $\boldsymbol{x}(0)=[1,0,-1]^{\rm T}$. The original state space equation can be expressed directly by the following MATLAB functions.

```
function y=c7mexp1x(t,x,k)
   if k<3, y=x(k+1);
   else
      y=-t^0.1*ml_func([1,1.545],-t)/ml_func([1,1.445],-t)...
         *exp(t)*x(1)*x(2)+exp(-2*t)-x(3)^2;
end, end
```

With the original differential equation defined, the equation can be solved directly using the following statements. The time required to solve it is only 0.0188 s, and the maximum error derived is as high as 0.0146.

```
>> alpha=[0.555, 0.445, 0.455]; h=0.01; tn=1; x0=[1; 0; -1];
   tic, [x,t]=nlfode_vec(@c7mexp1x,alpha,x0,h,tn); toc
   max(abs(x(:,1)-exp(-t)))
```

The good thing is that the algorithm is fast, so the accuracy can be improved by reducing the step size h. The measured elapsed time and maximum error for different step sizes are listed in Table 7.5. It can be seen that the time required is highly nonlinear because it involves the summation calculation in (7.28). It is not advisable to further reduce the step size to improve the accuracy. ◀

Table 7.5 Performance comparison of algorithms with different step sizes h

Step size h	0.01	0.005	0.001	0.0005	0.0001	0.00005	0.00001
Time elapse	0.0188	0.0198	0.0493	0.0774	0.7432	2.2098	46.581
Max. error	0.0146	0.0076	0.0016	7.847×10^{-4}	1.544×10^{-4}	7.870×10^{-5}	1.574×10^{-5}

7.2.4 An Algebraic Equation-Based FODE Solver

In Algorithm 7.3, the $x_i(t_k)$ term appears at both sides of the equation, so the previous algorithm iterates with some $x_i(t_{k-1})$ terms and finally approximates $x_i(t_k)$. Now reconsider the equation in (7.29), if the $x_i(t_{k-1})$ term is replaced back to $x_i(t_k)$, the form of each equation is the same, such that

7

$$x_i(t_k) - h^{\alpha_i} f_i\big(t_k, x_1(t_k), \cdots, x_n(t_k)\big) + \sum_{j=1}^{\min(L_0,k)} w_j^{(\alpha_i)} z_i(t_{k-j}) - x_i(0) = 0. \tag{7.31}$$

It can be seen that this generates n algebraic equations with n unknowns $x_1(t_k)$, $x_2(t_k)$, $\cdots$, and $x_n(t_k)$, so that for each moment t_k, the values of the state variables can be derived by solving the algebraic equations in such a way that the numerical solution of the original differential equation is finally derived.

Remarks Algebraic equation-based solver

(1) Solve the original differential equation using the loop structure.

(2) For each moment k, the value of the state variable can be derived by solving the algebraic equation.

(3) The summation expression in the equation is independent of the unknown variable $x_i(t_k)$, so only one summation in each step k is required. □

Based on the above considerations, the MATLAB program for solving the extended state space equation can be written under the new algorithm.

```
function [x,t]=nlfode_vec1(f,alpha,x0,h,tn,L0)
   arguments
      f, alpha(1,:), x0(:,1), h(1,1), tn(1,1), L0(1,1)=1e20;
   end
   n=length(x0); m=round(tn/h)+1; t=0; g=double(genfunc(1));
   x1=x0; ha=h.^alpha(:); z=zeros(n,m); tk=0;
   for i=1:n, W(i,:)=get_vecw(alpha(i),min(m,L0+1),g); end
   for k=2:m, tk=(k-1)*h; L=min(k-1,L0);
      for i=1:n
         F0(i,1)=W(i,2:L+1)*z(i,k-1:-1:k-L).'-x0(i);
      end
      F=@(x)x-f(tk,x).*ha+F0;
      x1=fsolve(F,x1); t=[t; tk]; z(:,k)=x1-x0;
```

```
    end
    x=(z+repmat(x0,[1,m])).'; t=t(:);
end
```

▶ Example 7.12

Consider again the extended FOSS equation from Example 7.10. Solve the equation again using the algorithm given here.

Solutions To solve the equation, an anonymous function describing the expression on the right-hand side of the equation should be created, as follows.

```
>> f=@(t,x)[(((x(2)-0.5)*(x(3)-0.3))^(1/5)+sqrt(t))/sqrt(pi);
            (x(1)-1)*gamma(2.2);
            (x(2)-0.5)*gamma(2.3)/gamma(2.2)];
```

Let $h = 0.001$, the solution process takes 45.34 s. However, the maximum error is as high as 0.0019. Therefore, the equation-based solution algorithm presented here does not work well in solving this benchmark problem.

```
>> alpha=[0.5,0.2,0.1]; x0=[1,0.5,0.3]; h=0.001; tn=10;
   tic, [x,t]=nlfode_vec1(f,alpha,x0,h,tn); toc
   xa=[t+1, t.^1.2+0.5, t.^1.3+0.3];
   err=x-xa; max(abs(err(:))), plot(t,x,'--',t,xa)
```

◀

▶ Example 7.13

Solve again the Caputo equation from Example 7.4.

Solutions This extended FOSS equation can be described directly by the following MATLAB function. Note that the format of this function is not the same as that in Example 7.11, where the standard form of function calculation is used to calculate the values of all state variables at the same time.

```
function y=c7mexp2m(t,x)
   y3=-t^0.1*ml_func([1,1.545],-t)/ml_func([1,1.445],-t)...
         *exp(t)*x(1)*x(2)+exp(-2*t)-x(3)^2;
   y=[x(2); x(3); y3];
end
```

The original Caputo equation can be solved by choosing the step size $h = 0.01$, when the time required is 0.416 s with a maximum error of 0.0012.

```
>> alpha=[0.555, 0.445, 0.455];
   h=0.01; tn=1; x0=[1; 0; -1];
   tic, [x,t]=nlfode_vec1(@c7mexp2m,alpha,x0,h,tn); toc
   max(abs(x(:,1)-exp(-t)))
```

By further reducing the step size h, the time elapse and maximum error can be measured for different step sizes, as shown in ◘ Table 7.6. In comparison, the present algorithm is slightly more efficient than Algorithm 7.3. Moreover, the relationship between the time

Table 7.6 Performance comparison of algorithms with different step size h

Step size h	**0.01**	0.005	0.001	0.0005	0.0001	0.00005
Time elapse	0.2885	0.5477	2.7510	5.4414	31.335	72.826
Max. error	0.0012	6.190×10^{-4}	1.304×10^{-4}	6.461×10^{-5}	1.286×10^{-5}	6.433×10^{-6}

elapse and the step length here is essentially linear, since the time elapse is closely related to the number of times the equation is solved. ◀

Attempting to solve the problem in Example 7.9 in this way would require calling the solution function for the algebraic equations 200,000 times, which is not realistically possible. Thus, a more efficient algorithm should be sought.

7

7.3 High-Precision Algorithm for Caputo Equations

As seen from the previously described predictor–corrector algorithm, this type of method has two drawbacks: first, the efficiency of the algorithm is not high; second, it is only suitable for solving commensurate-order differential equations. In addition, there is a potential drawback that the accuracy of the algorithm is not high. This section extends the high-precision algorithm for linear FODEs in ▶ Chap. 6 to the solution of nonlinear Caputo differential equations with the expectation that a high-precision solution algorithm can be derived.

This section first considers the genuine extended FOSS equation described in Definition 7.4 and proposes a special predictor–corrector method—decomposed into two parts, predictor and corrector, to first find a suitable predictor solution and search for the solution of the equation based on the predictor solution using the corrector method.

7.3.1 Predictor Equation

Consider again the nonlinear explicit Caputo equation as follows:

$$ {}_0^{\mathrm{C}}\mathscr{D}_t^{\alpha} y(t) = f\big(t, y(t), {}_0^{\mathrm{C}}\mathscr{D}_t^{\alpha_1} y(t), \cdots, {}_0^{\mathrm{C}}\mathscr{D}_t^{\alpha_{n-1}} y(t)\big). \tag{7.32} $$

Recall that the key point of the algorithm presented in ▶ Chap. 6 is the introduction of the Taylor auxiliary function $T(t)$, which allows the output signal to be decomposed into $y(t) = z(t) + T(t)$, where $z(t)$ is a function with zero initial conditions, and $T(t)$ is the Taylor auxiliary function defined in (6.54). The definition is given again below:

$$ T(t) = \sum_{k=0}^{q-1} \frac{y^{(k)}(0)}{k!} t^k = \sum_{k=0}^{q-1} \frac{y_k}{k!} t^k. \tag{7.33} $$

It can be seen from (7.32) that

$$ {}_{0}^{\mathrm{C}}\mathscr{D}_t^{\alpha} z(t) = f\big(t, z(t), {}_{0}^{\mathrm{C}}\mathscr{D}_t^{\alpha_1} z(t), \cdots, {}_{0}^{\mathrm{C}}\mathscr{D}_t^{\alpha_{n-1}} z(t)\big). \tag{7.34}$$

Furthermore, since ${}_{0}^{\mathrm{C}}\mathscr{D}_t^{\gamma} z(t) = {}_{0}^{\mathrm{RL}}\mathscr{D}_t^{\gamma} z(t)$, it follows that

$$ {}_{0}^{\mathrm{RL}}\mathscr{D}_t^{\alpha} z(t) = f\big(t, z(t), {}_{0}^{\mathrm{RL}}\mathscr{D}_t^{\alpha_1} z(t), \cdots, {}_{0}^{\mathrm{RL}}\mathscr{D}_t^{\alpha_{n-1}} z(t)\big). \tag{7.35}$$

Choose a step size h such that the first q points of the z_i function are zero and $y_i = T_i$, $i = 1, 2, \cdots, q$. This allows the original equation to be solved in a loop for each value of k. Assuming that z_k is known, to compute T_{k+1} of numerical solution, the predictor value of z_{k+1} can be denoted as $z_{k+1}^{\mathrm{p}} = z_k$ and substituting it into the right-hand side of the equation in (7.35). It can be seen that

$$ {}_{0}^{\mathrm{RL}}\mathscr{D}_t^{\alpha} z(t) = \hat{f}. \tag{7.36}$$

If $\hat{f}$ is regarded as a known function, then (7.36) is a one-term FODE, which yields its solution $\hat{z}_{k+1}$. If $||\hat{z}_{k+1} - z_{k+1}^{\mathrm{p}}|| < \epsilon$, where ϵ is a predetermined error tolerance, then the resulting $\hat{z}_{k+1}$ can be viewed as a numerical solution to the original equation; otherwise, let $z_{k+1}^{\mathrm{p}} = \hat{z}_{k+1}$ and continue the iterative arithmetic process. The predictor algorithm is summarized as follows.

Algorithm 7.4 Predictor algorithm

(1) Create an auxiliary function $T(t)$ and decompose $y(t) = z(t) + T(t)$ to create (7.35).
(2) Choose a step size h, let $y_i = T_i$ and $z_i = 0$ at $i = 1, 2, \cdots, q$.
(3) Let $k = q + 1$ to start the loop structure.
(4) Denote $z_{k+1}^{\mathrm{p}} = z_k$, compute ${}_{0}^{\mathrm{RL}}\mathscr{D}_t^{\alpha_i} z_{k+1}$, and substitute it into (7.35) to obtain (7.36) to find $\hat{z}_{k+1}$.
(5) If $||\hat{z}_{k+1} - z_{k+1}^{\mathrm{p}}|| < \epsilon$, then accept the solution $z_{k+1} = \hat{z}_{k+1}$, otherwise denote $z_{k+1}^{\mathrm{p}} = \hat{z}_{k+1}$ and continue the iteration process, until the solution is found.

Based on the above algorithm, the following MATLAB solver function `nlfep()` can be written.

```
function [y,t]=nlfep(fun,alpha,y0,tn,h,p,err)
   arguments, fun
      alpha(1,:), y0(1,:), tn(1,1), h(1,1){mustBePositive}
      p(1,1) {mustBePositiveInteger}, err(1,1) double=1e-10
   end
   m=ceil(tn/h)+1; t=(0:(m-1))'*h; ha=h.^alpha; z=0;
   [T,dT,w,d2]=aux_func(t,y0,alpha,p); y=z+T(1);
   for k=1:m-1
      zp=z(end); yp=zp+T(k+1); y=[y; yp]; z=[z; zp];
      [zc,yc]=repeat_funs(fun,t,y,d2,w,k,z,ha,dT,T);
      while abs(zc-zp)>err
```

```
        yp=yc; zp=zc; y(end)=yp; z(end)=zp;
        [zc,yc]=repeat_funs(fun,t,y,d2,w,k,z,ha,dT,T);
  end, end
  % a subfunction can be written to define the repeated code
  function [zc,yc]=repeat_funs(fun,t,y,d2,w,k,z,ha,dT,T)
     dy=zeros(1,d2-1);
     for j=1:(d2-1)
        dy(j)=w(1:k+1,j+1)'*z((k+1):-1:1)/ha(j+1)+dT(k,j+1);
     end, f=fun(t(k+1),y(k+1),dy);
     zc=((f-dT(k+1,1))*ha(1)-w(2:k+1,1)'*z(k:-1:1))/w(1,1);
     yc=zc+T(k+1);
  end
end
```

Among them, a supporting function `aux_func()` is used to reduce the duplication in this function and in the corrector algorithm that will be described later.

7

```
function [T,dT,w,d2]=aux_func(t,y0,alpha,p)
   an=ceil(alpha); y0=y0(:); q=length(y0); d2=length(alpha);
   m=length(t); g=double(genfunc(p));
   for i=1:d2, w(:,i)=get_vecw(alpha(i),m,g)'; end
   b=y0./gamma(1:q)'; T=0; dT=zeros(m,d2);
   for i=1:q, T=T+b(i)*t.^(i-1); end
   for i=1:d2
      if an(i)==0, dT(:,i)=T;
      elseif an(i)<q
         for j=(an(i)+1):q
            dT(:,i)=dT(:,i)+(t.^(j-1-alpha(i)))*...
                  b(j)*gamma(j)/gamma(j-alpha(i));
end, end, end, end
```

The syntax of the predictor solver is $[\boldsymbol{y}, \boldsymbol{t}]=$ `nlfep(fun,`$\boldsymbol{\alpha}$, $\boldsymbol{y}_0$, t_n, h, p, ϵ`)`, where `fun` is the function handle of the expression on the right-hand side of the original Caputo equation, the format of which will be demonstrated by examples later; the vector $\boldsymbol{y}_0$ is the initial condition vector; t_n is the termination time; h is the step size; p and ϵ are the order and the error tolerance of the algorithm, respectively. It must be noted that it is better to choose $p=1$ in step (4); too high an order p sometimes leads to large errors.

Remarks Illustration of the use of equation description functions

(1) The order vector $\boldsymbol{\alpha}=[\alpha, \alpha_1, \cdots, \alpha_{n-1}]$ contains all fractional orders at both sides of the equation.

(2) The differential equation description method: differential equation can be described not only by MATLAB function but also by anonymous function description. The variables describing the function are t (time column vector), $\boldsymbol{y}$ (output column vector), $\boldsymbol{D}$ (fractional-order derivative matrix), where the kth column of the $\boldsymbol{D}$ matrix $\boldsymbol{D}(:,k)$ represents the ${}_{t_0}^{\mathrm{C}}\mathscr{D}^{\alpha_k} y(t)$ function values.

(3) Dot operations should be used when describing differential equations. □

The resulting numerical solution may not be the numerical solution of the original equation, but the resulting solution can be used as the initial value for the corrector algorithm that will be described later.

▶ Example 7.14

Find the predictor solution of the differential equation in Example 7.11.

Solutions It can be seen that the order and initial value vectors are $\boldsymbol{\alpha} = [1.455, 0.555, 1]$ and $\boldsymbol{y}_0 = [1, -1]$. Introducing all the fractional-order derivative signals $d_1(t) = {}_0^{\mathrm{C}}\mathscr{D}_t^{0.555} y(t)$ and $d_2(t) = y'(t)$, the differential equation can be rewritten in the following standard form:

$$ {}_0^{\mathrm{C}}\mathscr{D}_t^{1.455} y(t) = -t^{0.1} \frac{\mathrm{E}_{1,1.545}(-t)}{\mathrm{E}_{1,1.445}(-t)} \mathrm{e}^t\, y(t)\, d_1(t) + \mathrm{e}^{-2t} - d_2^2(t). $$

In this way, the vectorized form description of the Caputo equation can be implemented by the anonymous function.

```
>> f=@(t,y,Dy)-t.^0.1.*ml_func([1,1.545],-t).*exp(t)./...
     ml_func([1,1.445],-t).*y.*Dy(:,1)+exp(-2*t)-Dy(:,2).^2;
```

The Caputo equation can be solved using the following statements, resulting in the predictor solution shown in Fig. 7.8. It can be seen that, for this example, the predictor solution is not exact, and sometimes the solution for $p = 2$ is even worse than that for $p = 1$. The solution process takes 0.03 s, yielding maximum errors of 0.0264 and 0.0364, respectively.

```
>> alpha=[1.455,0.555,1]; y0=[1,-1]; tn=1; h=0.01; err=1e-8; p=1;
   tic, [yp1,t]=nlfep(f,alpha,y0,tn,h,p,err); toc
   tic, [yp2,t]=nlfep(f,alpha,y0,tn,h,2,err); toc
   plot(t,yp1,t,yp2,'--',t,exp(-t),':')
   max(abs(yp1-exp(-t))), max(abs(yp2-exp(-t)))
```

◀

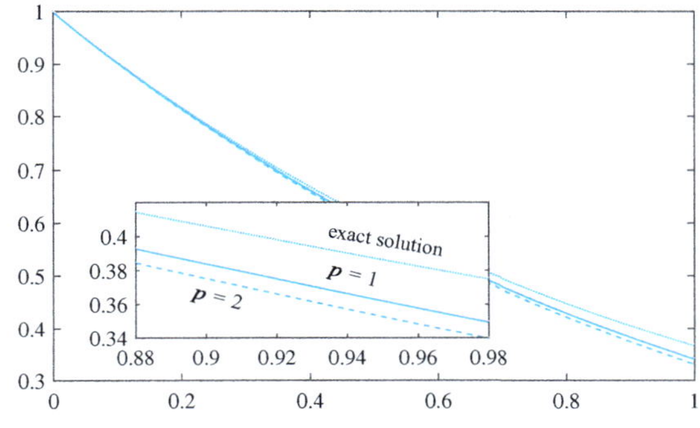

Fig. 7.8 Predictor solution of the equation

7.3.2 Corrector Solution Method

A better solution method—the corrector solution method—will be presented here. If the same step size h is chosen, and the resulting predictor solution $\boldsymbol{y}_\mathrm{p}$ is substituted into the right-hand side of (7.32), a one-term FODE can be derived. The solution of the equation is found using the iterative method until the solution is derived, and the vectorized algorithm for this idea is given below.

Algorithm 7.5 Corrector algorithm

(1) Assume that the solution $\boldsymbol{y}_\mathrm{p}$ of the predictor equation is known.
(2) Substitute $\boldsymbol{y}_\mathrm{p}$ into the right-hand side of (7.32), find a corrector solution $\hat{\boldsymbol{y}}$.
(3) If $||\hat{\boldsymbol{y}} - \boldsymbol{y}_\mathrm{p}|| < \epsilon$, accept the solution $\hat{\boldsymbol{y}}$, otherwise let $\boldsymbol{y}_\mathrm{p} = \hat{\boldsymbol{y}}$ and return to step (2) to continue iterating, until the solution of the equation is obtained.

Based on the above algorithm, the following MATLAB functions can be written, which can be used directly to solve arbitrary explicit Caputo equations.

```
function y=nlfec(fun,alpha,y0,yp,t,p,err)
   arguments
      fun, alpha(1,:), y0(:,1), yp(:,1), t(:,1)
      p(1,1){mustBePositiveInteger}, err(1,1) double=1e-10
   end
   h=t(2)-t(1); m=length(t); ha=h.^alpha;
   [T,dT,w,d2]=aux_func(t,y0,alpha,p);
   [z,y]=repeat_funs(fun,t,yp,T,d2,alpha,dT,ha,w,m,p);
   while norm(z)>err, yp=y; z=zeros(m,1);
      [z,y]=repeat_funs(fun,t,yp,T,d2,alpha,dT,ha,w,m,p);
   end
   % make the repeated code into a subfunction
   function [z,y]=repeat_funs(fun,t,yp,T,d2,alpha,dT,ha,w,m,p)
      for i=1:d2
         dyp(:,i)=glfdiff9(yp-T,t,alpha(i),p)+dT(:,i);
      end
      f=fun(t,yp,dyp(:,2:d2))-dyp(:,1); y=yp; z=zeros(m,1);
      for i=2:m, ii=(i-1):-1:1;
         z(i)=(f(i)*(ha(1))-w(2:i,1)'*z(ii))/w(1,1);
         y(i)=z(i)+yp(i);
   end, end
end
```

The syntax of the function is y=`nlfec(fun,`$\alpha, y_0, y_\mathrm{p}, t, p, \epsilon$`)`, where the input and output arguments are the same as those in `nlfep()` function. Similar to the algorithm presented earlier, since the predictor solution provides the initial values of the iteration of the corrector solution, it can also be replaced by a known vector, such as y_p=`ones(size(`t`))` skipping the predictor solution.

▶ Example 7.15

Solve again the Caputo equation from Example 7.11.

Solutions The previous example has yielded a predictor solution to the original equation. Choosing order $p = 2$ and step size $h = 0.01$, the total time elapse is 2.387 s, yielding a maximum error in the solution of 3.9350×10^{-5}. It can be seen that the corrector algorithm given here is still very efficient.

```
>> f=@(t,y,Dy)-t.^0.1.*ml_func([1,1.545],-t).*exp(t)./...
     ml_func([1,1.445],-t).*y.*Dy(:,1)+exp(-2*t)-Dy(:,2).^2;
   h=0.01; err=1e-10; tn=1; alpha=[1.455 0.555 1]; p=2; y0=[1,-1];
   tic, [yp,t]=nlfep(f,alpha,y0,tn,h,1,err); toc
   tic, [y2,t]=nlfec(f,alpha,y0,yp,t,p,err); toc
   max(abs(y2-exp(-t)))
```

Trying different step sizes and running the above code repeatedly, it is possible to measure the time elapse of the predictor function call, the time elapse of the corrector function call and the maximum computation error, as given by Table 7.7. It can be seen that although the time consumption is longer, around 1 minute, for $h = 0.0001$, the most accurate result is obtained, with maximum error as low as 6.0036×10^{-9}, which is significantly lower than all the previously presented algorithms.

If the predictor is skipped, and the predictor solution is set to a vector of ones, and execute again the following commands, the maximum error is 3.9709×10^{-7}, which is almost the same as the predictor–corrector method, but the time elapse increases significantly to 8.108 s. It is recommended to use the predictor function to derive the initial values when using this algorithm.

```
>> h=0.001; err=1e-10; t=[0:h:tn]'; yp=ones(size(t));
   tic, [y2,t]=nlfec(f,alpha,y0,yp,t,p,err); toc
   max(abs(y2-exp(-t)))
```

If $p = 3$ and the error tolerance is reduced to 10^{-14}, the following statements are repeatedly executed for different step sizes h, resulting in the measured data given in Table 7.8. It can be seen that if $h = 0.001$, the maximum error can be as low as 8.0543×10^{-10}, and the solution time is significantly lower than the best result for $p = 2$. If the p value is increased further, e.g., $p = 4$, the time elapse will increase significantly and the accuracy of the results will be further improved.

```
>> h=0.01; err=1e-14; p=3;
   tic, [yp,t]=nlfep(f,alpha,y0,tn,h,1,err); toc
```

Table 7.7 Solution efficiency for different step sizes

Step size h	0.01	0.005	0.001	0.0005	0.0001
Predictor time	0.0324	0.0554	0.192	0.369	2.895
Corrector time	2.355	2.791	3.613	5.586	61.635
Maximum error	3.9350×10^{-5}	7.6273×10^{-7}	3.9716×10^{-7}	1.2182×10^{-7}	6.0036×10^{-9}

Table 7.8 Solution efficiency for different step sizes with $p = 3$ and $p = 4$

Step size h	0.01	0.005	0.001	0.01	0.005	0.001
Order p	3	3	3	4	4	4
Time elapse	0.0476	0.0746	0.275	0.0448	0.0731	0.283
Corrector time	17.618	16.606	18.884	65.871	70.786	71.91
Maximum error	1.59×10^{-6}	1.55×10^{-7}	8.05×10^{-10}	8.93×10^{-9}	2.85×10^{-10}	3.85×10^{-13}

```
tic, [y2,t]=nlfec(f,alpha,y0,yp,t,p,err); toc
max(abs(y2-exp(-t)))
```

◀

Remarks Predictor–corrector solution of explicit nonlinear Caputo equations

(1) The algorithm here is much more efficient and accurate than the one given in ▶ Sect. 7.2.2.

(2) The algorithm here solves the explicit FODEs, not the FOSS equations.

(3) Compared with the previous algorithm, it does not require the differential equations to be commensurate-order. So it is applicable in a wider range.

(4) Since the theoretical accuracy of the algorithm is $O(h^p)$, the equations can be solved using slightly larger step sizes.

(5) The accuracy p of the algorithm is not recommended to be chosen arbitrarily. If it is too high, it will significantly increase the computational effort, and it is generally recommended to take $p = 2$ or $p = 3$. If α is large, it is recommended to increase further the value of p. □

7.4 Exercises

(1) Solve the following one-term FODE [4]:

$$ {}_0^{\mathrm{C}}\mathscr{D}_t^{\alpha} u(t) = \frac{\Gamma(5+\alpha)}{24} t^4 + t^{8+2\alpha} - u^2(t), $$

where $u(0) = u'(0) = u''(0) = 0$. The exact solution $u(t) = t^{4+\alpha}$. Selecting different values of α, find the numerical solutions, and assess the accuracy and solution efficiencies.

(2) Consider the one-term FODE with zero initial values in Example 7.5. Choosing different values of α (e.g., $\alpha = 0.4$, $\alpha = 1.6$, etc. [4]), solving the differential equation and evaluating the effect of the step size on the accuracy, time-consuming, etc., of the solution.

(3) The function `pepc_nlfode()` is written based on the equally spaced $\boldsymbol{t}$ vector. Equations (7.16)–(7.22) also gives the predictor–corrector method for unequally spaced $\boldsymbol{t}$. Write a general function to implement this algorithm.

(4) prove that the analytical solution of (7.3) is not $y(t) = \mathrm{e}^{-t}$.

(5) Consider the nonlinear FODE in Example 7.1 with nonzero initial value conditions $y(0) = 1$ and $y'(0) = -1$. Verify that $y(t) = \mathrm{e}^{-t}$ is a solution of the following equation by the functions `caputo9()` and `ml_func()` in the interval $[0, 2]$:

$$ {}_0^{\mathrm{C}}\mathscr{D}_t^{1.455} y(t) = -t^{0.1} \frac{\mathrm{E}_{1,1.545}(-t)}{\mathrm{E}_{1,1.445}(-t)} \mathrm{e}^t\, y(t)\, {}_0^{\mathrm{C}}\mathscr{D}_t^{0.555} y(t) + \mathrm{e}^{-2t} - \left[y'(t)\right]^2. $$

(6) Consider the system of FODEs in Example 7.4 with nonzero initial values of $x_1(0) = 1$, $x_2(0) = 0$, and $x_3(0) = -1$. Verify by `caputo9()` and `ml_func()` functions that $x_1(t) = \mathrm{e}^{-t}$, $x_2(t) = -t^{0.445}\mathrm{E}_{1,1.445}(-t)$, and $x_3(t) = -\mathrm{e}^{-t}$ are the solutions to the system of equations.

$$ \begin{cases} {}_0^{\mathrm{C}}\mathscr{D}_t^{0.555} x_1(t) = x_2(t) \\ {}_0^{\mathrm{C}}\mathscr{D}_t^{0.445} x_2(t) = x_3(t) \\ {}_0^{\mathrm{C}}\mathscr{D}_t^{0.455} x_3(t) = -t^{0.1} \dfrac{\mathrm{E}_{1,1.545}(-t)}{\mathrm{E}_{1,1.445}(-t)} \mathrm{e}^t x_1(t) x_2(t) + \mathrm{e}^{-2t} - x_3^2(t). \end{cases} $$

(7) Consider the following one-term nonlinear FODE with nonzero initial condition $y(0) = 1$. Verify that $y(t) = t + 1$ is the analytical solution of the equation. Function `pepc_nlfode()` is applied to solve the numerical solution in the interval $[0, 5]$.

$$ {}_0^{\mathrm{C}}\mathscr{D}_t^{0.5} y(t) = \sqrt{\frac{y(t) - 1}{\pi}} + \sqrt{\frac{t}{\pi}}. $$

(8) Consider the following system of FODEs with initial values of $x(0) = 1$ and $y(0) = 2$. Test that $x(t) = t^{0.7} + 1$ and $y(t) = t^{0.8} + 2$ are the analytical solutions of the equation. Use the function `nlfode_vec()` to find the numerical solution of the equation in the interval $[0, 10]$.

$$ \begin{cases} {}_0^{\mathrm{C}}\mathscr{D}_t^{0.5} x(t) = \dfrac{\Gamma(1.7)}{2\Gamma(1.2)} \left(\sqrt[15]{(x(t) - 1)^2 (y(t) - 2)^2} + t^{0.2} \right) \\ {}_0^{\mathrm{C}}\mathscr{D}_t^{0.7} y(t) = \dfrac{\Gamma(1.8)}{2\Gamma(1.1)} \left(\sqrt[7]{x(t) - 1} + t^{0.1} \right). \end{cases} $$

(9) The following equation is known to have the initial values $y(0) = 1$ and $y'(0) = 0$. Verify that $y(t) = t^2 + 1$ is the analytical solution of this equation. Find the numerical solution of the equation in the interval $[0, 10]$ by the functions `pepc_nlfode()` and `nlfep()`, respectively. Choose different step sizes and compare the computation time and error of the two functions.

$$ {}_0^{\mathrm{C}}\mathscr{D}_t^{1.8} y(t) = \frac{1}{\Gamma(1.2)} \left(\sqrt[10]{y(t) - 1} + \sqrt[5]{t} \right). $$

(10) The following equation has the initial value condition $y(0) = 1$. Verify that $y = t + 1$ is a solution of the equation. Use the function `nlfec()` to find the

numerical solution of the equation. Set the step size and order of the function to different values to compare the time consumption and error.

$$y'(t) + {}_0^{\rm C}\mathscr{D}_t^{0.5} y(t) + y(t) - 2\sqrt{\frac{y(t)-1}{\pi}} = t + 2.$$

(11) The following equation has an initial value of $y(0) = 1$. Verify that $y = \cos t$ is a solution to the equation by using the `nlfep()` and `nlfec()` functions to find the numerical solution of the equation in the interval $[0, 20]$. Choose different step sizes and orders and compare the computation time and error of the two functions.

$$ {}_0^{\rm C}\mathscr{D}_t^{0.8} y(t) = -\left[y^2(t) + \sin^2 t\right] {}_0^{\rm RL}\mathscr{D}_t^{-0.2} \sin t.$$

(12) The following equation has the initial values $y(0) = 3$ and $y'(0) = 0$. Test that $y(t) = t^2 + 3$ is the analytical solution of the equation and use the functions `nlfode_vec()` and `nlfode_vec1()` to find the numerical solution of the equation in the interval $[0, 2]$. Choose different parameters and compare the computation time and error of the two functions.

$$ {}_0^{\rm C}\mathscr{D}_t^{1.5} y(t) = \sqrt{\frac{8}{\pi}} \sqrt[4]{\Gamma(1.8)\,\Gamma(2.2)\,{}_0^{\rm C}\mathscr{D}_t^{1.2} y(t)\,{}_0^{\rm C}\mathscr{D}_t^{0.8} y(t)} + y(t) - t^2 - 3.$$

(13) The following equation has the initial value conditions $y(0) = 3$ and $y'(0) = -1$. Test that $y(t) = {\rm e}^{-1} + 2$ is the analytical solution of the equation. Use the functions `nlfode_vec()` and `nlfode_vec1()` to find the numerical solution of the equation in the interval $[0, 2]$. Choose different parameters and compare the computation time and error of the two functions.

$$ {}_0^{\rm C}\mathscr{D}_t^{1.7} y(t) = -\frac{t^{0.1}{\rm E}_{1,1.3}(-t)}{\left({\rm e}^{-t}+2\right){\rm E}_{1,1.2}(-t)}\, y(t)\, {}_0^{\rm C}\mathscr{D}_t^{0.8} y(t).$$

(14) The initial value conditions for the following equation are $y(0) = 1$ and $y'(0) = 0$. Test that $y(t) = \cosh t$ is the analytical solution of the equation and choose the appropriate function to find the numerical solution of the equation in the interval $[0, 2]$. Different parameters are chosen to compare the time consumption and error.

$$ {}_0^{\rm C}\mathscr{D}_t^{1.5} y(t) = t^{0.5}{\rm E}_{2,1.5}\left(t^2\right)\left[y^2(t) - \sinh^2 t\right].$$

(15) Consider the following FODE with the initial value condition $y(0) = 0$. Test that the analytical solution of the equation is $y(t) = 1 - \cos t^2$. Select the appropriate function to find the numerical solution of the equation in the interval $[0, 2]$. Choose different parameters and compare the computation time and error.

$$ {}_0^{\rm C}\mathscr{D}_t^{0.5} y(t) = 2\left[y^2(t) - 2y(t) + \sin^2 t^2 + 1\right] {}_0^{\rm RL}\mathscr{D}_t^{-0.5}\left(t \sin t^2\right).$$

References

1. Diethelm K (2010) The analysis of fractional differential equations: an application-oriented exposition using differential operators of Caputo type. Springer, New York
2. Petráš I (2011) Fractional-order nonlinear systems: modelling, analysis and simulation. Higher Education Press, Beijing
3. Xue DY, Bai L (2017) Benchmark problems for Caputo fractional-order ordinary differential equations. Fract Calculus Appl Anal 20(5):1305–1312
4. Wang ZQ, Cao JY (2015) Numerical algorithms and their error analysis in fractional-order integro-differential equations (in Chinese). Xi'an Jiaotong University Press, Xi'an

Block Diagram-Based Solutions of FODEs

Contents

D. Y. Xue and L. Bai, ***Fractional Calculus***, https://doi.org/10.1007/978-981-99-2070-9_8

Some numerical algorithms for nonlinear FODEs are presented in ▶ Chap. 7.

For more complicated systems, especially for complicated control systems, large-scale control systems are usually composed of several independent components, these components are usually connected to each other. If numerical solutions to such interconnected systems are expected, the entire system should be converted into a single explicit FODE. Otherwise the solutions cannot be found with the algorithms in ▶ Chap. 7.

Such a request is usually too demanding, since not all the complicated fractional-order systems can be expressed in a single explicit FODE model. Even though such a model can be established, the cost doing so may become too high, or the equations are too complicated to be solved. Therefore, for control systems, block diagram-based solution methods are more suitable for finding the numerical solutions of complicated nonlinear FODEs [1]. Under MATLAB framework, the more powerful Simulink environment can be adopted in solving related problems.

A dedicated Simulink blockset for fractional-order systems is designed by the author. The blocks to compute independently the fractional-order derivative signals are provided in the blockset. Also the blocks such as fractional-order PID controllers, FOTFs or FOTF matrices are provided. Details and internal structure programming are covered in detail in References [2, 3]. In ▶ Sect. 8.1, an introduction is made to the FOTF Toolbox. An introductory presentation on the modeling and simulation of differential equations in Simulink is provided in ▶ Sect. 8.2. The modeling and simulation of FODEs with zero initial conditions are discussed, and simulation result validation is demonstrated. In ▶ Sect. 8.3, block diagram-based modeling and simulation methods of FODEs with nonzero initial conditions are presented. Examples are used to demonstrate the rules and modeling procedures of explicit FODEs and FOSS models. Theoretically speaking, the Simulink models of nonlinear FODEs of any complexity can be established. In ▶ Sect. 8.4, the modeling and simulation of fractional-order control systems are also addressed.

8.1 Introduction of FOTF Toolbox and Blockset

For the convenience of fractional-order system research, a MATLAB toolbox named FOTF Toolbox is developed by the authors. The earliest toolbox was made up of a collection of some functions and Simulink models in 2002 [4]. In 2005, the object-oriented technique was adopted such that a FOTF class was defined. The toolbox was officially named the FOTF Toolbox [5, 6]. A new FOTF Toolbox was released in 2017, as a companion to the publication of the monograph [2]. The new versions include some high-precision algorithms, multivariable fractional-order control system support and a new Simulink blockset. The toolbox was regarded as one of the most widely used toolboxes in the fractional calculus and fractional-order control community worldwide [7]. With the development of this book, the most recent version of FOTF Toolbox is also released.

8.1.1 Input and Connections of Fractional-Order Transfer Functions

Fractional-order transfer function (FOTF) is the basic element of linear fractional-order control systems. The concepts of FOTF and FOTF matrix are presented here first, followed by the input and connection methods of fractional-order models.

Definition 8.1 Fractional-order transfer functions

The mathematical form of FOTF is

$$G(s) = \frac{b_1 s^{\gamma_1} + b_2 s^{\gamma_2} + \cdots + b_m s^{\gamma_m}}{a_1 s^{\eta_1} + a_2 s^{\eta_2} + \cdots + a_{n-1} s^{\eta_{n-1}} + a_n s^{\eta_n}} \mathrm{e}^{-Ts}, \tag{8.1}$$

where a_i, η_i, b_i, γ_i are all constants, T is the delay constant.

In normal cases, the FOTF in (8.1) can be described uniquely by four vectors and a scalar. It can be denoted as $(\boldsymbol{a}, \boldsymbol{\eta}, \boldsymbol{b}, \boldsymbol{\gamma}, T)$.

8

Definition 8.2 FOTF matrix

Similar to integer-order systems, the transfer function matrix model of a linear multivariable fractional-order system is

$$\boldsymbol{G}(s) = \begin{bmatrix} g_{11}(s) & \cdots & g_{1p}(s) \\ \vdots & \ddots & \vdots \\ g_{q1}(s) & \cdots & g_{qp}(s) \end{bmatrix}, \tag{8.2}$$

where $g_{ij}(s)$ is the univariate FOTF defined in (8.1).

A FOTF class is designed in FOTF Toolbox, allowing the users to describe the FOTF objects, and perform operations such as series, parallel and feedback connections. The related programming and ideas may refer to References [2, 3].

The following methods are allowed to input FOTF models with the facilities provided in the FOTF Toolbox.

(1) Use G=fotf($\boldsymbol{a}$,$\boldsymbol{n}_a$,$\boldsymbol{b}$,$\boldsymbol{n}_b$,T) command to input a FOTF object directly into MATLAB workspace, where $\boldsymbol{a} = [a_1, a_2, \cdots, a_n]$, $\boldsymbol{b} = [b_1, b_2, \cdots, b_m]$, $\boldsymbol{n}_a = [\eta_1, \eta_2, \cdots, \eta_n]$, and $\boldsymbol{n}_b = [\gamma_1, \gamma_2, \cdots, \gamma_m]$ can be used to input the coefficients and orders of the pseudo-polynomials in FOTF models, and the delay constant T. If there is no delay, this term can be omitted.

(2) If the numerator p_n and denominator p_d in pseudo-polynomial forms are known, G = `fotf(`p_d,p_n`)` can be used to create the FOTF object.

(3) Similar to the `tf()` function in Control System Toolbox, the s = `fotf('s')` command can be used to declare a Laplace operator s. Then simple MATLAB mathematical expressions can be made to input the FOTF objects or matrices.

(4) Command G = `fotf(`G_1`)` can be used to convert a G_1 model into an FOTF object, where G_1 can be a constant or a constant matrix, or an LTI object in Control

System Toolbox, which can be an integer-order transfer function or a state space model. It can also be an FOTF or an FOSS object defined in the FOTF Toolbox.

▶ Example 8.1

Input the following FOTF into MATLAB workspace.

$$G(s)=\frac{0.8s^{1.2}+2}{1.1s^{1.8}+1.9s^{0.5}+0.4}e^{-0.5s}.$$

Solutions The following commands can be used to input the above model.

```
>> G=fotf([1.1,1.9,0.4],[1.8,0.5,0],[0.8,2],[1.2,0],0.5)
```

With the `fotf()` function, the coefficient and order vectors can be assigned accordingly to the relevant fields in the object, such that a FOTF object can be established and displayed. An alternative method is to define the s operator, then input the whole FOTF model. The two methods are equivalent. For complicated FOTFs, the latter one is recommended.

```
>> s=fotf('s');
   G=(0.8*s^1.2+2)/(1.1*s^1.8+1.9*s^0.5+0.4)*exp(-0.5*s)
```

◀

▶ Example 8.2

Input the following FOTF matrix into MATLAB workspace.

$$\boldsymbol{G}(s)=\begin{bmatrix}\dfrac{1}{1.5s^{1.2}+0.7}e^{-0.5s} & \dfrac{2}{1.2s^{1.1}+1}e^{-0.2s}\\ \dfrac{3}{0.7s^{1.3}+1.5} & \dfrac{2}{1.3s^{1.1}+0.6}e^{-0.2s}\end{bmatrix}.$$

Solutions The four univariate FOTF objects should be entered first. Then with matrix input command, the FOTF matrix can be expressed in MATLAB workspace.

```
>> g1=fotf([1.5 0.7],[1.2 0],1,0,0.5);
   g2=fotf([1.2 1],[1.1 0],2,0,0.2);
   g3=fotf([0.7 1.5],[1.3 0],3,0);
   g4=fotf([1.3 0.6],[1.1 0],2,0,0.2); G=[g1,g2; g3,g4]
```

◀

Product and sum operations are supported for FOTF objects, such that the FOTF objects under series and parallel connections can be processed to find the overall model. Command G = `feedback`(G_1,G_2) is also supported in FOTF Toolbox to find the closed-loop model G, for the two FOTF objects under negative feedback connections.

▶ Example 8.3

Assume that under typical unity negative feedback connections, the plant and controller models are given, respectively, as

$$G(s) = \frac{0.8s^{1.2}+2}{1.1s^{1.8}+0.8s^{1.3}+1.9s^{0.5}+0.4},\quad G_c(s) = \frac{1.2s^{0.72}+1.5s^{0.33}}{3s^{0.8}},$$

find the closed-loop model.

Solutions The following commands can be used to enter the model into MATLAB workspace.

```
>> G=fotf([1.1,0.8 1.9 0.4],[1.8 1.3 0.5 0],[0.8 2],[1.2 0]);
   Gc=fotf([3],[0.8],[1.2 1.5],[0.72 0.33]); G1=feedback(G*Gc,1)
```

The obtained closed-loop model is

$$G_1(s) = \frac{0.96s^{1.59}+1.2s^{1.2}+2.4s^{0.39}+3}{3.3s^{2.27}+2.4s^{1.77}+0.96s^{1.59}+1.2s^{1.2}+5.7s^{0.97}+1.2s^{0.47}+2.4s^{0.39}+3}.$$

◀

8

8.1.2 Fractional-Order State Space Models

A fractional-order state space (FOSS) model is another important format for linear fractional-order systems. The conventional FOSS is defined in Definition 6.15. There are limitations in real applications [3]. Therefore a descriptor FOSS model is introduced.

Definition 8.3 Descriptor fractional-order state space model

The mathematical form of descriptor FOSS model is

$$\begin{cases} \boldsymbol{E}\mathscr{D}_t^{\alpha}\boldsymbol{x}(t) = \boldsymbol{A}\boldsymbol{x}(t)+\boldsymbol{B}\boldsymbol{u}(t) \\ \boldsymbol{y}(t) = \boldsymbol{C}\boldsymbol{x}(t)+\boldsymbol{D}\boldsymbol{u}(t), \end{cases} \tag{8.3}$$

where $\boldsymbol{x}(t) \in \mathbb{R}^{n\times 1}$ is the state variable vector, n is the order of the system. Matrices $\boldsymbol{E}\in\mathbb{R}^{n\times n}$, $\boldsymbol{A}\in\mathbb{R}^{n\times n}$, $\boldsymbol{B}\in\mathbb{R}^{n\times p}$, $\boldsymbol{C}\in\mathbb{R}^{q\times n}$ and $\boldsymbol{D}\in\mathbb{R}^{q\times p}$ are constant matrices. If $\boldsymbol{E}$ is an identity matrix, the system is in the form in Definition 6.15.

If matrices $\boldsymbol{A}$, $\boldsymbol{B}$, $\boldsymbol{C}$, $\boldsymbol{D}$, and $\boldsymbol{E}$ are known, the command $G=$ `foss(`$\boldsymbol{A},\boldsymbol{B},\boldsymbol{C},\boldsymbol{D},\alpha,T,\boldsymbol{E}$`)` can be used to input directly the model. If matrix $\boldsymbol{E}$ is omitted, the default $\boldsymbol{E}$ matrix is an identity one.

8.1.3 Analysis Functions for Linear Fractional-Order Systems

With contrast to the cases in Control System Toolbox in MATLAB, a series of modeling processing and system analysis functions are designed such that these functions can be called to fulfill corresponding tasks [3].

(1) **Model conversions**. FOTF objects can be converted into FOSS objects by function `foss()`, and the inverse conversion can be made with `fotf()` function. In addition, these two functions convert integer-order objects into fractional-order ones.

(2) **System properties analysis**. The stability of the system can be assessed with `isstable()` function. The controllability and observability can be tested with, respectively, `ctrb()` and `obsv()` functions. Function `norm()` can be used to compute fractional-order system norms. The syntaxes of these function are made as closely as their counterparts in the Control System Toolbox.

(3) **Time domain analysis**. The step response of fractional-order systems can be evaluated and plotted with `step()` function. Impulse response and arbitrary input response can be processed, respectively, with functions `impulse()` and `lsim()`.

(4) **Frequency domain analysis**. Like in the cases in integer-order systems, functions `bode()`, `nyquist()` and `nichols()` can be used to draw Bode diagrams, Nyquist plots and Nichols charts, respectively. The gain and phase margins can be evaluated with function `margin()`. For multivariable systems, functions `mfrd()` and `gershgorin()` can also be used in direct frequency domain analysis.

(5) **Root locus analysis**. The root locus of fractional-order systems can be drawn with function `rlocus()`. It should be noted that the method cannot be used to draw root loci for non-commensurate-order systems, or the system order is too high. The approximate root locus is recommended instead.

▶ Example 8.4

Assess the stability of the following FOTF, and compute the system norms.

$$G(s) = \frac{-2s^{0.63} - 4}{2s^{3.501} + 3.8s^{2.42} + 2.6s^{1.798} + 2.5s^{1.31} + 1.5}.$$

Solutions With the following commands, the overload function `isstable()` can be used to assess directly the system stability. Since the returned value is 1, the original system is stable.

```
>> a=[2,3.8,2.6,2.5,1.5]; na=[3.501,2.42,1.798,1.31,0];
   b=[-2,-4]; nb=[0.63,0]; G=fotf(a,na,b,nb);
   [key,alpha,a1,p]=isstable(G), p1=p.^(1/alpha)
```

The $\mathcal{H}_2$ and $\mathcal{H}_\infty$ norms of the system can be evaluated as 1.0122 and 8.6115.

```
>> n1=norm(G), n2=norm(G,inf)
```

◀

8.1.4 The FOTF Blockset

For ease of Simulink modeling, the new FOTF Blockset shown in Fig. 8.1 is developed. The blockset can be opened with the command `fotflib`. The commonly used blocks in the blockset are as follows:

(1) The kernel block is the Riemann–Liouville block, which is a masked block from Oustaloup or other filters. It can be used to evaluate the Riemann–Liouville fractional-order derivatives or integrals of its input signal. A method will be introduced later to acquire Caputo derivatives. Double click the block icon, a parameter dialog box is displayed, allowing the user to input information such as the filter orders and interested frequency bands. A listbox is provided to ask the user whether to append the block with a low-pass filter. Different filters are allowed, including discrete ones. Of course, from the comparative studies in ► Chap. 5, it is recommended to choose the default Oustaloup filter.

(2) An Implicit model block is masked version to implement the $(as+1)^{-\gamma}$ irrational transfer function model. The internal structure of the model is as shown in Fig. 5.25(b).

8

(3) Caputo vector block implements the vectorized Caputo derivative actions. It can be used directly in the modeling and simulation of Caputo state space systems.

In addition, the Approximate FOTF model block is used to describe the FOTF or FOTF matrix. Fractional-order PID controller block is used to represent fractional-order PID controller models, while in the Benchmark problem models group, a set of models for the benchmark problems are created. More details can be found in Appendix A.

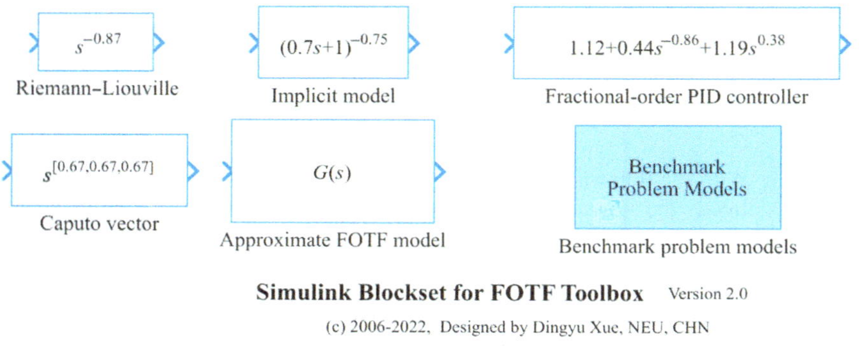

Fig. 8.1 FOTF Blockset (`fotflib.slx`)

8.2 Block Diagram-Based Solutions of FODEs with Zero Initial Conditions

Many nonlinear FODEs which cannot be easily solved with command-driven methods can be modeled and solved with Simulink. In this section, Simulink modeling rules and fundamental knowledge for ODEs are presented first. Then some examples in the modeling of FODEs are illustrated. The block diagram-based solution patterns are demonstrated. Simulation result validation methods are also presented.

8.2.1 Simulink Modeling Rules

Various blocks are provided in Simulink. It is not a simple matter to learn to model ODEs for the beginners. In Reference [8], a brief summary on Simulink modeling rules for ODEs is made. In actual ordinary differential equation modeling, the following rules and procedures can be adopted:

(1) To define two signals $x(t)$ and $x'(t)$, an integrator block can be introduced first. The output signal can be defined artificially as signal $x(t)$, and it is natural that the input signal is $x'(t)$. With such a method, the key signals for ODEs are defined. The initial value $x(0)$ can be assigned to the integrator block from its parameter dialog box.

(2) If two signals are equal, they can be joined by signal lines directly. This method is usually used in Simulink modeling to close simulation loops.

(3) For a given signal $u(t)$, if it is connected to a Gain block, and double the block to fill in the gain value k, the output signal thus configured is the $ku(t)$ signal.

(4) If algebraic operations are expected to the $v(t)$ signal, for instance, to compute $(v(t) + v^3(t)/3)\sin v(t)$, signal $v(t)$ can be connected to a Fcn or a MATLAB Function block. Double click the block, the expression `(u+u^3/3)*sin(u)` or a MATLAB function name can be specified in the parameter dialog box. Note that whatever the physical meaning of the input signal, it must be denoted as u in the Fcn block.

(5) Several signals can be fed into a Mux block, such that the output signal is a vector signal from these components. If a vector signal is fed into a Demux block, the scalar or small vector signals can be decomposed.

(6) In order to construct a delay signal $u(t-d)$ from signal $u(t)$, signal $u(t)$ can be fed into a Transport Delay block, and double click the block to enter parameter d, the output of the block is the expected delay signal.

(7) To acquire an interested signal, it can be connected into a Scope block or a Out block, such that the signal can be monitored and processed.

With the above modeling rules, the simulation models for ODEs can be created easily. There is no need to pre-process ODEs and convert them into standard forms. Direct modeling methods can be used to draw the ODEs in a Simulink model window. The simulation facilities provided in Simulink can be used to carry out simulation processes, so as to find the numerical solutions of the ODEs.

8.2.2 Simulink Environment Settings

Before introducing the simulation studies with Simulink, necessary parameter settings to the Simulink environment are recommended. It is suggested to set the necessary parameters in Simulink from the dialog boxes. Open a Simulink model window, and select Modeling pane, as shown in Fig. 8.2. Click the icon, a model parameter dialog box shown in Fig. 8.3 is opened. Two sets of parameters are suggested to be set again: (1) the solver parameter setting, (2) the input/output format setting [8]. The setting methods are presented next.

1. Solver Parameters Setting

Solver parameters are related to whether accurate solution results can be found. It is suggested to set the following two parameters:

(1) **Algorithm selection**. For reliable numerical computing, variable-step methods are recommended. This is the default setting in Simulink. In normal cases, the default numerical algorithm is set to Auto, and there is no need to alter this option. If the simulation process takes too long in certain problems, without yielding any simulation results, stiff ODE problems are probably encountered. In this case, the

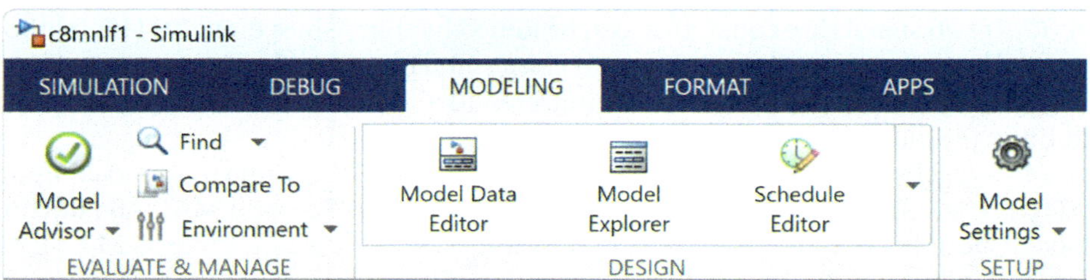

Fig. 8.2 Simulink model window toolbar

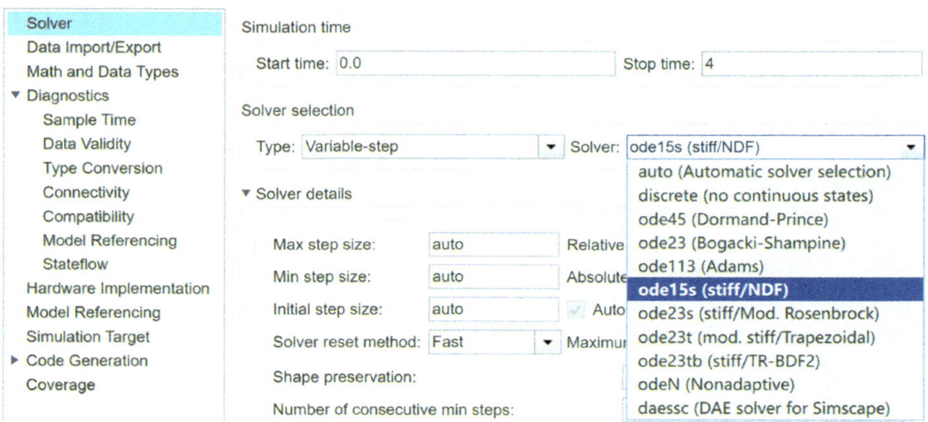

Fig. 8.3 Solver parameters setting

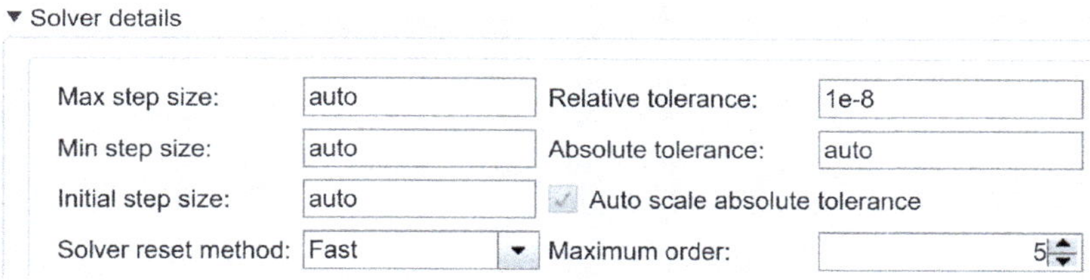

Fig. 8.4 Solver details setting

ode15(stiff/NDF) item from the Solvers listbox is recommended to speed up the simulation process. This option sometimes sacrifices slightly the accuracies in simulation.

(2) **Control parameters setting** [9]. Apart from the solver parameters, the Solver details part should be expanded, from which the Relative tolerance should be set to small values, to at least 10^{-8}. The default value in Simulink is 10^{-3}, implying one thousandth or relative error, which is far too large for accurate simulation solutions (Fig. 8.4).

2. Input and Output Format Setting

For the ease of visiting to the data in simulation results, it is recommended to set the output data format as follows. Select the Data Import/Export tag from the listbox on the left in Fig. 8.3, the Simulink parameter dialog box is automatically changed to the form in Fig. 8.5. It is suggested to uncheck the Single simulation output checkbox, and modify the default Dataset item in the Format listbox to Array. With such setting, simulation results are returned to MATLAB workspace in the reserved variables `tout` and `yout`. If the above setting is not made, it is rather complicated to extract the data from the simulation results.

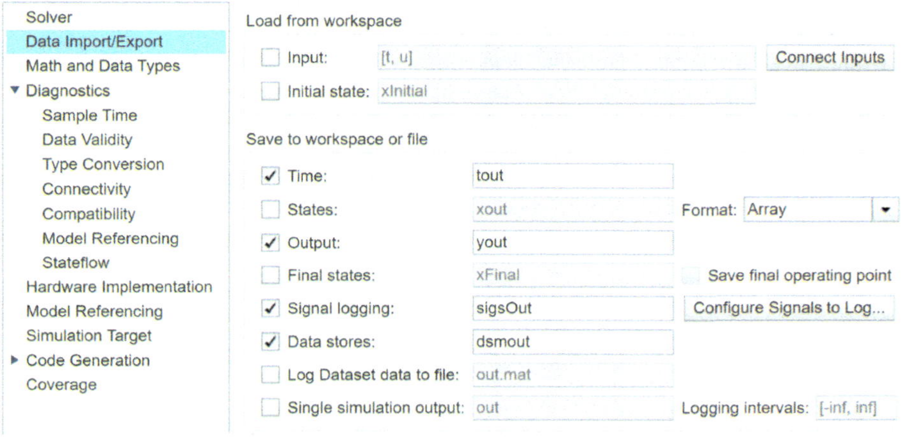

Fig. 8.5 Data import and export dialog box

8.2.3 Simulink Modeling and Solutions for FODEs

For the FODEs with zero initial conditions, the Riemann–Liouville block in the FOTF Blockset can be used to construct the key signals in Simulink modeling. Based on these key signals, the Simulink modeling for the entire fractional-order system can be established. The modeling procedures are demonstrated through the following examples.

► Example 8.5

Consider the linear FODE with zero initial conditions shown in Example 6.8.

$$\mathscr{D}_t^{3.5}y(t)+8\mathscr{D}_t^{3.1}y(t)+26\mathscr{D}_t^{2.3}y(t)+73\mathscr{D}_t^{1.2}y(t)+90\mathscr{D}_t^{0.5}y(t) = 30u'(t)+90\mathscr{D}_t^{0.3}u(t).$$

If the input signal is $u(t) = \sin t^2$, solve the FODE with Simulink.

Solutions For simplicity, the explicit form of the FODE is obtained first.

$$\mathscr{D}_t^{3.5}y(t) = -8\mathscr{D}_t^{3.1}y(t)-26\mathscr{D}_t^{2.3}y(t)-73\mathscr{D}_t^{1.2}y(t)-90\mathscr{D}_t^{0.5}y(t) + 30u'(t)+90\mathscr{D}_t^{0.3}u(t).$$

The following procedures can be used to set up the Simulink model.

(1) **Open a blank model window**. With the File → New → New Model menu item in MATLAB command window, a blank model window can be opened. The user can draw his own Simulink model in this window.

(2) **Define the key signals**. The key signals in FODEs refer to the signals such as the output signal $y(t)$ and its integer-order derivatives and the fractional-order ones involved. The integer-order derivatives can be defined directly from integer-order integrator chains. For this example, an integrator chain with three integrators can be constructed first, as shown in Fig. 8.6.

Apart from the integer-order derivatives, the signal $\mathscr{D}_t^{3.5}y(t)$ is also needed in the FODE. The order here is higher than the highest order in the integrator chain. Therefore a -0.5th-order Riemann–Liouville block can be connected to the $y'''(t)$ signal to build up signal $\mathscr{D}_t^{3.5}y(t)$. In addition, since the fractional-order derivatives $\mathscr{D}_t^{3.1}y(t)$, $\mathscr{D}_t^{2.3}y(t)$, $\mathscr{D}_t^{1.2}y(t)$ and $\mathscr{D}_t^{0.5}y(t)$ are needed in the FODE, these signals can be derived from the corresponding integer-order derivative signals, as shown in Fig. 8.7.

Double click a Riemann–Liouville block icon, a parameter dialog box in Fig. 8.8 is displayed. The filter parameters can be assigned in this dialog box. The parameters can either be specific values or variable names. If variable names are used, remember to assign the variables in MATLAB workspace, otherwise the Simulink model cannot be normally executed. In order to avoid the possible algebraic loop phenomena, the Is a low-pass filter required? checkbox can be checked. This checkbox sometimes may affect slightly the simulation precisions.

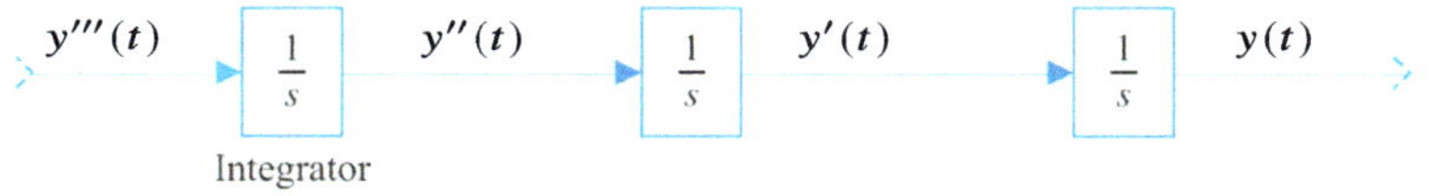

Fig. 8.6 Output signal $y(t)$ and its integer-order derivatives (c8mblk1a.slx)

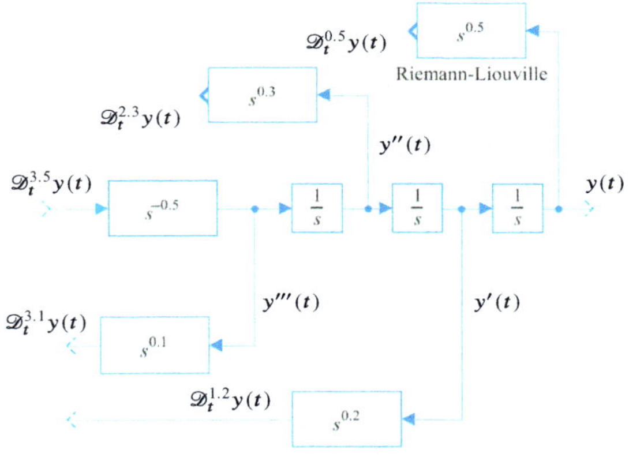

Fig. 8.7 The output signal $y(t)$ and other key signals (`c8mblk1b.slx`)

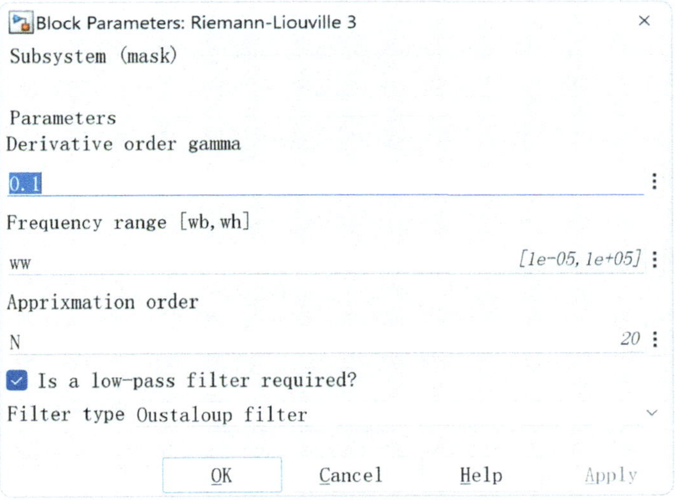

Fig. 8.8 Riemann–Liouville block parameter dialog box

For convenience, it is suggested to set Frequency range in a vector variable `ww`, and set Approximation order as variable N. The values of these two variables should be pre-specified and saved into the model.

(3) **Set up the complete Simulink model**. If all the key signals are defined, the low-level Simulink model can be constructed, as shown in Fig. 8.9. The parameters in the Riemann–Liouville block, `ww` and N, can be assigned first in MATLAB workspace, with the following commands.

```
>> ww=[1e-5 1e3]; N=15;
```

(4) **Start the simulation process**. Two methods can be used to start the simulation process. (a) Click the ▶ button in the toolbar in the Simulink model window. After simulation, two

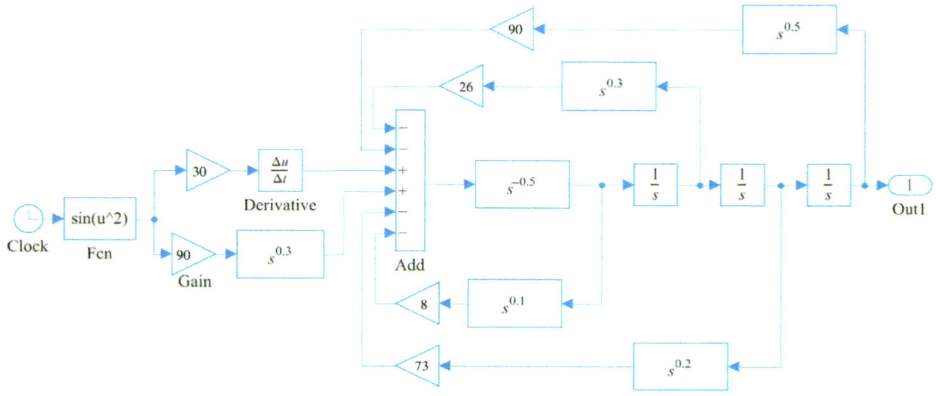

Fig. 8.9 Simulink model for the linear FODE (`c8mblk2.slx`)

8

variables `tout` and `yout` are returned to MATLAB workspace; (b) Call `sim()` function to start the simulation process, and the arguments t and y are returned.

```
>> [t,~,y]=sim('c8mblk3');
```

It is seen that with simulation methods, the numerical solutions are exactly the same as the ones obtained in Example 6.8. ◀

▶ Example 8.6

Since the model studied here is linear, a simpler way is to set up a Simulink model in Fig. 8.10. Before starting the simulation process, the following commands should be used to input the FOTF object into MATLAB workspace. The simulation result is also exactly the same.

```
>> b=[30,90]; nb=[1,0.3]; a=[1,8,26,73,90];
   na=[3.5,3.1,2.3,1.2,0.5]; G=fotf(a,na,b,nb);
```

◀

Fig. 8.10 A simpler Simulink model (`c8mblk3.slx`)

▶ Example 8.7

Consider again the linear FODE with zero initial conditions in Example 6.5.

$$\mathscr{D}_t^{1.2}y(t) + 5\mathscr{D}_t^{0.9}y(t) + 9\mathscr{D}_t^{0.6}y(t) + 7\mathscr{D}_t^{0.3}y(t) + 2y(t) = u(t).$$

If the input signal is a step one, it is known from ▶ Sect. 6.3 that, the exact solution of the FODE is

$$y_3(t) = \frac{1}{2} + \frac{1}{2}\mathrm{E}_{0.3}(-2t^{0.3}) - \mathrm{E}_{0.3}\left(-t^{0.3}\right) - t^{0.6}\mathrm{E}_{0.3,1.6}^{2}\left(-t^{0.3}\right) + t^{0.9}\mathrm{E}_{0.3,1.9}^{3}\left(-t^{0.3}\right).$$

Assess the impact of filter parameters on the Simulink results.

Solutions With the standard Simulink modeling procedures discussed earlier, the Simulink model in Fig. 8.11 can be established easily. If different filter parameters N and (ω_b, ω_h) are chosen, the simulation errors can be found, by comparing with the exact solutions, and the results are shown in Table 8.1.

It can be seen that if the frequency band of interest covers 6 decades, the filter order $N = 7$ may lead to very good results. If the number of decades is increased by 1, it is recommended to increase N by 1 or 2. In fact, since the simulation speed is rather fast, higher order filters can be selected for more accurate results.

```
>> ww=[1e-4 1e4]; T=[];         % the frequency band can be modified
   for N=5:18                   % try different orders
     [t,x,y]=sim('c8mblk5',0.5);
     y3=1/2+0.5*ml_func(0.3,-2*t.^0.3)-ml_func(0.3,-t.^0.3)...
        -t.^0.6.*ml_func([0.3,1.6,2],-t.^0.3)...
        +t.^0.9.*ml_func([0.3,1.9,3],-t.^0.3);
     e=norm(y-y3), T(N-4)=e;
   end
```

If the termination time is increased to $t_n = 1$, function `ml_func()` fails to get exact solutions. If Simulink model is adopted, the simulation interval can be increased significantly. It should be noted that if the termination time increases, the lower frequency bound ω_b is

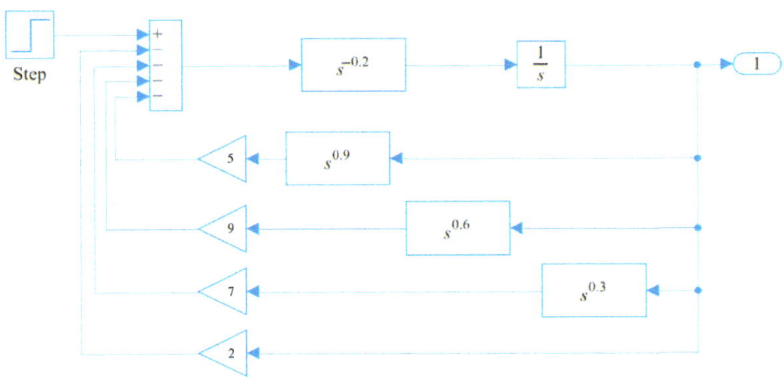

Fig. 8.11 Simulink model (c8mblk5.slx)

Table 8.1 Errors for different filter parameters

Order	Frequency band (ω_b, ω_h)				
N	$(10^{-3}, 10^3)$	$(10^{-4}, 10^4)$	$(10^{-4}, 10^3)$	$(10^{-5}, 10^3)$	$(10^{-5}, 10^5)$
5	0.0054	0.00802	0.00741	0.00733	0.08015
6	0.0047	0.00317	0.00521	0.00701	0.04668
7	0.0048	0.00196	0.00473	0.00512	0.01211
8	0.0048	0.00110	0.00478	0.00472	0.00570
9	0.0048	0.00071	0.00483	0.00476	0.00317
10	0.0049	0.00048	0.00483	0.00482	0.00416
11	0.0049	0.00042	0.00485	0.00483	0.00240
12	0.0049	0.00042	0.00487	0.00484	0.00120
13	0.0049	0.00041	0.00488	0.00486	0.00058
14	0.0049	0.00041	0.00489	0.00487	0.00044
15	0.0049	0.00040	0.00490	0.00488	0.00025
16	0.0049	0.00043	0.00490	0.00489	0.00017
17	0.0049	0.00043	0.00491	0.00490	0.00010
18	0.0050	0.00043	0.00491	0.00490	5.45×10^{-5}

8

increased. For instance, if the termination time is increased to $t_n = 500$, different orders and frequency bands can be tried, and the responses in Fig. 8.12 are obtained.

```
>> N=8; ww=[1e-3 1e3]; [t1,~,y1]=sim('c8mblk5',500);
   N=10; ww=[1e-4 1e3]; [t2,~,y2]=sim('c8mblk5',500);
   N=10; ww=[1e-5 1e3]; [t3,~,y3]=sim('c8mblk5',500);
   N=12; ww=[1e-6 1e3]; [t4,~,y4]=sim('c8mblk5',500);
   tic, N=16; ww=[1e-8 1e3]; [t5,~,y5]=sim('c8mblk5',500); toc
   plot(t1,y1,t2,y2,t3,y3,t4,y4,t5,y5)
```

It can be seen from the example that $\omega_b = 10^{-3}$ rad/s is not a good choice. If $\omega_b = 10^{-4}$ rad/s is selected, the approximation can be improved, however, errors can still be witnessed. For the simulation problem in such a large interval, $t_n = 500$, ω_b should by at least selected as 10^{-5} rad/s. Further increasing the order N, and increase ω_h, until satisfactory simulation results can be found. ◀

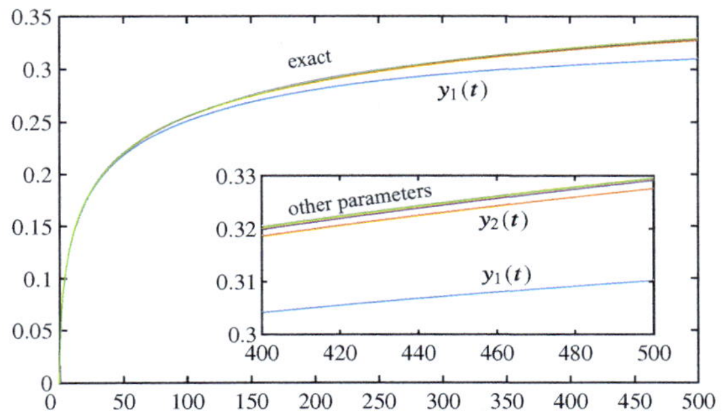

Fig. 8.12 Simulation results comparisons

▶ Example 8.8

Solve the nonlinear FODE with zero initial conditions.

$$\frac{3\mathscr{D}_t^{0.9}y(t)}{3+0.2\mathscr{D}_t^{0.8}y(t)+0.9\mathscr{D}_t^{0.2}y(t)}+\left|2\mathscr{D}_t^{0.7}y(t)\right|^{1.5}+\frac{4}{3}y(t)=5\sin 10t.$$

Solutions The explicit expression to the signal $y(t)$ can be obtained.

$$y(t)=\frac{3}{4}\left[5\sin 10t-\frac{3\mathscr{D}_t^{0.9}y(t)}{3+0.2\mathscr{D}_t^{0.8}y(t)+0.9\mathscr{D}_t^{0.2}y(t)}-\left|2\mathscr{D}_t^{0.7}y(t)\right|^{1.5}\right].$$

If can be seen from the explicit expression that the key signals in this model can be defined by the fractional-order derivatives such that the Simulink model in Fig. 8.13 can be established.

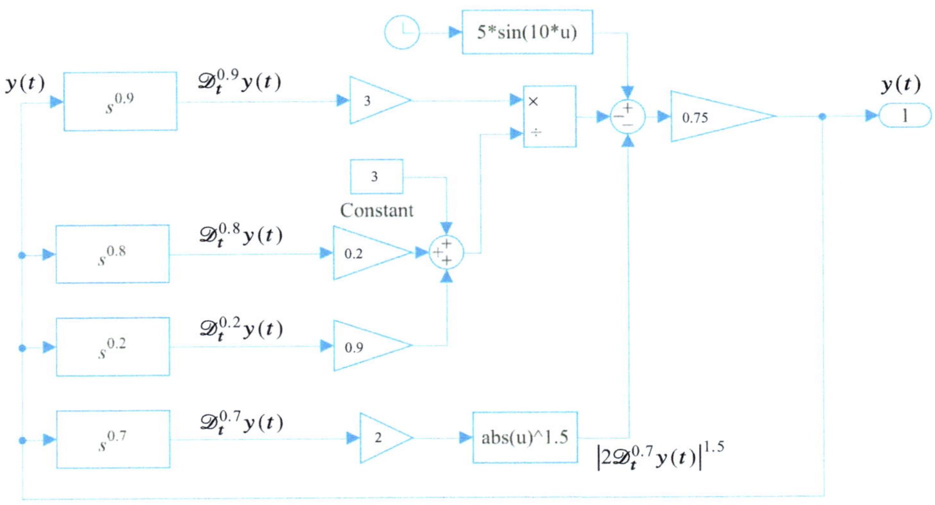

Fig. 8.13 Simulink model of the nonlinear FODE (`c8mnlf1.slx`)

Start simulation analysis to the system, the simulation results can be obtained, as shown in Fig. 8.14. The simulation results can be validated by altering filter parameters and Simulink model parameters.

```
>> [t,~,y]=sim('c8mnlf1',4); plot(t,y)
```

It should be noted that the Simulink modeling methods are not unique, even for the same FODEs. Consider the original FODE model, if it is written in another form

$$\mathscr{D}_t^{0.9}y(t)=\frac{3+0.2\mathscr{D}_t^{0.8}y(t)+0.9\mathscr{D}_t^{0.2}y(t)}{3}\left[5\sin 10t-\frac{4}{3}y(t)-\left|2\mathscr{D}_t^{0.7}y(t)\right|^{1.5}\right],$$

there will no longer by any dynamic terms in the denominator. An alternative Simulink model can be established as shown in Fig. 8.15, which is completely different from the one obtained earlier. The simulation results from the new model are exactly the same as the ones from the other model.

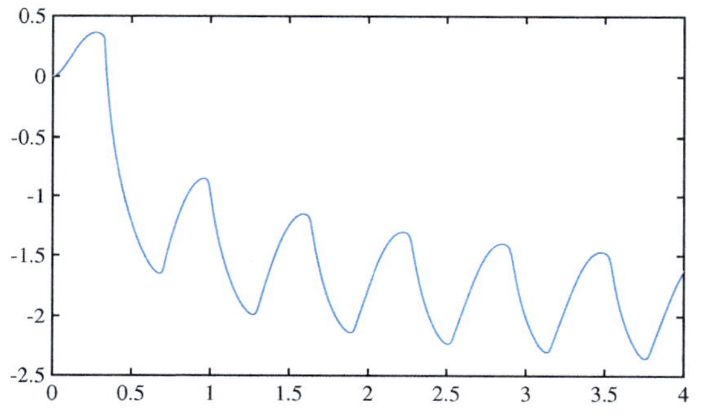

Fig. 8.14 System simulation results

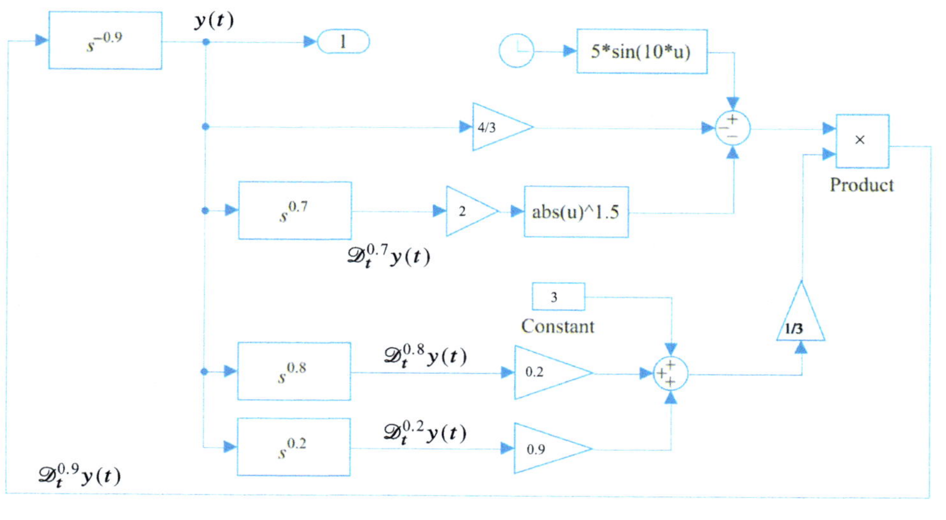

Fig. 8.15 A different Simulink model (c8mnlf2.slx)

For FODEs with complicated operations, there is a concise modeling method. The key signals can be packed into a vectorized signal. For instance, let $\boldsymbol{v} = [\mathscr{D}_t^{0.9}y(t), \mathscr{D}_t^{0.8}y(t), \mathscr{D}_t^{0.2}y(t), \mathscr{D}_t^{0.7}y(t), u(t)]$, then feed the vector into Fcn block to implement the complicated operations. The concise model thus established is as shown in Fig. 8.16. Start the simulation process, the same simulation results can be found. ◄

8.2.4 Validations of Numerical Solutions for Nonlinear FODEs

The validations discussed so far are based on the selection of different Oustaloup filter parameters and other parameters. In fact, the idea in ▶ Sect. 6.4.3 can also be used, such that the numerical solutions can be substituted back into the original equation to validate them. It seems that there are two difficulties doing these. (1) By default, variable-step algorithms are used in Simulink, while the functions such as `glfdiff()` are fixed-step ones. How can we find fixed-step simulation results? (2) The validations of nonlinear FODEs are not as simple as in their linear counterparts, as shown in ▶ Sect. 6.4.3. The two problems are explored in this section, such that a feasible validation approach is presented.

Although variable-step mechanism is used in Simulink, equally spaced time vector can be selected to return simulation results. A simple method is introduced here.

In Example 8.8, the command `[t,~,y]= sim('c8mnlf1',4)` was used, such that variable-step simulation results are returned. If a time vector $\boldsymbol{t}_0$ is assigned, the command `[t,~,y]=sim('c8mnlf1',t0)` can be used to compute simulation results on the $\boldsymbol{t}_0$ points. The returned $\boldsymbol{t}$ is identical to $\boldsymbol{t}_0$ vector. Therefore, a fixed-step vector $\boldsymbol{t}_0$ can be selected, and the equally spaced simulation results can be found. Note that, the default variable-step mechanism is still used in simulation, while the simulation results on the points in vector $\boldsymbol{t}_0$ are returned.

It is relatively simple to answer the second question. Dot operations can be used to evaluate expressions of any complexity. An example is used to demonstrate numerical solution validation methods for FODEs.

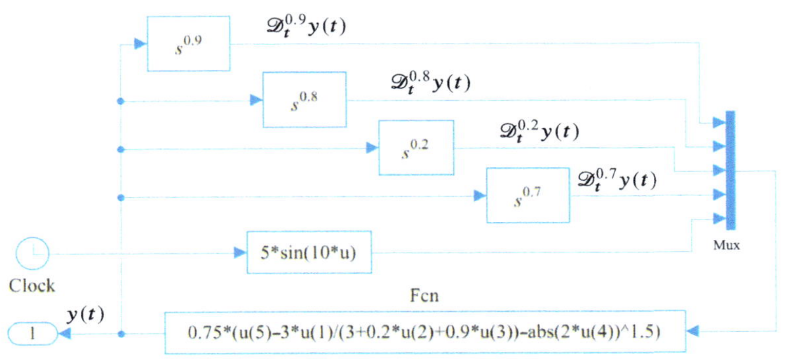

Fig. 8.16 A concise model for the nonlinear FODE (`c8mnlf3.slx`)

► **Example 8.9**

Consider again the nonlinear FODE in Example 8.8, solve it and validate results.

Solutions Select a fixed-step size $h = 0.002$, the following commands can be used to find the simulation results. It can be seen from the mathematical form of the FODE that 4 terms of Riemann–Liouville derivatives are used, with orders 0.9, 0.8, 0.2, and 0.7, respectively. Function `glfdiff()` can be called to evaluate the four fractional-order derivative signals. Then, dot operations are employed to compute the signals on the left- and right-hand sides. The curves on the two sides are shown in Fig. 8.17. It can be seen that apart from certain points where minor differences are visible, the two curves are the same. It means that the solutions obtained are correct.

```
>> t0=0:0.002:4; [t,~,y]=sim('c8mnlf1',t0);
   y1=glfdiff(y,t,0.9); y2=glfdiff(y,t,0.8);
   y3=glfdiff(y,t,0.2); y4=glfdiff(y,t,0.7);
   yy=3*y1./(3+0.2*y2+0.9*y3)+abs(2*y4).^1.5+4*y/3;
   uu=5*sin(10*t); max(abs(yy-uu)), plot(t,yy,t,uu,'--')
```

◄

8

8.3 Block Diagram-Based Solutions of Caputo Equations with Nonzero Initial Conditions

For Caputo equations, the output signal $y(t)$ or the input signal $u(t)$ may contain nonzero initial values. The modeling pattern in ► Sect. 8.2 cannot be used to establish the simulation models. Special methods must be introduced to set up the Simulink models. In this section, a new method is introduced to generate Caputo derivatives with Riemann–Liouville block. Based on this method, the modeling and simulation methods are proposed for nonlinear FODEs [1]. Theoretically speaking, the Simulink modeling method introduced in this section can be used to build models for Caputo equations of any complexity.

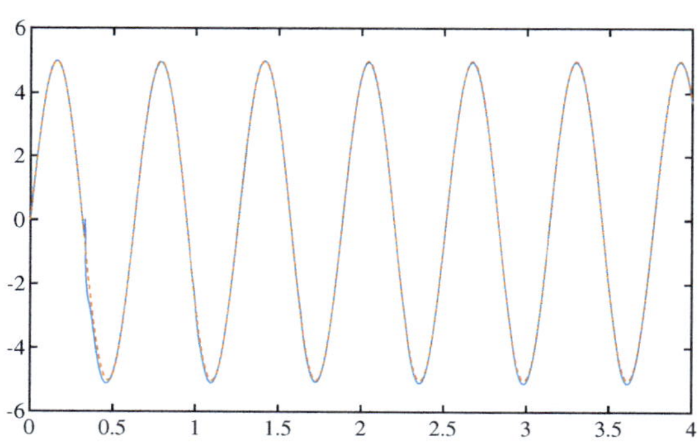

Fig. 8.17 Solution validations

8.3.1 Modeling and Solutions of Explicit Caputo Equations

Let us review the mathematical form of Caputo explicit FODE in ▸ Chap. 7. The mathematical form is given again next.

Definition 8.4 Explicit Caputo equation

The mathematical form of nonlinear explicit FODE is

$$ {}_{t_0}^{\rm C}\mathscr{D}_t^{\alpha} y(t) = f\left(t, y(t), {}_{t_0}^{\rm C}\mathscr{D}_t^{\alpha_1} y(t), {}_{t_0}^{\rm C}\mathscr{D}_t^{\alpha_2} y(t), \cdots, {}_{t_0}^{\rm C}\mathscr{D}_t^{\alpha_n} y(t)\right), \tag{8.4} $$

where $\alpha > \max(\alpha_1, \alpha_2, \cdots, \alpha_n) > 0$. The initial values of $y(t)$ and its integer-order derivatives are

$$ y(t_0) = y_0,\ y'(t_0) = y_1,\ y''(t_0) = y_2, \cdots,\ y^{(q-1)}(t_0) = y_q, \tag{8.5} $$

where $q = \lceil \alpha \rceil$. For simplicity, it is safe to assume that $\alpha_1 > \alpha_2 > \cdots > \alpha_n$.

As in the case discussed earlier, a chain composed of q integer-order integrators is composed, as shown in Fig. 8.18. The signal $y(t)$ and its integer-order derivatives are defined in this way. With the integrator chain, the initial values in (8.5) can be assigned accordingly into the integrator blocks.

With the integer-order derivative signals, it is natural to expect to compose Caputo derivative signals of signal $y(t)$. Consider the relationship between the Caputo and Riemann–Liouville derivatives, a compensating function can be introduced to define Caputo derivatives. Unfortunately, this method is not regarded as a good solution to the modeling problem [2]. Better solution methods are expected to solve the problems.

Review Theorem 3.18. For simplicity, the main conclusion of the theorem is summarized as follows: If $n = \lceil \gamma \rceil$, and $y(t)$ signal is nth-order differentiable, then

$$ {}_{t_0}^{\rm C}\mathscr{D}_t^{\gamma} y(t) = {}_{t_0}^{\rm RL}\mathscr{D}_t^{-(n-\gamma)} \left[y^{(n)}(t) \right]. \tag{8.6} $$

It can be seen from the theorem that Caputo derivative signal ${}_{t_0}^{\rm C}\mathscr{D}_t^{\gamma} y(t)$ can be generated by feeding integer-order derivative signal $y^{(n)}(t)$ into an $(n-\gamma)$th-order Riemann–Liouville integrator. For instance, in order to find the 2.6th-order Caputo derivative of signal $y(t)$, signal $y'''(t)$ can be fed into a 0.4th-order Oustaloup integrator. The $y'''(t)$ signal can be extracted from the integrator chain, and the 0.4th-order Oustaloup integrator can be built up from the Riemann–Liouville block.

Apart from Theorem 3.18, the following theorem is also very important in Caputo equation modeling. Based on this theorem, the simulation loop can be closed.

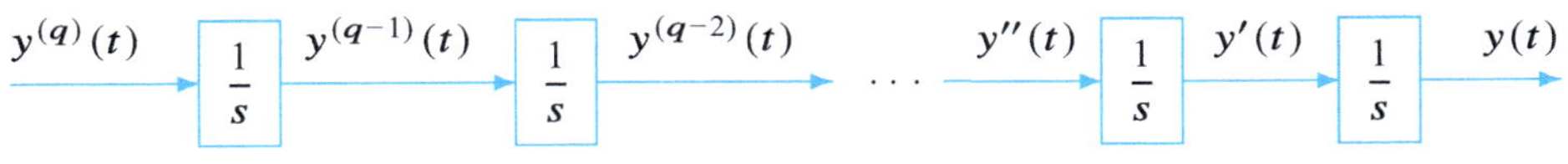

Fig. 8.18 Integer-order integrator chain

Theorem 8.1 Modeling theorem (II)

If $n = \lceil \gamma \rceil$, take $(n-\gamma)$th-order Riemann–Liouville derivatives to both sides of the formula in Theorem 3.18, then

$$ {}_{t_0}^{\mathrm{RL}}\mathscr{D}_t^{n-\gamma}\left[{}_{t_0}^{\mathrm{C}}\mathscr{D}_t^{\gamma}y(t)\right] = y^{(n)}(t). \tag{8.7}$$

It is seen from the theorem that, the nth-order derivative of signal $y(t)$ can be composed by feeding the Caputo derivative signal ${}_{t_0}^{\mathrm{C}}\mathscr{D}_t^{\gamma}y(t)$ into an $(n-\gamma)$th-order Riemann–Liouville differentiator. Again the Riemann–Liouville block can be used. For instance, if the integrator chain of a system is composed of three integrators, and the 2.6th-order Caputo derivative signal of $y(t)$ is already found, it can be fed into a 0.4th-order Riemann–Liouville block, such that signal $y'''(t)$ can be established. Therefore, such a signal can be connected directly to the input port of the left-most integrator block, such that the closed path in the Simulink model can be completed.

Since the fractional-order derivative actions mentioned here are all made through the Riemann–Liouville operator block, the block Riemann–Liouville is sufficient. Based on the above theorems, the following simple modeling algorithm is proposed.

8

Algorithm 8.1 Simple modeling algorithm

(1) The key fractional-order derivative signal ${}_{t_0}^{\mathrm{C}}\mathscr{D}_t^{\gamma}y(t)$ can be composed by feeding integer-order derivatives $y^{(n)}(t)$ in a Riemann–Liouville block, with the order $(\gamma-n)$. For instance, if 2.3th-order Caputo derivative is needed, the key signal $y^{(3)}(t)$ can be fed into a 0.7th Riemann–Liouville integrator block.

(2) For explicit FODEs, the Caputo derivatives can be fed into Riemann–Liouville blocks to restore integer-order derivatives, so as to complete the simulation closed path.

(3) If all these key signals are defined, it is easy to set up the Simulink model for the whole system.

The modeling method proposed here combined the integrator chain and Oustaloup filters to define Caputo derivatives. The modeling procedures are simple and straightforward. Examples are given next to demonstrate the modeling method.

▶ Example 8.10

Consider the benchmark problem in Example 4.9. Create a Simulink model for the FODE and assess the impact of Oustaloup filter parameters on the simulation results.

Solutions Since the highest order in the FODE is 0.7, an integer-order integrator with initial value 1 can be used to define the key signals $y(t)$ and $y'(t)$. A 0.3th-order Oustaloup differentiator is connected to $y'(t)$, such that signal ${}_{0}^{\mathrm{C}}\mathscr{D}_t^{0.7}y(t)$ is composed. The Clock block generates time t, and feeds it into a Fcn block, which describes the piecewise function $f(t)$. This signal is connected with the ${}_{0}^{\mathrm{C}}\mathscr{D}_t^{0.7}y(t)$ signal, such that the simulation model in Fig. 8.19 is established.

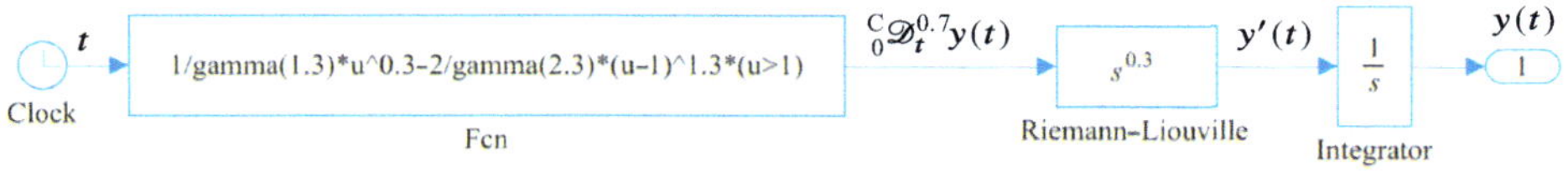

Fig. 8.19 Simulink model for benchmark problem 1 (`bp1_model.slx`)

Table 8.2 Impact of the filter parameters on the simulation results

Frequency bande	$(10^{-4}, 10^4)$	$(10^{-5}, 10^5)$	$(10^{-6}, 10^6)$	$(10^{-7}, 10^6)$	$(10^{-8}, 10^6)$
Order	21	25	29	35	41
Maximum error	7.6416×10^{-5}	6.0872×10^{-6}	3.3824×10^{-6}	1.2100×10^{-6}	4.7819×10^{-7}
Time elapse	0.579	0.589	0.583	0.604	0.562
Number of points	984	1057	1135	1141	1138

Selecting Oustaloup filter parameters, and initiate the simulation process, the simulation results can be obtained. Compared with the exact solutions, the information such as maximum error, time elapse and number of points can be measured. Select different filter parameters, the measured data can be obtained as shown in Table 8.2. It can be seen that the impact of the filter parameters on the time elapse and number of points are not significant. Normally speaking, the wider the interested frequency band, especially the smaller the value of ω_b, and the higher the order, the more accurate solutions can be found. It can be seen from the precision and speed that the method here is much better than the ones in Example 4.9.

```
>> N=11; ww=[1e-4 1e4];
   tic, [t,x,y]=sim('bp1_model'); toc, length(t)
   y0=t+1-(t-1).^2.*(t>1); err=y-y0; max(abs(err))
```

◀

▶ Example 8.11

Solve again the linear Caputo equation in Example 6.13 with Simulink.

$$y'''(t) + {}_0^C\mathscr{D}_t^{2.5}y(t) + y''(t) + 4y'(t) + {}_0^C\mathscr{D}_t^{0.5}y(t) + 4y(t) = 3\sqrt{2}\sin(t + 3\pi/4).$$

Assume that the initial values are $y(0) = 1$, $y'(0) = 0$ and $y''(0) = -1$, and the exact solution is $y(t) = \cos t$. The simulation interval is $t \in (0, 5000)$.

Solutions For the problems with such a large simulation interval, the fixed-step methods in Example 6.13 cannot be used at all. Variable-step methods such as Simulink modeling method can be used instead. The original FODE can be manually converted into the following explicit form:

$$y'''(t) = -{}_0^C\mathscr{D}_t^{2.5}y(t) - y''(t) - 4y'(t) - {}_0^C\mathscr{D}_t^{0.5}y(t) - 4y(t) + 3\sqrt{2}\sin(t + 3\pi/4).$$

With conventional modeling methods, three integrators can be introduced to build up the integrator chain, to define the key signals $y(t)$, $y'(t)$, $y''(t)$ and $y'''(t)$. With these four signals, Riemann–Liouville integrator blocks can be used to define the Caputo derivatives, such that the block diagram in Fig. 8.20 can be established. Note that, the key signal ${}_0^{\mathrm{C}}\mathscr{D}_t^{0.5}y(t)$ and ${}_0^{\mathrm{C}}\mathscr{D}_t^{2.5}y(t)$ are built upon Theorem 3.18. The modeling of the key signals is slightly different from the ones in Example 8.5.

Since the orders of the numerator and denominator in the Oustaloup integrator are the same, algebraic loops may be introduced in defining the ${}_0^{\mathrm{C}}\mathscr{D}_t^{0.5}y(t)$ signal, which means that within each ODE solution step, an algebraic equation must be solved once. Therefore, the solution process is very time consuming. If the Oustaloup filter parameters are set with the following commands, compared with the exact solution, it is found that the maximum error in the solution is 2.6990×10^{-11}, and the time elapse is 25.903 s.

```
>> N=15; ww=[1e-4 1e4];
   tic, [t,x,y]=sim('bp2_model',50); max(abs(y-cos(t))), toc
```

Double click the block who generates the ${}_0^{\mathrm{C}}\mathscr{D}_t^{0.5}y(t)$ signal, a parameter dialog box is opened, as shown in Fig. 8.8, from which there is an Is low-pass filter required? checkbox. If it is checked, a low-pass filter is appended automatically to the block, such that the total order of the denominator is higher than that of the numerator. In this case, the algebraic loop in the system is eliminated. Running the same commands, it is found that the simulation time is shortened to only 0.369 s, with maximum error of 3.6650×10^{-11}. It can be seen that if there exist algebraic loops, it may be eliminated by adding low-pass filters, such that the efficiency may be significantly raised. ◀

8

▶ Example 8.12

Solve again the nonlinear Caputo equations in Example 7.1 with the block diagram-based method. Compare the results with the methods in ▶ Chap. 7.

Solutions Since $q = \lceil 1.545 \rceil = 2$, two integer-order integrators can be used to make the integrator chain, and compose the key signals $y(t)$, $y'(t)$ and $y''(t)$. Assign accordingly the initial values $y(0) = 1$ and $y'(0) = -1$ to the integrators. The Simulink model in Fig. 8.21 can be constructed. In this model, signal ${}_0^{\mathrm{C}}\mathscr{D}_t^{0.555}y(t)$ is created by feeding signal $y'(t)$ into a Riemann–Liouville integrator block, such

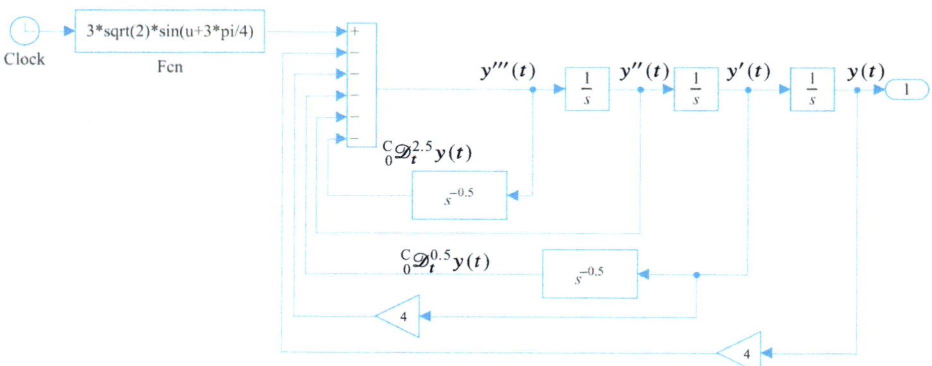

Fig. 8.20 Simulink model of the Caputo equation (bp2_model.slx)

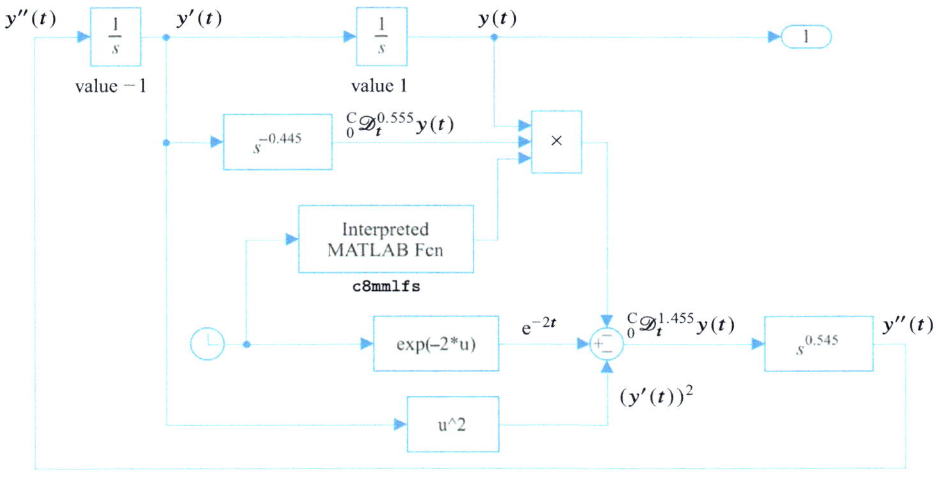

Fig. 8.21 AS new Simulink model (c8mexp2s.slx)

that the key ${}_0^{\mathrm{C}}\mathscr{D}_t^{0.555}y(t)$ signal is defined. This signal is composed by the use of Theorem 3.18.

How to make sure that the signal from the sum block is the expected ${}_0^{\mathrm{C}}\mathscr{D}_t^{1.455}y(t)$ signal? According to Theorem 8.1, further derivative actions can be made such that $y''(t)$ signal is generated. Then the signal can be linked to the $y''(t)$ at the input port of the left-most integrator, such that the simulation loop is completed. In this way, the output of the sum block is the expected ${}_0^{\mathrm{C}}\mathscr{D}_t^{1.455}y(t)$ signal.

To better describe the time domain function $t^{0.1}\,\mathrm{E}_{1,1.545}(-t)\,\mathrm{e}^t/\mathrm{E}_{1,1.445}(-t)$ in the first term on the right-hand side of the equation, an Interpreted MATLAB Fcn block is used, and the following MATLAB function should be written for the block.

```
function y=c8mmlfs(t)
   y=t^0.1*ml_func([1,1.545],-t)*exp(t)/ml_func([1,1.445],-t);
end
```

The file is saved to `c8mmlfs.m` and should be associated with the Interpreted MATLAB Fcn block. The method is double click the block, and fill in the file name `c8mmlfs` into the parameter dialog box, to set up the relationship.

The following parameters should be selected for the Oustaloup filter blocks. The simulation results can be obtained. Compared with the exact solution e^{-t}, the maximum error is 1.1451×10^{-4}, and time elapse is 0.4468 s.

```
>> N=11; ww=[1e-3 1e3];
   tic, [t,x,y]=sim('c8mexp2s'); toc, max(abs(y-exp(-t)))
```

Different frequency bands and filter orders can be tested for the filters. Some tests are made, and the measured results are shown in Table 8.3. It can be seen that large error comes mainly because t is large. Therefore, the value of ω_b usually determines the size of the maximum error. If ω_b is reduced, it is likely that the error is reduced. The value of ω_h affects the selection of minimum step size within Simulink mechanism. If the value is large, it may in turn increase the computational load. The higher the order of the filter, the

Table 8.3 Solution efficiencies under different parameters

Frequency bands	$(10^{-4}, 10^4)$	$(10^{-5}, 10^5)$	$(10^{-6}, 10^5)$	$(10^{-7}, 10^5)$	$(10^{-8}, 10^6)$
Order	21	23	25	27	31
Time elapse	0.4943	0.4194	0.5161	0.5153	0.6501
Number of points	1623	1762	1761	1759	1897
Maximum error	5.7446×10^{-5}	5.9895×10^{-6}	6.0452×10^{-6}	5.8927×10^{-6}	7.5992×10^{-7}

more accurate solutions to be found. The side effect is that the computational load is also increased. For this example, the impact of the latter two parameters is not very significant. Some extreme parameters are also tried: When the frequency band is $(10^{-9}, 10^7)$ rad/s, with order 39, the time elapse is 0.6456 s, with maximum error of 1.0321×10^{-7}, and the number of points is increased to 2033. Further modifying the filter parameters, no better performances are witnessed. Sometimes the simulation results even get worse.

Under the extreme parameters, if the termination time is increased to $t_{\rm n} = 2$ and $t_{\rm n} = 3$, the Simulink model can still be used in solving the FODE, with maximum error of 6.1451×10^{-7} and 3.4715×10^{-6}, respectively, and the time elapse is all around 0.9 s. Further increase the termination time, the simulation results may diverge.

8

```
>> N=39; ww=[1e-9 1e7];
   tic, [t,x,y]=sim('c8mexp2s',2); toc, max(abs(y-exp(-t)))
   tic, [t,x,y]=sim('c8mexp2s',3); toc, max(abs(y-exp(-t)))
```

◄

For this specific example, although the precision is much lower than the highest standard in ► Sect. 7.3, the speed of computation has significant advantages than the command-driven methods. Especially in the cases where the repeated execution of the solution process, the speed factor is more important.

8.3.2 Simulink Modeling of FOSS Models

In the previously discussed method, integrator chain is used to define the key signals, such that the whole simulation model can be established upon these key signals. For FOSS models, each state corresponds to an integrator block, such that the states as well as their first-order derivatives can be defined. Then, Theorem 3.18 can be used to define the Caputo derivatives of the state variables, such that the whole simulation model for FOSS expressions can be established. In this section, Simulink modeling and solution of ordinary FOSS expressions are discussed.

▶ Example 8.13

Consider the fractional-order Chua equation in Example 7.9. Create a Simulink model and find its solutions.

Solutions Three integer-order integrators can be introduced to define the three pairs of key signals, i.e., $x(t)$, $y(t)$, $z(t)$ and their first-order derivatives. From the first-order derivative signals, Theorem 3.18 can be used to define fractional-order derivatives such as ${}_0^{\rm C}\mathscr{D}_t^{\alpha_1}x(t)$. With all these signals, the Simulink model in Fig. 8.22 can be established directly. Therefore, the following commands can be issued to find the simulation results.

```
>> alpha=[0.93,0.99,0.92]; x0=[0.2; -0.1; 0.1];
   a=10.725; b=10.593; c=0.268; m0=-1.1726; m1=-0.7872;
   ww=[1e-4 1e3]; N=13; tic, [t,x,y]=sim('c8mchuasim',500); toc
   plot(y(:,1),y(:,2))
```

After 1.916 s of execution, the phase plane trajectory of the system can be obtained, and it is very close to the one obtained in Fig. 7.6. Therefore, the efficiency of such a method is much higher than those of the command-driven methods.
If a vectorized integrator block is used, the more concise model shown in Fig. 8.23 can be created. In the model, an Interpreted MATLAB Fcn block is written with the following MATLAB function file.

```
function Y=c8mchaosd(u)
   a=10.725; b=10.593; c=0.268; m0=-1.1726; m1=-0.7872;
   f=m1*u(1)+1/2*(m0-m1)*(abs(u(1)+1)-abs(u(1)-1));
   Y=[a*(u(2)-u(1)-f); u(1)-u(2)+u(3); -b*u(2)-c*u(3)];
end
```

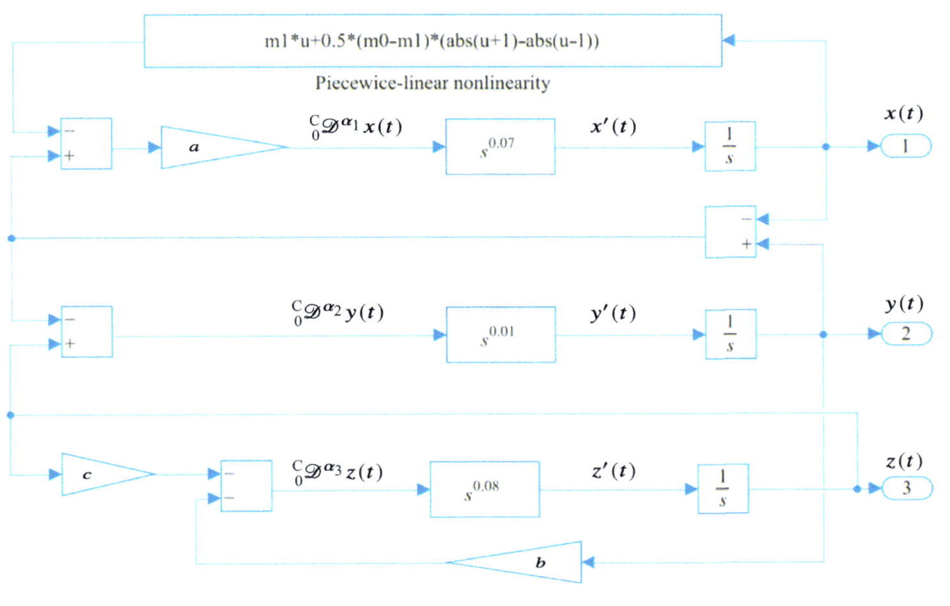

Fig. 8.22 Simulink model of Chua equations (`c8mchuasim.slx`)

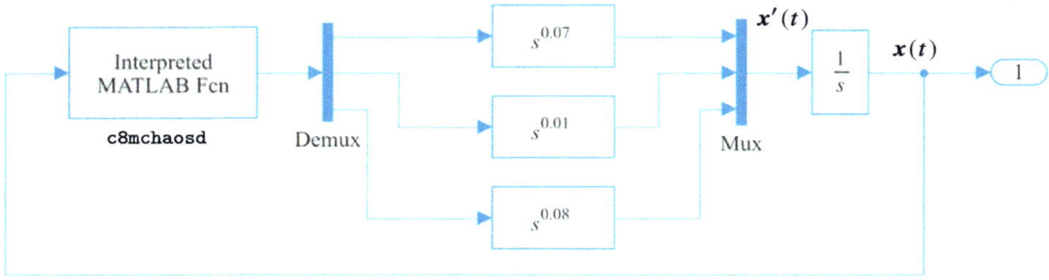

Fig. 8.23 Another model of Chua equations (c8mchaos.slx)

The following commands can be executed to simulate the system, and the results obtained are identical. The time elapse is increased to 5.705 s. Since the MATLAB function block is used, although the modeling looks much simpler, the speed is affected.

```
>> alpha=[0.93,0.99,0.92]; x0=[0.2; -0.1; 0.1];
   ww=[1e-4 1e3]; N=13; tic, [t,x,y]=sim('c8mchaos',500); toc
   plot(y(:,1),y(:,2))
```


It should be noted that in Reference [10], Simulink is used to implement Chua equations. The modeling scheme there exists problems, since the Oustaloup filter blocks and the integrator blocks are swapped, such that the theoretical foundation of the models in the references is not satisfied. ◂

In fact, the modeling method presented above is still rather complicated, since the vectorized modeling facilities in Simulink are not fully employed. Therefore, a Caputo vector block is provided in the FOTF Blockset. In the block, vectorized Caputo operator is established. Based on Theorem 8.1, the model is internally composed of Riemann–Liouville integrators, followed by a vectorized integrator block, such that vectorized modeling facilities are adopted, for the vectorized Caputo operator. It is especially suitable for modeling of FOSS expressions.

▸ Example 8.14

Compose a vectorized Chua equation model and compare the modeling difficulties and computational loads.

Solutions Vectorized method can be used to establish the model in Fig. 8.24 (a), where the internal structure of the Caputo vector block is shown in Fig. 8.24 (b). The internal structure of the block is established automatically by the corresponding MATLAB commands in the initialization pane of the masked block. In the model, $G(i)$ is the Oustaloup filter designed for the ith scalar input signal.

Double click the Caputo vector block icon, the parameter dialog box shown in Fig. 8.25 is opened, where dimension M is the number of channels in the vectorized input signal. Since there are three states, its value should be 3. The fractional orders in the vector should also be specified in the dialog box. The initial state vector as well as filter parameters should be provided in the dialog box.

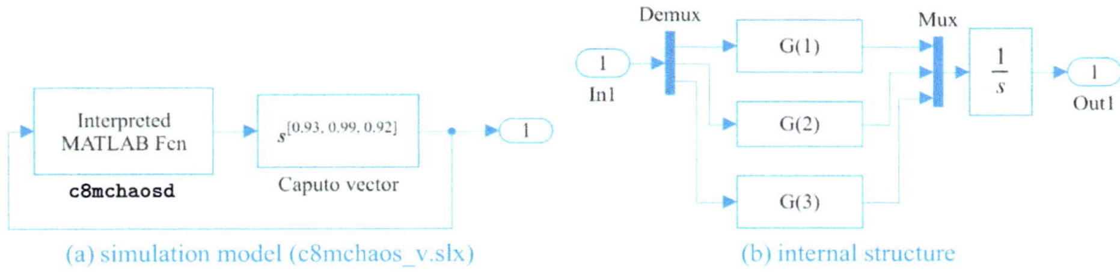

Fig. 8.24 Vectorized model of the Chua equations

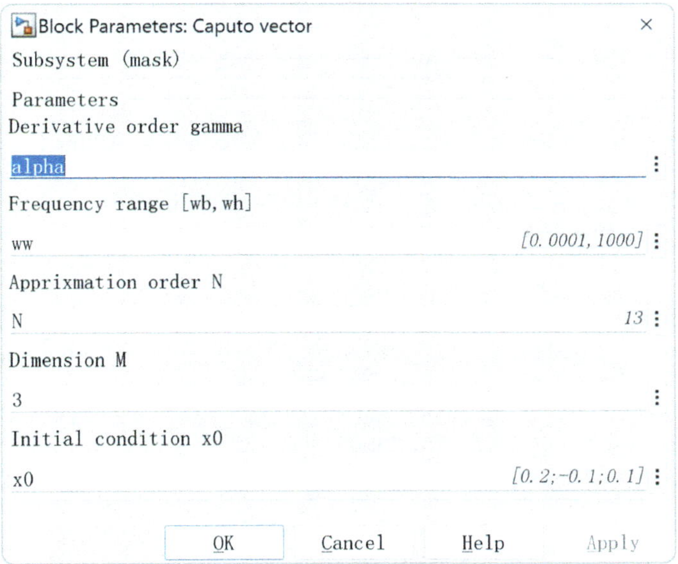

Fig. 8.25 Parameter dialog box of the Caputo vector block

The following commands can be used to specify Oustaloup filter parameters and initiate the simulation process. The simulation result obtained is the same, and the time elapse is 6.39 s, more than the other two models.

```
>> alpha=[0.93,0.99,0.92]; x0=[0.2; -0.1; 0.1];
   ww=[1e-4 1e3]; N=13; tic, [t,x,y]=sim('c8mchaos_v',500); toc
   plot(y(:,1),y(:,2))
```

◀

In the real modeling process, if conveniences are expected, the vectorized fractional-order operator block is recommended. The time elapse in simulation may also increase.

▶ Example 8.15

In the chaotic behavior analysis, most of the work assumed that the orders are very close to 1. If the three orders in Example 7.9 are all set to 0.8, find the simulation results in Chua equations.

Solutions If the value of α is set to 0.8 (note that, if a scalar is used, all the orders in the vector block are set to the same value), the following commands can be used to simulate the system. The phase plane trajectory is shown in Fig. 8.26, while the time elapse is 1.02s. For this specific example, if the orders are away from 1's, the chaotic phenomenon disappears.

```
>> alpha=0.8; x0=[0.2; -0.1; 0.1];
   ww=[1e-4 1e3]; N=13; tic, [t,x,y]=sim('c8mchaos_v',500); toc
   plot(y(:,1),y(:,2))
```

◀

Note that there are limitations in the Caputo vector block. All the fractional orders must satisfy $0 < \alpha_i \leq 1$. If the order exceeds this range, wrong results can be found. Therefore, if an order satisfies $\alpha_i > 1$, the following modeling strategy should be considered. This problem will be fully discussed in ▶ Sect. 8.3.3.

8

▶ Example 8.16

If the order is changed to $\alpha_3 = 1.08$ in Example 7.9, and $z'(0) = 0.1$, while other parameters are unchanged, solve the fractional-order Chua equations again.

Solutions It is obvious that the command-driven algorithms discussed in ▶ Chap. 7 cannot be used to solve this problem. The Caputo vector block cannot be used directly for this problem either. Observing the model `c8mchaos.slx`, a new integrator should be added to the $z(t)$ signal branch, such that $z''(t)$ is defined. The Simulink model in Fig. 8.27 is created. With the facilities in Simulink, the phase plane trajectory of $x(t)$ and $y(t)$ can be drawn, as shown in Fig. 8.28. Since the scales in the trajectory are too large, the trajectory is divergent, which implies that the system is unstable. ◀

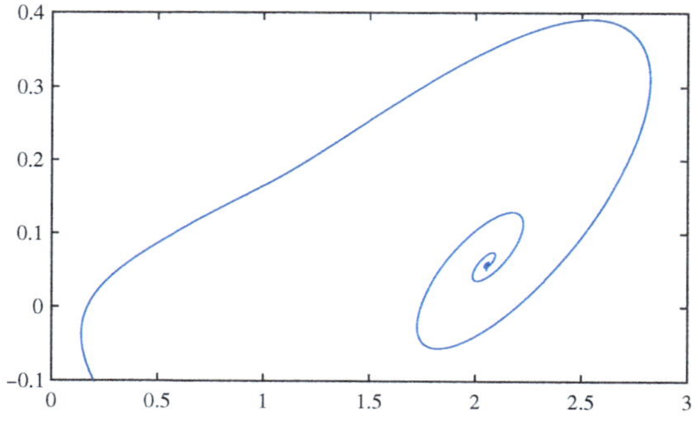

Fig. 8.26 0.8th-order response

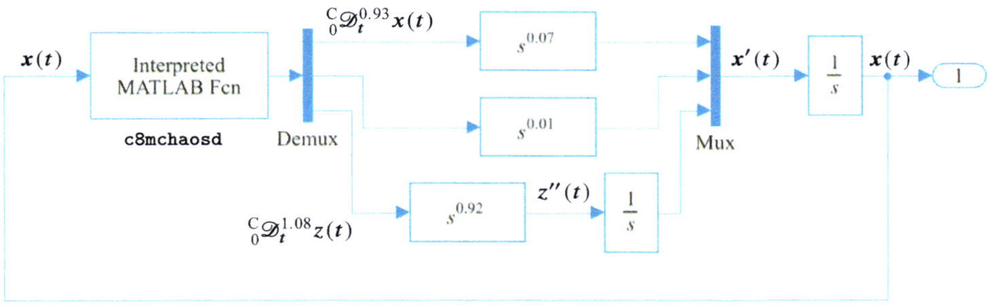

Fig. 8.27 Simulink model when $\alpha_3 > 1$ (c8mchaos2.slx)

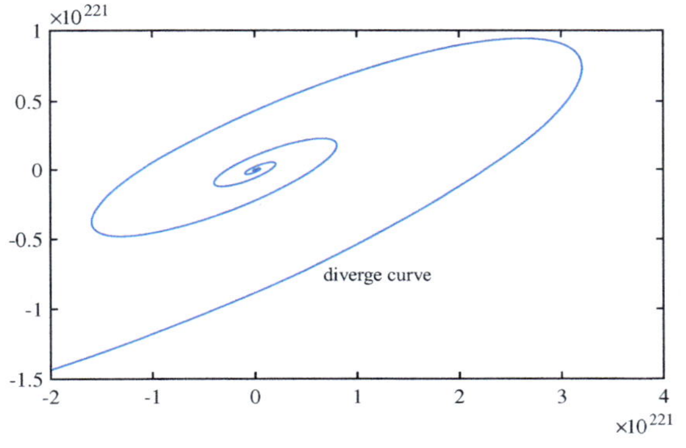

Fig. 8.28 Phase plane trajectory when $\alpha_3 > 1$

▶ Example 8.17

Since the exact solutions in the previous FODEs are not known, there is no way to assess the accuracies in the numerical solutions. The benchmark problem with known exact solutions is constructed [11].

$$\begin{cases} {}_0^{\mathrm{C}}\mathscr{D}_t^{0.5}x(t) = \dfrac{1}{\sqrt{\pi}}\left(\sqrt[5]{(y(t)-0.5)(z(t)-0.3)}+\sqrt{t}\right) \\ {}_0^{\mathrm{C}}\mathscr{D}_t^{0.2}y(t) = \Gamma(2.2)\big[x(t)-1\big] \\ {}_0^{\mathrm{C}}\mathscr{D}_t^{0.1}z(t) = \dfrac{\Gamma(2.3)}{\Gamma(2.2)}\big[y(t)-0.5\big], \end{cases}$$

where $x(0) = 1$, $y(0) = 0.5$ and $z(0) = 0.3$. The exact solutions are $x(t) = t + 1$, $y(t) = t^{1.2} + 0.5$ and $z(t) = t^{1.3} + 0.3$, and $0 \leqslant t \leqslant 10$. Solve the FOSS equations with simulation methods.

Solutions The **Caputo vector** block can be used to define the fractional-order derivatives on the left-hand side of the FODE. A Simulink model is then established as shown in Fig. 8.29, where the MATLAB function for the **Interpreted MATLAB Fcn** block is written as follows.

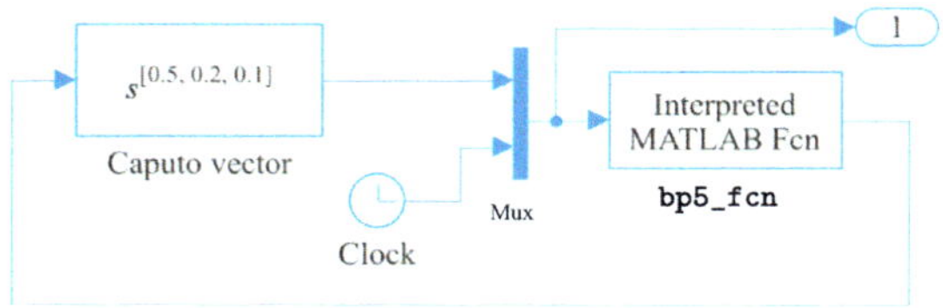

Fig. 8.29 Simulink model for the FOSS equations (bp5_model2.slx)

```
function y=bp5_fcn(u)
   y=[(((u(2)-0.5)*(u(3)-0.3))^(1/5)+sqrt(u(4)))/sqrt(pi);
      (u(1)-1)*gamma(2.2);
      (u(2)-0.5)*gamma(2.3)/gamma(2.2)];
   y=real(y);     % for unknown reasons, tiny imaginary part is introduced
end
```

The following commands can be used to solve numerically the FODEs. Compared with the exact solutions, it is found that the maximum error is 9.4733×10^{-5}. The time elapse is 36.38 s.

```
>> ww=[1e-6 1e5]; N=30;
   tic, [t,x,y]=sim('bp5_model2'); toc
   err=[y(:,1)-t-1, y(:,2)-t.^1.2-0.5, y(:,3)-t.^1.3-0.3];
   max(abs(err(:)))
```

If the simulation solver is set to ode15s, not the default ode45, and increase the frequency band to $(10^{-9}, 10^{7})$, with filter order $N = 59$, the time elapse is reduced to 1.181 s, and the maximum error is reduced to 9.4434×10^{-7}. ◀

▶ Example 8.18

Consider the benchmark problem in Example 4.9. Since the left-hand side of the equation is ${}_{0}^{\rm C}\mathscr{D}_{t}^{0.7}y(t)$, the Caputo vector block can be used to establish the simulation model, as shown in Fig. 8.30. The order vector in the block can be set to 0.7, while M is set to 1, and the initial value is set to 1, the problem can be solved again. The accuracies of the solution are similar to the ones obtained in Example 8.10, but the time elapse is almost doubled. Therefore, the efficiency is not high. ◀

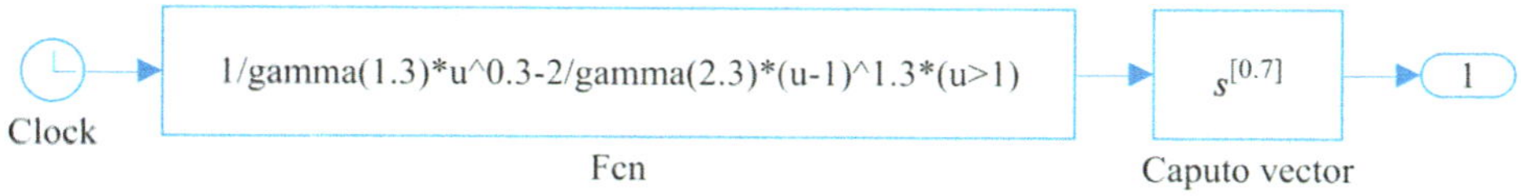

Fig. 8.30 Simulink model for the benchmark problem (bp1_model2.slx)

Remarks Simulink modeling of Caputo equations

(1) The Simulink modeling methods of Caputo equations are discussed in ▶ Sect. 8.3.1. They can be used directly in solving fractional-order Caputo equations, with satisfactory behaviors.

(2) If nonlinear FOSS equations are explored, the block Caputo vector can be adopted. It is not recommend to use such a block for scalar Caputo equations.

(3) Vectorized modeling method is used in the Caputo vector block. It is easy to use and the modeling is rather simple. The efficiency of such a vectorized block is not as efficient as the low-level models.

(4) The order in the block is restricted to $\alpha_i \in (0, 1]$. The cases where $\alpha_i > 1$ will be discussed later.

(5) The simulation results obtained should be validated. A specific suggestion is to alter Oustaloup filter or Simulink control parameters to cross-validate the results, until reliable simulation results are found. The validation methods in ▶ Sect. 8.2.4 can also be used. □

8.3.3 Handling of State Space Models with Order Higher Than 1

In certain problems, the case $\alpha_i > 1$ may be encountered in some FOSS equations. For instance, the modified FOSS model in Example 8.16. Difficulties may be encountered when Caputo vector block is used in vectorized modeling. Two solutions to this problem are explored here.

1. Low-Level Modeling Method

If a certain $\alpha_i > 1$, special treatment is needed, since the initial values of the first-order derivatives are also needed. Two consecutive integrators should be introduced to define the signals $x_i(t)$, $x_i'(t)$ and $x_i''(t)$, as in the case in Example 8.16. The Simulink model can then be established. The Simulink model is established upon Theorem 8.1, therefore, the results obtained are reliable. An example is provided next to illustrate the modeling method, and assess the impact of the filter parameters on the simulation efficiencies.

▶ Example 8.19

Modify slightly the benchmark problem in Example 8.17, the following FOSS equations can be constructed:

$$\begin{cases} {}_0^{\rm C}\mathscr{D}_t^{0.5}x(t) = \dfrac{2}{\sqrt{\pi}}\left[\sqrt[5]{(y(t)-0.3)(z(t)-0.5x(t))}+\sqrt{t}\right] \\ {}_0^{\rm C}\mathscr{D}_t^{0.1}y(t) = \dfrac{\Gamma(1.4)}{\Gamma(1.3)}\left(\dfrac{x(t)-1}{2}\right)^{0.3} \\ {}_0^{\rm C}\mathscr{D}_t^{1.5}z(t) = \dfrac{\Gamma(3.1)}{\Gamma(1.6)}\sqrt{(y(t)-0.3)^3}, \end{cases}$$

where $x(0) = 1$, $y(0) = 0.3$, $z(0) = 0.5$ and $z'(0) = 1$. The exact solutions are $x(t) = 2t+1$, $y(t) = t^{0.4}+0.3$ and $z(t) = t^{2.1}+t+0.5$. Find the numerical solutions with Simulink.

Solutions Since in the last equation, the 1.5th-order derivative of $z(t)$ is involved, while the order is larger than 1, an extra integrator should be introduced to define the $z''(t)$ signal. The property ${}_0^{\rm RL}\mathscr{D}_t^{0.5}\left[{}_0^{\rm C}\mathscr{D}_t^{1.5}z(t)\right] = z''(t)$ can be used, and similar to the modeling method in Example 8.16, the Simulink model in Fig. 8.31 can be established directly. The MATLAB function written for the Interpreted MATLAB Fcn block is listed as follows:

```
function y=bp5a_fcn(u)
   y=[2/sqrt(pi)*(((u(2)-0.3)*(u(3)-0.5*u(1)))^(1/5)+sqrt(u(4)));
      gamma(1.4)/gamma(1.3)*((u(1)-1)/2)^0.3;
      gamma(3.1)/gamma(1.6)*sqrt((u(2)-0.3)^3)];
   y=real(y);     %extract the real part
end
```

To solve the FOSS equations, the following commands can be issued, and the curves of the three variables $x(t)$, $y(t)$ and $z(t)$ are obtained, as shown in Fig. 8.32. The exact solutions are also superimposed on the plots. It can be seen that since the solutions are very accurate, the differences between the solutions and the exact ones cannot be distinguished from the plots.

8

```
>> ww=[1e-5, 1e5]; N=20;
   tic, [t,~,y]=sim('bp5a_model',10); toc
   y0=[2*t+1, t.^0.4+0.3, t.^2.1+t+0.5];
   max(max(abs(y-y0))), plot(t,y,t,y0,'--')
```

Different parameters for the Oustaloup filters are tested, and the measured data are provided in Table 8.4. Compared with other tables, the errors in this one is larger, since the magnitude of signal $z(t)$ is large. Although the signal magnitude at $t = 10$ is 136.3925, the maximum errors obtained here are still very large. ◄

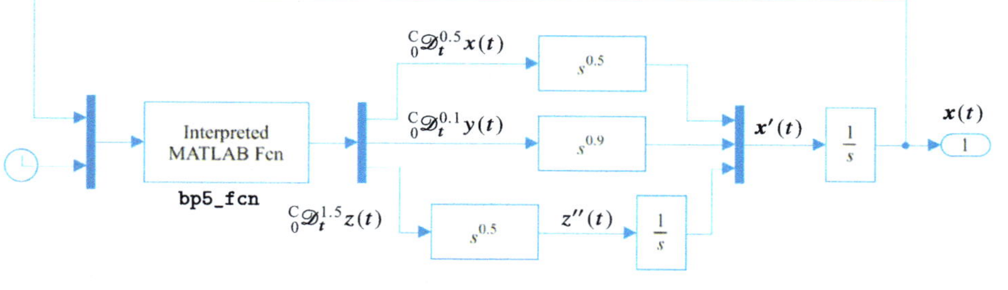

Fig. 8.31 Simulink model when the order is larger than 1 (bp5a_model.slx)

Table 8.4 The solution efficiencies under different filter parameters

Frequency bands	$(10^{-4}, 10^4)$	$(10^{-5}, 10^5)$	$(10^{-6}, 10^5)$	$(10^{-7}, 10^6)$	$(10^{-8}, 10^6)$	$(10^{-9}, 10^6)$
Order	21	23	25	27	31	43
Time elapse	0.880	0.895	0.432	0.565	0.625	0.678
Number of points	1079	1242	1242	1433	1477	1450
Maximum error	0.0859	0.0088	0.0020	0.0020	0.0020	0.0020

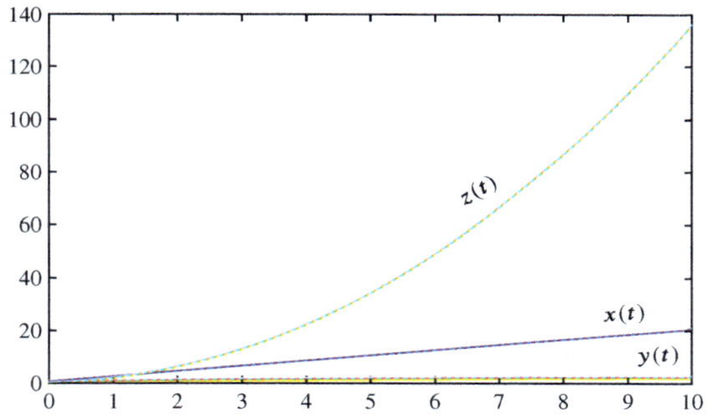

Fig. 8.32 Time responses of the state variables

For this specific example, if the frequency band is further enlarged, the precisions cannot be increased. Especially the value ω_h cannot be further increased, otherwise simulation process cannot be completed.

2. State Augmentation Method

It can be seen from the previous example that the low-level model procedures are still too complicated. An alternative modeling method can be introduced. The vectorized modeling method is revisited. It is quite natural to introduce augmented state variables, such that the orders in the state space models do not exceed 1. Therefore, the Caputo vector block can still be employed in vectorized modeling. An example is given next to demonstrate the modeling method.

▶ Example 8.20

Consider again the FOSS equations in Example 8.19. Build a Simulink model with vectorized modeling method.

Solutions Since in the original state space expression, the derivative order of $z(t)$ is larger than 1, it can be decomposed into two states, such that new state variables are assigned. For instance, let $v_1(t) = x(t)$, $v_2(t) = y(t)$, $v_3(t) = z(t)$ and $v_4(t) = z'(t)$. The new FOSS equations can be established.

$$\begin{cases} {}_0^{\mathrm{C}}\mathscr{D}_t^{0.5} v_1(t) = \dfrac{2}{\sqrt{\pi}}\left[\sqrt[5]{(v_2(t)-0.3)(v_3(t)-0.5v_1(t))} + \sqrt{t}\right] \\ {}_0^{\mathrm{C}}\mathscr{D}_t^{0.1} v_2(t) = \dfrac{\Gamma(1.4)}{\Gamma(1.3)}\left(\dfrac{v_1(t)-1}{2}\right)^{0.3} \\ {}_0^{\mathrm{C}}\mathscr{D}_t^{1} v_3(t) = v_4(t) \\ {}_0^{\mathrm{C}}\mathscr{D}_t^{0.5} v_4(t) = \dfrac{\Gamma(3.1)}{\Gamma(1.6)}\sqrt{(v_2(t)-0.3)^3}. \end{cases}$$

The initial values of the augmented states are $v_1(0) = 1$, $v_2(0) = 0.3$, $v_3(0) = 0.5$ and $v_4(0) = 1$. Therefore, with Caputo vector block, vectorized modeling method can be used to establish the Simulink model in Fig. 8.33. The following MATLAB function can be written:

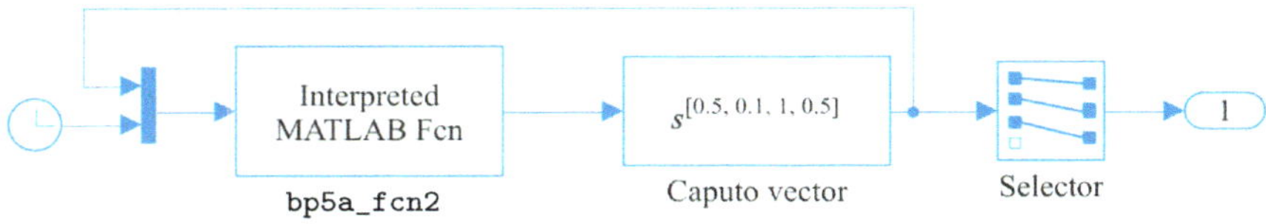

Fig. 8.33 Vectorized model (bp5a_model2.slx)

```
function y=bp5a_fcn2(u)
   y=[2/sqrt(pi)*(((u(2)-0.3)*(u(3)-0.5*u(1)))^(1/5)+sqrt(u(5)));
      gamma(1.4)/gamma(1.3)*((u(1)-1)/2)^0.3;
      u(4);
      gamma(3.1)/gamma(1.6)*sqrt((u(2)-0.3)^3)];
   y=real(y);    %extract the real part
end
```

It can be seen that the model constructed in this way is much simpler than the one in Example 8.19. The following commands can be used to initiate the simulation process, and the results obtained are exactly the same as the ones in Example 8.19. In this model, a Selector block is used to extract the first three signals from the state vector and connect them to the output port.

```
>> alpha=[0.5, 0.3, 1, 0.5]; x0=[1; 0.3; 0.5; 1];
   ww=[1e-5, 1e5]; N=20;
   tic, [t,~,y]=sim('bp5a_model2',10); toc
   y0=[2*t+1, t.^0.4+0.3, t.^2.1+t+0.5];
   max(max(abs(y-y0))), plot(t,y,t,y0,'--')
```

◄

8.4 Simulation of Fractional-Order Feedback Control Systems with Simulink

For linear fractional-order systems, FOTF or FOTF matrix are the most fundamental elements, An FOTF block is provided in the FOTF Blockset. Also, a fractional-order PID controller block is provided. With these blocks, Simulink models for linear fractional-order systems can be established easily. The time responses of the closed-loop systems can be easily obtained under Simulink environment.

8.4.1 Fractional-Order Transfer Function Block

An Approximate FOTF model block is provided in the FOTF Blockset, see Fig. 8.1, to describe FOTFs. The essentials of such a block are to build up a linear high integer-order transfer function with filters such as Oustaloup filter. It can be used to approximate the behaviors of FOTFs.

Double click the block, a parameter dialog box shown in Fig. 8.34 is opened. The FOTFs or FOTF matrices can be specified in the block. A high integer-order transfer function model can be generated automatically with filters such as Oustaloup ones, and it can be used in the modeling and simulation of fractional-order systems.

Two methods can be used to input FOTF models.

(1) In the first four edit boxes, the four vectors can be specified. Also the information such as the frequency band of interest and filter order can be specified. Such a block can then be used in a Simulink model.

(2) The FOTFs or FOTF matrices can be defined in MATLAB workspace first, in the ways demonstrated in Examples 8.1 and 8.2. The model names can then be filled into the first edit box of the parameter dialog box, such that the block can be used to represent the models.

A fractional-order PID controller block is also provided in FOTF Blockset, and the controller parameters can also be specified in the block, such that it can be used in Simulink modeling.

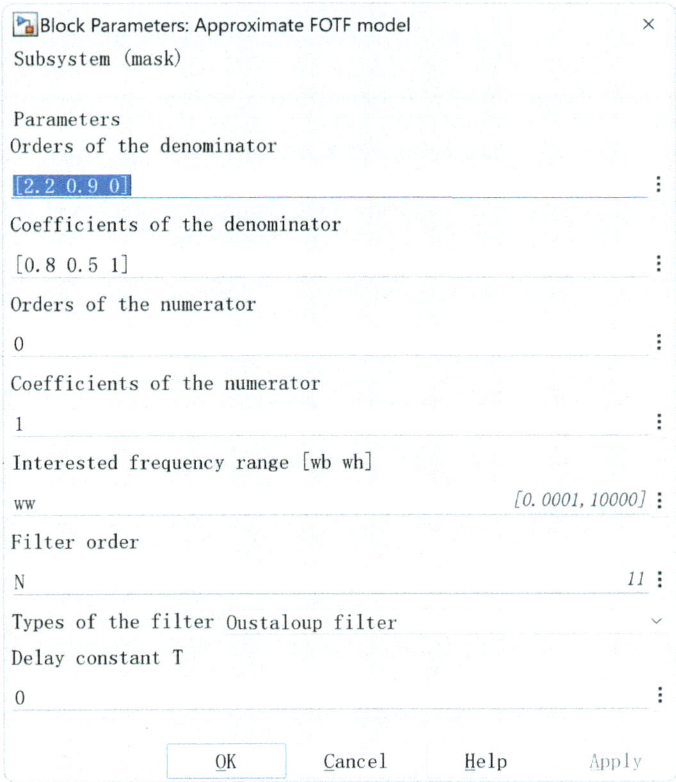

Fig. 8.34 Parameter dialog box of FOTF block

8.4.2 Fractional-Order PID Controllers and Closed-Loop System Simulation

With the two important blocks, examples are used in this section to demonstrate the modeling and simulation of fractional-order PID control systems.

► Example 8.21

If a FOTF plant with delay and PID controller in the typical negative feedback control system are known [2], draw the closed-loop step response.

$$G(s) = \frac{1}{0.8s^{2.2} + 0.5s^{0.9} + 1} e^{-s}, \quad G_c(s) = 0.45966 + \frac{0.5761}{s^{0.99627}} + 0.49337s^{1.3792}.$$

Solutions Before establishing the simulation model, the plant model without delay should be described first in MATLAB workspace.

```
>> s=fotf('s'); G=1/(0.8*s^2.2+0.5*s^0.9+1);
```


The Approximate FOTF model block can be copied into the Simulink model window, such that the simulation model in Fig. 8.35 can be established. The fractional-order PID controller parameters are also specified in the model, by double clicking the controller block. The controller parameters can be specified in the dialog box shown in Fig. 8.36. The parameters can be modified in the parameter dialog box.

The parameters of the fractional-order PID controller can also be entered into MATLAB workspace, with the command G_c =fopid(K_p,K_i,K_d,λ,μ). The controller's variable name G_c can be specified into the first edit box in the parameter dialog box in Fig. 8.36.

Initiate the simulation process, the closed-loop step response of the system can be obtained, as shown in Fig. 8.37. The response with numerical inverse Laplace transform is also obtained, and superimposed on the plot. It can be seen that the two curves are very close.

```
>> tic, [t,~,y]=sim('c8mfpid',20); toc
   G='exp(-s)/(0.8*s^2.2+0.5*s^0.9+1)';
   Gc='(0.45966+0.5761/s^0.99627+0.49337*s^1.3792)';
   [t1,y1]=INVLAP_new([G,'*',Gc],0,20,200,1,'1/s');
   plot(t,y,t1,y1,'--')
```

◄

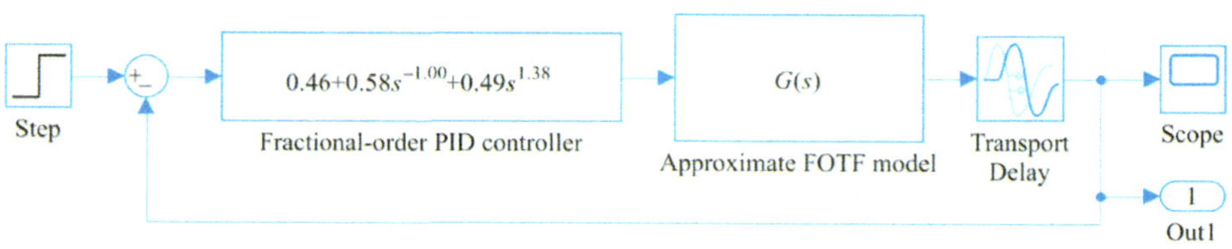

Fig. 8.35 Simulink model for fractional-order PID control system (`c8mfpid.slx`)

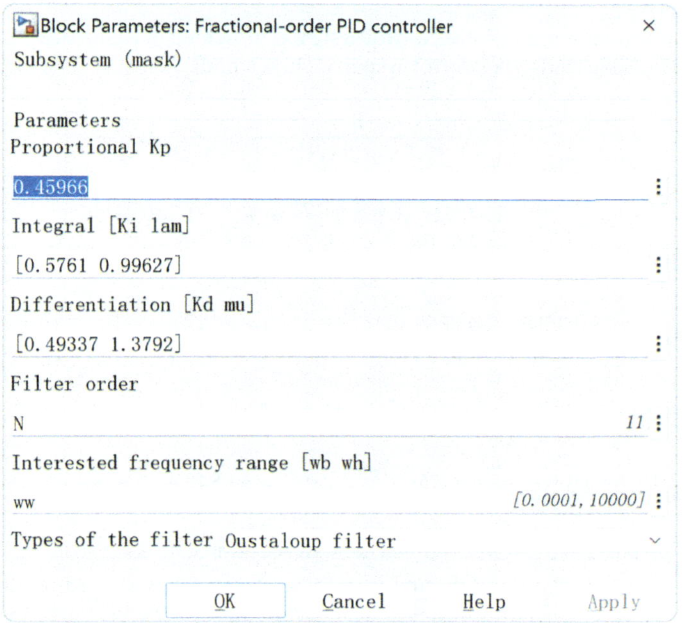

Fig. 8.36 Parameter dialog box of the fractional-order PID controller

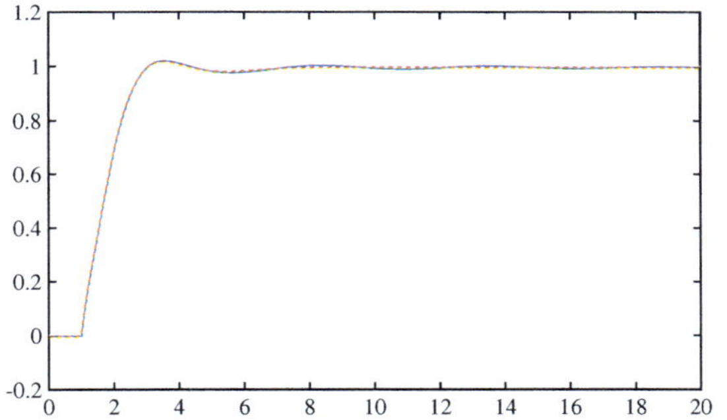

Fig. 8.37 Closed-loop step response

8.4.3 Simulation of Multivariable Systems

The Approximate FOTF model block in FOTF Blockset also supports the direct modeling of FOTF matrices. The transfer function matrix in MATLAB workspace can be embedded into the block, such that a multivariable model with filters such as Oustaloup filters can be used to represent the transfer function matrix. Therefore, if FOTF matrices are known, they can be used in Simulink modeling, such that the modeling and simulation analysis of multivariable systems can be carried out. An example is given next to demonstrate the modeling of multivariable control systems.

► Example 8.22

A fractional-order multivariable control system is presented in Reference [3]. The structure of a multivariable feedback control system is shown in Fig. 8.38. The plant model is

$$G(s) = \begin{bmatrix} \dfrac{1}{1.35s^{1.2}+2.3s^{0.9}+1} & \dfrac{2}{4.13s^{0.7}+1} \\ \dfrac{1}{0.52s^{1.5}+2.03s^{0.7}+1} & -\dfrac{1}{3.8s^{0.8}+1} \end{bmatrix},$$

and the compensating matrices are

$$K_\mathrm{p} = \begin{bmatrix} -0.4596 & -0.8881 \\ -0.8503 & 0.5262 \end{bmatrix}, \quad K_\mathrm{d}(s) = \begin{bmatrix} -1/(2.5s+1) & 0 \\ 0 & -1/(s+1) \end{bmatrix}.$$

The fractional-order controller $G_\mathrm{c}(s)$ is a diagonal matrix, whose diagonal elements are

$$c_1(s) = 10.7003 + \frac{2.9743}{s^{0.86736}} + 15s^{0.7876}, \quad c_2(s) = 14.848 + \frac{10.1421}{s^{0.81932}} + 14.6848s^{0.7355}.$$

8

Draw the step response of the multivariable closed-loop control system.

Solutions The step response of the multivariable system means the two input channels driven the system individually with unit step signals and observe the outputs of the system. To solve the problem, the plant model, two compensators, and two fractional-order controllers can be entered into MATLAB workspace first. The simulation model in Fig. 8.39 can be established.

```
>> s=fotf('s'); g1=1/(1.35*s^1.2+2.3*s^0.9+1);
   g2=2/(4.13*s^0.7+1); g4=-1/(3.8*s^0.8+1);
   g3=1/(0.52*s^1.5+2.03*s^0.7+1); G=[g1,g2; g3,g4];
```

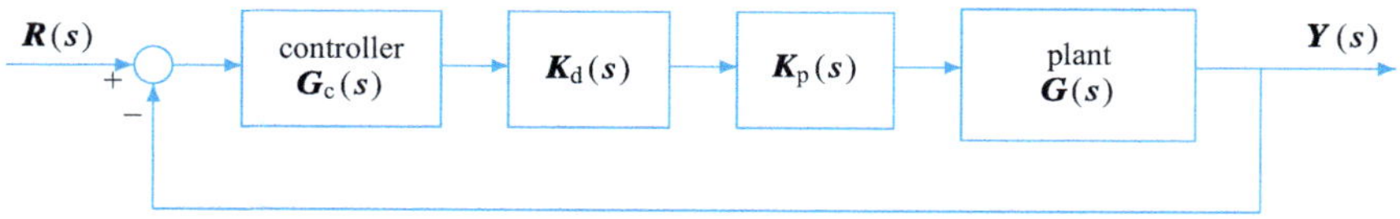

Fig. 8.38 Block diagram of a typical multivariable control system

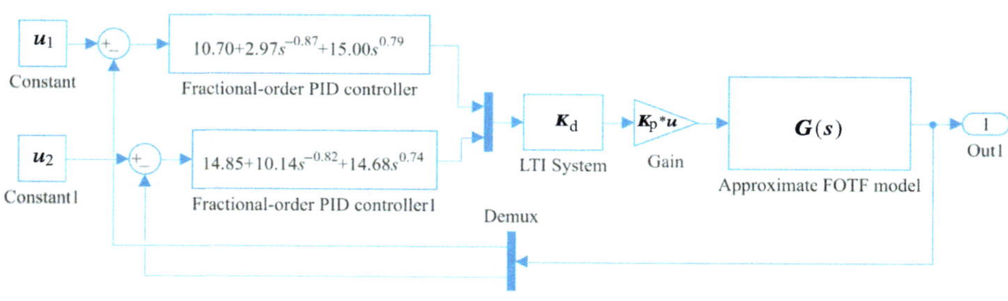

Fig. 8.39 Simulink model of a multivariable system (`c8mmimo.slx`)

```
Kp=[-0.4596,-0.8881; -0.8503,0.5262];
c1=10.7003+2.9743*s^-0.86736+15*s^0.7876;
c2=14.848+10.1421*s^-0.81932+14.6848*s^0.7355;
s=tf('s'); Kd=[-1/(2.5*s+1), 0; 0, -1/(s+1)];
```

With the simulation model, the following commands can be used to draw directly the closed-loop step responses. Assume that $u_1=1$ and $u_2=0$, the two output signals can be obtained, which are the outputs driven by the first input alone can be obtained. Assume then $u_1=0$ and $u_2=1$, the closed-loop step responses under the second input signal can be found. The closed-loop step responses can be obtained, as shown in Fig. 8.40.

```
>> u1=1; u2=0; [t1,~,y1]=sim('c8mmimo'); % first signal acts alone
   u1=0; u2=1; [t2,~,y2]=sim('c8mmimo'); % second signal acts alone
   subplot(221), plot(t1,y1(:,1)), ylim([-0.1 1.1])
   subplot(223), plot(t1,y1(:,2)), ylim([-0.1 1.1])
   subplot(222), plot(t2,y2(:,1)), ylim([-0.1 1.1])
   subplot(224), plot(t2,y2(:,2)), ylim([-0.1 1.1])
```

◀

8.5 Exercises

(1) Assume that the plant model is $G(s) = 1/(s^2+3s+2)$, while the controller is a discrete PI controller $G_c(z) = 3.86+0.263/(z-1)$, with sample time $T = 0.1$ s, can you write out the explicit differential equation of the control system? Why?

(2) Consider the simulation model in Example 8.5. Why the signal $y'''(t)$ is connected to an integrator to create signal ${}_0^{\mathrm{C}}\mathscr{D}_t^{3.5}y(t)$, while when compute ${}_0^{\mathrm{C}}\mathscr{D}_t^{3.1}y(t)$, signal $y'''(t)$ should be connected to a differentiator? Is it possible to connect the signals to differentiators? Why?

(3) Consider the following FODE with initial value $y(0) = 0$. Validate that signal $y(t) = \sin^2 t$ is the solution of the following equation. Establish a Simulink

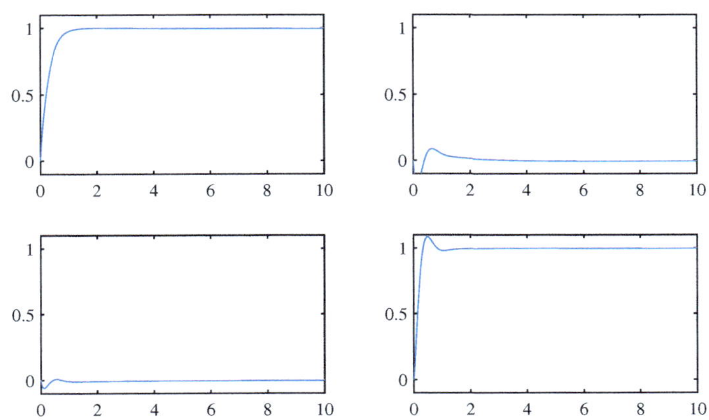

Fig. 8.40 Step responses of the multivariable closed-loop system

model to find the numerical solution in the interval [0, 100]. Select filter parameters and observe the impact of these parameters on the solution errors.

$$ {}_0^{\rm C}\mathscr{D}_t^{0.8} y(t) = {}_0^{\rm RL}\mathscr{D}_t^{-0.2} \sin 2t. $$

(4) For the given initial values $y(0) = y'(0) = 0$, consider the following Riemann–Liouville differential equation.

$$ y''(t) + {}_0^{\rm RL}\mathscr{D}_t^{1.8} y(t) = \frac{1}{\Gamma(t)} + {}_0^{\rm RL}\mathscr{D}_t^{-0.2}\left(\frac{1}{\Gamma(t)}\right). $$

Validate that $y(t) = \displaystyle\int_0^t \frac{t-\tau}{\Gamma(\tau)}{\rm d}\tau$ is the exact solution of the equation. Establish a Simulink model, find the numerical solution in the [0, 10] interval, and validate the solution.

(5) The initial conditions $y(0) = 2$, $y'(0) = 0$ and $y''(0) = -2$ are known for the linear FODE.

$$ y'''(t) + {}_0^{\rm C}\mathscr{D}_t^{2.7} y(t) + 2y''(t) + 2y'(t) + {}_0^{\rm C}\mathscr{D}_t^{0.7} y(t) + y(t) = -2\sqrt{2}\sin\left(t + \frac{\pi}{4}\right). $$

Validate that the analytical solution is $y = 2\cos t$. Establish a Simulink model in the interval [0, 100]. Select filter parameters and analyze their impact on the simulation errors.

(6) For the following Caputo equation with initial values $y(0) = 1$ and $y'(0) = 0$.

$$ {}_0^{\rm C}\mathscr{D}_t^{1.5} y(t) + y'(t) - t^{0.5}{\rm E}_{2,1.5}\left(t^2\right) y^2(t) = \sinh t - t^{0.5}{\rm E}_{2,1.5}\left(t^2\right)\sinh^2 t. $$

Validate that $y(t) = \cosh t$ is the exact solution of the equation. Establish a Simulink model and find its numerical solutions in the interval [0, 2].

(7) Consider the following Caputo equation:

$$ y'(t) + y^2(t)\,{}_0^{\rm C}\mathscr{D}_t^{0.8} y(t) + \sin^2 t\,{}_0^{\rm C}\mathscr{D}_t^{0.8} y(t) + y(t) = \sqrt{2}\sin\left(t + \frac{3\pi}{4}\right) - {}_0^{\rm RL}\mathscr{D}_t^{-0.2} \sin t. $$

The initial condition is $y(0) = 1$. Validate that $y = \cos t$ is the exact solution. Establish a Simulink model, and find the numerical solutions in the [0, 100] interval. Select filter parameters and observe their impact on simulation results.

(8) Consider the following Caputo equation with the initial condition $y(0) = 0$:

$$ {}_0^{\rm C}\mathscr{D}_t^{0.8} y(t) = \begin{cases} \dfrac{1}{\Gamma(1.2)} t^{0.2}, & 0 \leqslant t \leqslant 1 \\[2ex] \dfrac{1}{\Gamma(1.2)} t^{0.2} - \dfrac{2}{\Gamma(2.2)}(t-1)^{1.2}, & t > 1. \end{cases} $$

Validate that the exact solution of the equation is

$$y(t) = \begin{cases} t, & 0 \leqslant t \leqslant 1 \\ t - (t-1)^2, & t > 1. \end{cases}$$

Establish a Simulink model, and find it numerical solution in the interval [0, 2].

(9) For the following Caputo equation with initial conditions $x_1(0) = 1, x_2(0) = 2$ and $x_3(0) = 3$, validate that $x_1(t) = t^{0.7}+1, x_2(t) = t^{0.8}+2$ and $x_3(t) = t^{0.9}+3$ are the exact solutions. Establish a Simulink model, and find its numerical solutions in the interval [0, 20].

$$\begin{cases} {}_0^{\rm C}\mathscr{D}_t^{0.5} x_1(t) = \dfrac{\Gamma(1.7)}{2\Gamma(1.2)} \left(\sqrt[8]{(x_1(t)-1)(x_3(t)-3)} + t^{0.2}\right) \\ {}_0^{\rm C}\mathscr{D}_t^{0.7} x_2(t) = \dfrac{\Gamma(1.8)}{2\Gamma(1.1)} \left(\sqrt[7]{x_1(t)-1} + t^{0.1}\right) \\ {}_0^{\rm C}\mathscr{D}_t^{0.8} x_3(t) = \dfrac{\Gamma(1.9)}{\Gamma(1.1)} \sqrt[8]{x_2(t)-2}. \end{cases}$$

(10) The initial conditions of a Caputo equation are $y(0) = 1$ and $y'(0) = -1$, validate that the exact solution of the equation is $y(t) = {\rm e}^{-t}$. Establish a Simulink model, and compute the numerical solutions in the interval [0, 3]. Select filter parameters, and analyze their impact on the solution errors.

$${}_0^{\rm C}\mathscr{D}_t^{1.8} y(t) = -t^{0.1}\,{\rm e}^t \frac{{\rm E}_{1,1.2}(-t)}{{\rm E}_{1,1.1}(-t)} y(t)\, {}_0^{\rm C}\mathscr{D}_t^{0.9} y(t) + y^2(t) - {\rm e}^{-2t}.$$

(11) For the Caputo equation with initial conditions $y(0) = 1$ and $y'(0) = 0$, and validate the function $y(t) = t^2 + 1$ is the exact solution. Establish a Simulink model and compute the numerical solutions in the [0, 10] interval. Select filter parameters and analyze their impact on the solution errors.

$${}_0^{\rm C}\mathscr{D}_t^{1.5} y(t) = \sqrt{\frac{8}{\pi}} \sqrt[4]{\Gamma(1.7)\,\Gamma(2.3)\, {}_0^{\rm C}\mathscr{D}_t^{1.3} y(t)\, {}_0^{\rm C}\mathscr{D}_t^{0.7} y(t)} + y'(t) - 2t.$$

(12) For the `bp5a_fcn2.m` file on page 334, why there are tiny complex quantities occur?

(13) In Example 8.11, it is claimed that the command-driven method cannot be used in the solutions in large time interval. Do you know why?

(14) Consider the problem in Example 8.13. In the simulation model Reference [10], the sequential order of each Oustaloup filter and integrator block is different from the ones presented here. The sequential order of these blocks is very important. Consider again the benchmark problem in Example 8.17. If the sequential orders in Reference [10] are used to establish a simulation model, whether correct simulation results can be found? Why?

(15) In constructing Simulink models for FODEs, especially when describing Caputo equations, an integrator chain is recommended to define the key signals. Can we use a differentiator chain instead? Why?

(16) Use a block to express the following FOTF in Simulink:

$$G(s) = \frac{7s^{1.8} + 42s^{1.5} + 106s^{1.2} + 147s^{0.9} + 120s^{0.6} + 55s^{0.3} + 11}{s^{2.1} + 8s^{1.8} + 27s^{1.5} + 50s^{1.2} + 55s^{0.9} + 36s^{0.6} + 13s^{0.3} + 2}.$$

(17) Consider the multivariable model in Example 8.22. If the gain of the plant $G(s)$ is increased by 50% or decreased by 50%, and the controller parameters are the same, compute the closed-loop system response and observe the control results.

References

1. Bai L, Xue DY (2018) Universal block diagram based modeling and simulation schemes for fractional-order control systems. ISA Trans 82:153–162
2. Xue DY (2017) Fractional-order control systems – Fundamentals and numerical implementations. De Gruyter, Berlin
3. Xue DY (2018) Fractional calculus and fractional-order control (in Chinese). Science Press, Beijing
4. Xue DY, Chen YQ (2004) MATLAB solutions of higher advanced applicational mathematical problems (in Chinese). Tsinghua University Press, Beijing
5. Xue D, Chen YQ, Atherton DP (2007) Linear feedback control –analysis and design with MATLAB. SIAM Press, Philadephia
6. Chen YQ, Petráš I, Xue DY (2009) Fractional order control - a tutorial. In: Proceedings of the American control conference. St Louis, pp 1397–1411
7. Li Z, Liu L, Dehghan S, et al. (2017) A review and evaluation of numerical tools for fractional calculus and fractional order control. Int J Control 90(6):1165–1181
8. Xue DY (2022) Modeling ang simulation with Simulink. De Gruyter Press, Berlin
9. Xue DY (2022) MATLAB and Simulink in action: programming, computing and simulation (in Chinese). Tsinghua University Press, Beijing
10. Petráš I (2011) Fractional-order nonlinear systems: modelling, analysis and simulation. Higher Education Press, Beijing
11. Xue DY, Bai L (2017) Benchmark problems for Caputo fractional-order ordinary differential equations. Fract Calculus Appl Anal 20(5):1305–1312

8

Numerical Solutions of Special Fractional-Order Differential Equations

Contents

D. Y. Xue and L. Bai, *Fractional Calculus*, https://doi.org/10.1007/978-981-99-2070-9_9

Command-driven methods for nonlinear Caputo equations are introduced in ▶ Chap. 7, and the methods are mainly used in solving explicit equations and FOSS equations. In ▶ Chap. 8, block diagram-based methods are introduced for these FODEs, with wider application fields.

Some special FODEs are discussed in particular in this chapter. In ▶ Sect. 9.1, a command-driven high-precision algorithm is introduced for implicit Caputo equations. Block diagram-based methods are also introduced for implicit Caputo equations, with even better performance. In ▶ Sect. 9.2, fractional-order delay differential equations (FODDEs) are discussed, and the equations with nonzero history functions are studied with block diagram-based approach. In ▶ Sect. 9.3, an optimization-based method and an algebraic equation-based method are proposed for boundary value problems in FODEs. In ▶ Sect. 9.4, numerical algorithms and implementations are presented for time-fractional partial differential equations (TFPDEs).

9.1 Implicit Fractional-Order Differential Equations

The solution commands and block diagrams discussed so far are for explicit fractional-order differential equations (FODEs). If a differential equation cannot be converted into its explicit form, those methods cannot be used. In this section, discussions are made on solving various implicit Caputo equations.

9.1.1 A High-Precision Matrix Algorithm for Caputo Equations

The definition of implicit Caputo equations is presented first, followed by numerical solution methods for the equations. A high-precision numerical algorithm is proposed, with a general-purpose MATLAB solver.

Definition 9.1 Implicit FODEs

The general form of a nonlinear implicit Caputo equation is

$$F\big(t, y(t), {}_{t_0}^{\;\mathrm{C}}\mathscr{D}_t^{\alpha_1} y(t), \dots, {}_{t_0}^{\;\mathrm{C}}\mathscr{D}_t^{\alpha_n} y(t)\big) = 0. \tag{9.1}$$

It is safe to assume that α_i is a rising sequence, and denote $q = \lceil \alpha_n \rceil$. The given initial values are $y^{(q-1)}(0) = c_{q-1}, y^{(q-2)}(0) = c_{q-2}, \dots, y(0) = c_0$.

As presented earlier, signal $y(t)$ can still be decomposed as $y(t) = z(t) + T(t)$, where the definition of $T(t)$ is in (7.33), while signal $z(t)$ has zero initial conditions. If the fractional calculus definition is changed into Riemann–Liouville definition, the equation below is still satisfied.

$$F\big(t, z(t), {}_{t_0}^{\mathrm{RL}}\mathscr{D}_t^{\alpha_1} z(t), \dots, {}_{t_0}^{\mathrm{RL}}\mathscr{D}_t^{\alpha_n} z(t)\big) = 0. \tag{9.2}$$

Recall the pth-order matrix algorithm for evaluating the Riemann–Liouville derivatives, the original function $y(t)$ can be decomposed into $u(t) + v(t)$, with

$$u(t)=\sum_{k=0}^{p} c_k(t-t_0)^k, \tag{9.3}$$

where $c_k=y_k/k!$. Therefore, the α_ith-order derivative can be evaluated directly from

$${}_{t_0}^{\rm RL}\mathscr{D}_t^{\alpha_i}y(t)=\frac{1}{h^{\alpha_i}}\boldsymbol{W}^{\alpha_i}\boldsymbol{v}+\sum_{k=0}^{p}c_k\frac{\Gamma(k+1)}{\Gamma(k+1-\alpha_i)}(t-t_0)^{k-\alpha_i}, \tag{9.4}$$

where $\boldsymbol{W}^{\alpha_i}$ is a lower triangular matrix composed of w_i coefficients. For simplicity, denote $\boldsymbol{B}_i=\boldsymbol{W}^{\alpha_i}/h^{\alpha_i}$.

Note that to evaluate the α_ith-order Caputo derivative to a polynomial signal, $(\lceil\alpha_i\rceil+1)$th integer-order derivative should be evaluated first, and then fractional-order integrals can be taken, such that the polynomial signal whose order lower than $\lceil\alpha_i\rceil$ vanish directly, where (9.4) can be rewritten as

$${}_{t_0}^{\rm RL}\mathscr{D}_t^{\alpha_i}y(t)=\frac{1}{h^{\alpha_i}}\boldsymbol{W}^{\alpha_i}\boldsymbol{v}+\sum_{k=\lceil\alpha_i\rceil+1}^{q}\frac{c_k\,\Gamma(k+1)}{\Gamma(k+1-\alpha_i)}(t-t_0)^{k-\alpha_i}. \tag{9.5}$$

For an equally spaced time vector $\boldsymbol{t}=[0,h,2h,\dots,mh]$, the following nonlinear algebraic equation can be constructed

$$\boldsymbol{f}\big(\boldsymbol{t},\boldsymbol{B}_1\boldsymbol{v},\boldsymbol{B}_2\boldsymbol{v},\cdots,\boldsymbol{B}_n\boldsymbol{v}\big)=\boldsymbol{0}. \tag{9.6}$$

The equation can be solved directly with the MATLAB solver `fsolve()`. The following MATLAB code implements the algorithm, where $\boldsymbol{B}_i$ and matrix $\boldsymbol{d}_u$ can be passed to the function $\boldsymbol{f}$ as additional arguments, with $\boldsymbol{B}_i$ represented by a 3D array.

```
function [y,t]=nlfode_mat(f,alpha,y0,tn,h,p,yp)
  arguments
     f, alpha(1,:) double, y0(:,1) double, tn(1,1) double
     h(1,1) double, p(1,1){mustBePositiveInteger}
     yp(:,1) double=zeros(ceil(tn/h)+1,1);
  end
  alfn=ceil(alpha); m=ceil(tn/h)+1;
  t=(0:(m-1))'*h; d1=length(y0); d2=length(alpha);
  B=zeros(m,m,d2); g=double(genfunc(p));
  for i=1:d2, w=get_vecw(alpha(i),m,g);
     B(:,:,i)=rot90(hankel(w(end:-1:1)))/h^alpha(i);
  end
  c=y0./gamma(1:d1)'; du=zeros(m,d2);
  u=0; for i=1:d1, u=u+c(i)*t.^(i-1); end
  for i=1:d2
     if alfn(i)==0, du(:,i)=u;
     elseif alfn(i)<d1
        for k=(alfn(i)+1):d1
           du(:,i)=du(:,i)+(t.^(k-1-alpha(i)))*c(k)...
```

```
                    *gamma(k)/gamma(k-alpha(i));
   end, end, end
   v=fsolve(f,yp,[],t,u,B,du); y=u+v;
end
```

The syntax of the solver is $[y,t]$=`nlfode_mat(fun,`α`,`y_0`,`t_n`,`h`,`p`,`y_p`)`, where `fun` is the function handle describing the implicit FODEs, whose format will be demonstrated next through examples. The row vector $\boldsymbol{\alpha}$ contains all the related orders where $\boldsymbol{\alpha} = [\alpha_1, \alpha_2, \ldots, \alpha_m]$. The vector $\boldsymbol{y}_0$ is the vector of initial conditions. The arguments h and t_n are scalars for step size and termination time, respectively. The integer p is the order of the algorithm, and $p \le q$. The optional vector $\boldsymbol{y}_p$ is the initially predicted initial values. If the argument is not provided, a default zero vector is adopted.

In order to solve the implicit FODEs, a MATLAB function or an anonymous function must be written to describe the implicit FODEs. The format of the function is rather special, the input arguments are $(\boldsymbol{v}, \boldsymbol{t}, \boldsymbol{u}, \boldsymbol{B}, \boldsymbol{d}_u)$. When describing signal y, it should be denoted as $\boldsymbol{u}+\boldsymbol{v}$ instead. When describing the term ${}_0^{\mathrm{C}}\mathscr{D}_t^{\alpha_i}y(t)$ in the FODEs, denote it as $\boldsymbol{d}_u(:,i) + \boldsymbol{B}(:,:,i)*\boldsymbol{v}$.

The function `fsolve()` is the kernel of this solver. In the user-defined function describing the implicit FODEs, the variables $\boldsymbol{t}$, $\boldsymbol{u}$, $\boldsymbol{B}$ and $\boldsymbol{d}_u$ are passed to function `fsolve()` as additional parameters, which are implemented internally in the `nlfode_mat()` solver. The solver also applies to the solutions of explicit FODEs. Examples are presented next to demonstrate the solutions of explicit and implicit Caputo equations with such a solver.

9

► Example 9.1

Solve again the explicit Caputo equation in Example 7.11 with the new solver.

Solutions The Caputo equation in Example 7.3 is an explicit one. It should be converted into an implicit one before it can be solved with this solver. The conversion in this example is easy, just simply by moving all the terms on the right-hand side of the equation to the left. The following implicit form can be obtained directly from the original FODE:

$$ {}_0^{\mathrm{C}}\mathscr{D}_t^{1.455}y(t) + t^{0.1}\frac{\mathrm{E}_{1,1.545}(-t)}{\mathrm{E}_{1,1.445}(-t)}\,\mathrm{e}^t y(t)\,{}_0^{\mathrm{C}}\mathscr{D}_t^{0.555}y(t) - \mathrm{e}^{-2t} + \left[y'(t)\right]^2 = 0. $$

Selecting the order vector $\boldsymbol{\alpha} = \left[1.455, 0.555, 1\right]$, an anonymous function can then be used to describe the implicit FODE. The term ${}_0^{\mathrm{C}}\mathscr{D}_t^{1.455}y(t)$ should be represented as $\boldsymbol{d}_u(:,1)+\boldsymbol{B}(:,:,1)*\boldsymbol{v}$ in the anonymous function, since the order 1.455 is the first term in vector $\boldsymbol{\alpha}$. The subsequent terms are also represented similarly. It can be seen that the following anonymous function can be constructed directly. Note that, dot operations should be used whenever necessary in the anonymous function.

```
>> alpha=[1.455,0.555,1]; y0=[1,-1]; tn=1;
   f=@(v,t,u,B,du)(du(:,1)+B(:,:,1)*v)+t.^0.1.*exp(t)...
         .*ml_func([1,1.545],-t)./ml_func([1,1.445],-t)...
         .*(v+u).*(du(:,2)+B(:,:,2)*v)-exp(-2*t)+...
         (du(:,3)+B(:,:,3)*v).^2;
```

Selecting different step sizes h, and try, respectively, $p = 1$ and $p = 2$, the time elapse and maximum errors can be measured, as shown in Table 9.1. It can be seen that, for each step size, the computational loads for $p = 1$ and $p = 2$ are similar, but the maximum error is reduced significantly. Unfortunately, the efficiency of this solver is much lower than the solver `nlfec()`, even much lower the block diagram-based method.

```
>> h=0.01;
   tic, p=1; [y1,t]=nlfode_mat(f,alpha,y0,tn,h,p); toc
   tic, p=2; [y2,t]=nlfode_mat(f,alpha,y0,tn,h,p); toc
   max(abs(y1-exp(-t))), max(abs(y2-exp(-t)))
```

◄

It should be noted that the solver applies to implicit FODEs, therefore it is not fair to compare it with the explicit solvers. In order to assess fairly the efficiency of implicit FODE solvers, the authors designed a benchmark problem [1]. An example is given next to demonstrate the high-precision solver for implicit FODEs.

► Example 9.2

Solve the following implicit FODE [1, 2]:

$$\begin{aligned}&{}_0^{\mathrm{C}}\mathscr{D}_t^{0.2}y(t)\,{}_0^{\mathrm{C}}\mathscr{D}_t^{1.8}y(t)+{}_0^{\mathrm{C}}\mathscr{D}_t^{0.3}y(t)\,{}_0^{\mathrm{C}}\mathscr{D}_t^{1.7}y(t)\\ &=-\frac{t}{8}\left[\mathrm{E}_{1,1.8}\left(-\frac{t}{2}\right)\mathrm{E}_{1,1.2}\left(-\frac{t}{2}\right)+\mathrm{E}_{1,1.7}\left(-\frac{t}{2}\right)\mathrm{E}_{1,1.3}\left(-\frac{t}{2}\right)\right],\end{aligned}$$

where $y(0) = 1$, $y'(0) = -1/2$, and the exact solution is $y(t) = \mathrm{e}^{-t/2}$, with $t \in (0, 10)$.

Solutions It has been pointed out that since the algorithm here involves the solutions of large-scale algebraic equations with function `fsolve()`, where the number of unknowns equals to the length of vector $\boldsymbol{t}$, the computational load is relatively large. It is not suitable for solving FODEs over relatively large time intervals. In this example, the numerical solution problem in the interval $t \in (0, 2)$ is studied.

Before solving the implicit FODE, it should be converted first into the standard form. For this example, the standard form can be found by moving all the terms on the right-hand side of the equation to the left.

$$\begin{aligned}&{}_0^{\mathrm{C}}\mathscr{D}_t^{0.2}y(t)\,{}_0^{\mathrm{C}}\mathscr{D}_t^{1.8}y(t)+{}_0^{\mathrm{C}}\mathscr{D}_t^{0.3}y(t)\,{}_0^{\mathrm{C}}\mathscr{D}_t^{1.7}y(t)+\\ &\frac{t}{8}\left[\mathrm{E}_{1,1.8}\left(-\frac{t}{2}\right)\mathrm{E}_{1,1.2}\left(-\frac{t}{2}\right)+\mathrm{E}_{1,1.7}\left(-\frac{t}{2}\right)\mathrm{E}_{1,1.3}\left(-\frac{t}{2}\right)\right]=0.\end{aligned}$$

Table 9.1 The impacts of the step sizes and orders

Step size h	0.01	0.01	0.005	0.005	0.001	0.001
Order p	1	2	1	2	1	2
Time elapse	2.4976	2.2347	13.068	13.374	368.034	356.121
Maximum error	0.0090	0.0010	0.0046	3.8442×10^{-4}	9.7603×10^{-4}	3.5270×10^{-5}

Table 9.2 The impacts of the step sizes and orders

Step sizes and orders	$h = 0.01, p = 1$	$h = 0.01, p = 2$	$h = 0.005, p = 1$	$h = 0.005, p = 2$
Maximum error	0.0036	8.6190×10^{-4}	0.0019	3.6895×10^{-4}
Time elapse	6.4692	7.5325	38.025	36.688

To solve the implicit FODE, the following anonymous function should be written:

```
>> alpha=[0.2 1.8 0.3 1.7]; y0=[1,-1/2]; tn=2;
   f=@(v,t,u,B,du)(du(:,1)+B(:,:,1)*v).*(du(:,2)+B(:,:,2)*v)+...
                  (du(:,3)+B(:,:,3)*v).*(du(:,4)+B(:,:,4)*v)+...
            1/8*t.*(ml_func([1,1.8],-t/2).*ml_func([1,1.2],-t/2)+...
                 ml_func([1,1.7],-t/2).*ml_func([1,1.3],-t/2));
```

With the function handle f, the solver `nlfode_mat()` can be called to solve directly the implicit FODE. Different h and p combinations are tested, and the solutions can be found. The maximum errors and time elapse are recorded, as shown in Table 9.2. Note that the step size h should not be set to very small values, otherwise the computational load may be too heavy to accomplish the tasks.


```
>> h=0.01; p=2;
   tic, [y1,t]=nlfode_mat(f,alpha,y0,tn,h,p); toc
   max(abs(y1-exp(-t/2)))
```

If one selects $h = 0.01$, and selects $t_n = 10$ and $p = 2$, the time elapse is 272.084 s, with maximum error of 0.0117. Therefore, this solver is not quite suitable for solving general implicit FODEs.◀

9.1.2 Block Diagram-Based Solutions of Implicit FODEs

The block diagram-based modeling and simulation methods studied so far are for explicit FODEs. If implicit ones are encountered in the research, special modeling and simulation methods should be established. For the general form of implicit FODE in Definition 9.1, a block diagram-based solution method is proposed, and examples are used to demonstrate the modeling and simulation issues.

Algorithm 9.1 Simulink modeling of implicit FODEs

(1) If the equation has q initial conditions, an integrator chain in Fig. 8.18 can be configured first to define the key signals $y(t), y'(t), \ldots, y^{(q)}(t)$.

(2) The q initial values can be assigned into the corresponding integrators, such that the initial conditions of the output signal $y(t)$ and its integer-order derivatives are satisfied.

(3) The fractional-order derivative signals can be constructed by feeding the integer-order derivative signals to the Oustaloup filters.
(4) Construct the left-hand side signal in implicit FODE in (9.1) and feed it into an **Algebraic Constraint** block, its output is connected to a $(q-\alpha_n)$th Oustaloup filter, the output of the filter can be connected to the $y^{(q)}(t)$ port, to complete the closed-loop in the Simulink model.

A specific example is given next to demonstrate the modeling of implicit FODEs.

▶ Example 9.3

Consider again the example in Example 9.2. Solve the implicit FODE with Simulink.

Solutions The implicit Caputo equation should be converted into the standard form.

$$ {}_0^{\rm C}\mathscr{D}_t^{0.2}y(t)\,{}_0^{\rm C}\mathscr{D}_t^{1.8}y(t)+{}_0^{\rm C}\mathscr{D}_t^{0.3}y(t)\,{}_0^{\rm C}\mathscr{D}_t^{1.7}y(t) + \frac{t}{8}\left[{\rm E}_{1,1.8}\left(-\frac{t}{2}\right){\rm E}_{1,1.2}\left(-\frac{t}{2}\right)+{\rm E}_{1,1.7}\left(-\frac{t}{2}\right){\rm E}_{1,1.3}\left(-\frac{t}{2}\right)\right]=0. $$

Based on the modeling procedures, the key signals $y(t)$, $y'(t)$ and $y''(t)$ are defined first, then the Caputo derivative signals ${}_0^{\rm C}\mathscr{D}_t^{0.2}y(t)$, ${}_0^{\rm C}\mathscr{D}_t^{0.3}y(t)$, ${}_0^{\rm C}\mathscr{D}_t^{1.7}y(t)$ and ${}_0^{\rm C}\mathscr{D}_t^{1.8}y(t)$ can be defined. With these signals, the left-hand side of the implicit FODE can be modeled, and the signal can be fed into the **Algebraic Constraint** block. The output of the block can be defined as ${}_0^{\rm C}\mathscr{D}_t^{1.8}y(t)$, if it is fed into a 0.2th order differentiator to generate the derivative signal $y''(t)$. With such a signal, it can be linked directly to the $y''(t)$ signal, defined in the integrator chain. The complete Simulink model for the implicit Caputo equation can be established as shown in Fig. 9.1. Since the initial values of the system are already defined in the integrator chain, the fractional-order differentiators can be simply modeled with Oustaloup filters. The function designed for the **Interpreted MATLAB Fcn** block is

```
function y=bp4_fcn(t)
   y=1/8*t*(ml_func([1,1.8],-t/2)*ml_func([1,1.2],-t/2)+...
            ml_func([1,1.7],-t/2)*ml_func([1,1.3],-t/2));
end
```

Selecting the following Oustaloup filter parameters, the numerical solution of the implicit FODE can be found, with a maximum error of 3.8182×10^{-5}, and time elapse of 164.21s. Compared with other models, the time elapse is much longer. This is because that there exists an algebraic loop in the model, such that algebraic equation should be solved once in every simulation step.

```
>> ww=[1e-5 1e5]; N=30; tic, [t,x,y]=sim('bp4_model'); toc
   max(abs(y-exp(-t/2)))
```

For different Oustaloup filter parameters, the above commands can be executed repeatedly, and the measured data in Table 9.3 can be found. It is seen that is $\omega_{\rm b}$ is selected too small, the solution precision can be increased, while the time elapse is also increased.
Since there exists an algebraic loop, blocks $s^{-0.2}$ and $s^{-0.3}$ can be set to the ones with low-pass filter. Unfortunately although algebraic loop can be avoided, the errors thus generated

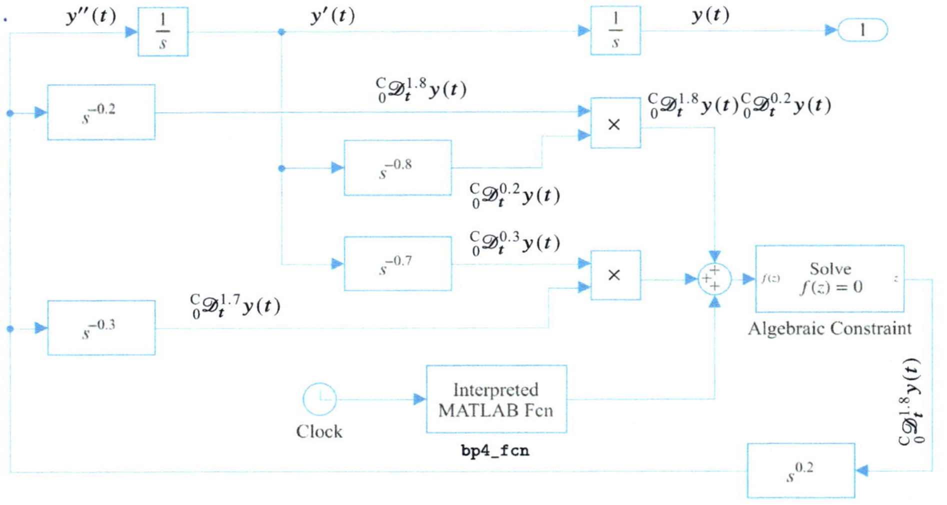

Fig. 9.1 Simulink model of the implicit FODE (`bp4_model.slx`)

Table 9.3 The impacts of frequency bands and orders

Frequency bands	$(10^{-3}, 10^{3})$	$(10^{-4}, 10^{4})$	$(10^{-5}, 10^{5})$	$(10^{-6}, 10^{6})$
Order N	11	30	30	35
Maximum error	0.0024	3.2104×10^{-4}	3.8182×10^{-5}	4.0166×10^{-6}
Time elapse	6.444	33.566	164.21	1705.82
Number of points	10639	54065	254488	2227969

are too large to be accepted. Therefore, for this example, the algebraic loop in the model should be maintained. ◀

9.1.3 Stiff ODE-Based Method

If the computational load is too high in solving a certain ODE, alternative ODE solvers can be tested. For instance, the solver can be set to ode15s, which is designed for solving stiff ODEs. For stiff ODEs, normally the number of points is significantly reduced, without sacrificing too much the solution precision. The computational load is usually significantly reduced. An example is used next to demonstrate the stiff ODE solvers in handling implicit FODEs.

▶ Example 9.4

A Simulink model is established in solving the implicit FODE in Example 9.3, and different filter parameters are tested. When the frequency band increases, the computational load increases significantly. Use stiff ODE solvers, and see whether high-precision solutions can be found.

Solutions With the `set_param()` function in Simulink, the simulation control parameters can be set, without redrawing the whole Simulink model. With the following commands, the simulation algorithm can be set to ode15s. The solver can also be selected in the dialog box in ◘ Fig. 8.4.

```
>> set_param('bp4_model','Solver','ode15s')
```

Different frequency bands and order combinations are tested, and with the repeated execution of the following statements, the measured data in ◘ Table 9.4 are obtained. It can be seen that with the stiff ODE solver, and with suitable choice of filter parameters, the solution precision is promoted significantly, and the time elapse is much lower than the previously described FODE solvers.

```
>> ww=[1e-6 1e6]; N=35; tic, [t,x,y]=sim('bp4_model'); toc
   max(abs(y-exp(-t/2))), length(t)
```

◀

9.1.4 The Fitting with the Implicit Block

Although some frequency response data-based filter design methods are presented in ▶ Chap. 5, it is extremely difficult to model a general irrational system with filters, and the fitting qualities are not satisfactory. Therefore for ordinary irrational systems, if the system can be described by transfer function models, numerical Laplace transform methods can be adopted, to carry out simulation analysis of the systems. Unfortunately, if the irrational transfer function is only a part of the whole system, simulation in Simulink framework should be explored.

In this section, only a small category of irrational transfer function model, that is $G(s)=(\tau s+1)^{-\nu}$, is considered. An Implicit model block is provided in the FOTF Blockset, which implements the continuous approximation model in ▶ Sect. 5.5.1 and the discrete fitting models in ▶ Sect. 5.6.3. Double click the Implicit model block,

◘ **Table 9.4** The impact of frequency band and orders with stiff ODE solvers

Frequency bands	$(10^{-6}, 10^{6})$	$(10^{-7}, 10^{7})$	$(10^{-8}, 10^{8})$	$(10^{-8}, 10^{8})$
Order N	35	39	43	51
Maximum error	4.0163×10^{-6}	6.8395×10^{-7}	9.8850×10^{-7}	8.8621×10^{-8}
Time elapse	9.2335	22.5766	44.885	44.538
Number of points	5458	11953	20156	15197

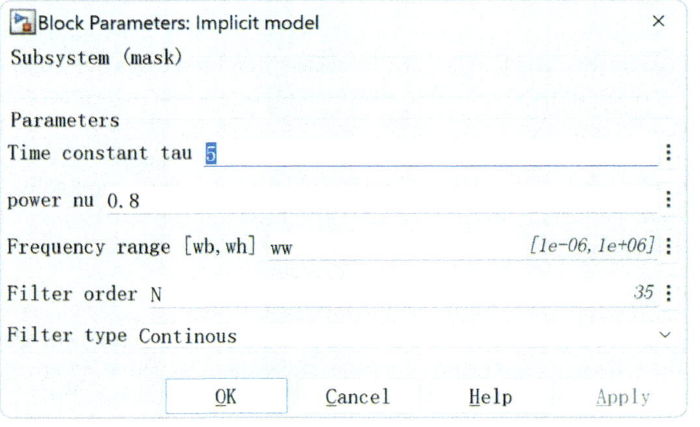

Fig. 9.2 Dialog box of Implicit model

a dialog box shown in Fig. 9.2 is opened. The values τ and ν can be assigned. In addition, the continuous or discrete filters can be selected from the Filter type listbox.

If the continuous filter is selected, the parameters N and `ww` in the Oustaloup filter should also be assigned. If discrete filter is selected, the order N is also required, while in the Frequency range edit box, sample time should be assigned.

A very important Laplace transform formula is presented in Theorem 6.4:

$$\mathscr{L}^{-1}\left[\frac{s^{\alpha\gamma-\beta}}{(s^\alpha+a)^\gamma}\right]=t^{\beta-1}\mathrm{E}_{\alpha,\beta}^{\gamma}\left(-at^\alpha\right). \tag{9.7}$$

With this formula, many properties regarding implicit models can be derived. For instance, let $\alpha=1$, $\gamma=\nu$, $a=1/\tau$ and $\beta=\nu$, the impulse response of the implicit block can be obtained.

$$\mathscr{L}^{-1}\left[\frac{(1/a)^\nu}{(s/a+1)^\nu}\right]=\frac{t^{\nu-1}}{a^\nu}\mathrm{E}_{1,\nu}^{\nu}\left(-\frac{t}{a}\right)=\tau^\nu t^{\nu-1}\mathrm{E}_{1,\nu}^{\nu}\left(-\tau t\right). \tag{9.8}$$

If $\alpha=1$, $\gamma=\nu$, $a=1/\tau$ and $\beta=\nu+1$ are selected, the step response of the implicit model can be formulated.

$$\mathscr{L}^{-1}\left[\frac{(1/a)^\nu}{s(s/a+1)^\nu}\right]=\frac{t^{\nu}}{a^\nu}\mathrm{E}_{1,1+\nu}^{\nu}\left(-\frac{t}{a}\right)=(\tau t)^\nu\mathrm{E}_{1,1+\nu}^{\nu}\left(-\tau t\right). \tag{9.9}$$

An example is presented next to study the impacts of filter parameters on the fitting qualities.

► Example 9.5

Draw the unity step response curve of the irrational model $G(s)=1/(5s+1)^{0.8}$.

Solutions It is seen from (9.9) that, when $\tau=1/5$ and $\nu=0.8$, the step response can be written as

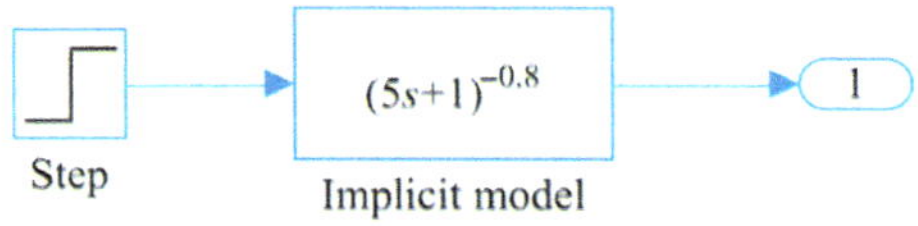

Fig. 9.3 Simulink model of the implicit block (`c9mimp3.slx`)

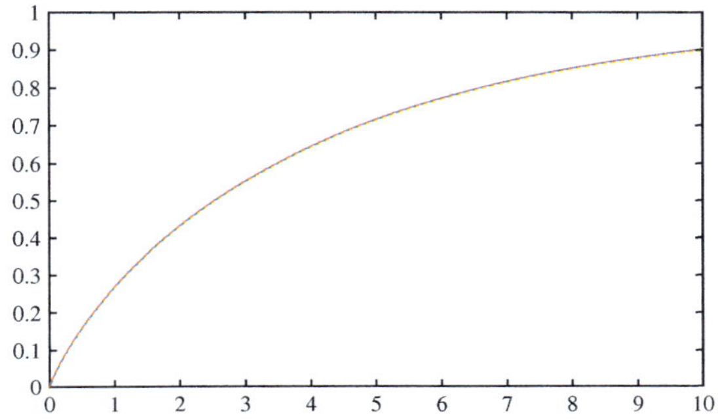

Fig. 9.4 Step responses comparisons

$$\mathscr{L}^{-1}\left[\frac{1}{s(5s+1)^{0.8}}\right]=\frac{t^{0.8}}{5^{0.8}}\,\mathrm{E}_{1,1.8}^{0.8}\left(-\frac{t}{5}\right).$$

With the **Implicit model** block in the FOTF Blockset, a Simulink model in Fig. 9.3 can be established. The frequency band of interest can be represented by variable `ww`, and express the order with variable N, such that the step response can be found with the following commands. The results are shown in Fig. 9.4. The maximum error of the model is 3.5694×10^{-4}. It can be seen that the two curves cannot be distinguished from the plot.

```
>> ww=[1e-4,1e4]; N=11; tic, [t,~,y]=sim('c9mimp3',10); toc
   y0=5^-0.8*t.^0.8.*ml_func([1,1.8,0.8],-t/5);
   max(abs(y-y0)), length(t), plot(t,y0,t,y,'--')
```

To assess quantitatively the impacts of the filter parameters to the fitting results, some combinations of frequency bands and orders are chosen, and the measured data are presented in Table 9.5. It can be seen that with suitable selection of filter parameters, accurate solutions can be found. ◀

▶ Example 9.6

Consider the model in Example 9.5, assess the fitting quality of the discrete filters.

Solutions Double click the **Implicit model** block, select the **Discrete** item from the **Filter type** listbox, the discrete filters in Example 5.22 can be assigned. The filter parameters N and sample time ω_b can be assigned in the dialog box. With different sample time and order combinations, the filters can be designed with the following statements, and compare the simulation results. It is found that when the sample time is 10^{-4}, and the filter order $N=8$, the maximum error is 0.0011, and time elapse if 0.2308 s, with 4741 points computed. It is

Table 9.5 The impacts of continuous filter parameters

Frequency bands	$(10^{-4}, 10^{4})$	$(10^{-4}, 10^{4})$	$(10^{-5}, 10^{5})$	$(10^{-6}, 10^{6})$
Order N	11	31	31	35
Maximum error	3.5694×10^{-4}	2.4363×10^{-4}	2.7595×10^{-5}	4.3734×10^{-6}
Time elapse	1.3197	1.7536	4.2930	29.1493
Number of points	19262	36252	157206	1480694

not possible to further improving the fitting quality, by raising the orders or reducing the sample time. If the filter is over-designed, the final results may diverge.

```
>> ww=[1e-4 1e3]; N=10; tic, [t,~,y]=sim('c9mimp3',10); toc
   y0=5^-0.8*t.^0.8.*ml_func([1,1.8,0.8],-t/5);
   max(abs(y-y0)), length(t), plot(t,y0,t,y,'--')
```

◀

9

Apart from the precision issues, large-scale matrix problems may also be encountered. If the number of points, $N = t_{\rm n}/h$, is selected too large, the $N \times N$ large-scale matrix appears in the discrete filter design, and it is not possible to design the filter. It is concluded from the real testing that the continuous filters are recommended to approximate irrational blocks.

9.2 Solutions of Fractional-Order Delay Differential Equations

In the FODEs studied so far, all the signals happen at exactly the same time. If the signal values in the past are involved, e.g., if $y(t-0.1)$ signal appears, it can be regarded as the signal value of $y(t)$ 0.1 s earlier, also known as 0.1 s of transport delay in the signal $y(t)$. Such an equation is known as fractional-order delay differential equation (FODDE).

There are very few literatures so far in exploring genuine FODDEs, while the delay phenomena exist in many FODEs. For instance, in fractional-order control systems, if there exist delays in the plant, the related FODEs are FODDEs. In this section, discussions are made on the numerical solutions of FODDEs.

9.2.1 Design of Benchmark Problems

Since in the existing literatures, there is no valuable benchmark problems for testing FODDEs, it is not possible to comment fairly on the behaviors of FODDE solvers. In order to bridge the gap, some of the FODEs are modified, and two benchmark problems are designed such that fair comparisons on different FODDE solvers can be made [3].

▶ Example 9.7

Consider the Caputo equation in Reference [1].

$${}_0^{\mathrm{C}}\mathscr{D}_t^{\sqrt{2}}y(t) = -t^{1.5-\sqrt{2}}\frac{\mathrm{E}_{1,3-\sqrt{2}}(-t)}{\mathrm{E}_{1,1.5}(-t)}\,\mathrm{e}^t y(t)\,{}_0^{\mathrm{C}}\mathscr{D}_t^{0.5}y(t) + \mathrm{e}^{-2t} - \left[y'(t)\right]^2,$$

where $y(0) = 1$ and $y'(0) = -1$, with exact solution of $y(t) = \mathrm{e}^{-t}$. Process the FODE, and construct Caputo delay differential equations.

Solutions For the FODDE model, it is hard to convert the fractional-order derivative signals into delay terms. Therefore the original signal and its integer-order derivatives should be processed instead. If the expected exact solution is still $y(t) = \mathrm{e}^{-t}$, the squared term of first-order derivative signal can be modified as $[y'(t-0.1)]^2 = \mathrm{e}^{0.2-2t}$. Therefore, to modify this term, the e^{-2t} term in the front should be canceled. Therefore the delay Caputo equation can be constructed as

$${}_0^{\mathrm{C}}\mathscr{D}_t^{\sqrt{2}}y(t) = -t^{1.5-\sqrt{2}}\frac{\mathrm{E}_{1,3-\sqrt{2}}(-t)}{\mathrm{E}_{1,1.5}(-t)}\,\mathrm{e}^t y(t)\,{}_0^{\mathrm{C}}\mathscr{D}_t^{0.5}y(t) + \mathrm{e}^{-2t} - \mathrm{e}^{-0.2}\left[y'(t-0.1)\right]^2. \tag{9.10}$$

For $t < 0$, the history functions are $y(t) = \mathrm{e}^{-t}$ and $y'(t) = -\mathrm{e}^{-t}$. ◀

▶ Example 9.8

Consider again the benchmark problem in Example 9.7. Design a benchmark problem with two delay signals.

Solutions The delay term $y'(t-0.1)$ in Example 9.7 is reserved, with slight modifications on its representations, such that $\mathrm{e}^{-2t} - \mathrm{e}^{-0.2}\left[y'(t-0.1)\right]^2$ term is changed into $\mathrm{e}^{-2t+0.2} - \left[y'(t-0.1)\right]^2$, without affecting the exact solutions.

Now we can consider modifying the $y(t)$ in the first term into $y(t-0.2)$. To ensure that the solution is still $y(t) = \mathrm{e}^{-t}$, the new term should be canceled with the e^t term in the front. That is to find $g(t)$, such that $g(t)y(t-0.2) = 1$. Therefore, $g(t) = \mathrm{e}^{t-0.2}$, and the benchmark problem of FODDE with two delay terms can be constructed directly as follows:

$${}_0^{\mathrm{C}}\mathscr{D}_t^{\sqrt{2}}y(t) = -t^{1.5-\sqrt{2}}\frac{\mathrm{E}_{1,3-\sqrt{2}}(-t)}{\mathrm{E}_{1,1.5}(-t)}\,\mathrm{e}^{t-0.2}y(t-0.2)\,{}_0^{\mathrm{C}}\mathscr{D}_t^{0.5}y(t) + \mathrm{e}^{0.2-2t} - \left[y'(t-0.1)\right]^2, \tag{9.11}$$

with history functions $y(t) = \mathrm{e}^{-t}$ and $y'(t) = -\mathrm{e}^{-t}$, when $t \le 0$. The exact solution is $y(t) = \mathrm{e}^{-t}$. ◀

9.2.2 Modeling of Nonzero History Functions

History functions are the output function $y(t)$ when $t < t_0$. Since Simulink environment is used in this section, examples are used to demonstrate how to use Simulink to handle history functions.

► Example 9.9

Consider the benchmark problem in Example 9.7. Observing mainly the signal $y'(t-0.1)$. If $t < 0$, the history functions are $y(t) = \mathrm{e}^{-t}$ and $y'(t) = -\mathrm{e}^{-t}$. How can we describe correctly the signal $y'(t-0.1)$ in Simulink?

Solutions Since $y'(t-0.1)$ signal is a delayed one from signal $y'(t)$, it is quite natural to connect the $y'(t)$ into a Transport Delay block, and assign the delay constant in block to 0.1, such that the model in Fig. 9.5a can be established. In fact, the signal thus built up does not consider at all the case when $t < 0$, such that in the first 0.1 s, the output signal remains as 0, not the expected history function $y'(t) = -\mathrm{e}^{-t}$.
If the impact of the history functions is taken into account, a Switch block should be introduced, to treat, respectively, the signal $y'(t-0.1)$ in the cases within and after 0.1 s. Based on such an idea, the simulation model in Fig. 9.5b should be created instead, where the input signal is $y'(t)$. Note that (1) the threshold of the switch block is set to 0.1, i.e., to activate the delay block after 0.1 s. Before that time, the history function of t is used instead. (2) The Fcn block process the translated function of t, where it should not be expressed as `-exp(-u)`, but `-exp(0.1-u)`. Therefore, the model thus created is the correct one, with the considerations of the history functions. Such a delay signal can be used directly in FODDE modeling and simulation. ◄

9

9.2.3 Solutions of FODDEs

The block diagram-based modeling methodologies for FODDEs are considered in this section. The block diagram of the FODDE model should be established first, and the delay signals should be treated with delay blocks and history function blocks. Therefore, the entire Simulink model can be created. With the powerful system simulation facilities provided in Simulink, the numerical solutions of the FODDEs can be found. In this section, examples are used to demonstrate the modeling and simulation methods for FODDEs.

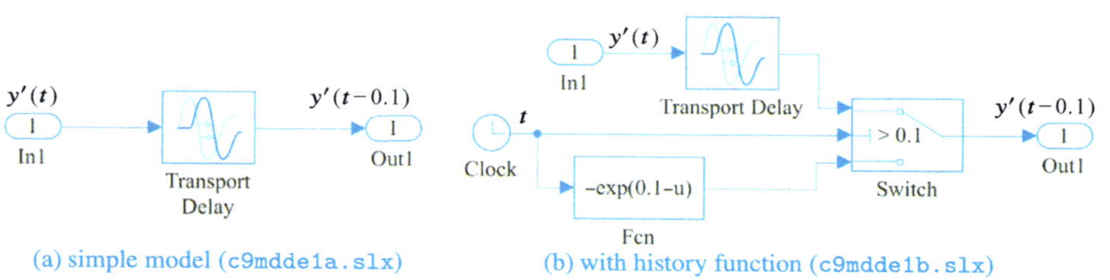

Fig. 9.5 Handling history functions

▶ Example 9.10

Solve the FODDE in Example 9.7, and assess the impact of the parameters on the simulation results.

Solutions Consider the FODDE model. In fact, the block diagram in Example 8.12 can be used as the initial model. The delay signal with nonzero history function should be constructed according to the suggestions in Fig. 9.5b, such that the simulation model in Fig. 9.6 can be constructed. The MATLAB function for the corresponding block is written as follows.

```
function y=bp8_fcn(t)
   y=t^(1.5-sqrt(2))*exp(t)*...
      ml_func([1,3-sqrt(2)],-t)./ml_func([1,1.5],-t);
end
```

Different combinations of `ww` and N are tried, and the measured data in Table 9.6. It is found that to have fast simulation, relatively accurate results can be found within a very short-time interval. To carry out accurate simulation, the maximum error may be reduced to a level as low as 10^{-8}.

```
>> ww=[1e-4 1e4]; N=15;                     %set different parameters
   tic, [t,~,y]=sim('bp8_model'); toc      %initiate simulation process
   length(t), max(abs(y-exp(-t)))          %compute the maximum error
```

◀

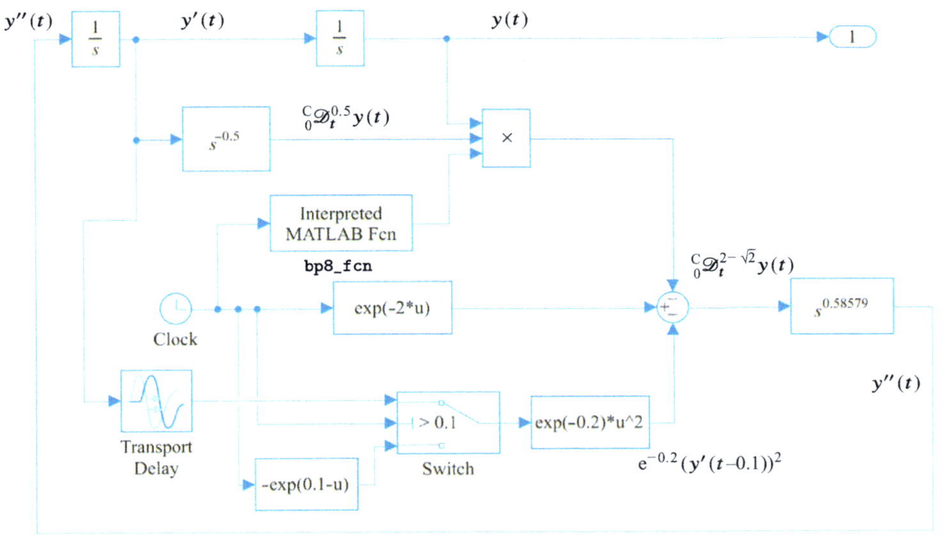

Fig. 9.6 Simulink model of the FODDE (`bp8_model.slx`)

Table 9.6 The impacts of filter parameters on simulation results

Frequency bands	$(10^{-3}, 10^3)$	$(10^{-4}, 10^4)$	$(10^{-5}, 10^5)$	$(10^{-6}, 10^6)$	$(10^{-7}, 10^7)$
Order	11	15	25	29	39
Maximum error	2.4420×10^{-4}	2.4332×10^{-5}	2.5392×10^{-6}	3.2158×10^{-7}	2.6870×10^{-8}
Time elapse	4.180	6.518	7.182	11.553	13.833
Number of points	8150	15717	28410	38360	40972

► Example 9.11

Consider now the FODDE has two delay signals

$$ {}_0^{\mathrm{C}}\mathscr{D}_t^{\sqrt{2}}y(t)=-t^{1.5-\sqrt{2}}\frac{\mathrm{E}_{1,3-\sqrt{2}}(-t)}{\mathrm{E}_{1,1.5}(-t)}\,\mathrm{e}^{t-0.3}y(t-0.3)\,{}_0^{\mathrm{C}}\mathscr{D}_t^{0.5}y(t)+\mathrm{e}^{0.2-2t}-\left[y'(t-0.1)\right]^2. $$

When $t<0$, the history functions are $y(t)=\mathrm{e}^{-t}$ and $y'(t)=-\mathrm{e}^{-t}$. Find the numerical solutions of the FODDE in the interval $0\leqslant t\leqslant 2$. The exact solution of the FODDE is $y(t)=\mathrm{e}^{-t}$.

9

Solutions If two delay signals are involved, these signals can be constructed separately, according to the method in Fig. 9.5b, such that the Simulink model in Fig. 9.7 can be created. The MATLAB function written for the Interpreted MATLAB Fcn block is

```
function y=bp8a_fcn(t)
   y=t^(1.5-sqrt(2))*exp(t-0.3)*...
       ml_func([1,3-sqrt(2)],-t)./ml_func([1,1.5],-t);
end
```

Selecting the values ww and N as follows, the time elapse is 4.33s, and the number of points computed is 19850. The maximum error is 2.3759×10^{-5}. To get more accurate simulation results, the two parameters can be tuned.

```
>> ww=[1e-4 1e4]; N=15;                    % set the filter parameters
   tic, [t,~,y]=sim('bp8a_model'); toc     % initiate simulation
   length(t), max(abs(y-exp(-t)))          % find themaximum error
```

◄

► Example 9.12

Consider the problem in Example 9.10. Assume that the history functions are all zero. The FODDE is modified into the following form:

$$ {}_0^{\mathrm{C}}\mathscr{D}_t^{\sqrt{2}}y(t)=-t^{1.5-\sqrt{2}}\frac{\mathrm{E}_{1,3-\sqrt{2}}(-t)}{\mathrm{E}_{1,1.5}(-t)}\,\mathrm{e}^{t}\,y(t)\,{}_0^{\mathrm{C}}\mathscr{D}_t^{0.5}y(t-0.2)+\mathrm{e}^{-2t}-\left[y'(t-1)\right]^2. $$

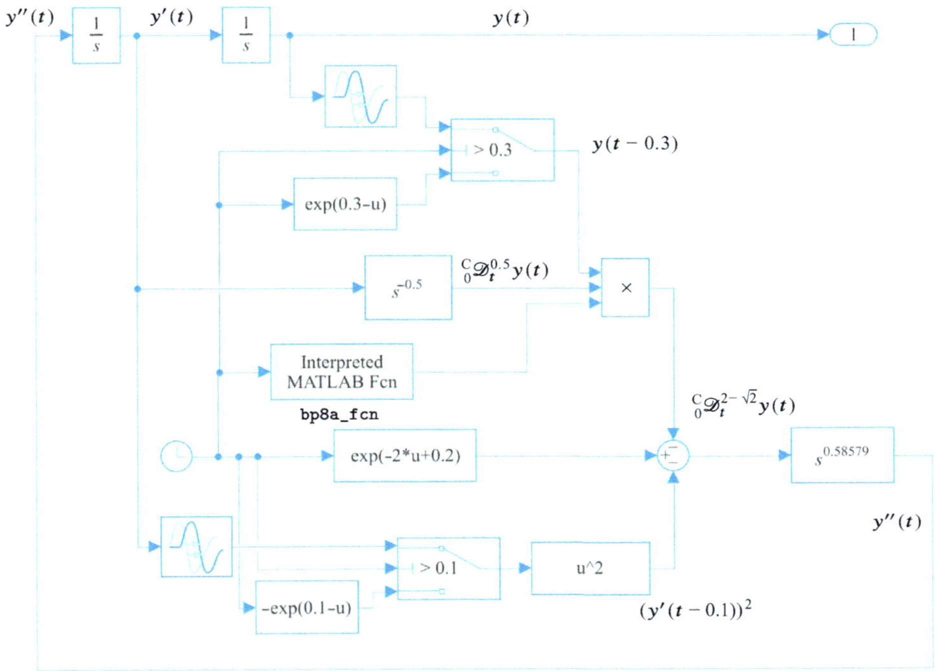

Fig. 9.7 The FODDE with two delay signals (`bp8a_model.slx`)

Solve the equation and validate the results.

Solutions Assume that the history functions are all zero, the function $y(t) = \mathrm{e}^{-t}$ is no longer the exact solution of the problem. There is no use to assess the accuracy by comparing this function. How can we validate the numerical solutions? Different filter parameters and Simulink solver parameters can be used to solve the differential equations. If the two results in two runs are identical, the solutions can be accepted, otherwise the parameters should be modified and validations should be made again.

Since the history functions of all these signals are 0, the signals can be fed directly into the Transport Delay blocks to find the delay signals. There is no longer any need for switch blocks to describe the history functions. Therefore the modeling process is much more simplified. The Simulink model in Fig. 9.8 can be created, where two different sets of values are assigned to the Oustaloup filters and draw simulation results. It can be seen that the two sets of simulation results are exactly the same, implying that the simulation results are correct.

```
>> ww=[1e-5,1e5]; N=20; [t1,~,y1]=sim('c9mdde3');
   ww=[1e-8,1e6]; N=41; [t2,~,y2]=sim('c9mdde3');
   plot(t1,y1,t2,y2,'--')
```

◀

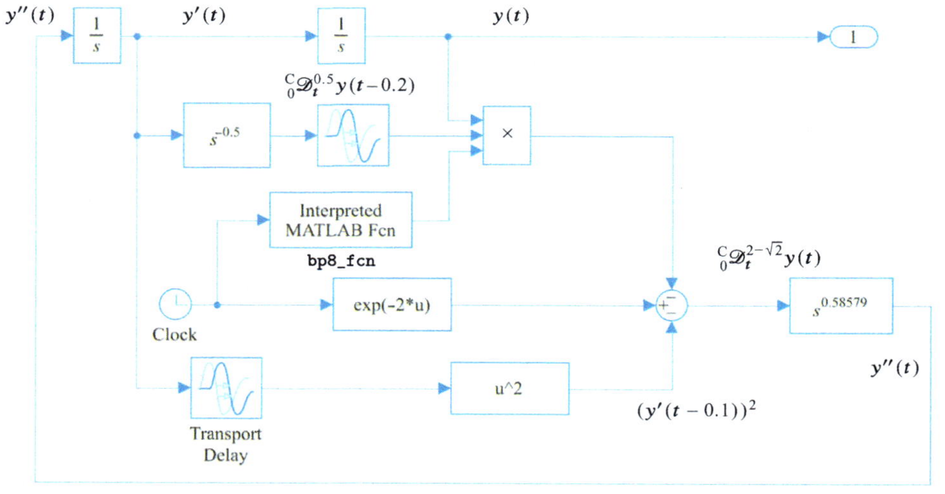

Fig. 9.8 FODDE with zero history functions (`c9mdde3.slx`)

9.3 Boundary Value Problems in FODEs

In the previous discussions on the numerical solutions of various FODE problems, it is always assumed that the function values of $y(t)$ and its integer-order derivatives at time t_0 are given. These problems are also known as initial value problems (IVPs). In certain FODEs, if the initial values of some state variables are known, also the relationship between the initial and final values of some states are known, or satisfy certain requirements, the problems are classified as boundary value problems (BVPs). In this section, the mathematical forms of BVPs are presented first, followed by the introduction of the shooting methods. Finally the combination of numerical optimization methods and algebraic equation solutions method with shooting methods are presented in solving BVPs.

9.3.1 Mathematical Forms of BVPs

The FODE descriptions of BVPs are identical to the ones in ▶ Chap. 8. That is, the FODEs can either be the explicit models in Definition 7.1, or in the extended state space models in Definition 7.4. Even, they can be described as implicit FODEs or FODDEs.

Consider first the explicit FODE model. Since the given conditions here are not the same as the ones in IVPs, the mathematical definition of the BVPs of explicit FODEs is presented next.

Definition 9.2 BVPs of explicit FODEs

The mathematical form of an explicit FODE is

$$ {}_{t_0}^{\mathrm{C}}\mathscr{D}_t^{\alpha} y(t) = f\left(t, y(t), {}_{t_0}^{\mathrm{C}}\mathscr{D}_t^{\alpha_1} y(t), {}_{t_0}^{\mathrm{C}}\mathscr{D}_t^{\alpha_2} y(t), \cdots, {}_{t_0}^{\mathrm{C}}\mathscr{D}_t^{\alpha_n} y(t)\right), \tag{9.12} $$

where $\alpha > \max(\alpha_1, \alpha_2, \ldots, \alpha_n) > 0$, and $q = \lceil \alpha \rceil$. The boundary value conditions of function $y(t)$ and its derivatives are

$$ \psi_i(y(t_0), y'(t_0), \ldots, y^{(q-1)}(t_0), y(t_{\mathrm{n}}), y'(t_{\mathrm{n}}), \ldots, y^{(q-1)}(t_{\mathrm{n}})) = 0, \tag{9.13} $$

where $\psi_i(\cdot)$ is an arbitrary nonlinear function $i = 1, 2, \ldots, q$. The solution interval is $t \in [t_0, t_{\mathrm{n}}]$.

For simplicity, only the boundary values of the function $y(t)$ and its integer-order derivatives are considered. In fact, the methods in this section apply also to the cases of fractional-order boundary value conditions. This is not discussed in this section.

For FOSS models, there are also boundary value problems. The general form is given next.

Definition 9.3 BVPs for fractional state space models

The mathematical form of the nonlinear FOSS model is

$$ {}_{t_0}^{\mathrm{C}}\mathscr{D}_t^{\boldsymbol{\alpha}} \boldsymbol{x}(t) = \boldsymbol{f}\left(t, \boldsymbol{x}(t)\right), \tag{9.14} $$

where $\boldsymbol{\alpha} = [\alpha_1, \alpha_2, \ldots, \alpha_n]$, and $q = \lceil \max(\alpha_i) \rceil$. The given boundary value conditions of the states $\boldsymbol{x}(t)$ are

$$ \psi_i(\boldsymbol{x}(t_0), \boldsymbol{x}(t_{\mathrm{n}})) = 0, \tag{9.15} $$

where $i = 1, 2, \ldots, q$. The solution interval is $t \in [t_0, t_{\mathrm{n}}]$, $\psi_i(\cdot)$ is an arbitrary nonlinear function.

For the third benchmark problem in Reference [1], slight modification can be made to construct the following boundary value problems.

▶ Example 9.13

The explicit FODE is given by

$$ {}_0^{\mathrm{C}}\mathscr{D}_t^{\sqrt{2}} y(t) = -t^{1.5-\sqrt{2}} \frac{\mathrm{E}_{1,3-\sqrt{2}}(-t)}{\mathrm{E}_{1,1.5}(-t)} \mathrm{e}^t y(t)\, {}_0^{\mathrm{C}}\mathscr{D}_t^{0.5} y(t) + \mathrm{e}^{-2t} - \left[y'(t)\right]^2, $$

with exact solution $y(t) = \mathrm{e}^{-t}$. Two BVPs can be constructed as
(1) The known boundary value conditions are $y(0) = 1$ and $y(2) = \mathrm{e}^{-2}$.
(2) The boundary value conditions are $y(2) = \mathrm{e}^{-2}$ and $y'(2) - y(0) = -\mathrm{e}^{-2} - 1$. ◀

9.3.2 Optimization and Algebraic Equation Modeling in Shooting Methods

Shooting methods are commonly used BVP methods. As discussed earlier, if the initial values of FODEs are known, command-driven or block diagram-based methods can be used to find the numerical solutions of FODEs. Therefore, the following ideas can be conceived: The unknown initial values can be assigned first, with the known IVP solvers, the numerical solutions of the FODEs can be found. There are of course differences between the solutions and the expected boundary value conditions. The differences can then be used to modify the assigned initial values so as to find better numerical solutions. With repeated modifications of the given initial values, it is possible to find consistent ones, such that the numerical solutions satisfying the given boundary value conditions.

Since this idea is similar to target shooting to adjust repeatedly the angle of a gun so as to find a suitable gun angle to hit the target. This type of methods is referred to as shooting methods.

With ideas in the shooting methods, two actual modeling methods can be conceived: (1) to build up an optimization-based method, (2) to build up an algebraic equation-based method. If the given boundary value conditions have m initial values, it is safe to assume that the first m conditions are initial values, they can be assigned in the corresponding integrators in the key signals in the Simulink model. The remaining $q-m$ boundary value conditions can be used to set up the optimization or algebraic equation model.

Definition 9.4 The optimization model

The sum of the absolute values of the remaining boundary value condition errors can be selected as the objective function

$$\min_{\boldsymbol{x}} \sum_{i=m+1}^{q} \left|\psi_i\big(y(t_0), y'(t_0), \dots, y^{(q-1)}(t_0), y(t_{\rm n}), y'(t_{\rm n}), \dots, y^{(q-1)}(t_{\rm n})\big)\right|, \tag{9.16}$$

where the decision variables $\boldsymbol{x}$ can be selected as the unknowns of the remaining ones in $y(t_0), y'(t_0), \dots, y^{(q-1)}(t_0)$.

Definition 9.5 Algebraic equation model

The remaining boundary value conditions are simultaneous algebraic equations

$$\psi_i\big(y(t_0), y'(t_0), \dots, y^{(q-1)}(t_0), y(t_{\rm n}), y'(t_{\rm n}), \dots, y^{(q-1)}(t_{\rm n})\big) = 0, \tag{9.17}$$

where $i = m+1, m+2, \dots, q$, and the unknowns are the unknown ones in the remaining of $y(t_0), y'(t_0), \dots, y^{(q-1)}(t_0)$.

Remarks The solution models

(1) In the optimization model, the sum of absolute errors is used. However many scholars tend to select the sum of squared errors as the objective functions, since the derivatives can be made so as to find the analytical solutions. Unfortunately, there are not at all any analytical solutions in this problem. A squared sum sometimes may lead to premature solutions. Thus the sum of absolute errors is recommended.

(2) In the algebraic equation model, the equations can alternatively be selected as the absolute values of the boundary conditions. While in the numerical solution methods, the absolute values may mislead the searching directions. Therefore it is not suggested to use absolute values in constructing the algebraic equations. □

▶ Example 9.14

Consider the two BVPs constructed in Example 9.13. Write the objective functions for them.

Solutions If the initial values $y(0)$ and $y'(0)$ are known, the boundary values $\hat{y}(2)$ and $\hat{y}'(2)$ can be found by numerical FODE solvers. Therefore the following two cases can be considered, such that the original BVPs can be converted into numerical optimization problems. The key issues in the conversion include the selection of decision variables, and how to write out the objective functions. The optimization modeling problems are, respectively, explored for the two BVPs.

(1) Since the initial value $y(0)=1$ is definitely known, the solution is to find the consistent $y'(0)$, such that the result $\hat{y}(2)$ obtained in simulation approaches $y(2) = \mathrm{e}^{-2}$ as close as possible. This is to introduce a decision variable $x=y'(0)$, such that the objective function $f(x)=|\hat{y}(2)-\mathrm{e}^{-2}|$ is minimized.

(2) Since $y(0)$ and $y'(0)$ are all unknown, two decision variables $x_1 = y(0)$ and $x_2 = y'(0)$ should be introduced, such that the two boundary value conditions are satisfied. Therefore the optimization problem below can be set up

$$\min_{\boldsymbol{x}} \left|\hat{y}(2)-\mathrm{e}^{-2}\right| + \left|\hat{y}'(2)-x_1+\mathrm{e}^{-2}+1\right|.$$

◀

9.3.3 Fast Restart Mode in Simulink

In normal situations, no matter how the simulation process is invoked - by calling `sim()` function, or by clicking the ▶ button in Simulink interface, the first thing to do in Simulink is the model compiling process. The compiling process is not quite time consuming, if it is executed only once. However in the optimization process, the simulation process should be invoked repeatedly. If the model is compiled each time, the total time elapse is significant. If there is a mechanism which bypasses the compiling process, the efficiency in the optimization process can be boosted.

Fast restart mode in Simulink can be utilized. If there is no change in the model structure, with only the parameters in the model changed, the compiling process can be skipped in such a mode. The procedures in fast restart mode setting and objective function descriptions are summarized as follows:

(1) Open the Simulink model and issue the following command, the fast restart mode is activated.

```
>> set_param(gcs,'FastRestart','on')
```

(2) In writing the MATLAB file to describe the objective function, the updated values of the decision variables should be assigned into Simulink model workspace, not MATLAB workspace. Specific objective function writing methods and optimization process will be demonstrated later through examples.

(3) When calling `sim()` function to initiate simulation process in Simulink, only one returned argument `out` is allowed. Such a variable is a structured one, whose `tout` and `yout` members store the time and output signals.

When the objective function is written, the functions `fminsearch()` or `fminunc()` can be used to find the optimum values of the decision variables. When the optimization process is complete, the optimum decision variables can be used to restart the whole simulation process, so as to find the numerical solutions of the BVPs. In this section, an algebraic equation solution-based method can also be used to solve the boundary value problems. Some of the BVP problems which cannot be solved with optimization methods can be explored with the new solver.

9.3.4 Solutions of BVPs

In the previous section, the methods converting the BVPs into numerical optimization problems and algebraic equation solution problems are introduced, and the implementing tactics are introduced. Since in the equation solution of optimization processes, the numerical solutions of FODE should be called tens or even hundreds of times, the efficiency and accuracy of the FODE solvers should be ensured. It is obvious that it is hard for the command-driven solvers in ▶ Chap. 7 to satisfying such a request. Block diagram-based FODE solvers should be adopted instead. It is not necessary to select too demanding filter parameters, since the time elapse issues should also be considered. Some examples are presented next to demonstrate the solution methods of BVPs for FODEs.

▶ Example 9.15

Solve the first BVP in Example 9.13.

Solutions To solve the BVP, a corresponding Simulink model should be created first. Since the model is quite similar to the one in Example 8.12, the Simulink model in ◘ Fig. 9.9 can be established. More importantly, the initial values of the two integrators should be set to variables a and b, and let $b = 1$ for later use. Therefore, the value of variable a should be updated within the MATLAB function describing the objective function.
The MATLAB function written for the Interpreted MATLAB Fcn block is

```
function y=bp6_fcn(u)
   y=u^(1.5-sqrt(2))*exp(u)*...
        ml_func([1,3-sqrt(2)],-u)./ml_func([1,1.5],-u);
end
```

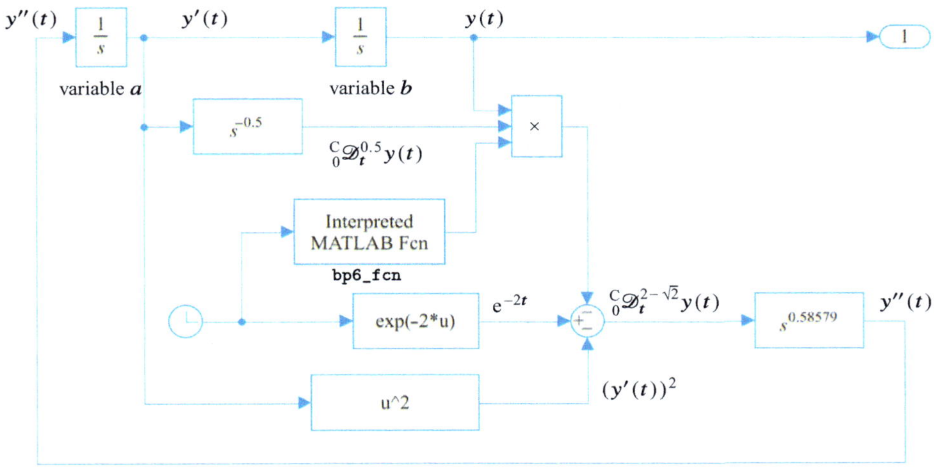

Fig. 9.9 New Simulink model (bp6_model.slx)

In normal cases, for a specific a value, that is $y'(0)$, it may lead singularities to the whole problem, which causes abnormal termination of the solution process, with error messages. A trial structure is used here. If there are errors occur, the objective function value is assigned alternatively into very large values, to avoid singularities, such that the solution process can be executed. Based on such an idea, the following objective function can be written.

```
function y=c9mbvp1(x)
   W=get_param(gcs,'ModelWorkspace'); assignin(W,'a',x)
   try
      txy=sim('bp6_model.slx'); y0=txy.yout;
      y=abs(y0(end)-exp(-2));
   catch, y=10; end
end
```

The model workspace W is obtained first in the objective function. The decision variable is written to variable a in the model workspace. Start the simulation process, the final value can be found and written into the objective function. Therefore, the optimization process can be executed, to find the consistent initial value $x = -0.9999999726$, which is very close to the theoretical value of -1. The time elapse of the solution process is 36.61 s. Starting from the value, the actual final value error is 2.5326×10^{-9}.

```
>> ww=[1e-7,1e7]; N=39;
   bp6_model; set_param(gcs,'FastRestart','on');      % fast restart mode
   tic, x=fminunc(@c9mbvp1,rand(1)), toc, c9mbvp1(x)
   set_param(gcs,'FastRestart','off');                % close the mode
```

If the algebraic equation-based method is used, the following function can be used to describe the algebraic equation.

```
function y=c9mbvp1a(x)
   W=get_param(gcs,'ModelWorkspace'); assignin(W,'a',x)
   try, txy=sim('bp6_model.slx'); y0=txy.yout; y=y0(end)-exp(-2);
   catch, y=10; end
end
```

The following commands can be used to solve directly the following differential equation, with the consistent initial value of $x = -0.9999999727$, very close to the theoretical value of -1. The entire process takes about 125.43s, with final value error of 3.7068×10^{-10}. It can be seen that for this example, the optimization method has higher efficiency, since the time elapse is less than the algebraic equation-based solver.

```
>> bp6_model; set_param(gcs,'FastRestart','on');     %fast restart mode
   tic, x=fsolve(@c9mbvp1a,rand(1)), toc, c9mbvp1a(x)
   set_param(gcs,'FastRestart','off');                %close the mode
```

◀

▶ Example 9.16

Solve the second BVP in Example 9.13.

Solutions Compared with the previous example, the initial values in the two integrators should be assigned, such that they can be set to variables a and b. Since signal $y'(t)$ is involved, the Simulink model should be modified, by adding an extra output port, as shown in Fig. 9.10. The new model is saved into the file `bp6a_model.slx`.
In fact, it is seen from a numerical solution viewpoint that the second BVP has one more decision variables. With reference to the objective function in Example 9.14, the new objective function is written.

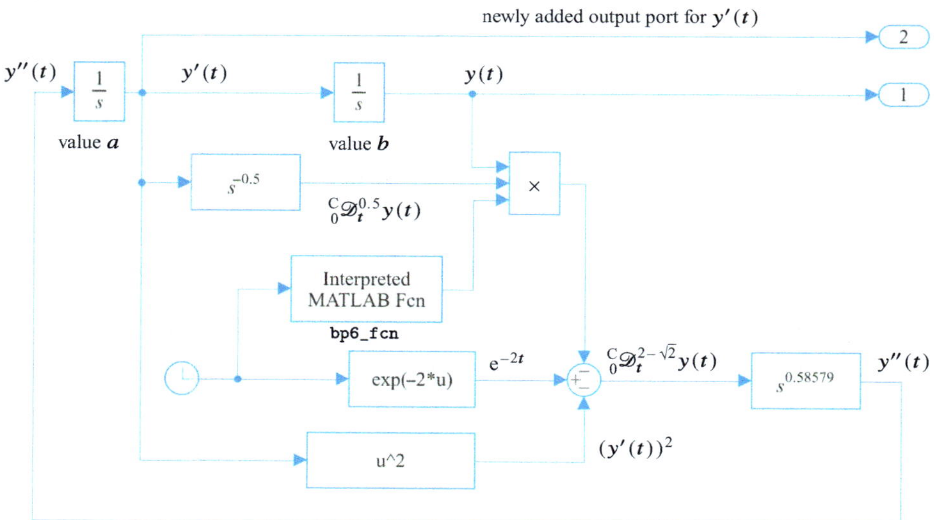

Fig. 9.10 New Simulink model (bp6a_model.slx)

```
function y=c9mbvp2(x)
   W=get_param(gcs,'ModelWorkspace');
   assignin(W,'b',x(1)), assignin(W,'a',x(2))
   try
      txy=sim('bp6a_model.slx'); y0=txy.yout;
      y=abs(y0(end,1)-exp(-2))+abs(y0(end,2)-y0(1,1)+exp(-2)+1);
   catch, y=10; end
end
```

The following commands are used to solve the optimization problem. Unfortunately, after several executions, the impact of initial values on the optimization process is too heavy. The correct solutions cannot be found in most of the time. More than 10 times are executed, each takes about 80 to 120 s, and exact solutions are not found. The best objective function value found is 0.0043. Therefore, it is concluded that the optimization-based method failed.

```
>> ww=[1e-7,1e7]; N=39; x=rand(2,1);
   bp6a_model; set_param(gcs,'FastRestart','on');
   tic, x=fminunc(@c9mbvp2,x), toc, c9mbvp2(x)
   set_param(gcs,'FastRestart','off');
```

Now consider the algebraic equation-based solver. The following function can be written to describe the simultaneous algebraic equations.

```
function y=c9mbvp2a(x)
   W=get_param(gcs,'ModelWorkspace');
   assignin(W,'b',x(1)), assignin(W,'a',x(2))
   try
      txy=sim('bp6a_model.slx'); y0=txy.yout;
      y=[y0(end,1)-exp(-2); y0(end,2)-y0(1,1)+exp(-2)+1];
   catch, y=[10; 10]; end
end
```

Therefore the following commands can be used to solve directly the algebraic equations.

```
>> bp6a_model; set_param(gcs,'FastRestart','on');        %fast restart mode
   tic, [x,f,c,d]=fsolve(@c9mbvp2a,rand(2,1)), toc
   c9mbvp2a(x), set_param(gcs,'FastRestart','off');     %close the mode
```

The consistent initial values $b = 1.000001178730345$ and $a = -1.000000605777505$ are found. The errors for the two boundary values are, respectively, -1.0602×10^{-10} and -2.2265×10^{-10}, with time elapse 75.32s. The statements in the paragraph are executed repeatedly more than 10 times, and the solutions are all very close. For this particular problem, the algebraic equation-based method is more reliable. In ordinary cases, the algebraic equation solution process calls about 60 to 80 runs of the simulation process. It can be imagined that when the command-driven solvers in ► Chap. 7 are used as the kernel, the computational load may be too heavy to be actually used in BVP solutions. ◄

▶ Example 9.17

Consider the implicit FODE in Example 9.2. Assume that the exact solution is still $y(t) = \mathrm{e}^{-t/2}$, and the final values satisfy the conditions $y(4) = \mathrm{e}^{-2}$ and $y'(4) = -0.5\,\mathrm{e}^{-2}$. Use algebraic equation solution method to solve again the final value implicit FODE problem.

Solutions Assume the model `c9mimps.slx` in Fig. 9.1 is used. The initial values of the two integrators are assigned to variables a and b, respectively. In addition, connect the $y'(t)$ signal to the second output port. The new model `bp7a_model.slx` can be created, as shown in Fig. 9.11. Therefore, the following MATLAB function can be written to describe the algebraic equations for the boundary value conditions.

```
function y=c9mbvp3(x)
   W=get_param(gcs,'ModelWorkspace');
   assignin(W,'b',x(1)), assignin(W,'a',x(2))
   try
      txy=sim('bp7a_model.slx'); y0=txy.yout;
      y=[y0(end,1)-exp(-2); y0(end,2)+0.5*exp(-2)];
   catch, y=[10; 10]; end
end
```

The following commands can be used to solve the boundary value problem of the implicit FODE. It should be noted that since random initial search points are assigned, the following algebraic equation solution commands may sometimes fail. In this case, the initial values should be chosen again, until a solution can be found.


```
>> ww=[1e-5 1e5]; N=27; x=rand(2,1);
   bp7a_model; set_param(gcs,'FastRestart','on');
```

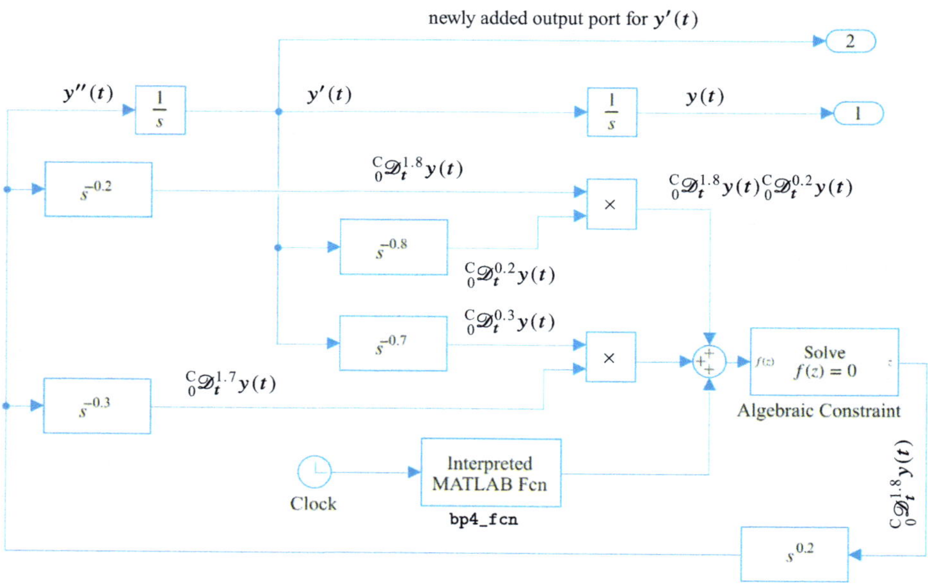

Fig. 9.11 Simulink model of the implicit FODE (`bp7a_model.slx`)

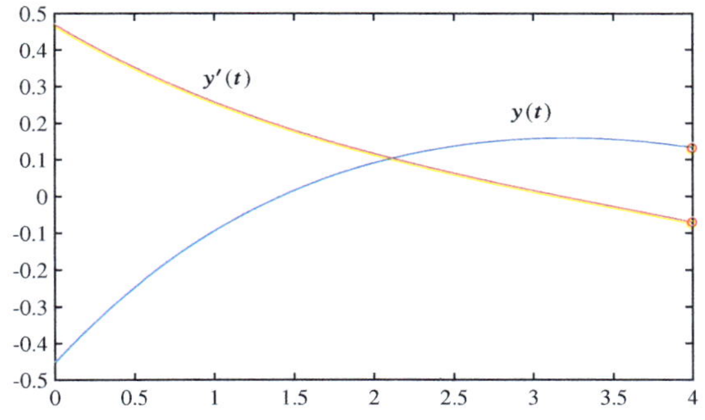

Fig. 9.12 An alternative implicit FODE solution

```
tic, x=fsolve(@c9mbvp3,x), toc, c9mbvp3(x)
set_param(gcs,'FastRestart','off');
```

With the above commands, apart from the known exact analytical solution, another solution is also found, as shown in Fig. 9.12, indicating that the solution is not unique. The equivalent initial values are $b = -0.4506$ and $a = 0.4731$, with error of 10^{-8}. The time elapse is about 70 s.

```
>> plot(tout,yout,[4 4],[exp(-2),-0.5*exp(-2)],'o')
```

◀

9.4 Numerical Solutions of Time-Fractional Partial Differential Equations

Fractional-order partial differential equations (FOPDEs) are attractive research fields. Complicated behaviors such as power-law are often encountered in FOPDEs, which are used widely in many fields such as physics, chemistry, biology, fluid mechanics, anomalous diffusion and control systems. Usually the analytical solutions to FOPDEs cannot be found. Even in very few cases where exact solutions can be found, complicated special functions are involved and cannot be evaluated easily. Therefore, numerical solution algorithms receive significant attentions in FOPDEs and many algorithms are proposed. For instance, the difference methods [5–7], finite difference methods [8, 9], spectral methods [10, 11] and so on.

For one-dimensional FOPDEs, difference methods are effective ones. For different fractional orders, the difference formats are also different. Specifically, α is used to denote the fractional order, the difference formulas for $0 < \alpha < 1$ and $1 < \alpha < 2$ are different [7]. In Reference [12], an algorithm with the identical difference format is proposed to solve the equations, and there is no need to consider separately the fractional orders.

Since the initial values in Caputo fractional-order derivatives have specific physical meanings, Caputo definitions are often considered for the fractional-order derivatives. Therefore the following FOPDE form is considered.

Definition 9.6 Time-fractional partial differential equation

The mathematical form of one-dimensional time-fractional partial differential equation (TFPDE) is

$$\begin{cases} {}_{t_0}^{\mathrm{C}}\mathscr{D}_t^{\alpha} y(x,t) = a\dfrac{\partial^2 y(x,t)}{\partial x^2} + f(x,t) \\ y(x,t_0) = \phi_1(x),\ \dfrac{\partial}{\partial t}y(x,t_0) = \phi_2(x), \cdots, \dfrac{\partial^{n-1}}{\partial t^{n-1}}y(x,t_0) = \phi_n(x) \\ y(x_0,t) = \psi_1(t),\ y(x_M,t) = \psi_2(t),\ x \in [x_0, x_M],\ t \in [t_0, t_N]. \end{cases} \tag{9.18}$$

where a and α are positive real numbers, $n = \lceil \alpha \rceil$, $\phi_i(x)$ are initial functions, while $\psi_1(t)$ and $\psi_2(t)$ are boundary value conditions.

Under theoretic framework, very large value of n can be assigned for initial value problems, however in real problems if $n > 2$, the system is unstable. Therefore at most two initial conditions, $\phi_1(x)$ and $\phi_2(x)$, are considered.

Definition 9.7 Generating mesh grids

The space interval $[x_0, x_M]$ is divided equally into M subintervals (x_i, x_{i+1}), $i = 0, 1, \ldots, M-1$, whose step size is $h = (x_M - x_0)/M$. The time interval $[t_0, t_N]$ is divided equally into N subintervals (t_j, t_{j+1}), $j = 0, 1, \ldots, N-1$, with step size $\tau = (t_N - t_0)/N$. Cut-off rules parallel to x and t axes can be made, to generate mesh grids in the interested region, denoted as (x_i, t_j). The exact solution at the mesh grids are denoted as $y(x_i, t_j)$, while the numerical solutions are denoted as y_j^i.

9

If the x axis step size is h, the following difference formula can be adopted to compute the second-order derivatives, whose accuracy is $o(h^2)$.

$$\frac{\partial^2}{\partial x^2} y(x_i, t_j) = \frac{1}{h^2}\big[y(x_{i-1}, t_j) - 2y(x_i, t_j) + y(x_{i+1}, t_j)\big]. \tag{9.19}$$

Computing the Caputo derivative on the left-hand side of (9.18), where an axillary function is still adopted, the zero initial condition and nonzero initial condition parts in $y(x, t)$ can be separated. The axillary function can be chosen as

$$\hat{y}(x_i, t) = \sum_{k=0}^{n-1} \frac{1}{k!}\phi_{k+1}(x_i)t^k. \tag{9.20}$$

Therefore the solution of the equation can be decomposed into $y(x_i, t) = \hat{y}(x_i, t) + \bar{y}(x_i, t)$, where the initial conditions of $\hat{y}(x_i, t)$ and $y(x_i, t)$ are identical. The initial conditions of $\bar{y}(x_i, t)$ are 0, such that ${}_{t_0}^{\mathrm{C}}\mathscr{D}_t^{\alpha}\bar{y}(x_i, t) = {}_{t_0}^{\mathrm{RL}}\mathscr{D}_t^{\alpha}\bar{y}(x_i, t)$. Since $\alpha > n-1$, ${}_{t_0}^{\mathrm{C}}\mathscr{D}_t^{\alpha}\hat{y}(x_i, t) = 0$. Therefore it is found that

$${}_{t_0}^{\mathrm{C}}\mathscr{D}_t^{\alpha} y(x_i, t) = {}_{t_0}^{\mathrm{C}}\mathscr{D}_t^{\alpha}\hat{y}(x_i, t) + {}_{t_0}^{\mathrm{C}}\mathscr{D}_t^{\alpha}\bar{y}(x_i, t) = {}_{t_0}^{\mathrm{RL}}\mathscr{D}_t^{\alpha}\bar{y}(x_i, t). \tag{9.21}$$

The methods in ▶ Chap. 3 can be used to evaluate directly ${}_{t_0}^{\rm C}\mathscr{D}_t^{\alpha}y(x_i,t)$, such that

$$ {}_{t_0}^{\rm C}\mathscr{D}_t^{\alpha}y(x_i,t)={}_{t_0}^{\rm RL}\mathscr{D}_t^{\alpha}\bar{y}(x_i,t)=\frac{1}{\tau^\alpha}\sum_{k=0}^{j}w_{j-k}\bar{y}(x_i,t_k). \tag{9.22}$$

In each mesh grid line $x=x_i$, the axillary function can be constructed

$$\hat{y}(x_i,t)=\begin{cases}\psi_1(t), & i=0\\ \psi_2(t), & i=M\\ \displaystyle\sum_{k=0}^{n-1}\frac{1}{k!}\phi_k(x_i)t^k, & 0<i<M.\end{cases} \tag{9.23}$$

The solution $y(x_i,t)$ in equation (9.18) can be decomposed as

$$y(x_i,t)=\hat{y}(x_i,t)+\bar{y}(x_i,t). \tag{9.24}$$

Substituting (9.24) and (9.21) into (9.18), the equation (9.24) on the mesh grid (x_i,t_j) is converted to

$$ {}_{t_0}^{\rm RL}\mathscr{D}_t^{\alpha}\bar{y}(x_i,t_j)=a\frac{\partial^2}{\partial x^2}\bar{y}(x_i,t_j)+a\frac{\partial^2}{\partial x^2}\hat{y}(x_i,t_j)+f(x_i,t_j). \tag{9.25}$$

Substituting (9.19) and (9.22) into (9.25), it is found that

$$\begin{aligned}\frac{1}{\tau^\alpha}\sum_{k=0}^{j}w_{j-k}\bar{y}(x_i,t_k)=&\frac{a}{h^2}\big[\bar{y}(x_{i-1},t_j)-2\bar{y}(x_i,t_j)+\bar{y}(x_{i+1},t_j)\big]+\\ &\frac{a}{h^2}\big[\hat{y}(x_{i-1},t_j)-2\hat{y}(x_i,t_j)+\hat{y}(x_{i+1},t_j)\big]+f(x_i,t_j).\end{aligned} \tag{9.26}$$

Multiply both sides of (9.26) simultaneously by τ^α, then

$$\sum_{k=0}^{j}w_{j-k}\bar{y}(x_i,t_k)=r\big(\bar{y}(x_{i-1},t_j)-2\bar{y}(x_i,t_j)+\bar{y}(x_{i+1},t_j)\big)+\hat{f}_j^i, \tag{9.27}$$

where $r=a\tau^\alpha/h^2$, $\hat{f}_j^i=r\big(\hat{y}(x_{i-1},t_j)-2\hat{y}(x_i,t_j)+\hat{y}(x_{i+1},t_j)\big)+\tau^\alpha f(x_i,t_j)$. Denote $\bar{y}_j^i=\bar{y}(x_i,t_j)$, it is found that

$$\sum_{k=0}^{j}w_{j-k}\bar{y}_k^i=r\big(\bar{y}_j^{i-1}-2\bar{y}_j^i+\bar{y}_j^{i+1}\big)+\hat{f}_j^i, \tag{9.28}$$

where since $\bar{y}_j^{i-1}$, $\bar{y}_j^i$ and $\bar{y}_j^{i+1}$ are unknowns, and move them to the left-hand side of the equal sign, it is found that

$$-r\bar{y}_j^{i-1}+(1+2r)\bar{y}_j^i-r\bar{y}_j^{i+1}=\hat{f}_j^i-\sum_{k=0}^{j-1}w_{j-k}\bar{y}_k^i, \tag{9.29}$$

such that the matrix form can be written as

$$\begin{bmatrix} 1+2r & -r & & \\ -r & 1+2r & r & \\ & \ddots & \ddots & \ddots \\ & & -r & 1+2r \end{bmatrix}\begin{bmatrix} \bar{y}_j^1 \\ \bar{y}_j^2 \\ \vdots \\ \bar{y}_j^{M-1} \end{bmatrix}=\begin{bmatrix} \hat{f}_j^1-\sum_{k=0}^{j-1}w_{j-k}\bar{y}_k^1 \\ \hat{f}_j^2-\sum_{k=0}^{j-1}w_{j-k}\bar{y}_k^2 \\ \vdots \\ \hat{f}_j^M-\sum_{k=0}^{j-1}w_{j-k}\bar{y}_k^M \end{bmatrix}. \tag{9.30}$$

Since in the tri-diagonal coefficient matrix, there are $r=a\tau^\alpha/h^2>0$, the matrix is a diagonal dominant one, thus the equation has a unique solution. Substituting the solution in (9.30) into (9.24), the solution of the TFPDE can be found.

It is shown that the accuracy of the algorithm is $O(\tau^{1+\alpha}, \tau^\alpha h^2)$ [12]. The algorithm and procedures for TFPDEs are summarized as follows.

Algorithm 9.2 Solutions of TFPDE

(1) From the initial and boundary value conditions in (9.18), the axillary function $\hat{y}(x_i, t)$ in (9.23) can be composed.

(2) The solution $y(x_i, t)$ in (9.18) is decomposed into $\hat{y}(x_i, t)+\bar{y}(x_i, t)$. The equation in (9.18) can be converted into the form in (9.25).

(3) With (9.22) and (9.19), the fractional-order and second-order derivatives in (9.25) can be found, to find (9.30).

(4) Find the numerical solution $\bar{y}(x, t)$ from (9.30). Substituting it into (9.24,) the numerical solution of (9.18) can be found.

Remarks Special treatment of algebraic equations

Equation (9.30) is an $\boldsymbol{Ax}=\boldsymbol{b}$ type linear algebraic equation. Various methods can be used to solve it, including the `inv(A)*b` and `A\b` methods, as well as the chasing method for the tri-diagonal equations. In fact, since matrix $\boldsymbol{A}$ is a constant one, and the linear equation is solved repeatedly, a higher efficiency method is to work out its inverse matrix $\boldsymbol{A}_0=\boldsymbol{A}^{-1}$, so that in later equation solutions, `A0*b` can be used directly to find the solutions. □

Based on the above algorithm, a TFPDE solver can be written as

```
function [U,X,T]=ftpde_sol(a,f,alpha,L,T0,phi1,phi2,psi1,psi2,m,n)
   arguments
      a(1,1) double, f, alpha(1,1){mustBePositive}
```

```
      L(1,:) double, T0(1,:) double
      phi1(1,1), phi2, psi1(1,1), psi2(1,1)
      m(1,1){mustBeInteger}=100; n(1,1){mustBeInteger}=100;
   end
   if isscalar(L), L=[0,L]; end, dx=(L(2)-L(1))/m;
   if isscalar(T0), T0=[0,T0]; end, dt=(T0(2)-T0(1))/n;
   x=L(1)+(0:m)'*dx; t=T0(1)+(0:n)'*dt;     %set up the segments
   [T,X]=meshgrid(t,x); F=f(X,T);           %set up the mesh grids
   r=a*(dt^alpha)/dx/dx; U=zeros(m+1,n+1);
   v=phi1(x); V=v; I=2:m;                   %initial conditions
   if alpha>1, v1=phi2(x)*dt; V=[V zeros(m+1,n)]; end
   if alpha>1, for i=1:n, v=v+v1; V(:,i+1)=v; end, end
   if alpha>1, dV=zeros(m+1,n+1);
      for j=2:n+1
         dV(I,j)=r*V(I-1,j)-2*r*V(I,j)+r*V(I+1,j);
      end
   else, dV(I)=r*V(I-1)-2*r*V(I)+r*V(I+1); dV=[dV(:); 0]; end
   F=F*(dt^alpha)+dV; r0=r*ones(m-2,1);
   g=genfunc(1); w=get_vecw(alpha,n+1,double(g))';
   A0=(2*r+1)*eye(m-1)-diag(r0,1)-diag(r0,-1); A0=inv(A0);
   U(1,:)=psi1(t)-V(1,:)'; U(m+1,:)=psi2(t)-V(m+1,:)';
   for j=1:n                     %set up the tri-diagonal equation
      b1=[r*U(1,j+1); zeros(m-3,1); r*U(m+1,j+1)];
      b1=b1-U(2:m,1:j)*w(j+1:-1:2)+F(2:m,j+1);
      U(2:m,j+1)=A0*b1;          %use inverse to solve the equation
   end, U=U+V;                   %find the solution matrix
end
```

The syntax is [U,X,T]=`ftpde_sol`(a,f,α,L,T,ϕ_1,ϕ_2,ψ_1,ψ_2,m,n), where a, α, $\boldsymbol{L} = [x_0, x_n]$, $\boldsymbol{T} = [t_0, t_m]$ are the same as the ones in the formulas. If only scalars are provided in the latter two arguments, it implies that $x_0 = 0$ or $t_0 = 0$. Argument f is the function handle, while ϕ_1 and ϕ_2 are initial function handles. If $\alpha \leq 1$, ϕ_2 can be set to an empty matrix. Arguments ψ_1 and ψ_2 are the handles of boundary functions, Arguments m and n are, respectively, the segmentation numbers on x and t axes, with default values of 100. The returned argument $\boldsymbol{U}$ is the solution matrix of $y(x, t)$, while $\boldsymbol{X}$ and $\boldsymbol{T}$ are, respectively, the mesh grid matrices of x and t axes.

▶ Example 9.18

Solve numerically the following slow-diffusion TFPDE [12]:

$$ {}_0^{\mathrm{C}}\mathscr{D}_t^{0.8} y(x,t) = \frac{\partial^2 y(x,t)}{\partial x^2} - \mathrm{e}^{x+0.3t} + 0.3\, t^{0.2} \mathrm{e}^x\, \mathrm{E}_{1,1.2}(0.3t), $$

where $x \in [0, 1]$, $t \in [0, 1]$. The initial values are $y(x, 0) = \mathrm{e}^x$ and the boundary value conditions are, respectively, $y(0, t) = \mathrm{e}^{0.3t}$ and $y(1, t) = \mathrm{e}^{1+0.3t}$. The exact solution of the equation is $y(x, t) = \mathrm{e}^{x+0.3t}$.

Solutions Compared with the TFPDE in Definition 9.6, it is found that $a = 1$, $\alpha = 0.8$, $L = T = 1$, $f(x, t) = -e^{x+3t} + 0.3\,t^{0.2}e^x\,E_{1,1.2}(0.3t)$, $\phi_1(x) = e^x$. Since $\alpha = 0.8 < 1$, there no need to assign $\phi_2(x)$. The boundary value conditions are $\psi_1(t) = e^{0.3t}$ and $\psi_2(t) = e^{1+0.3t}$. Divide x and t axes into 15 and 100 mesh grids, the following commands can be issued to solve directly the equation, and the solution surface is shown in Fig. 9.13.

```
>> a=1; alpha=0.8; L=1; T0=1; m=15; n=100;
   f=@(x,t)-exp(x+0.3*t)+...          %anonymous function with dot operations
           0.3*(t.^0.2).*exp(x).*ml_func([1,1.2],0.3*t);
   phi1=@(x)exp(x); psi1=@(t)exp(0.3*t); psi2=@(t)exp(1+0.3*t);
   [U,X,T]=ftpde_sol(a,f,alpha,L,T0,phi1,'',psi1,psi2,m,n);
   surf(X,T,U)
```

Since the exact solution is known, the exact solution can be found and the errors from the numerical solutions can be found. The maximum error is 6.0399×10^{-4}, and the error surface is shown in Fig. 9.14. It can be seen from the error that when t is very small, the error is large, as in the case in Fig. 4.2. To reduce the error, the step size in x axis should be reduced.

```
>> f0=@(x,t)exp(x+0.3*t);                          %exact solution
   U0=f0(X,T); err=U-U0; max(abs(err(:)))          %find the maximum error
   surf(X,T,U-U0)                                  %draw error surface
```

9

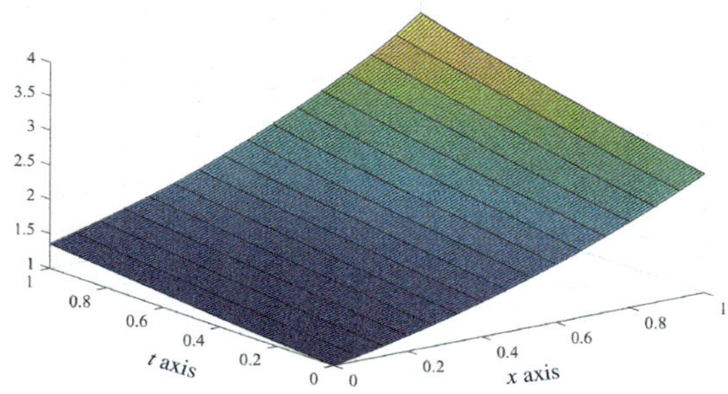

Fig. 9.13 Solution surface

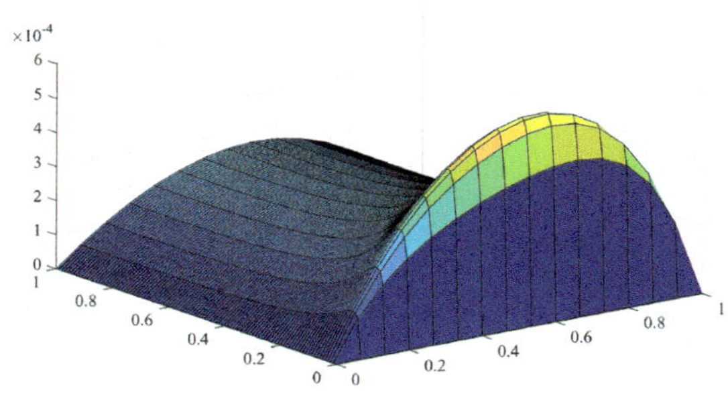

Fig. 9.14 The error surface

Table 9.7 Solution efficiencies for different values of m

m	10	20	30	40	50	100
Time elapse	0.0065	0.0272	0.0647	0.164	0.375	14.979
Maximum error	6.502×10^{-4}	2.196×10^{-4}	1.128×10^{-4}	6.946×10^{-5}	4.738×10^{-5}	1.405×10^{-5}

Different values of m are tested, and let $n = m^2$, the errors are measured, as shown in Table 9.7. It is seen that with the increase of m, the precision is also increasing. When $m = 100$, the time elapse also significantly increases, since $n = 10000$. Large matrix of 101×10001 is involved, such that the algebraic equation is solved 10000 times. If n is not too large, the algorithm here is rather highly efficient.
Apart from 3D surface display, animation method can also be used to display the solutions [13], with the following commands.

```
>> [U,X,T]=ftpde_sol(a,f,alpha,L,T0,phi1,'',psi1,psi2,120,100);
   figure(gcf), h=plot(x,U(:,1)), ylim([0,4])
   for i=2:size(U,2), h.YData=U(:,i); drawnow, end
```

◀

▶ Example 9.19

Solve the following hyper-diffusion TFPDE [12].

$$ {}_0^{\rm C}\mathscr{D}_t^{1.7} y(x,t) = \frac{\partial^2 y(x,t)}{\partial x^2} - {\rm e}^{x+0.5t} + 0.25\, t^{0.3} {\rm e}^x\, {\rm E}_{1,1.3}(0.5t), $$

where $x \in [0,1]$, $t \in [0,1]$. The initial values are $y(x,0) = {\rm e}^x$ and $y'(x,0) = 0.5\,{\rm e}^x$, while the boundary functions are $y(0,t) = {\rm e}^{0.5t}$ and $y(1,t) = {\rm e}^{1+0.5t}$. The exact solution of the equation is $y(x,t) = {\rm e}^{x+0.5t}$.

Solutions Compared with the one in Example 9.18, the order α is larger than 1. Therefore besides $y(x,0)$, the function $y'(x,0)$ is also needed. Compared with the TFPDE standard form, it is seen that $\alpha = 1.7$, $a = 1$ and $L = T = 1$, and the function $f(x,t) = -{\rm e}^{x+0.5t} + 0.25\, t^{0.3} {\rm e}^x\, {\rm E}_{1,1.3}(0.5t)$, with initial value functions $\phi_1(x) = {\rm e}^x$ and $\phi_2(x) = 0.5{\rm e}^x$. The boundary functions are $\psi_1(t) = {\rm e}^{0.5t}$ and $\psi_2(t) = {\rm e}^{1+0.5t}$. The following commands can be used to solve the TFPDE.

```
>> a=1; alpha=1.7; L=1; T0=1; m=10; n=m^2;
   f=@(x,t)-exp(x+0.5*t)+...
           0.25*(t.^0.3).*exp(x).*ml_func([1,1.3],0.5*t);
   phi1=@(x)exp(x); phi2=@(x)0.5*exp(x);
   psi1=@(t)exp(0.5*t); psi2=@(t)exp(1+0.5*t); tic
   [U,X,T]=ftpde_sol(a,f,alpha,L,T0,phi1,phi2,psi1,psi2,m,n); toc
   f0=@(x,t)exp(x+0.5*t); err=U-f0(X,T); max(abs(err(:)))
```

Table 9.8 Solution efficiencies for different m's when $\alpha > 1$

m	10	20	30	40	50	100
Time elapse	0.0132	0.0243	0.0749	0.175	0.395	15.138
Maximum error	0.0013	3.481×10^{-4}	1.566×10^{-4}	8.865×10^{-5}	5.697×10^{-5}	1.438×10^{-5}

It can be seen that the solver here can be used to solve accurately the TFPDEs. If different values of n are chosen, the time elapse and maximum errors are then measured, as shown in Table 9.8. It can be seen that the maximum error is similar to the cases in $\alpha < 1$, indicating that the algorithm can also be used to solve the problems when $\alpha > 1$. ◄

► Example 9.20

Consider the TFPDE in Reference [14].

$$ {}_0^{\mathrm{C}}\mathscr{D}_t^{\alpha} u(x,t) = \frac{\partial^2 u(x,t)}{\partial x^2} + \frac{2}{\Gamma(3-\alpha)} t^{2-\alpha} \sin 2\pi x + 4\pi^2 t^2 \sin 2\pi x, $$

9

where $\alpha = 0.5$, with initial conditions $\phi(x) = 0$. Zero boundary value conditions were also assumed in the original example, where $u(0,t) = u(1,t) = 0$, and the exact solution is $u(x,t) = t^2 \sin 2\pi x$. In fact, by assigning the solution interval to $x \in [0.25, 0.75]$, the nonzero boundary value conditions can also be constructed from the exact solutions, which are $u(0.25,t) = t^2$ and $u(0.75,t) = -t^2$. If $t \in [0,1]$, solve the TFPDE.

Solutions The algorithms in Reference [14] can only be used for zero boundary value problems, the nonzero boundary value condition here cannot be solved with the algorithms. With the new solver `ftpde_sol()`, the nonzero boundary value condition problem can be solved directly, with the solution surface shown in Fig. 9.15. Meanwhile, the boundary functions are superimposed on the surface. It can be seen that the left and right x axis boundaries match the surface very well. The maximum error is 7.8492×10^{-4}.

```
>> a=1; alpha=0.5; L=[0.25 0.75]; T0=1; m=15; n=ceil(m^2/2);
   f=@(x,t)2/gamma(3-alpha)*t.^(2-alpha).*sin(2*pi*x)+...
           4*pi^2*t.^2.*sin(2*pi*x);
   f0=@(x,t)t.^2.*sin(2*pi*x);                    %exact solution
   psi1=@(t)t.^2; psi2=@(t)-t.^2; phi1=@(x)zeros(size(x)); tic
   [U,X,T]=ftpde_sol(a,f,alpha,L,T0,phi1,'',psi1,psi2,m,n); toc
   surf(X,T,U), U0=f0(X,T); err=U-U0; max(abs(err(:)))  %maximum error
   t=T(1,:); x ones(size(t)); x1=0.25*x; x2=0.75*x;  %boundary functions
   hold on, plot3(x1,t,psi1(t),'r',x2,t,psi2(t),'r'),hold off
```

Selecting different values of m, and let $n = \lceil m^2/2 \rceil$, the errors are measured, as shown in Table 9.9. It can be seen that the accuracy of the numerical solution is rather high. ◄

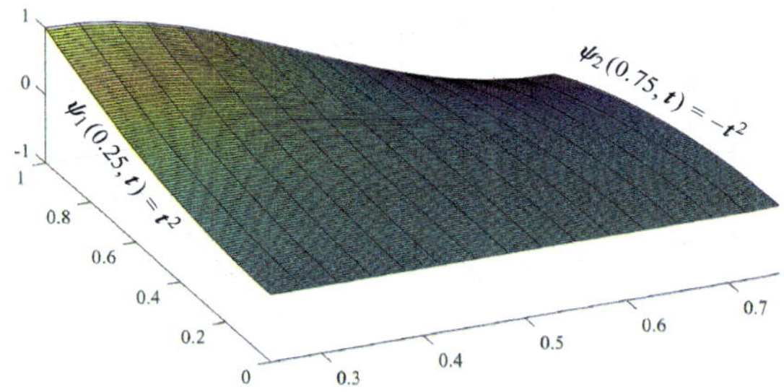

Fig. 9.15 Solution surface

Table 9.9 Solution efficiencies for different m's

m	15	30	45	60	100	120
Time elapse	0.0143	0.0392	0.1872	0.4448	4.5363	10.933
Maximum error	7.849×10^{-4}	1.966×10^{-4}	8.766×10^{-5}	4.930×10^{-5}	1.775×10^{-5}	1.233×10^{-5}

9.5 Exercises

(1) Consider the Algebraic Constraint block in Fig. 9.3. Its output is connected to an $s^{0.2}$ block. What happens to the system, if it is connected to an $s^{0.3}$ block? Are the simulation results affected by such settings. What are the actual signals in each branch? What happens if the block is changed into $s^{1.2}$ or $s^{1.3}$? Why? Validate your conclusions with simulation experiments.

(2) If the block $s^{-0.5}$ is swapped with the transport delay block in Fig. 9.8, is the simulation result affected? Why? Validate your conclusions with simulation experiments.

(3) Consider the following FODDE [3]

$$y'''(t) + {}_0^{\rm C}\mathscr{D}_t^{2.5}y(t) + y''(t) + 4y'(t) + {}_0^{\rm C}\mathscr{D}_t^{0.5}y(t) + 4y(t) = 3\sqrt{2}y(t-7\pi/4),$$

and the history function is $y(t) = \sin t + \cos t$ when $t \leqslant 0$, and the exact solution is $y(t) = \sqrt{2}\sin(t+\pi/4)$, find the numerical solutions of the FODDE model, and assess the precision and efficiencies.

(4) Based on the model in Example 9.8, try to introduce the third delay term such a new benchmark problem can be created. Use simulation results to solve the corresponding FODDE.

(5) The Fcn block is used in Fig. 9.6, and the function is assigned to `-exp(0.1-u)`. If it is modified to `-exp(-u)`, what may happen? Why?

(6) For the fractional-order delay state space model [3]

$$\begin{cases} {}_0^{\rm C}\mathscr{D}_t^{0.5}x(t) = \dfrac{1}{\sqrt{\rm p}}\left(\sqrt[6]{(y(t)-2)(z(t)-3)}+\sqrt{x(t-1)}\right) \\ {}_0^{\rm C}\mathscr{D}_t^{0.2}y(t) = \Gamma(2.2)x(t-1) \\ {}_0^{\rm C}\mathscr{D}_t^{0.6}z(t) = \dfrac{\Gamma(2.8)}{\Gamma(2.2)}x^{1.2}(t-1), \end{cases}$$

with initial conditions $x(0) = 1$, $y(0) = 2$ and $z(0) = 3$, and history function $x(t) = t+1$ when $t \leqslant 0$. The exact solutions are $x(t) = t+1$, $y(t) = t^{1.2}+2$ and $z(t) = t^{1.8} + 3$. Find the numerical solutions of the FODDE within $t \in (0, 10)$, and assess the precision and efficiencies.

(7) Solve numerically the FODDE for $t \in (0, 10)$.

$$\begin{aligned} &{}_0^{\rm C}\mathscr{D}_t^{0.2}y(t)\,{}_0^{\rm C}\mathscr{D}_t^{1.8}y(t) + {}_0^{\rm C}\mathscr{D}_t^{0.3}y(t)\,{}_0^{\rm C}\mathscr{D}_t^{1.7}y(t) + {\rm e}^{-0.1}y(t-0.2) - {\rm e}^{-t/2} \\ &= -\frac{t}{8}\left[{\rm E}_{1,1.8}\left(-\frac{t}{2}\right){\rm E}_{1,1.2}\left(-\frac{t}{2}\right) + {\rm E}_{1,1.7}\left(-\frac{t}{2}\right){\rm E}_{1,1.3}\left(-\frac{t}{2}\right)\right], \end{aligned}$$

where $y'(0) = -1/2$, with history function $y(t) = {\rm e}^{-t/2}$ when $t \leqslant 0$. The exact solution is $y(t) = {\rm e}^{-t/2}$.

(8) Consider the BVP in Example 9.15. If fast restart mode is not used, assess the efficiencies in the solutions.

9

(9) Consider the state space model in Example 8.9. If the given boundary value conditions are $x(0) = 1$, $x(10) - y(10) = 10^{0.4} - 207/10$, $z'(10) - z(10) = -79\times 10^{0.1} - 10$ and $y(10)-y(0) = 10^{0.4}$, and the exact solutions are $x(t) = 2t+1$, $y(t) = t^{0.4}+0.3$ and $z(t) = t+t^{2.1}+1$, find the numerical solutions, and assess the precisions and efficiencies.

(10) Consider the linear FODE in Example 8.11. If the boundary value conditions are $y''(2\pi)-y''(0) = 0$, $y'(0) = 0$ and $y'(2\pi)-y'(0) = 0$, with known exact solution of $y(t) = \cos t$, solve the FODE and assess the precisions and efficiencies.

(11) Solve the following TFPDE [14]

$$\frac{\partial^\alpha u(x,t)}{\partial t^\alpha} = \frac{\partial^2 u(x,t)}{\partial x^2} + x^2(x-1)^2\cos t - \frac{2\sin t}{\Gamma(3-\alpha)}x^{2-\alpha} + \frac{12\sin t}{\Gamma(4-\alpha)}x^{3-\alpha} - \frac{24\sin t}{\Gamma(t-\alpha)}x^{4-\alpha},$$

where $\alpha = 1.2$ and $\alpha = 1.6$. The initial conditions are $\phi(x) = 0$, and boundary value conditions $u(0,t) = u(1,t) = 0$. The exact solution is $u(x,t) = x^2(x-1)^2\sin t$. Assess the impact of mesh grid parameters to the solutions.

(12) Compose a TFPDE model with nonzero initial conditions and nonzero boundary value conditions from the following equation [15]:

$${}_0^{\rm C}\mathscr{D}_t^{\lambda}f(x,t) = \frac{\partial^2 f(x,t)}{\partial x^2} + \sin\pi x{\rm E}_\lambda(t^\lambda) + \pi^2\sin\pi x\big({\rm E}_\lambda(t^\lambda)-1\big),$$

with the exact solution $f(x,t) = \sin\pi x\big({\rm E}_\lambda(t^\lambda)-1\big)$. Find the numerical solutions to this TFPDE.

(13) From the exact solution in Exercise (11) a TFPDE model with nonzero initial values and nonzero boundary value conditions. Find the numerical solutions and assess their precisions and efficiencies.

(14) Consider again the TFPDE in Example 9.20. If $\alpha = 1.5$, and an extra initial condition is $\phi_2(x) = 0$, find the numerical solution and assess its numerical solutions and efficiencies.

References

1. Xue DY, Bai L (2017) Benchmark problems for Caputo fractional-order ordinary differential equations[J]. Frac Calc Appl Anal 20(5):1305–1312
2. Bai L, Xue DY (2018) Universal block diagram based modeling and simulation schemes for fractional-order control systems. ISA Trans 82:153–162
3. Bai L, Xue DY (2022) Block diagram modeling and simulation for delay fractional-order differential equations[C]. In: Proceedings of Chinese control and decision conference. Hefei, China, pp 137–142
4. Xue DY (2022) Modeling ang simulation with Simulink[M]. De Gruyter Press, Berlin
5. Gao GH, Sun ZZ (2011) A compact finite difference scheme for the fractional sub-diffusion equations[M]. J Comput Phys 230:586–595
6. Gao GH, Sun ZZ (2013) The finite difference approximation for a class of fractional sub-diffusion equations on a space uncounded domain[J]. J Comput Phys 236:443–460
7. Sun ZZ, Gao GH (2015) Finite difference algorithms in fractional-order differential equations (in Chinese)[M]. Science Press, Beijing
8. Zhuang P, Liu FW, Turner I et al (2016) Galerskin finite element method and error analysis for the fractional cable equations[J]. Numer Algorithms 72(2):447–466
9. Jiang YJ, Ma YJ (2011) High-order finite element methods for time-fractional partial differential equations[J]. J Comput Appl Math 235(11):3285–3290
10. Liu FW, Zhuang PH, Liu QX (2015) Numerical algorithms and their applications in fractional-order partial differential equations (in Chinese)[M]. Science Press, Beijing
11. Li X, Xu C (2009) A space-time spectral method for the time-fractional diffusion equations [J]. SIAM J Numer Anal 47(3):2108–2131
12. Bai L, Xue DY, Meng L (2021) A numerical algorithm for time-fractional partial differential equations (in Chinese)[J]. J Math Pract Theory 51(12):245–251
13. Xue DY (2022) MATLAB and Simulink in action: Programming, computing and simulation (in Chinese)[M]. Tsinghua University Press, Beijing
14. Wang ZQ, Cao JY (2015) Numerical algorithms and their error analysis in fractional-order integro-differential equations (in Chinese)[M]. Xi'an Jiaotong University Press, Xi'an
15. Doha EH, Zaky MA, Abdelkawy MA (2019) Spectral methods within fractional calculus. In: Baleanu D, Lopes AM (eds) Handbook of fractional calculus with applications, vol 8. Applications in Engineering, Life and Social Sciences, Part B[M]. De Gruyter, Berlin

Supplementary Information

X. Dingyü and B. Lu, *Fractional Calculus*, https://doi.org/10.1007/978-981-99-2070-9

Appendix A: Benchmark Problems for Fractional-order Differential Equations

The solution of FODEs is an important foundation of fractional calculus and its applications. There are various numerical methods for solving FODEs in the literature. However, it is difficult to compare the advantages and disadvantages of the algorithms, since the application examples of each algorithm are very different, which makes it difficult for the users to choose the suitable algorithm for their applications. In many FODE examples, the power functions are deliberately chosen as analytical solutions. This makes the original differential equation satisfy the zero initial conditions to a large extent, avoiding the generality of the nonzero initial value Caputo equation, so that even if the proposed algorithm can produce good numerical solutions, the results are not convincing.

With this in mind, the authors constructed some benchmark problems with analytical solutions [1, 2] in an attempt to fill the gaps in the literature in these areas. Using the high-precision numerical methods presented in this book with Simulink block diagram solutions, the numerical solutions of these problems should be the most accurate numerical solutions currently available in a double-precision data structure. Their accuracy is usually several orders of magnitude higher than that of any other existing algorithms.

Several benchmark problems are listed in ▶ Sect. A.1. Some of the models are modified slightly compared to those in Reference [1]. The proofs of these benchmark problems are given. A set of Simulink models for the benchmark problems is given in ▶ Sect. A.2.

A.1 Mathematical Descriptions and Proofs in Benchmark Problems

A.1.1 Initial Value Problems in FODEs

Definition A.1 Benchmark problem 1

A one-term FODE with piecewise function is

$$ {}_0^{\mathrm{C}}\mathscr{D}_t^{0.7}y(t) = \begin{cases} \dfrac{1}{\Gamma(1.3)}t^{0.3}, & 0 \leqslant t \leqslant 1 \\ \dfrac{1}{\Gamma(1.3)}t^{0.3} - \dfrac{2}{\Gamma(2.3)}(t-1)^{1.3}, & t > 1, \end{cases} \tag{A.1} $$

where $0 \leqslant t \leqslant 2$, with initial condition $y(0) = 1$. The analytical solution of the FODE is also a piecewise function

$$ y(t) = \begin{cases} t+1, & 0 \leqslant t \leqslant 1 \\ t+1-(t-1)^2, & t > 1. \end{cases} \tag{A.2} $$

Proof The easiest way to prove that an analytical solution satisfies an equation is to substitute the analytical solution back into the equation and see whether the two sides of the equation are the same.

According to the properties given in (3.62), all the orders satisfy $\lceil\alpha\rceil \leqslant \lambda$. So the following formula can be used:

$$ {}_0^{\rm C}\mathscr{D}_t^{\alpha} t^{\lambda} = \frac{\Gamma(\lambda+1)}{\Gamma(\lambda+1-\alpha)} t^{\lambda-\alpha}. \tag{A.3}$$

If $t \leqslant 1$, it is found from (A.3) that, since $y(t) = t+1$, the term 1 vanishes when Caputo derivative is taken. Therefore, to find Caputo derivative with respect to t, we have $\lambda = 1$ and $\alpha = 0.7$. It follows that

$$ {}_0^{\rm C}\mathscr{D}_t^{0.7} y(t) = \frac{\Gamma(2)}{\Gamma(2-0.7)} t^{1-0.7} = \frac{1!}{\Gamma(1.3)} t^{0.3} = \frac{1}{\Gamma(1.3)} t^{0.3}. $$

Thus, the first segment of the piecewise function satisfies (A.1), and $y(0) = 1$ satisfies the known initial value.

Now look at $t > 1$. Since $y(t) = t+1-(t-1)^2$, so for the $(t-1)^2$ term, $\lambda = 2$, so

$$ {}_0^{\rm C}\mathscr{D}_t^{0.7} y(t) = \frac{t^{0.3}}{\Gamma(1.3)} - \frac{\Gamma(2+1)}{\Gamma(3-0.7)}(t-1)^{2-0.7} = \frac{t^{0.3}}{\Gamma(1.3)} - \frac{2}{\Gamma(2.3)}(t-1)^{1.3}. $$

It can be seen that the second segment also satisfies (A.1). Therefore, the $y(t)$ function given by (A.2) is the solution of the Caputo equation in (A.1). □

Remarks Although this equation is indeed not a decent FODE, and it is better not to use the term "FODE" to demonstrate your own algorithm for solving FODEs, since the right-hand side of the equation is a given function, independent of $y(t)$, as using such an example would have a negative effect. As a numerical calculus calculation, this problem can be compared as a benchmark problem. □

Definition A.2 Benchmark problem 2

Linear FODE with nonzero initial conditions

$$ y'''(t) + {}_0^{\rm C}\mathscr{D}_t^{4-\sqrt{\pi}} y(t) + y''(t) + 4y'(t) + {}_0^{\rm C}\mathscr{D}_t^{2-\sqrt{\pi}} y(t) + 4y(t) = 6\cos t, \tag{A.4}$$

with the initial values $y(0) = 1$, $y'(0) = 1$, $y''(0) = -1$, and $0 \leqslant t \leqslant 5000$. The exact solution of the equation is $y(t) = \sqrt{2}\sin(t+\pi/4)$.

Proof Consider the property in (3.66)

$$ {}_0^{\rm C}\mathscr{D}_t^{\alpha} \sin\lambda t = \frac{({\rm j}\lambda)^m}{2{\rm j}} t^{\gamma}\left[{\rm E}_{1,\gamma+1}({\rm j}\lambda t) - (-1)^m {\rm E}_{1,\gamma+1}(-{\rm j}\lambda t)\right], \tag{A.5}$$

where $m = \lceil\alpha\rceil$ and $\gamma = m-\alpha$. If $y(t) = \sqrt{2}\sin(t+\pi/4)$, then by (A.5), ${}_0^{\rm C}\mathscr{D}_t^{4-\sqrt{\pi}}y(t)$ term satisfies $\lambda = 1$, $m = 3$, $\gamma = \sqrt{\pi}-1$, while function ${}_0^{\rm C}\mathscr{D}_t^{2-\sqrt{\pi}}$ satisfies $m = 1$, $\lambda = 1$ and $\gamma = \sqrt{\pi}-1$. Since γ is the same in both terms, the difference between these terms is $\mathrm{j}^2 = -1$ times. Therefore, it is seen from (A.4) that ${}_0^{\rm C}\mathscr{D}_t^{4-\sqrt{\pi}}y(t)$ term and ${}_0^{\rm C}\mathscr{D}_t^{2-\sqrt{\pi}}y(t)$ term cancels completely.

The remaining integer-order part of the left-hand side of the equation can be directly calculated and normalized using the following MATLAB statements. Also, the initial conditions are validated.

```
>> syms t, y(t)=sqrt(2)*sin(t+pi/4);
   simplify(diff(y,3)+diff(y,2)+4*diff(y)+4*y)
   y(0), y1=diff(y); y1(0), y2=diff(y1); y2(0).
```

It can be seen that the remaining part on the left-hand side happens to be $6\cos t$, the same as the one on the right-hand side. The three initial conditions are also satisfied. It is shown that the analytical solution of the FODE is $y(t) = \sqrt{2}\sin(t+\pi/4)$. □

Remarks The difficulty of this problem is that all fixed-step algorithms are difficult to guarantee the accuracy of the algorithm, because the termination time is too large, and too small a step size may lead to too much computation, which is difficult to implement by general computers. □

Definition A.3 Benchmark problem 3

Nonlinear FODE with nonzero conditions.

$$ {}_0^{\rm C}\mathscr{D}_t^{\sqrt{2}}y(t) = -t^{1.5-\sqrt{2}}\frac{\mathrm{E}_{1,3-\sqrt{2}}(-t)}{\mathrm{E}_{1,1.5}(-t)}\,\mathrm{e}^t y(t)\,{}_0^{\rm C}\mathscr{D}_t^{0.5}y(t) + \mathrm{e}^{-2t} - \left[y'(t)\right]^2, \tag{A.6}$$

where $y(0) = 1$, $y'(0) = -1$, and $0 \leqslant t \leqslant 2$. The analytical solution is $y(t) = \mathrm{e}^{-t}$. Larger time intervals can be tried, such as the numerical solution in $t \in (0, 3)$, or even larger intervals can be tested.

Proof With the important property ${}_0^{\rm C}\mathscr{D}_t^{\alpha}\mathrm{e}^{\lambda t} = \lambda^m t^\gamma \mathrm{E}_{1,\gamma+1}(\lambda t)$ in Theorem 3.6, where $m = \lceil\alpha\rceil$ and $\gamma = m-\alpha$. If $y(t) = \mathrm{e}^{-t}$, the two fractional-order terms are as follows:

(1) In ${}_0^{\rm C}\mathscr{D}_t^{\sqrt{2}}y(t)$, $m = 2$, $\gamma = 2-\sqrt{2}$, $\lambda = -1$. It is found that ${}_0^{\rm C}\mathscr{D}_t^{\sqrt{2}}y(t) = (-1)^2 t^{2-\sqrt{2}}\mathrm{E}_{1,3-\sqrt{2}}(-t)$.

(2) In ${}_0^{\rm C}\mathscr{D}_t^{0.5}y(t)$, $m = 1$, $\gamma = 0.5$. It is found that ${}_0^{\rm C}\mathscr{D}_t^{0.5}y(t) = (-1)^1 t^{0.5}\mathrm{E}_{1,1.5}(-t)$.

Substituting (2) into the right-hand side of (A.6), then

$$ -t^{1.5-\sqrt{2}}\frac{\mathrm{E}_{1,3-\sqrt{2}}(-t)}{\mathrm{E}_{1,1.5}(-t)}\mathrm{e}^t\mathrm{e}^{-t}(-1)^1 t^{0.5}\mathrm{E}_{1,1.5}(-t) + \mathrm{e}^{-2t} - \left[\mathrm{e}^{-t}\right]^2 = t^{2-\sqrt{2}}\mathrm{E}_{1,3-\sqrt{2}}(-t). $$

It is obvious that the right-hand side is the same as ${}_0^{\rm C}\mathscr{D}_t^{\sqrt{2}}y(t)$. In addition, if $y(t)={\rm e}^{-t}$, it is found that $y(0)=1$ and $y'(0)=-1$. It is proved that the analytical solution of (A.6) is $y(t)={\rm e}^{-t}$. □

Remarks Nonlinear non-commensurate-order differential equation

(1) The design of this benchmark problem is inspired by the example in Reference [3.], but with added difficulty. The differential equation is not convertible to commensurate-order differential equations.

(2) The similar differential equations of this problem are solved in Examples 7.11, 8.12 and 7.15. Algorithms vary in accuracy and speed, and many of them are no longer suitable for solving this problem. Other methods can be used to solve this similar problem and yield fair results for algorithm comparison.

(3) If the termination time is chosen to be 3 s or even 4 s, will the numerical solution of the equation still be obtained? □

Definition A.4 Benchmark problem 4

Implicit FODE with nonzero initial conditions

$$\begin{aligned}&{}_0^{\rm C}\mathscr{D}_t^{0.2}y(t)\,{}_0^{\rm C}\mathscr{D}_t^{1.8}y(t)+{}_0^{\rm C}\mathscr{D}_t^{0.3}y(t)\,{}_0^{\rm C}\mathscr{D}_t^{1.7}y(t)\\&=-\frac{t}{8}\left[{\rm E}_{1,1.8}\left(-\frac{t}{2}\right){\rm E}_{1,1.2}\left(-\frac{t}{2}\right)+{\rm E}_{1,1.7}\left(-\frac{t}{2}\right){\rm E}_{1,1.3}\left(-\frac{t}{2}\right)\right],\end{aligned}\tag{A.7}$$

where $y(0)=1$, $y'(0)=-1/2$ and $0\leqslant t\leqslant 5$. The analytical solution is $y(t)={\rm e}^{-t/2}$.

Proof If $y={\rm e}^{-t/2}$, consider first the term ${}_0^{\rm C}\mathscr{D}_t^{0.2}y(t)\,{}_0^{\rm C}\mathscr{D}_t^{1.8}y(t)$. Again, by Theorem 3.6, it follows that $\lambda=-1/2$. For ${}_0^{\rm C}\mathscr{D}_t^{0.2}y(t)$, $m=1$ and $\gamma=0.8$. It follows that

$${}_0^{\rm C}\mathscr{D}_t^{0.2}y(t)=\left(-\frac{1}{2}\right)^1 t^{0.8}{\rm E}_{1,1.8}\left(-\frac{t}{2}\right),$$

and in ${}_0^{\rm C}\mathscr{D}_t^{1.8}y(t)$, $m=2$ and $\gamma=0.2$. Thus

$${}_0^{\rm C}\mathscr{D}_t^{1.8}y(t)=\left(-\frac{1}{2}\right)^2 t^{0.2}{\rm E}_{1,1.2}\left(-\frac{t}{2}\right).$$

Clearly, the product of these two terms is identical to the first term on the right-hand side of the equation. Similarly, it can be shown that the second product term, and therefore the equation itself, is satisfying the implicit FODE. If the analytical solution of the equation is $y(t)={\rm e}^{-t/2}$, it is clear that the initial value condition $y(0)=1$ and $y'(0)=-1/2$ also satisfy the given conditions. Thus, $y(t)={\rm e}^{-t/2}$ is the analytical solution of this implicit equation. □

Definition A.5 Benchmark problem 5

Nonlinear FOSS equations 1.

$$\begin{cases} {}_0^{\mathrm{C}}\mathscr{D}_t^{0.5}x(t) = \dfrac{1}{\sqrt{\pi}}\left(\sqrt[5]{(y(t)-0.5)(z(t)-0.3)} + \sqrt{t}\right) \\ {}_0^{\mathrm{C}}\mathscr{D}_t^{0.2}y(t) = \Gamma(2.2)\big[x(t)-1\big] \\ {}_0^{\mathrm{C}}\mathscr{D}_t^{0.1}z(t) = \dfrac{\Gamma(2.3)}{\Gamma(2.2)}\big[y(t)-0.5\big], \end{cases} \tag{A.8}$$

where $x(0) = 1$, $y(0) = 0.5$ and $z(0) = 0.3$. It is known that the analytical solutions of the equations are $x(t) = t+1$, $y(t) = t^{1.2}+0.5$ and $z(t) = t^{1.3}+0.3$, when $0 \leqslant t \leqslant 10$.

Proof Suppose $x(t) = t + 1$, $y(t) = t^{1.2} + 0.5$ and $z(t) = t^{1.3} + 0.3$, and obviously, the initial conditions $x(0) = 1$, $y(0) = 0.5$ and $z(0) = 0.3$ are satisfied. Now verify the three equations one by one. Substituting these three functions into the right-hand side of the first equation, it is seen that

$$\frac{1}{\sqrt{\pi}}\left(\sqrt[5]{(y(t)-0.5)(z(t)-0.3)} + \sqrt{t}\right) = \frac{1}{\sqrt{\pi}}\left(\sqrt[5]{t^{1.2}t^{1.3}} + \sqrt{t}\right) = 2\sqrt{\frac{t}{\pi}}. \tag{A.9}$$

Substitute them to (A.3), since ${}_0^{\mathrm{C}}\mathscr{D}_t^{0.5}t^0 = 0$ and $\Gamma(3/2) = \sqrt{\pi}/2$, it is found that

$${}_0^{\mathrm{C}}\mathscr{D}_t^{0.5}x(t) = {}_0^{\mathrm{C}}\mathscr{D}_t^{0.5}t + {}_0^{\mathrm{C}}\mathscr{D}_t^{0.5}t^0 = \frac{\Gamma(2)}{\Gamma(1.5)}\sqrt{t} = 2\sqrt{\frac{t}{\pi}}.$$

It is seen that the solution is the same as the one in (A.9), thus proving that the analytical solution satisfies the first equation.

Substituting the analytical solutions into the second equation, it is deduced that the left-hand side of the equation is equal to the right-hand side, so the second equation holds.

$${}_0^{\mathrm{C}}\mathscr{D}_t^{0.2}y(t) = {}_0^{\mathrm{C}}\mathscr{D}_t^{0.2}t^{1.2} + 0.5\,{}_0^{\mathrm{C}}\mathscr{D}_t^{0.2}t^0 = \frac{\Gamma(2.2)}{\Gamma(2)}t = \Gamma(2.2)t = \Gamma(2.2)\big[x(t)-1\big].$$

Substituting the analytical solutions into the third equation also proves that the equation holds. Therefore, the analytical solution satisfies the original equation.

$${}_0^{\mathrm{C}}\mathscr{D}_t^{0.1}z(t) = {}_0^{\mathrm{C}}\mathscr{D}_t^{0.1}t^{1.3} + 0.3\,{}_0^{\mathrm{C}}\mathscr{D}_t^{0.1}t^0 = \frac{\Gamma(2.3)}{\Gamma(2.2)}t^{1.2} = \frac{\Gamma(2.3)}{\Gamma(2.2)}\big[y(t)-0.5\big].$$

□

Definition A.6 Benchmark problem 5a

Nonlinear state space equations 2.

$$\begin{cases} {}_0^{\mathrm{C}}\mathscr{D}_t^{0.5}x(t) = \dfrac{2}{\sqrt{\pi}}\left[\sqrt[5]{(y(t)-0.3)(z(t)-0.5x(t))}+\sqrt{t}\right] \\ {}_0^{\mathrm{C}}\mathscr{D}_t^{0.1}y(t) = \dfrac{\Gamma(1.4)}{\Gamma(1.3)}\left(\dfrac{x(t)-1}{2}\right)^{0.3} \\ {}_0^{\mathrm{C}}\mathscr{D}_t^{1.5}z(t) = \dfrac{\Gamma(3.1)}{\Gamma(1.6)}\sqrt{(y(t)-0.3)^3}, \end{cases} \tag{A.10}$$

where $x(0) = 1$, $y(0) = 0.3$, $z(0) = 0.5$ and $z'(0) = 1$. The analytical solutions are $x(t) = 2t + 1$, $y(t) = t^{0.4} + 0.3$, $z(t) = t^{2.1} + t + 0.5$.

Remarks FOSS equations

The difficulty of this FOSS equations is that the order of the last term is 1.5, which is greater than 1. □

Proof The left-hand side of the first equation is first derived separately from the analytical solution function.

$${}_0^{\mathrm{C}}\mathscr{D}_t^{0.5}x(t) = {}_0^{\mathrm{C}}\mathscr{D}_t^{0.5}(2t+1) = 2\,{}_0^{\mathrm{C}}\mathscr{D}_t^{0.5}t = 4\sqrt{\frac{t}{\pi}}.$$

In fact, since $x(t)$ contains only the integer-powers of t, it can be derived directly from the `caputosym()` function, giving the same result as above.

```
>> syms t positive
   x(t)=2*t+1; L1=caputosym(x,sym(1/2),0,t).
```

Since $y(t)$ and $z(t)$ contain non-integer-powers of t, the `caputosym()` function is powerless and the left-hand side of the latter two equations can only be derived by hand.

$${}_0^{\mathrm{C}}\mathscr{D}_t^{0.1}y(t) = {}_0^{\mathrm{C}}\mathscr{D}_t^{0.1}(t^{0.4}+0.3) = {}_0^{\mathrm{C}}\mathscr{D}_t^{0.1}t^{0.4} = \frac{\Gamma(1.4)}{\Gamma(1.3)}t^{0.3} \tag{A.11}$$

$${}_0^{\mathrm{C}}\mathscr{D}_t^{1.5}z(t) = {}_0^{\mathrm{C}}\mathscr{D}_t^{1.5}(t^{2.1}+t+0.5) = {}_0^{\mathrm{C}}\mathscr{D}_t^{1.5}t^{2.1} = \frac{\Gamma(3.1)}{\Gamma(1.6)}t^{0.6}. \tag{A.12}$$

Using MATLAB's symbolic computing, substitute the analytical solution into the right-hand side of each of the three equations and simplify.

```
>> syms t positive; P=sym(pi);
   x(t)=2*t+1; y=t^0.4+0.3; z(t)=t^2.1+t+0.5;
   R1=2/sqrt(P)*(((y-0.3)*(z-sym(0.5)*x))^(1/5)+sqrt(t))
   R2a=((x-1)/2)^sym(0.3), R3a=sqrt((y-sym(0.3))^3).
```

The following results can be derived (the coefficients of the terms R_{2a} and R_{3a} are already identical to (A.10)).

$$R_1 = 4\sqrt{\frac{t}{\pi}},\ R_{2a} = t^{0.3},\ R_{3a} = t^{0.6}.$$

The result obtained is exactly the same as the left-hand side of the equation. Moreover, it is clear that the initial conditions given can be proved directly by the three analytical solutions. So it is proved that this benchmark problem is correct. □

A.1.2 Boundary Value Problems for FODEs

Definition A.7 Benchmark problem 6

Boundary value problem 1 of nonlinear FODE is

$$ {}_0^{\mathrm{C}}\mathscr{D}_t^{\sqrt{2}}y(t) = -t^{1.5-\sqrt{2}}\frac{\mathrm{E}_{1,3-\sqrt{2}}(-t)}{\mathrm{E}_{1,1.5}(-t)}\,\mathrm{e}^t y(t)\,{}_0^{\mathrm{C}}\mathscr{D}_t^{0.5}y(t) + \mathrm{e}^{-2t} - \left[y'(t)\right]^2, \tag{A.13}$$

where $y(0) = 1$, $y(2) = \mathrm{e}^{-2}$ and $0 \leqslant t \leqslant 2$. The exact solution is $y(t) = \mathrm{e}^{-t}$.

Proof The proof of the Caputo equation is given in ► Definition A.3. The following proof of the boundary values follows: If the analytical solution of the equation is $y(t) = \mathrm{e}^{-t}$, it is clear that $y(0) = 1$ and $y(2) = \mathrm{e}^{-2}$. This proves that the analytical solution of the equation is valid. □

Definition A.8 Benchmark problem 6a

Boundary value problem 2 of nonlinear FODE is

$$ {}_0^{\mathrm{C}}\mathscr{D}_t^{\sqrt{2}}y(t) = -t^{1.5-\sqrt{2}}\frac{\mathrm{E}_{1,3-\sqrt{2}}(-t)}{\mathrm{E}_{1,1.5}(-t)}\,\mathrm{e}^t y(t)\,{}_0^{\mathrm{C}}\mathscr{D}_t^{0.5}y(t) + \mathrm{e}^{-2t} - \left[y'(t)\right]^2, \tag{A.14}$$

where $y(2) = -2$, $y'(2) - y(0) = -\mathrm{e}^{-2} - 1$, and $0 \leqslant t \leqslant 2$, with exact solution $y(t) = \mathrm{e}^{-t}$.

Proof Substituting the boundary value conditions to the equation to prove the theorem. □

Definition A.9 Benchmark problem 7

Boundary value problem of implicit FODE.

$$\begin{aligned} &{}_0^{\mathrm{C}}\mathscr{D}_t^{0.2}y(t)\,{}_0^{\mathrm{C}}\mathscr{D}_t^{1.8}y(t) + {}_0^{\mathrm{C}}\mathscr{D}_t^{0.3}y(t)\,{}_0^{\mathrm{C}}\mathscr{D}_t^{1.7}y(t) \\ &= -\frac{t}{8}\left[\mathrm{E}_{1,1.8}\left(-\frac{t}{2}\right)\mathrm{E}_{1,1.2}\left(-\frac{t}{2}\right) + \mathrm{E}_{1,1.7}\left(-\frac{t}{2}\right)\mathrm{E}_{1,1.3}\left(-\frac{t}{2}\right)\right], \end{aligned} \tag{A.15}$$

where $y(0) = 1$, $y(5) = \mathrm{e}^{-5/2}$ and $0 \leqslant t \leqslant 5$, with exact solution $y(t) = \mathrm{e}^{-t/2}$.

Proof The proof of the implicit Caputo equation is given in ▶ Definition A.4. The following proves the boundary values: If the analytical solution of the equation is $y(t) = \mathrm{e}^{-t/2}$, it is clear that $y(0) = 1$ and $y(5) = \mathrm{e}^{-5/2}$. Thus proving that $y(t) = \mathrm{e}^{-t/2}$ is the analytical solution of the problem. □

Remarks Selection of termination time
The termination time is intentionally set to 5 instead of 10 here because the value of $\mathrm{e}^{-10/2} = 0.0067$ is too small to be discriminative for the numerical solution of the boundary value problem. □

Definition A.10 Benchmark problem 7a

Boundary value problem of implicit FODE

$$\begin{aligned}&{}_0^{\mathrm{C}}\mathscr{D}_t^{0.2}y(t)\,{}_0^{\mathrm{C}}\mathscr{D}_t^{1.8}y(t)+{}_0^{\mathrm{C}}\mathscr{D}_t^{0.3}y(t)\,{}_0^{\mathrm{C}}\mathscr{D}_t^{1.7}y(t)\\&=-\frac{t}{8}\left[\mathrm{E}_{1,1.8}\left(-\frac{t}{2}\right)\mathrm{E}_{1,1.2}\left(-\frac{t}{2}\right)+\mathrm{E}_{1,1.7}\left(-\frac{t}{2}\right)\mathrm{E}_{1,1.3}\left(-\frac{t}{2}\right)\right],\end{aligned} \tag{A.16}$$

where $y(5) = \mathrm{e}^{-2.5}$, $y'(5) = -\mathrm{e}^{-2.5}/2$ and $0 \leqslant t \leqslant 5$. The analytical solution of the equation is $y(t) = \mathrm{e}^{-t/2}$.

Proof Clearly, the final value given can be verified by $y(t) = \mathrm{e}^{-t/2}$. However, by Example 9.17, the solution of the equation is not unique. □

A.1.3 Fractional-Order Delay Differential Equations

Definition A.11 Benchmark problem 8

Nonlinear FODDE [2].

$${}_0^{\mathrm{C}}\mathscr{D}_t^{\sqrt{2}}y(t) = -t^{1.5-\sqrt{2}}\frac{\mathrm{E}_{1,3-\sqrt{2}}(-t)}{\mathrm{E}_{1,1.5}(-t)}\,\mathrm{e}^t y(t)\,{}_0^{\mathrm{C}}\mathscr{D}_t^{0.5}y(t)+\mathrm{e}^{-2t}-\mathrm{e}^{-0.2}\left[y'(t-0.1)\right]^2, \tag{A.17}$$

where $y(0) = 1$ and $y'(0) = -1$. When $t < 0$, the history function is $y(t) = \mathrm{e}^{-t}$, the interval is $0 \leqslant t \leqslant 2$. The analytical solution is $y(t) = \mathrm{e}^{-t}$.

Proof With the proof in Definition A.3, it can be seen that, if the last two right-hand sides of the equation in (A.17) are satisfied. Then the solution satisfies the original differential equation. Obviously, if $y = \mathrm{e}^{-t}$, then $\left[y'(t-0.1)\right]^2 = \mathrm{e}^{0.2-2t}$, which exactly cancels the previous term, therefore, the analytical solution of this delayed differential equation is $y(t) = \mathrm{e}^{-t}$. □

Remarks The impact of history functions

The solutions of FODDE is closely related to the history functions. For the same differential equations, if the history functions are different, the results are also different. □

Definition A.12 Benchmark problem 8a

Nonlinear FODDE with two delay constants.

$$ {}_0^{\mathrm{C}}\mathscr{D}_t^{\sqrt{2}}y(t)=-t^{1.5-\sqrt{2}}\frac{\mathrm{E}_{1,3-\sqrt{2}}(-t)}{\mathrm{E}_{1,1.5}(-t)}\,\mathrm{e}^{t-0.3}y(t-0.3)\,{}_0^{\mathrm{C}}\mathscr{D}_t^{0.5}y(t)+\mathrm{e}^{0.2-2t}-\left[y'(t-0.1)\right]^2, \tag{A.18} $$

where $y(0) = 1$ and $y'(0) = -1$. When $t < 0$, the history function is $y(t) = \mathrm{e}^{-t}$. Find the numerical solution in the interval $0 \leqslant t \leqslant 2$. The exact solution is $y(t)=\mathrm{e}^{-t}$.

Proof It is easy to show that $\mathrm{e}^{t-0.3}y(t-0.3)=1$ for this definition. In fact, the proof is straightforward by substituting $y(t)=\mathrm{e}^{-t}$. □

A.2 A Simulink Group for Benchmark Problems

The new version of the FOTF Toolbox specifically provides a blockset of benchmark problems. Double-clicking the Benchmark problem models icon in the FOTF Blockset (see Fig. 8.1) will open the blockset shown in Fig. A.1.

The block group gives the Simulink solution models for all the benchmark problems listed earlier, and many of them also provide pre-set good `ww` and N parameters.

The benchmark problem blockset can also be opened directly via the command `bp_models` without going through the FOTF Blockset. The details of the Simulink models for each benchmark problem are provided in Table A.1. The core of the

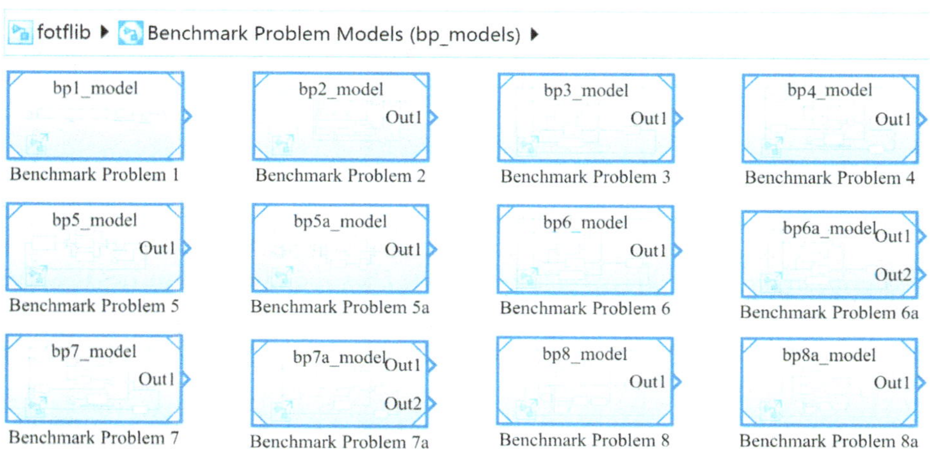

Fig. A.1 Group of benchmark problems for the FOTF Blockset

Table A.1 Simulink models for benchmark problems provided in the blockset

No.	Model names	Model for reference	Recommended parameters	Other descriptions
1	`bp1_model`	Fig. 8.19	$(10^{-8}, 10^6), 41$	Vectorized version of `bp1_model2`, but not recommended for real use
2	`bp2_model`	Fig. 8.20	$(10^{-7}, 10^6), 41$	
3	`bp3_model`	Fig. 9.9	$(10^{-9}, 10^7), 39$	Similar to `bp6_model`, let $b = 1$ and $a = -1$
4	`bp4_model`	Fig. 9.1	$(10^{-6}, 10^5), 30$	
5	`bp5_model2`	Fig. 8.29	$(10^{-9}, 10^7), 59$	Vectorized version of `bp5_model`
5a	`bp5a_model`	Fig. 8.31	$(10^{-9}, 10^6), 43$	Vectorized version of `bp5a_model2`
6	`bp6_model`	Fig. 9.9	$(10^{-7}, 10^7), 39$	Unknown a, and let $b = 1$
6a	`bp6a_model`	Fig. 9.10	$(10^{-7}, 10^7), 39$	Unknown parameters a and b
7	`bp7_model`	Fig. 9.1	$(10^{-5}, 10^5), 27$	Same to `bp4_model`, but known a
7a	`bp7a_model`	Fig. 9.11	$(10^{-5}, 10^5), 27$	Unknown parameters a and b
8	`bp8_model`	Fig. 9.6	$(10^{-7}, 10^7), 39$	
8a	`bp8a_model`	Fig. 9.7	$(10^{-7}, 10^7), 39$	

models is the Oustaloup filter, and the suggested ω_b, ω_h and N parameters are given in the table, although users can try other parameters themselves to see if they yield better results.

A.3 Exercises

(1) Prove the benchmark problem in Definition A.1 with `caputosym()`

(2) Define the problem in Definition A.1 as a function of zero initial values. Modify the model to a nonzero initial value problem and test various methods to find the numerical solution of the problem.

(3) If $y(t) = \sqrt{2}\sin(t + \pi/2)$, try to prove ${}_0^{\mathrm{C}}\mathscr{D}_t^{4-\sqrt{\pi}}y(t) + {}_0^{\mathrm{C}}\mathscr{D}_t^{2-\sqrt{\pi}}y(t) = 0$ by means of symbolic computing. The equation can also be verified by means of the formulation.

(4) Consider the following delay differential equation, known as $-1 \leqslant t \leqslant 0$ when the history function is $y(t) = t + 1$.

$$y'(t) + {}_0^{\mathrm{C}}\mathscr{D}_t^{0.5} y(t) + 2y(t) = 2\sqrt{\frac{y(t-1)}{\pi}} + 2t + 3.$$

Prove that $y(t) = t + 1$ is the analytical solution of the equation for $t > 0$. Build a Simulink model and find the numerical solution on $[0, 10]$.

(5) Consider the method of constructing the explicit FODE in Definition A.3. Solve the differential equation numerically using the MATLAB command-driven method and the block diagram-based method introduced in this book and see if the resulting solution still satisfies the differential equation.

(6) In the two FODDEs given, the delay terms act on both the $y(t)$ and $y'(t)$ terms. Modify the differential equation in Definition A.12 to act with delay on ${}_0^{\mathrm{C}}\mathscr{D}_t^{0.5} y(t)$ such that the analytical solution of the differential equation remains $y(t) = \mathrm{e}^{-t}$. Build a simulation model and verify that the numerical solution of this differential equation still has a high accuracy.

(7) Construct the FODDE from the implicit differential equation given in Definition A.4. Build a simulation model to demonstrate the reliability of the algorithm using simulation methods.

(8) Consider the implicit FODE model in Definition A.9. If the termination time of the simulation is set to $t_{\mathrm{n}} = 10$, and the final value $y(10) = \mathrm{e}^{-10/2}$ is known, can the boundary value problem be solved numerically by executing the simulation model. Why?

(9) Since there are too few known functions to derive an analytical solution of the Caputo derivative, the analytical solution functions listed here are mostly exponential and power functions, making the constructed analytical solutions monotonic and lacking in challenge to the algorithm. Try to construct some benchmark problems with non-monotone analytical solutions?

(10) The suggested values of `ww` and N parameters for each model of the benchmark problem blockset are given in Table A.1. In fact, the solution accuracy can be further adjusted on the basis of these suggested values. Modify the parameters of some of the models to see whether the solution accuracy and efficiency can be further improved.

(11) When constructing the benchmark problem blockset of the FOTF Toolbox, two models for vectorized solution of the state space equations, `bp5_model2` and `bp5a_model2`, are not included. Add these two vectorized simulation models to the blockset in the same form.

References

1. Xue D Y, Bai L. Benchmark problems for Caputo fractional-order ordinary differential equations[J]. Fractional Calculus and Applied Analysis, 2017, 20(5): 1305~1312.
2. Bai L, Xue D Y. Block diagram modeling and simulation for delay fractional-order differential equations[C]. Proceedings of Chinese Control and Decision Conference. Hefei, China, 2022, 137~142.
3. Diethelm K. The analysis of fractional differential equations: an application-oriented exposition using differential operators of Caputo type[M]. New York: Springer, 2010.

Appendix B: Inverse Laplace Transforms for Fractional-order and Irrational Functions

B.1 Commonly Used Special Functions in Fractional Calculus

Due to the difficulties of fractional-order integral itself, special functions should be introduced. Some of them are listed in Table B.1. More details of the general special functions can be found in ► Chap. 2.

Table B.1 The commonly used special functions

Special functions	Definitions
Bessel function	$J_\nu(t)$ is the solution of the equation $t^2y'' + ty' + (t^2 - \nu^2)y = 0$
Beta function	$B(z, m) = \int_0^1 t^{m-1}(1-t)^{z-1}dt,\ \mathbb{R}(m) > 0, \mathbb{R}(z) > 0$
Incomplete Beta	$B_x(z, m) = \dfrac{1}{B(z, m)}\int_0^x t^{z-1}(1-t)^{m-1}dt,\ \mathbb{R}(m) > 0,$
Function	$\mathbb{R}(z) > 0, 0 \leqslant x \leqslant 1$
Incomplete Gamma function	$\Gamma(z, \alpha) = \dfrac{1}{\Gamma(\alpha)}\int_0^z e^{-t}t^{\alpha-1}dt, \quad z \geqslant 0$
Dawson function	$\mathrm{daw}(t) = e^{-t^2}\int_0^t e^{\tau^2}d\tau$
Gamma function	$\Gamma(z) = \int_0^\infty e^{-t}t^{z-1}dt,\ \mathbb{R}(z) > 0$
Hermite polynomial	$H_n(t) = e^{t^2}\dfrac{d^n}{dt^n}e^{-t^2}$
Mittag-Leffler function	$E_{\alpha,\beta}^{\gamma,q}(z) = \sum_{k=0}^{\infty}\dfrac{(\gamma)_{kq}}{\Gamma(\alpha k+\beta)}\dfrac{z^k}{k!},\ \mathbb{R}(\alpha), \mathbb{R}(\beta), \mathbb{R}(\gamma) > 0$, and $q \in \mathbb{N}$
Special cases in M-L functions	$E_{\alpha,\beta}^{\gamma}(z) = E_{\alpha,\beta}^{\gamma,1}(z),\ E_{\alpha,\beta}(z) = E_{\alpha,\beta}^{1}(z),\ E_\alpha(z) = E_{\alpha,1}(z)$
Pochhammer symbol	$(\gamma)_k = \gamma(\gamma+1)(\gamma+2)\cdots(\gamma+k-1) = \dfrac{\Gamma(k+\gamma)}{\Gamma(\gamma)}$
Complementary error function	$\mathrm{erfc}(t) = \dfrac{2}{\sqrt{\pi}}\int_t^\infty e^{-\tau^2}d\tau = 1 - \mathrm{erf}(t)$
Generalized Bessel function	$I_\nu(t) = j^{-\nu}I_\nu(jt)$
Error function	$\mathrm{erf}(t) = \dfrac{2}{\sqrt{\pi}}\int_0^t e^{-\tau^2}d\tau$

B.2 Inverse Laplace Transform Table

Table B.2 gives the inverse Laplace transformation tables [1, 2] for functions associated with fractional-order and irrational operations.

Table B.2 Table of inverse Laplace transformations of fractional-order and irrational operations

$F(s)$	$f(t)=\mathscr{L}^{-1}[F(s)]$	$F(s)$	$f(t)=\mathscr{L}^{-1}[F(s)]$
$\dfrac{s^{\alpha\gamma-\beta}}{(s^\alpha+a)^\gamma}$	$t^{\beta-1}\mathrm{E}^{\gamma}_{\alpha,\beta}\left(-at^\alpha\right)$	$\dfrac{1}{s^n\sqrt{s}},\ n=1,2,\cdots$	$\dfrac{2^n t^{n-1/2}}{1\cdot3\cdot5\cdot\cdots\cdot(2n-1)\sqrt{\pi}}$
$\arctan\dfrac{k}{s}$	$\dfrac{1}{t}\sin kt$	$\dfrac{k}{s^2+k^2}\coth\dfrac{\pi s}{2k}$	$\lvert\sin kt\rvert$
$\ln\dfrac{s^2-a^2}{s^2}$	$\dfrac{2}{t}(1-\cosh at)$	$\dfrac{1}{s\sqrt{s}}\mathrm{e}^{-k\sqrt{s}}$	$2\sqrt{\dfrac{t}{\pi}}\mathrm{e}^{-k^2/(4t)}-k\,\mathrm{erfc}\left(\dfrac{k}{2\sqrt{t}}\right)$
$\ln\dfrac{s^2+a^2}{s^2}$	$\dfrac{2}{t}(1-\cos at)$	$\dfrac{\mathrm{e}^{-k\sqrt{s}}}{\sqrt{s}\,(a+\sqrt{s})}$	$\mathrm{e}^{ak}\mathrm{e}^{a^2t}\mathrm{erfc}\left(a\sqrt{t}+\dfrac{k}{2\sqrt{t}}\right)$
$\dfrac{(1-s)^n}{s^{n+1/2}}$	$\dfrac{n!}{(2n)!\sqrt{\pi t}}\mathrm{H}_{2n}\left(\sqrt{t}\right)^{\dagger}$	$\dfrac{1}{\sqrt{s+b}\,(s+a)}$	$\dfrac{1}{\sqrt{b-a}}\mathrm{e}^{-at}\mathrm{erf}\left(\sqrt{(b-a)t}\right)$
$\dfrac{1}{\sqrt{s^2+a^2}}$	$\mathrm{J}_0(at)$	$\dfrac{(1-s)^n}{s^{n+3/2}}$	$-\dfrac{n!}{(2n+1)!\sqrt{\pi}}\mathrm{H}_{2n+1}\left(\sqrt{t}\right)$
$\dfrac{1}{\sqrt{s^2-a^2}}$	$\mathrm{I}_0(at)$	$\dfrac{(a-b)^k}{(\sqrt{s+a}+\sqrt{s+b})^{2k}}$	$\dfrac{k}{t}\mathrm{e}^{-(a+b)t/2}\,\mathrm{I}_k\left(\dfrac{a-b}{2}t\right),\ k>0$
$\dfrac{1}{(s+a)^\alpha}$	$\dfrac{t^{\alpha-1}}{\Gamma(\alpha)}\mathrm{e}^{-at}$	$\dfrac{\sqrt{s+2a}-\sqrt{s}}{\sqrt{s+2a}+\sqrt{s}}$	$\dfrac{1}{t}\mathrm{e}^{-at}\mathrm{I}_1(at)$
$\dfrac{1}{s^\alpha+a}$	$t^{\alpha-1}\mathrm{E}_{\alpha,\alpha}\left(-at^\alpha\right)$	$\dfrac{\sqrt{s+2a}-\sqrt{s}}{\sqrt{s+2a}+\sqrt{s}}$	$\dfrac{1}{t}\mathrm{e}^{-at}\mathrm{I}_1(at)$
$\dfrac{1}{s}\mathrm{e}^{-k/s}$	$\mathrm{J}_0\left(2\sqrt{kt}\right)$	$\dfrac{(\sqrt{s^2+a^2}-s)^\nu}{\sqrt{s^2+a^2}}$	$a^\nu\mathrm{J}_\nu(at),\ \nu>-1$
$\dfrac{1}{s\sqrt{s}}\mathrm{e}^{-k/s}$	$\dfrac{1}{\sqrt{\pi k}}\sin 2\sqrt{kt}$	$\dfrac{1}{(s^2-a^2)^k}$	$\dfrac{\sqrt{\pi}}{\Gamma(k)}\left(\dfrac{t}{2a}\right)^{k-1/2}\mathrm{I}_{k-1/2}(at)$
$\dfrac{1}{\sqrt{s}}\mathrm{e}^{k/s}$	$\dfrac{1}{\sqrt{\pi t}}\cosh 2\sqrt{kt}$	$\dfrac{(\sqrt{s^2-a^2}+s)^\nu}{\sqrt{s^2-a^2}}$	$a^\nu\mathrm{I}_\nu(at),\ \nu>-1$

† Hermite polynomial $\mathrm{H}_n(x)=\displaystyle\sum_{k=0}^{[n/2]}\frac{(-1)^k n!}{k!(n-2k)!}(2x)^{n-2k}$, the solution of the differential equations $y''(x)-2xy'(x)+2ny(x)=0$ or $[\mathrm{e}^{-x^2}y'(x)]'+2n\mathrm{e}^{-x^2}y(x)=0$. The recursive formula is alternatively expressed as $\mathrm{H}_0(x)=1$, $\mathrm{H}_{n+1}(x)=2x\mathrm{H}_n(x)-n\mathrm{H}_{n-1}(x)$.

Table B.2 (continued)

$F(s)$	$f(t)=\mathscr{L}^{-1}[F(s)]$	$F(s)$	$f(t)=\mathscr{L}^{-1}[F(s)]$
$\frac{1}{\sqrt{s}}\mathrm{e}^{-k/s}$	$\frac{1}{\sqrt{\pi t}}\cos 2\sqrt{kt}$	$\frac{1}{(\sqrt{s^2+a^2})^k}$	$\frac{\sqrt{\pi}}{\Gamma(k)}\left(\frac{t}{2a}\right)^{k-1/2}\mathrm{J}_{k-1/2}(at)$
$\ln\frac{s-a}{s-b}$	$\frac{1}{t}\left(\mathrm{e}^{bt}-\mathrm{e}^{at}\right)$	$(\sqrt{s^2+a^2}-s)^k$	$\frac{ka^k}{t}\mathrm{J}_k(at),\ k>0$
$\frac{a}{s(s^\alpha+a)}$	$1-\mathrm{E}_\alpha\left(-at^\alpha\right)$	$\frac{1}{s+\sqrt{s^2+a^2}}$	$\frac{\mathrm{J}_1(at)}{at}$
$\frac{s^\alpha}{s(s^\alpha+a)}$	$\mathrm{E}_\alpha\left(-at^\alpha\right)$	$\frac{1}{\sqrt{s+a}\sqrt{s+b}}$	$\mathrm{e}^{-(a+b)t/2}\,\mathrm{I}_0\left(\frac{a-b}{2}t\right)$
$\frac{1}{s}\mathrm{e}^{-k\sqrt{s}}$	$\mathrm{erfc}\left(\frac{k}{2\sqrt{t}}\right)$	$\frac{1}{(s+\sqrt{s^2+a^2})^N}$	$\frac{N\mathrm{J}_N(at)}{at},\ N>0$
$\frac{1}{\sqrt{s}}\mathrm{e}^{-k\sqrt{s}}$	$\frac{1}{\sqrt{\pi t}}\mathrm{e}^{-k^2/(4t)}$	$\frac{1}{s\sqrt{s}}\mathrm{e}^{-\sqrt{s}}$	$2\sqrt{\frac{t}{\pi}}\mathrm{e}^{-1/(4t)}-\mathrm{erfc}\left(\frac{1}{2\sqrt{t}}\right)$
$\frac{1}{s^\alpha(s-a)}$	$t^\alpha\mathrm{E}_{1,1+\alpha}(at)$	$\frac{\mathrm{e}^{-\sqrt{s}}}{\sqrt{s}(\sqrt{s}+1)}$	$\mathrm{e}^{t+1}\mathrm{erfc}\left(\sqrt{t}+\frac{1}{2\sqrt{t}}\right)$
$\frac{1}{\sqrt{s}}$	$\frac{1}{\sqrt{\pi t}}$	$\frac{s^\alpha}{s-a}$	$-t^\alpha\mathrm{E}_{1,1-\alpha}(at),\ 0<\alpha<1$
$\frac{1}{s\sqrt{s}}\mathrm{e}^{k/s}$	$\frac{1}{\sqrt{\pi k}}\sinh 2\sqrt{kt}$	$\frac{b^2-a^2}{(s-a^2)(\sqrt{s}+b)}$	$\mathrm{e}^{a^2t}[b-a\,\mathrm{erf}(a\sqrt{t})]-b\mathrm{e}^{b^2t}\mathrm{erfc}(b\sqrt{t})$
$\frac{1}{s\sqrt{s}}$	$2\sqrt{\frac{t}{\pi}}$	$\frac{1}{s^\nu}\mathrm{e}^{k/s}$	$\left(\frac{t}{k}\right)^{(\nu-1)/2}\mathrm{I}_{\nu-1}\left(2\sqrt{kt}\right)$
$\frac{1}{\sqrt{s+1}}$	$\frac{\mathrm{e}^{-t}}{\sqrt{\pi t}}$	$\frac{1}{s^\nu}\mathrm{e}^{-k/s}$	$\left(\frac{t}{k}\right)^{(\nu-1)/2}\mathrm{J}_{\nu-1}\left(2\sqrt{kt}\right),\ \nu>0$
$\mathrm{e}^{-k\sqrt{s}}$	$\frac{k}{2\sqrt{\pi t^3}}\mathrm{e}^{-k^2/(4t)}$	$\sqrt{s-a}-\sqrt{s-b}$	$\frac{1}{2\sqrt{\pi t^3}}\left(\mathrm{e}^{bt}-\mathrm{e}^{at}\right)$
$\frac{1}{\sqrt{s}(s-1)}$	$\mathrm{e}^t\mathrm{erf}\left(\sqrt{t}\right)$	$\frac{\sqrt{s+2a}-\sqrt{s}}{\sqrt{s}}$	$a\mathrm{e}^{-at}[\mathrm{I}_1(at)+\mathrm{I}_0(at)]$
$\frac{1}{\sqrt{s}(s+1)}$	$\frac{2}{\sqrt{\pi}}\mathrm{daw}\left(\sqrt{t}\right)$	$\frac{\sqrt{s}}{s+1}$	$\frac{1}{\sqrt{\pi t}}-\frac{2}{\sqrt{\pi}}\mathrm{daw}\left(\sqrt{t}\right)$
$\frac{1}{\sqrt{s}(s+a^2)}$	$\sqrt{t}\,\mathrm{E}_{1,3/2}\left(-a^2t\right)$	$\frac{s}{(s-a)\sqrt{s-a}}$	$\frac{1}{\sqrt{\pi t}}\mathrm{e}^{at}(1+2at)$
$\frac{\sqrt{s}}{s+a^2}$	$\frac{1}{\sqrt{t}}\mathrm{E}_{1,1/2}\left(-a^2t\right)$	$\frac{1}{\sqrt{s}+a}$	$\frac{1}{\sqrt{\pi t}}-a\mathrm{e}^{a^2t}\mathrm{erfc}\left(a\sqrt{t}\right)$
$\frac{1}{s\sqrt{s+1}}$	$\mathrm{erf}\left(\sqrt{t}\right)$	$\frac{\sqrt{s}}{s-a^2}$	$\frac{1}{\sqrt{\pi t}}+a\mathrm{e}^{a^2t}\mathrm{erf}\left(a\sqrt{t}\right)$
$\frac{1}{\sqrt{s}(s-a^2)}$	$\frac{1}{a}\mathrm{e}^{a^2t}\mathrm{erf}\left(a\sqrt{t}\right)$	$\frac{1}{\sqrt{s}(s+a^2)}$	$\frac{2}{a\sqrt{\pi}}\mathrm{e}^{-a^2t}\int_0^{a\sqrt{t}}\mathrm{e}^{\tau^2}\mathrm{d}\tau$

Table B.2 (continued)

$F(s)$	$f(t)=\mathscr{L}^{-1}[F(s)]$	$F(s)$	$f(t)=\mathscr{L}^{-1}[F(s)]$
$\dfrac{1}{\sqrt{s}(\sqrt{s}+a)}$	$\mathrm{e}^{a^2t}\mathrm{erfc}\left(a\sqrt{t}\right)$	$\dfrac{s\sqrt{s}}{s+1}$	$2\sqrt{\dfrac{t}{\pi}}-\dfrac{2}{\sqrt{\pi}}\mathrm{daw}\left(\sqrt{t}\right)$
$\dfrac{\sqrt{s}}{s-1}$	$\dfrac{1}{\sqrt{\pi t}}+\mathrm{e}^t\mathrm{erf}\left(\sqrt{t}\right)$	$\dfrac{k!}{\sqrt{s}\pm\lambda}$	$t^{(k-1)/2}\mathrm{E}_{1/2,1/2}^{(k)}\left(\mp\lambda\sqrt{t}\right),\ \mathbb{R}(s)>\lambda^2$
$\dfrac{1}{s^\alpha}$	$\dfrac{t^{\alpha-1}}{\Gamma(\alpha)}$	$\dfrac{s^{\alpha-1}}{s^\alpha\pm\lambda}$	$\mathrm{E}_\alpha\left(\mp\lambda t^\alpha\right),\ \mathbb{R}(s)>\lvert\lambda\rvert^{1/\alpha}$

$F(s)$	$f(t)=\mathscr{L}^{-1}[F(s)]$
$\dfrac{1}{\sqrt{s(s+a)}(\sqrt{s+a}+\sqrt{s})^{2\nu}}$	$\dfrac{1}{a^\nu}\mathrm{e}^{-at/2}\mathrm{I}_\nu\left(\dfrac{a}{2}t\right),\ k>0$
$\dfrac{\Gamma(k)}{(s+a)^k(s+b)^k}$	$\sqrt{\pi}\left(\dfrac{t}{a-b}\right)^{k-1/2}\mathrm{e}^{-(a+b)t/2}\mathrm{I}_{k-1/2}\left(\dfrac{a-b}{2}t\right)$
$\dfrac{1}{\sqrt{s(s+a)}(\sqrt{s+a}+\sqrt{s})^{2\nu}}$	$\dfrac{1}{a^\nu}\mathrm{e}^{-at/2}\mathrm{I}_\nu\left(\dfrac{a}{2}t\right),\ k>0$
$\dfrac{1}{\sqrt{s(s+a)}(\sqrt{s+a}+\sqrt{s})^{2\nu}}$	$\dfrac{1}{a^\nu}\mathrm{e}^{-at/2}\mathrm{I}_\nu\left(\dfrac{a}{2}t\right),\ k>0$
$\dfrac{1}{\sqrt{s^2+a^2}(s+\sqrt{s^2+a^2})^N}$	$\dfrac{\mathrm{J}_N(at)}{a^N}$
$\dfrac{1}{\sqrt{s^2+a^2}(s+\sqrt{s^2+a^2})}$	$\dfrac{\mathrm{J}_1(at)}{a}$
$\dfrac{b^2-a^2}{\sqrt{s}(s-a^2)(\sqrt{s}+b)}$	$\mathrm{e}^{a^2t}\left[\dfrac{b}{a}\mathrm{erf}\left(a\sqrt{t}\right)-1\right]+\mathrm{e}^{b^2t}\mathrm{erfc}\left(b\sqrt{t}\right)$
$\dfrac{a\mathrm{e}^{-k\sqrt{s}}}{s(a+\sqrt{s})}$	$-\mathrm{e}^{ak}\mathrm{e}^{a^2t}\mathrm{erfc}\left(a\sqrt{t}+\dfrac{k}{2\sqrt{t}}\right)+\mathrm{erfc}\left(\dfrac{k}{2\sqrt{t}}\right)$
$\dfrac{1}{\sqrt{s+a}(s+b)\sqrt{s+b}}$	$t\mathrm{e}^{-(a+b)t/2}\left[\mathrm{I}_0\left(\dfrac{a-b}{2}t\right)+\mathrm{I}_1\left(\dfrac{a-b}{2}t\right)\right]$
$\dfrac{\mathrm{e}^{-\sqrt{s}}}{\sqrt{s}+1}$	$\dfrac{\mathrm{e}^{-1/(4k)}}{\sqrt{\pi t}}-\mathrm{e}^{t+1}\mathrm{erfc}\left(\sqrt{t}+\dfrac{1}{2\sqrt{t}}\right)$
$\dfrac{\mathrm{e}^{-\sqrt{s}}}{s(\sqrt{s}+1)}$	$\mathrm{erfc}\left(\dfrac{1}{2\sqrt{t}}\right)-\mathrm{e}^{t+1}\mathrm{erfc}\left(\sqrt{t}+\dfrac{1}{2\sqrt{t}}\right)$

References

1. Magin R L. Fractional calculus in bioengineering[M]. Redding: Begell House Publishers, 2006.
2. Chen Y Q, Petráš I, Vinagre B M. A list of Laplace and inverse Laplace transforms related to fractional-order calculus[OL]. [April 29, 2024] ► http://people.tuke.sk/ivo.petras/foc_laplace.pdf, 2007.

Appendix C: FOTF Toolbox Functions and Models

C.1 Fundamental Computing Functions

In order to cooperate with the introduction of the book, the author has written a large number of MATLAB programs. These programs can be directly called to solve the relevant problems in this book. The page number given below is the page number of the source program list. Certain programs do not have the page number, the reader can read the program code in the FOTF Toolbox.

C.1.1 Special Functions and Other Supporting Functions

Function names	Function descriptions	Pages
beta_c()	Compute Beta function with complex arguments	31
caputosym()	Compute analytical solution of Caputo derivatives	86
common_order()	Find the base order from a set of orders	
fdcoef()	Compute finite difference coefficients	134
fdiffcom()	Common function for fractional-order derivative	65
fotf2sym()	Convert FOTF object to symbolic expression	
fence_shadow()	Fractional-order integral interpretation: shadow on the fence	94
fmincon_global()	Global solver for constrained optimization problems	
fminunc_global()	Global solver for unconstrained optimization problems	
funmsym()	Analytical solution of matrix functions	
gamma_c()	Compute Gamma function with complex arguments	28
kronsum()	Compute Kronecker sum	
mfd2frd()	Conversion of multivariable frequency response data types	
mittag_leffler()	Analytical solution of Mittag-Leffler functions	39
ml_func()	Numerical solution of Mittag-Leffler functions	40
more_sols()	Find all roots of nonlinear matrix equations	256
more_vpasols()	Symbolic version of more_sols() function	
num_diff()	High-precision solution of fractional-order derivatives	130
pade_app()	Padé approximation of rational transfer functions	
paderm()	Padé approximation of independent numerator and denominator orders	
riemannsym()	Analytical solution of Riemann–Liouville derivatives	83
svec2sl()	Internal modeling of the Caputo vector block	

C.1.2 Numerical Computing in Fractional-Order Derivatives

Function names	Function descriptions	Pages
caputo()	Compute Caputo derivatives, not recommended for use	89
caputo9()	High-precision $O(h^p)$ Caputo computing, recommended	124
genfunc()	Generating function coefficients evaluation from symbolic computing	107
get_vecw()	Computing $O(h^p)$ high-precision weighting coefficients	116
glfdiff0()	Compute Grünwald–Letnikov derivative, not recommended	64
glfdiff()	Computing $O(h)$ Grünwald–Letnikov fractional-order derivatives	66
glfdiff2()	Uncompensate $O(h^p)$ Grünwald–Letnikov derivatives, not recommended	115
glfdiff9()	$O(h^p)$ high-precision Grünwald–Letnikov derivative, recommended	118
glfdiff9_mat()	Matrix implementation of glfdiff9(), not recommended	124
glfdiff_fft()	FFT-based $O(h^p)$ GL derivative, not recommended	110
glfdiff_mat()	Matrix version of glfdiff() function	71
glfdiff_mem()	Grünwald–Letnikov derivative with short-time memory effect	73
rlfdiff()	Riemann–Liouville derivative computing, not recommended	81

C.1.3 Filter Design

Function names	Function descriptions	Pages
carlson_fod()	Carlson filter design	147
charef_fod()	Charef filter design	176
charef_opt()	Optimal Charef filter design	182
contfrac0()	Continued fraction approximation of irrational systems	144
dfod2()	Discrete filter for fractional-order operators	188
iir_pade()	IIR filter design using Padé approximation	190
irid_fod()	Impulse response invariant-based IIR filter design	192
irid_folpf()	Impulse response invariant-based irrational model filter	192
matsuda_fod()	Matsuda–Fujii filter design	150
new_fod()	Modified Oustaloup filter design	160
opt_app()	High-order optimal integer-order filter design	167
ousta_fod()	Standard Oustaloup filter design	155
srid_fod()	Step response invariant IIR filter design	192

C.1.4 Solutions of Linear FODEs

Function names	Function descriptions	Pages
caputo_ics()	Equivalent initial conditions, called by fode_caputo9() Function	236
fode_caputo0()	Simple solver for Caputo equations with nonzero initial conditions	231
fode_caputo9()	$O(h^p)$ numerical solver for Caputo equations	236
fode_sol()	Closed-form solution of FODE with zero initial conditions	220
fode_solm()	Matrix version of fode_sol(), not recommended	225
fode_sol9()	$O(h^p)$ version of fode_sol() function	226
ml_step3()	Step responses of three-term linear FODE	208

C.1.5 Solutions of Nonlinear FODEs

Function names	Function descriptions	Pages
ftpde_sol()	Time fractional differential equation solver	372
INVLAP_new()	Closed-loop response solution with numerical Laplace transform	246
nlfode_mat()	Matrix-based implicit FODE solver	281
nlfode_vec()	Nonlinear extended FOSS equation solver	281
nlfode_vec1()	Another version of nlfode_vec(), not recommended	286
nlfec()	$O(h^p)$ corrector solver for nonlinear FODE	292
nlfep()	$O(h^p)$ predictor solver for nonlinear FODE	289
pepc_nlfode()	One-term FODE solver with predictor–corrector method	274

C.2 Object-Oriented Program Design

In order to make the analysis and design of linear fractional-order control systems more convenient, two classes—FOTF class (FOTF) and fractional-order state space class (FOSS)—are designed after the control system toolbox of MATLAB. Each class creates a number of overload functions whose call format is as consistent as possible with its control system toolbox prototype, making it easier to use both the classes [1, 2]. The low-level class of the FOTF class has been replaced with the ppoly class [3], making the program more stable.

C.2.1 FOTF Class for Fractional-Order Transfer Functions

Function names	Function descriptions
base_order()	Find base order of a FOTF object
bode( )	Draw Bode diagram for a FOTF object

Function names	Function descriptions
`diag( )`	Construction diagonal FOTF matrix
`display( )`	Display a FOTF object
`eig( )`	Computing the poles of a FOTF object
`eq( )`	Judge whether two FOTF objects equal or not
`exp()`	Input delay term to a FOTF object
`feedback()`	Compose feedback system from two FOTF objects
`foss_a()`	Convert FOTF object into a extended FOSS object
`fotf()`	Constructing a FOTF object
`fotf2cotf()`	Convert a FOTF object into commensurate-order form
`fotf2foss()`	Low-level function to convert FOTF into FOSS, called by `foss()`
`fotfdata()`	Extracting all the fields from a FOTF object
`freqresp()`	Low-level frequency response computing of a FOTF object
`high_order()`	High-order approximation of a FOTF object
`impulse()`	Impulse response
`inv()`	Inverse of a FOTF object, not recommended in multivariable systems
`isstable()`	Judge the stability of a FOTF object
`iszero()`	Judge whether a FOTF object is zero or not
`lsim()`	Time response for arbitrary input response
`margin()`	Compute gain and phase margin of a FOTF object
`maxdelay()`	Maximum delay in the multivariable FOTF matrices
`mfrd()`	Frequency response
`minus()`	Subtracting two FOTF objects
`mldivide()`	Left division of a FOTF object
`mpower()`	Power evaluation of a FOTF object
`mrdivide()`	Right division of a FOTF object
`mtimes()`	Product of two FOTF objects
`nichols()`	Draw Nichols chart
`norm()`	Evaluate $\mathcal{H}_2$ and $\mathcal{H}_\infty$ norms of a FOTF object
`nyquist()`	Draw Nyquist plot
`plus()`	Adding up two FOTF objects
`residue()`	Partial fraction expansion
`rlocus()`	Root locus of a SISO FOTF object
`sigma()`	Multivariable singular value plot
`simplify()`	Simplification of a FOTF object
`step()`	Step response plotting
`uminus()`	Unary minus of a FOTF object

C.2.2 FOSS Class for Fractional-Order State Space Models

Function names	Function descriptions
bode()	Bode diagram plotting
coss_aug()	State augmentation of a FOSS object
ctrb()	Generating a controllability judgement matrix
display()	Display a FOSS object
eig()	Compute poles of a FOSS object
eq()	Judge whether two FOSS objects equal or not
feedback()	Feedback connection of two FOSS objects
foss()	Create a FOSS object
foss2fotf()	Low-level function to convert a FOSS object to a FOTF object
impulse()	Impulse response of a FOSS object
inv()	Inverse of a FOSS object
isstable()	Judge the stability of a FOSS object
lsim()	Compute arbitrary time response
margin()	Gain and phase margins computation
mfrd()	Multivariable frequency response
minreal()	Minimum realization of a FOSS object
minus()	Subtracting two FOSS objects
mpower()	Power computing of a FOSS object
mtimes()	Multiplying two FOSS objects
nichols()	Draw Nichols chart
norm()	Evaluate $\mathcal{H}_2$ and $\mathcal{H}_\infty$ norms of a FOSS object
nyquist()	Plotting Nyquist plot
obsv()	Constructing observability judgement matrix
order()	Compute the order of a FOSS object
plus()	Adding up two FOSS objects
rlocus()	Root locus of a SISO FOSS object
size()	Find the number of states, inputs and outputs from a FOSS object
ss_extract()	Extract integer-order state space model from a FOSS object
step()	Step response of a FOSS object
uminus()	Unary minus of a FOSS object

C.3 Simulink Models

The FOTF Toolbox builds a Simulink blockset with the startup command `fotflib`, which includes the necessary blocks for modeling fractional-order systems. In addition to this blockset, several reusable general Simulink model frameworks are built that can simulate a class of fractional-order control systems.

Here we introduce the basic blocks of the FOTF Blockset and present a few representative models of reusable simulation systems.

C.3.1 FOTF Blockset in Simulink

Model names	Model descriptions
`fotflib`	Type `fotflib` command to open FOTF Blockset
Riemann–Liouville	Use integer-order block to evaluate various of Riemann–Liouville filters, including Oustaloup filter, Matsuda–Fujii filter, modified Oustaloup filter, impulse response invariant discrete filters, etc.
Caputo vector	Vectorized Caputo operator block, suitable for state space equations
Implicit model	$(as+1)^{-\gamma}$ block, with continuous or discrete implementation
Approximate FOTF model	Filter implementation of FOTFs and FOTF matrices
Fractional PID controller	Fractional-order PID controller block
Benchmark problem models	Benchmark problem model group, a series implementation of benchmark models with pre-assigned parameters. Also use `bp_models` to open

C.3.2 Important Reusable Fractional-Order Models

Model names	Model descriptions	Pages
`c8mblk2.slx`	Linear FODE with zero initial conditions	312
`c8mblk3.slx`	Transfer function model of linear FODE with zero initial conditions	312
`bp1_model.slx`	Fractional-order integral evaluation	321
`bp2_model.slx`	Linear Caputo differential equation model	322
`c8mexp2s.slx`	Linear Caputo equation model	323
`c8mchaos.slx`	Low-level model of Caputo state space model	326
`c8mchaos_v.slx`	Vectorized implementation of Caputo state space model	327
`bp5_model.slx`	Low-level model of Caputo state space model with order greater than 1	332
`bp5a_model.slx`	Vectorized Caputo state space model with order greater than 1	333
`c8mpid.slx`	Fractional-order feedback control system model	336
`c8mmimo.slx`	Multivariable fractional-order control system	338
`bp4_model.slx`	Implicit Caputo equation	350
`bp8_model.slx`	Caputo delay models with nonzero history conditions	357

References

1. Xue DY (2017) Fractional-order control systems—fundamentals and numerical implementations [M]. De Gruyter, Berlin
2. Xue DY (2018) Fractional calculus and fractional-order control (in Chinese)[M]. Science Press, Beijing
3. Xue DY, Pan F (2023) MATLAB and Simulink in action: programming, scientific computing and simulation [M]. Springer, Singapore

Index

F

G

H

I

K

L

M

N

O

P

SPRINGER NATURE

GPSR Compliance

The European Union's (EU) General Product Safety Regulation (GPSR) is a set of rules that requires consumer products to be safe and our obligations to ensure this.

If you have any concerns about our products, you can contact us on ProductSafety@springernature.com

In case Publisher is established outside the EU, the EU authorized representative is:

Springer Nature Customer Service Center GmbH
Europaplatz 3
69115 Heidelberg, Germany

The manufacturer's authorised representative in the EU is Springer Nature Customer Service Centre GmbH, Europaplatz 3, 69115 Heidelberg, Germany. If you have any concerns regarding our products, please contact ProductSafety@springernature.com

Printed and bound by CPI Group (UK) Ltd, Croydon, CR0 4YY
01/10/2026
02225907-0020